Fisheries Economics
Volume I

International Library of Environmental Economics and Policy

General Editors: Tom Tietenberg and Wendy Morrison

Titles in the Series

Fisheries Economics
Volume I

Collected Essays

Edited by

Lee G. Anderson

College of Marine Studies, University of Delaware, USA

Routledge
Taylor & Francis Group

LONDON AND NEW YORK

First published 2002 by Ashgate Publishing

Reissued 2018 by Routledge
2 Park Square, Milton Park, Abingdon, Oxon, OX14 4RN
711 Third Avenue, New York, NY 10017

Routledge is an imprint of the Taylor & Francis Group, an informa business

Notice:
Product or corporate names may be trademarks or registered trademarks, and are used only for identification and explanation without intent to infringe.

Publisher's Note
The publisher has gone to great lengths to ensure the quality of this reprint but points out that some imperfections in the original copies may be apparent.

Disclaimer
The publisher has made every effort to trace copyright holders and welcomes correspondence from those they have been unable to contact.

A Library of Congress record exists under LC control number: 00030623

ISBN 13: 978-1-138-72315-3 (hbk)
ISBN 13: 978-1-138-70898-3 (pbk)
ISBN 13: 978-1-315-19318-2 (ebk)

Contents

Disaggregated Models

PART III   REGULATION

Acknowledgements

The editor and publishers wish to thank the following for permission to use copyright material.

Academic Press for the essays: Colin W. Clark and Gordon R. Munro (1975), 'The Economics of Fishing and Modern Capital Theory: A Simplified Approach', *Journal of Environmental Economics and Management*, **2**, pp. 96–106. Copyright © 1975 Academic Press, reprinted by permission of the publisher. All rights reserved.

American Economic Association for the essays: Ralph Turvey (1964), 'Optimization and Suboptimization in Fishery Regulation', *The American Economic Review*, **54**, pp. 64–76; Ronald N. Johnson and Gary D. Libecap (1982), 'Contracting Problems and Regulation: The Case of the Fishery', *The American Economic Review*, **72**, pp. 1005–22.

Blackwell Publishers for the essays: Parzival Copes (1972), 'Factor Rents, Sole Ownership and the Optimum Level of Fisheries Exploitation', *The Manchester School of Economics and Social Studies*, **41**, pp. 145–63; J.R. Gould (1972), 'Externalities, Factor Proportions and the Level of Exploitation of Free Access Resources', *Economica*, **39**, pp. 383–402.

Econometric Society for the essay: Colin W. Clark, Frank H. Clarke and Gordon R. Munro (1979), 'The Optimal Exploitation of Renewable Resource Stocks: Problems of Irreversible Investment', *Econometrica*, **47**, pp. 25–47.

Fisheries and Oceans for the essays: Milner B. Schaefer (1957), 'Some Considerations of Population Dynamics and Economics in Relation to the Management of the Commercial Marine Fisheries', *Journal of the Fisheries Research Board of Canada*, **14**, pp. 669–81; Anthony Scott (1979), 'Development of Economic Theory on Fisheries Regulation', *Journal of the Fisheries Research Board of Canada*, **36**, pp. 725–41; J.A. Crutchfield (1979), 'Economic and Social Implications of the Main Policy Alternatives for Controlling Fishing Effort', *Journal of the Fisheries Research Board of Canada*, **36**, pp. 742–52.

Kluwer Academic Publishers for the essay: R. Quentin Grafton (1996), 'Individual Transferable Quotas: Theory and Practice', *Reviews in Fish Biology and Fisheries*, **6**, pp. 5–20. With kind permission of Kluwer Academic Publishers.

NRC Research Press for the essay: Colin W. Clark (1980), 'Towards a Predictive Model for the Economic Regulation of Commercial Fisheries', *Canadian Journal of Fisheries and Aquatic Sciences*, **37**, pp. 1111–29.

Series Preface

The *International Library of Environmental Economics and Policy* explores the influence of economics on the development of environmental and natural resource policy. In a series of twenty five volumes, the most significant journal essays in key areas of contemporary environmental and resource policy are collected. Scholars who are recognized for their expertise and contribution to the literature in the various research areas serve as volume editors and write an introductory essay that provides the context for the collection.

Volumes in the series reflect the broad strands of economic research including 1) Natural and Environmental Resources, 2) Policy Instruments and Institutions and 3) Methodology. The editors, in their introduction to each volume, provide a state-of-the-art overview of the topic and explain the influence and relevance of the collected papers on the development of policy. This reference series provides access to the economic literature that has made an enduring contribution to contemporary and natural resource policy.

TOM TIETENBERG
WENDY MORRISON
General Editors

Introduction

Economic Aspects of Fisheries Utilization

The purpose of this collection is twofold. The first is to describe the history of how the economic theory of fisheries exploitation developed. The second is to provide coverage of the important topics in the field. At the same time, the book is aimed at two audiences. It is hoped that it will be a useful addition to the libraries of academic and other professional economists, as well as supplementary text for natural resources, fisheries economics, or fisheries management classes.

The range of inquiry by those who call themselves fisheries economists is very wide and not all of it can be covered in a single volume. For the most part, the focus here will be on the traditional economic problem of optimal resource allocation. With free and unlimited access to a fishery, the incentive structure is such that the actions of private actors will lead to inefficient production. It is possible to state the conditions under which dynamic efficiency can be obtained and to suggest regulation programs which help achieve efficiency. This focus may appear to ignore the fact that economic efficiency is not the prime, or even a secondary, objective of actual management programs in many parts of the world. However, the focus here is not how fisheries agencies do work, or even necessarily, how they should work. The subject is the lack of efficiency under open access and the specification of exactly what dynamic efficiency entails. This means that work by economists and other students of fisheries management on multi-objective management, the special problems associated with artisinal fisheries, and other interesting and important issues will not be covered.

Space constraints will also preclude coverage of many other areas of economic research that are directly or indirectly related to fisheries utilization and management. Perhaps the most important is the valuation of recreational fisheries by contingent valuation, hedonic pricing, and the travel cost method. However, each of these is covered in a separate volume in this series. Other topics which will not be covered are demand and market studies, international trade in seafood products, and aquaculture. Even with the constrained focus, the selection of essays was an extremely difficult task, and many important and interesting essays were, by necessity, not included. To partially compensate, an extended bibliography is attached, but there are no claims that it includes all relevant citations. For more references, the interested reader is referred to an online bibliography maintained by John Ward and sponsored by National Marine Fisheries Service, Office of Science and Technology, Fisheries Statistics and Economics Division at *http://www.st.nmfs.gov/st1/econ/pubs.html* .

I am grateful to Ragnar Arnason, Keith Criddle, Bruce Rettig, Ken Roberts, and Ralph Townsend for providing suggestions on which articles to include and especially to Trond Bjørndal, Jon Conrad, and Rögnvaldur Hannesson for making suggestions and for providing comments on a preliminary version of the introductory essay. I would also like to thank the many subscribers to FISHFOLK who responded to my request for suggestions for citations for the extended bibliography.

Introduction

To set the stage, consider the following *brief* description of what I believe is the essence of modern fisheries economics. Fish stocks are living capital which, when combined with other inputs, are capable of producing food and recreational experiences on a potentially continuing basis. Using a standard welfare economics approach, economically efficient utilization of fish stocks requires obtaining the time path of harvests that will maximize the net present value of production, taking into account the full opportunity cost of all inputs involved. For fisheries management purposes, it is imperative that the focus on the time path of harvest (an output measure) does not detract from the necessity of using the most efficient combinations of inputs. Because both the net annual value of production and the recruitment, mortality, and growth of individuals in fish stocks can depend upon where, when, how, and at what size the fish of any one stock are captured, the determination of the optimal utilization paths can be an enormously difficult problem. It is made more demanding because the production functions and population dynamics are stochastic and not subject to complete specification given existing knowledge.

Due to the incentives facing individual operators (given the lack of property rights in the living capital which is the fish stock), it is highly unlikely that the optimal time path of harvest and input utilization will be achieved in an unregulated fishery. Profit maximizing individual participants will base entry and annual operating decisions on private returns and private costs. While such behaviour is rational from the private point of view, the effects of their harvest on the current harvest of others and upon the future status of the stocks are not being considered. This means that the marginal cost of harvest in any period, including foregone current production of other goods and services and foregone discounted net gains from future harvests will be higher than the current production value of the harvest. If left unregulated, the time path of open access use will result in stock sizes which are too low (in terms of reproduction and marginal productivity effects on harvest) and a set of fleets and processing plants that is too large and will likely consist of the wrong mix of variable and fixed inputs.

From a strict economic efficiency point of view, the goal of management is to find the time path of harvest and input usage starting with the status quo stock size and set of man-made capital that will maximize the net present value of output. And the full problem must consider the cost of implementing the management structure and selecting and enforcing the regulations which cause the private individuals to mimic the decisions that would be made if they were fully responsible for the short and long term costs of their decisions.

Because the solution of the optimal time path problem is so difficult, and because of political realities in real world management, a great deal of the theoretical and applied work on fisheries regulation focuses attention on developing programs which will achieve biological goals of management without causing the inefficient use of inputs. It has been shown that traditional regulations such as gear restrictions, area closures, and so on which proscribe individual activity, but do not restrict access, may be able to increase expected stock size as long as they increase the cost at which fishing effort can be produced. However, they are not likely to be biologically effective in the long run as participants find ways they can, at a cost, increase effective effort given the regulation constraints. This can lead to a never-ending battle between regulators and participants leading to further cost inefficiencies. Total allowable catch programs (TACs) may be biologically effective, provided they can be enforced, but they cause a race for fish and higher costs than necessary. However, access control programs, especially those which focus on outputs and create some aspects of property rights, can obtain biological objectives and potentially create incentives to reduce or minimize the average cost of fish harvest.

These points and the associated nuances and subtleties are described in greater detail in the collected readings. Taken together the first five essays contain the core of the received analysis of the economics of fishing utilization. The remaining essays contain refinements, extensions, and provide practical applications of the theory. The remainder of this essay will provide more detail on the above summary of basic principles. The majority of the discussion will be in the context of synthesizing and relating the analysis of the five core essays. There will also be some discussion of the reasons for and the insights from the refinements and extensions.

The Basics

The first three essays (Gordon, Scott, and Schaefer) are among the first essays written on what is now called fisheries economics, and provide a good description of the basic insights that are fundamental to the field, especially the problem of open access inefficiencies.[1] However, from a historical perspective, it is interesting to note just exactly what is and what is not contained in the essays. For example, the standard sustainable revenue and cost curve analysis which is frequently associated with Gordon and is often used to describe the static operation of a fishery, is not found in the Gordon essay although its evolution can be traced through the Gordon, Scott, and Schaefer essays. Also, while Scott presents the basics of optimal use through time, the more formal analysis using the maximum principle, which now forms the current state of knowledge, was not published until 15 years later, primarily by Clark although others contributed as well. See Brown (1974), Burt and Cummings (1969, 1970, and 1977), and Plourde (1970, 1971). Crutchfield and Zellner (1962) studied dynamic utilization with a calculus of variations approach.

Gordon (Chapter 1) describes the inefficiency of open access[2] utilization in terms of average and marginal returns to a fishery assuming that the unit cost of effort is constant (see his Figure 1). He stresses that this is a long run analysis and that the diminishing returns to effort are due to the effect that catch will have on stock size. He assumes that there is a linear relationship between catch and effort in the short run. He notes, however, that the basic results follow with a nonlinear short term production function. As will be demonstrated below, the shape of the short run production function is important in determining the tractability of the bioeconomic model and in the types of inefficiencies introduced by open access utilization.

He demonstrates that the level of effort that will maximize seasonal net revenues is where the (value of) marginal product is equal to the marginal cost of effort. He then shows that this is not an equilibrium because at that point the average return to effort, the amount that would be earned by a new entrant, is higher than the m«"ginal cost of effort. The open access equilibrium will occur where the value of the average product is equal to average cost. Following the traditional Ricardian analysis of farm lands of different productivity, he goes one step further and shows that with fishing grounds of different productivity, not only will there be over-exploitation on each ground, but also there will likely be a misallocation of effort between grounds because the value of the marginal output will not be the same on each ground.

This demonstration of the difference between open access and optimal utilization set a pattern that has been followed in scores of essays up to this day. The analysis may be different because of different assumptions about population dynamics, the price of fish, the number of species involved, the time horizon and so on but the story is the same. A set of equations can be derived to show how individuals will choose to produce, assuming they maximize private returns, and that set will not correspond to the conditions necessary to maximize total discounted net returns.

Gordon also gives considerable emphasis to the limited mobility of fishermen and its effect on their opportunity income and argues that 'overfishing' in the real world may be exacerbated by these low opportunity costs. Scott replies that this is a red herring. Open access inefficiency will result regardless of whether opportunity cost is high or low. However, later essays demonstrate that malleability does have an effect on the dynamic time paths of open access and optimal utilization (see below).

Readers may be well advised to skip the last section of the Gordon essay which contains an early bioeconomic equilibrium analysis. Although the model is internally consistent, Gordon assumes that the equilibrium stock size is inversely related to the amount of harvest, which is to say that sustainable yield increases as stock size decreases to zero. The received sustainable static analysis in terms of revenues and costs is displayed by Schaefer (Chapter 3) in his Figure 4. This analysis is based on the logistic growth curve, where sustainable yield reaches a maximum when the stock is one half its virgin size (see below). The analysis is presented in terms of total revenue and cost as well as average revenue and cost. Gordon's Figure 1, which uses average and marginal curves, is analogous to the bottom of Schaefer's Figure 4, although Schaefer provides an explicit analysis of long term diminishing returns to effort. As mentioned, Schaefer uses the logistic growth curve, which has been the basis for explaining population dynamics in a very large proportion of the received literature on fisheries economics.

Using the notation of the Clark and Munro essay[3], the Schaefer model can be expressed as follows. Let annual net growth of a fish stock (the net effect of recruitment, individual growth, and natural mortality) equal

$$F(X) = k_1 X(M - X), \tag{1}$$

where X is stock size in biomass, k_1 is a constant, and M is the ecological carrying capacity.[4] This produces a symmetrical curve where growth equals zero at $X = 0$ and $X = M$, and maximum growth, which is the maximum sustainable yield, occurs at $X = M/2$. Call the stock size which generates the maximum growth, X_{msy}. Let annual harvest, h, equal

$$h = h(E, X) = bEX \tag{2}$$

where b is the catchability coefficient and E equals effort. Sustainable yield requires that harvest equals growth

$$bEX = k_1 X(M - X)$$

Solving for X obtains

$$X = M - (b/k_1)E \tag{3}$$

Substituting (3) into (2) produces the sustainable yield curve

$$Y = bE(M - (b/k_1)E) \tag{4}$$

The short run production function is linear in E but the sustainable yield curve is nonlinear due to the long term effect of effort on stock size. It produces the symmetrical shaped total sustainable revenue curve.

Schaefer uses his Figure 4 to show that depending on the price-cost ratio, the equality of average returns and average cost (the open access equilibrium condition) can possibly lead to an open access equilibrium in the downward sloping portion of the sustainable yield curve. In that situation, the equilibrium stock size will be less than X_{msy}. The point where marginal revenue equals marginal cost (the condition for a maximum sustained net revenue) will always occur in the upward sloping portion of the sustainable yield curve. This gives the impression that the optimal stock size will always be greater than X_{msy}. The dynamic analysis of Clark and Munro shows that this may not necessarily be the case.

The Scott essay (Chapter 2) confirms the primary results of Gordon, although unfortunately it uses the same biological assumption. In addition, it makes a significant contribution by distinguishing between short and long run analysis and by introducing the basics of optimal utilization through time, which is presented more formally and eloquently by Clark and Munro using techniques not available to Scott. Scott also presents an early, but incomplete, disaggregated analysis. He switches the frame of reference from effort to output, and says that given the fixed factors on a boat, the marginal cost of catching an extra fish at the individual firm level will be an increasing function of output. This places the concept of diminishing returns more in terms of typical microeconomic analysis and away from the primarily biological aspects of the aggregate yield function. The supply curve of the fishery will be the sum of the supply curves of existing boats, and Scott states that the short run equilibrium will occur when the demand for fish, which he assumed to be horizontal, intersects the aggregate supply curve. Scott then turns his attention to long term optimum utilization. He makes the interesting point that if a sole owner were to take over this disaggregated fishery for *one season only*, there would be no incentive to switch from the open access equilibrium point because the boats are operating where marginal cost is equal to price. He notes that this conclusion will not hold if there are external diseconomies between firms. The significance of this point with respect to the types of inefficiencies introduced by open access will be discussed below. However, if the sole owner has a long term horizon, there will be incentives to change the operation by taking into account the effect of production this year on potential profit in future years through changes in stock size. He formalizes this by introducing the concept of user cost, and using his Figure 2 shows that to achieve optimal utilization, the change in net revenue that is achieved by the last unit of catch in any period must be compared with the change in the present value of net revenues which results from taking that unit of fish out of the water.

In order to analyze optimal long term utilization, Scott had to step away from the sustainable yield curves. This is an important point. The early essays devote considerable space developing the sustainable yield curve, and while it is a useful pedagogical device for understanding the open access bioeconomic equilibrium, it is not useful in explaining how the equilibria are achieved or in looking at optimization. This is because it is implicitly assumed that with movements along the sustainable yield curve, the stock size instantaneously adjusts with changes in effort. This is not the case, and this adjustment time, along with the adjustments in effort, is very critical in explaining equilibria and optimization. The maximization of sustainable net revenues is not the same as the maximization of the present value of net revenue unless the discount rate is equal to zero.

The Clark and Munro essay (Chapter 4) provides a much more detailed look at the dynamic optimization problem. However, because they use harvest as the control variable instead of effort, it can be difficult to make straightforward comparisons to the Gordon–Schaefer analysis. Comparisons are also difficult because of the assumed biological equilibrium in the Gordon–Schaefer model. The following discussion presents the Clark and Munro model in terms of effort. This provides for a straightforward comparison to the earlier work and demonstrates the significance of the linearity assumptions.

Assume a short run production function $h(E, X)$ with diminishing returns to both effort and stock (that is, h_E and h_X are positive and h_{EE} and h_{XX} are negative), a nonlinear total social cost of effort function $c(E)$, where $c_E > 0$ and $c_{EE} > 0$, and a constant price, P. Note that these are more general assumptions than those used in both the Gordon–Schaefer and the Clark and Munro models. Comparisons with the more specific assumptions will be made in the course of the analysis.

There are two possible explanations as to why the total cost function can be nonlinear. See Anderson (Chapter 35). First, inputs used in the production of effort can have positively sloped supply curves which will cause the cost of producing each additional unit of effort to increase. Call this Case 1. Let $\gamma(E)$ represent the marginal cost of producing effort where $\gamma_E > 0$. In this case, $c(E) = \int_0^E \gamma(E)dE$. Assuming perfect competition in the production of effort, the total payments necessary to generate any level of effort will equal $\gamma(E)E$ and rents to intra-marginal factors of production will equal $\gamma(E)E - \int_0^E \gamma(E)dE$. A second reason for a nonlinear total cost curve is congestion among vessels where the unit cost of producing effort is a function of total effort. Call this Case 2. Let $\Phi(E)$ represent the increasing unit cost function where $\Phi_E > 0$. In this case both the social cost of effort, $c(E)$, and payments necessary to generate a given level of effort will equal $\Phi(E)E$.

Following the logic of open access utilization, with a given stock size, in any period, effort will expand until the return per unit of effort is equal to the cost of producing the last unit. (Remember that this analysis is in terms of the short run yield curve and not the sustainable yield curve.) For the two cases this can be represented as:

    Case 1 $Ph/E - \gamma(E) = 0$ (5a)

    Case 2 $Ph/E - \Phi(E) = 0$ (5b)

Given the assumptions, average revenue decreases and the cost of the effort increases with E. Therefore, there will be a positive E for which the above equations hold as long as returns are greater than costs for low levels of effort. In fact, there will be a solution if either average revenue is decreasing *or* the cost of effort is increasing. (How this system can reach a bioeconomic equilibrium is the subject of the Smith essay described below.) Note that using the assumptions in the Gordon–Schaefer model, short run revenue and cost are both linear functions of E, which means that the short run equilibrium level of effort is indeterminate. Or more realistically, if E_{max} is the maximum amount of effort that the existing fleet can produce, then with linear costs and revenues, the open access fishery will produce E_{max} when average return per unit of effort is greater than or equal to the cost of producing it, but it will produce no effort when returns are less than costs. These niceties are obscured by using the sustainable yield curve.

Equations (5a) and (5b) summarize the Gordon–Schaefer open access conditions for the *short run*. They will provide a useful basis for comparison to the equations which follow from the Clark and Munro dynamic analysis, which considers short run decisions taking the long run point of view. The population dynamics are introduced in a different way from that in deriving the sustainable yield curve.

With a discount rate of δ, the net present value of harvest can be represented as

$$NPV = \int_0^\infty e^{-\delta t}[Ph\{E(t), X(t)\} - C\{E(t)\}]dt \tag{6}$$

The problem is to find the time path of annual effort that will maximize this function over an infinite time horizon subject to the constraint on how stock size will change over time.

$$dX(t)/dt = F\{X(t)\} - h\{E(t), X(t)\} \tag{7}$$

The current value Hamiltonian for this maximization problem is

$$H = [Ph\{E(t), X(t)\} - c\{E(t)\}] + \lambda(t)[F\{X(t)\} - h\{E(t), X(t)\}]$$

The $\lambda(t)$ term is the unknown adjoint variable which can be interpreted as the current value of a unit of stock *in situ*. It represents the gains that would be achieved at a point in time by an incremental increase in stock size. Two of the conditions required for a maximum to the problem are:

$$\partial H / \partial E = 0 \quad \text{for all t.}$$

$$d\lambda / dt = \delta\lambda - \partial H / \partial X$$

The first equation specifies the optimal level of E in any period while the second describes the required time rate of change of the adjoint variable. Since the equations must hold simultaneously, E must be chosen through time, taking into account the current value of λ and also the effect it will have on future values of λ.

Evaluating the first condition yields (the t subscripts are suppressed):

$$Ph_E - c' - \lambda h_E = 0.$$

Or $$(P - \lambda)h_E - c' = 0 \tag{8}$$

This provides a statement of the conditions which must hold in every period (at all points in time to be precise) in order to maximize NPV. As opposed to open access, where effort will enter until the average current revenue is equal to the cost of the last unit of effort, optimal utilization requires that the production of the last unit of effort, valued by the net effect it will have on NPV, be balanced against the cost of producing that unit of effort at every point in time. To be specific, P is the value gained in current revenue for putting a unit of fish on the dock, but λ is the present value of future net revenue lost by taking that unit of fish out of the water.

Substituting the appropriate expression for c', (8) can be expressed for the two cases as follows:

Case 1 $(P-\lambda)h_E - \gamma(E) = 0$ (8a)

Case 2 $(P-\lambda)h_E - \Phi(E) - E\Phi' = 0$ (8b)

Note that in Case 1, if $h_E = h/E$ as assumed by Gordon and Schaefer, then the only difference between open access and optimal utilization is the addition of the λ term. See (5a) and (5b). Scott was correct when he said that a sole owner would make no changes in production if he/she only had control for a single period. But it can be seen that in the more general model, there can be short run concerns for the reasons Scott noted. When h_E does not equal h/E there is an intra-period interaction between individual operators. An extra unit of output by one producer will decrease production by all others. Whether increasing costs will cause intra-period inefficiencies depends on the case. In Case 1, the last unit of effort to enter under open access pays the full social cost of production. The changes in costs borne by other producers are transfer payments to intra-marginal inputs. However in Case 2, the full cost of the last unit of effort is the cost of producing the last unit, $\Phi(E)$, plus the increase in real cost imposed on other producers, $E\Phi'$.

The second required condition can be evaluated as follows:

$$d\lambda/dt = \delta\lambda - Ph_X - \lambda(F' - h_X) \qquad (9)$$

This equation and the equation for the net rate of stock growth (7) are a coupled system of two differential equations which determine the optimal trajectories of λ and X. See the Appendix in the Clark and Munro essay for more on the mathematics of solving these equations. As noted, the solutions become extremely difficult if not impossible given the general assumptions about diminishing returns to effort and nonlinear cost functions. The equations are useful for providing insights into the economics of dynamic use, but their application is difficult. Some of the essays presented below demonstrate how this compli-cated principle can be applied to real world fisheries, but for now consider the economic meaning of the equations. The trajectory of stock size can reach a point where the stock will remain constant.[5] In that case, $d\lambda/dt$ will equal zero. Rearranging (9) at that point yields:

$$\delta\lambda = (P-\lambda)h_X + \lambda F' \qquad (9a)$$

Bearing in mind that λ represents the value of a unit of fish in the water, this has a straightforward interpretation. The right-hand side is the value of the marginal product of a unit of stock in the water. The first term is the extra harvest which results from increasing stock size, h_X, evaluated at the net price of fish. The second term is the change in growth of the stock evaluated at the unit value of stock in the water. The sum of these values must equal the annual return that would be earned by an asset equal in value to a unit of stock in the water.

From (8) it can be shown that

$$\lambda = [Ph_E - c'] / h_E.$$

Substituting this in (9a) yields:

$$F' + c' h_X / [Ph_E - c'] = \delta \tag{10}$$

This is the equivalent of Clark's golden rule equation when effort rather than harvest is the control variable. The second term is called the marginal stock effect because it takes into account the effect that stock size has on the cost of producing a unit of fish. Note that if h_X equals zero, which is to say that annual production does not vary with stock size, this equation reduces to

$$F' = \delta$$

The optimal stock size occurs where the slope of the growth function is equal to the discount rate. The slope of the growth function can be interpreted as the own rate of return to the stock because it shows the extra annual physical return that can be obtained by allowing for a unit increase in stock size. This means that in this case, the optimal stock size will be less than X_{msy} because the slope of the growth curve is positive only in this range. Therefore, based on the growth of stock alone, it never makes sense to operate at a stock size above X_{msy} because the own rate of return will be negative. This is contrary to the result of the static model where the optimum always occurs at a stock size greater than X_{msy}. However, when h_X is positive, the optimal equilibrium stock size is where the slope of the growth curve equals $\delta - c' h_X / [Ph_E - c']$. If $c' h_X / [Ph_E - c'] > \delta$, the optimal stock size will be greater than X_{msy}.

Elsewhere Clark (1973b) compares the extreme cases where δ equals zero and infinity. If δ is zero, the optimal stock size is the one that follows from the static model and this is true regardless of the assumptions about the shapes of the short run production function and/or the cost curve. Given the assumptions of the Gordon–Schaefer model, the optimal stock size at an infinite discount rate will be the open access stock size. With an infinite interest rate, future returns are of no concern which is analogous to a sole owner with control for one year only. But this is not true under the more general assumptions. While an infinite discount rate will mean that the user cost in terms of long run changes in stock size will be ignored, optimal utilization will still require a correction for the intra-period inefficiencies of open access.

In commenting on the difference between a static and a dynamic formulation of the optimal utilization problem, Turvey (1964, p. 75) agreed that the trade-off between current and future returns should be considered, but he concluded that 'all this complicates matters, but introduces no interesting new principles.' One might be tempted to agree when considering things from an equilibrium point of view. The difference between a static optimum stock size and a dynamic optimum equilibrium stock size can be very small for interest rates in the relevant range. But that misses the point. The real issue is that solving the fishery management problem must be viewed as finding the appropriate time path of effort and stock size. This is indeed an interesting and very important new principle. When looking at an overexploited fishery, solving for the dynamically optimal stock size is only part of the problem. The real issue is how the investment program (that is, reduction in harvest) should be set up so as to achieve that stock size.

The optimal approach path and how it relates to the assumption of a short term linear production function can be described in more detail by taking a closer look at the first order condition for effort in any period.

$$\lambda = [Ph_E - c']/h_E.$$

With a linear production function, h_E is constant for any level of X, and with a linear cost equation c' is constant as well. In that case the right-hand side of the above will be a constant for any level of X. This situation leads to the so-called bang-bang approach to the dynamically optimal stock size. If stock size is such that $\lambda > [Ph_E - c'] / h_E$, a fish in the water will always add more to the PV of net revenue than will harvesting a fish in the current period. Therefore effort should be zero and the stock should be allowed to grow as fast as it can. When the stock size is such that the inequality is reversed, a fish on the dock is always worth more than a fish in the water, and so it makes sense to take as much fish as is physically possible. In terms of the above, effort should equal E_{max} and the fish stock should be reduced as fast as possible. When the stock is pushed to the optimum size, further investment or disinvestment in the stock is not warranted, and E should be set such that catch is equal to growth.

With nonlinearity in production or cost (or with a downward sloping demand curve), the approach path to the optimal stock size is asymptotic. As effort expands in any period, the marginal return will decrease. Therefore even when the stock size is low, it may make sense to harvest some fish. Effort should be expanded until net current returns equal the user cost. Catch will be less than growth, and so there will still be investment in stock size but there will be an optimal annual combination of fish for current use and fish for investment in the stock. In the opposite case where current stock size is higher than the optimum, catch must be greater than growth so that the stock can be reduced. However, the amount by which catch should exceed growth will depend upon current net returns which are a diminishing function of effort. Therefore, the optimal rate of reduction in stock size will be less than in the linear case. Clark and Munro discuss both kinds of approach paths.

But the problem of optimal utilization is more complicated than this because the above analysis does not distinguish between the current operating cost of producing effort and the capital cost of building the boats which produce the effort. The amount of capital, which is to say the size of the fleet, is not considered except that there is some (constant) maximum amount of effort that can be produced in any period. Of course, optimal utilization in the sense of maximizing NPV must consider capital costs as well. The problem is to determine the optimal size of the fleet through time and then to decide how much effort to produce with the existing fleet at any point in time. What makes the problem difficult is that vessels last for many periods but the actual production decision is made on an annual basis. If the fleet is too small, it will constrain current output, but if it is too large it represents an ex-post waste of resources. Of course what is too large or too small can only be measured relative to the current stock size which can change over time.

The Clark, Clarke, and Munro essay (Chapter 5) addresses this issue using a sophisticated and rather complicated model which assumes linear production and cost functions. The intuitive results can be explained quite easily, however, using the above discussion as a frame of reference.[6] The basic point is that fishing capital investment decisions must be based on returns inclusive of capital costs while current operating decisions are based on operating costs only. They show that, given their assumptions, there is a unique optimum stock and fishing capital size combination, which if maintained will maximize the NPV of production, taking into account the cost of building and maintaining the fishing capital which is assumed to depreciate at a known rate. Fishing capital is designated by K and is defined in terms of the maximum amount of annual effort it can produce. Let the optimum combination be represented by X^* and K^*. The interesting point of course is the shape of the optimal time paths of K, X, and E as the fishery moves to X^* and K^*. In this case, the adjustments will not be bang-bang even with the linearity assumptions. In the simple model, the time path depended only on the size of the initial X. In this case the time path depends upon the relative size of the initial combination of X and K. This is shown in their Figure 2.

The basics of the issues under consideration can best be introduced by discussing first the case of an unexploited stock with no fleet. The optimum path will necessitate an initial K greater than K^* to allow for the required reduction of X. The size of the initial K will depend upon the difference between the initial X and X^*. Given the initial X and K, catch will be greater than growth and the stock size will fall. At the same time, K will decrease because of depreciation. The difference in this case is that even if X falls below X^*, the optimum amount of effort will not necessarily fall to zero. Even though it does not make sense to invest in new capital at these stock sizes, it may make sense to operate existing capital if current returns are above the sum of operating costs and the marginal user cost. The optimum path will have both K and X decreasing as long as the effort from K produces a catch greater than growth. Eventually K will decrease such that X will begin to grow again. The optimal path will eventually arrive at a point where $K < K^*$ and $X = X^*$. Given the assumptions in the essay, at this point K should be increased to K^*.

The cases for stock rebuilding will follow the same general trend except in cases where X is very low relative to stock size. At very low stock sizes, it may make sense to stop fishing completely because current returns may be lower than the sum of operating costs and the marginal user cost. However, the critical stock size for determining whether a complete cessation of fishing is necessary will depend upon the existing K. Up to a point, the larger the K, the larger will be the critical X. At low fleet sizes it is appropriate to fish at lower stock sizes because, given the depreciation rate, the long term effects on stock size and hence NPV will be less.

Refinements

Demand, Supply, and Inputs

The three essays in this section, all published before the Clark and Munro essay, expand upon the basic principles of open access and static optimal use by showing how the analysis will change by using more realistic assumptions about price, costs, and other variables. They are important because they show that the frequently stated conclusion that open access exploitation will dissipate all rents is not quite accurate. They also demonstrate that different participants can view open access and optimal utilization differently depending upon their relative comparative advantages. Turvey (Chapter 6) introduces consumer and producer surplus as well as the controlling of the size of fish at first capture. He shows that even ignoring the surpluses, it is necessary to choose an optimal combination of effort and size at first capture in order to insure that the gains to the fishery are maximized. Although he acknowledges that the passage of time will be an important element in gaining benefits from changing the size at first capture, the analysis is not explicitly dynamic. He also shows that maximizing the net value of output with increasing costs of effort and downward sloping demand curves means that the maximand is not simply rent to the fishery as in the simple Gordon model, but the sum of consumer surplus, producer surplus, and rent to the fish stock.

Copes (Chapter 7) improves upon the Turvey analysis by presenting the information in graphs which are comparable to standard demand and supply curves. His backward-bending long run supply curve, which he describes in more detail in a 1970 essay, shows the possibility of multiple equilibria and demonstrates how a fishery can be sustainable even at low catch and stock levels if the price of output is high enough. This curve can also be used to demonstrate that returns in the form of consumer surplus will be achieved even in open access and that moving from an open access to an optimum position can lead to a decrease in consumer surplus. He also considers rents to factors of production due to differences in skill levels and, using diagrams, demonstrates Turvey's point that for optimal utilization it is necessary to use marginal cost net of factor rents. Here again he shows that all factor rents are not dissipated under open access and that the size of these rents can vary depending upon what sort of management objectives were used. This makes explicit the commonly noted phenomenon that various segments of a fishery can have different private incentives when discussing fisheries regulation. While a movement to an optimum situation could increase the total net gains, some sectors can lose and some can win.

Gould (Chapter 8) refines the analysis by moving beyond the concept of a conglomerate input (effort) and considers a model where aggregate output is a function of two variable inputs and the size of the fish stock. He shows that under open access not only will there be an incorrect level of total output, but also that the factor proportions chosen by individual profit maximizing participants will not minimize the cost of producing that level of output. The results follow from an assumption of an aggregate production function which shows diminishing returns to both variable inputs for any given stock size. It turns out that the average return to each variable input, the parameter used by participants to choose the level of input use, is greater than the marginal return, which is the parameter that must be used in order to obtain the optimum input mix.

Unfortunately all of these essays focus on the maximization of sustainable net revenue not the net present value of output and hence they do not consider user cost and dynamic adjustment.

Extinction

The issue of stock extinction both in an open access and an optimally controlled fishery has received considerable attention in the literature. Part of this had to do with the simplicity of the Gordon–Schaefer model which always generates a positive open access equilibrium stock size. Gould (Chapter 9) gives a more general analysis and shows that if the growth function goes to zero at positive stock sizes or if there are diminishing returns to effort, it is possible to show that an open access fishery may result in species extinction. Clark (Chapter 10) demonstrates the same thing by formulating a cost function where cost per unit of fish does not go to infinity as stock size approaches zero. He then shows that extinction is possible in open access if the price of fish is greater than the cost of catching the last fish. He goes on to show that under the same conditions, the maximization of NPV will also result in extinction if the interest rate is sufficiently high. This assumes that the value of the fish stock derives exclusively from the ability to provide food or raw materials. The policy conclusions do not necessarily hold if there are ecological interdependencies between stocks or if the stock has an intrinsic value through the provision of viewing opportunities, such as is the case with whales. Other essays not included in this collection add more to this general conclusion. Clark (1973b) provides more details on the conditions for optimal extinction; Cropper, Lee, and Pannu (1979) show that in the nonlinear case the size of the initial stock size is crucial in determining whether extinction is optimal. Clark and Munro (1978) look at extinction in the non-autonomous case where important parameters do not remain constant over time.

Disaggregated Models

The essay by Smith (Chapter 11) is a very important contribution to the literature on fisheries economics for two reasons. First, it produced a disaggregated model which allowed for specific consideration of activities at the individual firm or vessel level. The model also directly considered many issues that were studied independently in earlier works, including size at first capture, diminishing returns to effort at the seasonal level, and congestion externalities among vessels. The essay demonstrates very clearly the differences between open access and static optimal utilization taking all of these things into consideration. The essay is also the first to look at open access using short run dynamics rather than using the sustained yield curve as a frame of reference. This allows for the second new insight which is a detailed analysis of how a fishery will operate as it moves toward the bioeconomic equilibrium which had been taken more or less for granted in earlier works. Smith shows how existing private operators will operate so as to maximize profits given their capital and the size of the fish stock. Over time the fish stock will change according to the difference between catch and stock growth and the number of participants will vary according to periodic profits. A bioeconomic equilibrium will be obtained when growth equals catch and net returns equal annual fixed costs. The trajectory of the fishery toward that equilibrium depends upon the relative rates of change of fleet and stock size. In certain cases an equilibrium can be achieved; in others there can be permanent oscillations around a potential equilibrium point, or the stock can be pushed to extinction. Combining these results with the Clark and Munro analysis, it is clear that it is necessary to consider time paths of effort (output) and stock size when analyzing both open access and optimal utilization. The sustainable revenue curve does not allow for a complete examination.

Anderson (Chapter 12) expands upon the disaggregated analysis of Smith and presents a more detailed look at the relationship between firm and aggregate behaviour in a fishery in a way that is directly comparable to standard micro-economic analysis. The analysis differs from Smith's in that effort is viewed as the control variable rather than output. This allows for a direct comparison to earlier works and it also demonstrates that while vessels can directly control the amount of effort they produce, they can only indirectly control their vessel's catch rate: that depends upon the relationship between the activities of the entire fleet and short run harvest (that is, on whether the aggregate short run harvest function is linear or nonlinear), and on stock size. The mathematical analysis shows why the effects of changes in fleet size and in effort per boat on changes in annual catch, on changes in stock size, and on changes in real vessel costs must be considered in order to maximize net fishery wide returns. By looking at the way that effort is produced, the analysis also shows that optimal regulation must consider the number of vessels and output per vessel rather than just focusing on aggregate effort.

Regulation

Many of the essays in this collection focus on the difference between open access and optimal utilization and in the process provide some valuable information for designing regulation. The essays in this section focus more directly on the economic aspects of regulation. The essay by Scott (Chapter 13) summarizes some of the lessons offered by earlier essays and provides a detailed analysis of taxes and fishing rights. Crutchfield (Chapter 14) summarizes the differences between different types of regulations and compares them according to a number of different criteria. Clark (Chapter 15) provides a more rigorous analysis by providing a formal analysis of how individual participants react to various sorts of regulations. The bottom line to all of these discussions is that most traditional regulation methods such as gear restrictions, closed seasons, closed areas, and so on, will directly cause the cost of producing effort to go up because of the adverse effect on input choice, but in the long run, they will likely lose some of their efficacy in protecting the stock because of the way participants can react and adjust to the regulations. Total quotas (TACs), to the extent that they can be enforced, will not lose biological efficacy, but they do indirectly cause costs to increase by encouraging or exacerbating the race to fish.

On the other hand, controlling access by providing limited property rights in the form of individual transferable quotas (ITQs), will provide incentives to harvest efficiently. In essence, ITQs are based on the one traditional regulation that has the potential to achieve biological goals, but in addition, they produce incentives to correct for inefficiencies that would normally be present. Grafton (Chapter 16) presents a detailed analysis of ITQs in theory and practice. Arnason (Chapter 17) describes the workings of individual transferable quotas and also shows how the information provided by the market value for quota shares can help managers set the appropriate time path of quotas. Both essays show that ITQs provide incentives for the efficient harvest of the quota in any period, and Arnason shows that they provide a low cost method of determining the optimal time path of harvest. Boyce (Chapter 18) confirms the first point, but shows that ITQs, as traditionally implemented, cannot correct for intra-season inefficiencies caused by vessel congestion or other phenomena which will cause the aggregate annual harvest function to be nonlinear. Copes (Chapter 19) asserts that 'it is perhaps best to consider the individual quota a generally attractive management device, except where circumstances leave it vulnerable to serious problems' (p. 289). He then describes situations where ITQs may not work because of the peculiarities of a particular fishery. Townsend (Chapter 20) provides a survey of the effects of rights-based fishing. He finds that the evidence supports some of the theoretical predictions but contradicts others. Johnson and Libecap (Chapter 21) provide a slightly different picture of regulation by looking at the motivation for different parts of a heterogenous industry to favour different types of regulation.

Extensions

Multispecies Models

The problem of open access and optimal utilization is more complex than that illustrated in a model with one fleet independently harvesting a single stock. In most cases there is the simultaneous harvest of multiple stocks by different fleets and there can be technological or biological interrelationships between them. A technological relationship exists when one fleet inflicts mortality on stocks harvested by another fleet. A biological relationship exists when the population dynamics of the different stocks are related through predator–prey relationships or competition for habitat or food. Anderson (Chapter 22) describes the open access and optimal utilization of fisheries in these cases. It is shown that both the open access equilibrium and the dynamic optimum can occur with one or both of the fisheries operating, depending upon the types and strengths of the interdependencies. For optimal utilization, the marginal current net revenue from a unit of effort from either fleet must be balanced against the effect it will have on future net revenues of both fleets. Boyce (Chapter 23) looks at the technological interdependence (by-catch) issue in more detail giving particular attention to current real world regulation problems. By-catch is frequently regulated by setting total allowable catches (TACs) for the different species. However, this can cause problems if the by-catch ratios do not equal the ratios of the TACs because the TAC on one fishery will be reached before the TAC on another is achieved. Boyce considers open access and optimal policy choices under different conditions including prohibition of landing by-catch and the by-catch stock having existence value but no commercial value. See also Hannesson (1983) and Ward (1994).

International Utilization

The optimal utilization of a fish stock by more than one country poses some very interesting problems. Put simply, the derivation of an optimal time path of effort (harvest) implicitly assumes that the management agency can control all effort directed toward the stock. With unilateral management, this is not the case. In essence, the fall-back position that the government can correct for the lack of property rights does not hold. The problem becomes one of how individual governments behave, or should behave, when devising their own management strategies and/or negotiating with other countries to establish joint management rules. Questions of distribution are just as important as questions of efficiency. While a country can presumably control its own effort through time (absent international agreement), it has no authority over the effort of another country. This leads to the possibility that some or all of the potential gains from unilateral effort reductions can be dissipated through increases in effort by other countries. Game theory suggests that in a two country case, each government should choose its annual effort based on the estimated effort in the other country. Levhari and Mirman (Chapter 24) analyze this problem and show that if two countries each try to maximize its own welfare, taking into account the actions of the other, a long term equilibrium can be achieved. Moreover, assuming the countries are identical, they demonstrate that this equilibrium will occur at a lower stock level than would be optimal if the two countries formed a cooperative venture. Munro (Chapter 25) takes the problem one step further and studies the possibility of agreements on long term mutual harvesting strategies. The issues at stake are the share of fish given to each country and the weight given to the social discount rate, effort cost, and consumer preferences of each country when the harvest path is determined. The basis for the negotiations is a status quo distribution which, implicitly, is similar to the one that would follow from the Levhari and Mirman analysis. Agreement is based upon the principle that neither country will be worse off than they would be with no agreement. Hannesson (Chapter 26) extends the analysis by considering more than two countries in situations of both shared and straddling stocks, and considers threat strategies where strict adherence to the agreement cannot be guaranteed. He finds that the possibility of achieving a cooperative solution is good if nations perceive the 'game' as being repeating indefinitely (which seems reasonable for a renewable resource) and adopt retaliatory measures for those who deviate from the agreement, although the conclusion is sensitive to the number of nations involved and the differences in costs among nations. Other essays on this topic include Anderson (1975b), Vislie (1987), and Fischer and Mirman (1992 and 1996).

Uncertainty

The use of deterministic models simplifies the analysis but obviously causes potentially important aspects of open access and optimal use to be ignored. There is considerable uncertainty in both the biological and the economic aspects of fisheries. However, using expected utility analysis it is possible to rigorously analyze the effects of uncertainty on the conclusions of deterministic models. Andersen and Sutinen (Chapter 27) provide a useful summary of stochastic bioeconomic models. But as they point out, the effects of uncertainty on the theoretical conclusions are rarely unambiguous and, in some instances, differ little from deterministic results. It is not possible to make sweeping generalizations. Hannesson and Steinshamn (Chapter 28) provide a good example of analyzing fisheries with fluctuating stock sizes and at the same time provide information on the choice of constant effort or constant quota programs to promote stability, a question that has received much attention in real world management circles. They present a stochastic theoretical model and make it operational by using a recruitment function that is governed by a sinusoidal function. They show that a concave revenue function implies that a constant quota policy will allow for a higher expected profit, but a stock-dependent unit cost of landed fish implies that a constant effort policy is superior. For more detail on the expected utility approach to addressing uncertainty, see the references cited in the Andersen and Sutinen essay, Charles (1983a, 1983b), and Clark and Kirkwood (1986).

An interesting new area of inquiry resulting from management efforts to cope with uncertainty involves a spatial analysis of fish stocks (see below as well) and the use of marine reserves or marine protected areas. A marine reserve program would prohibit fishing in a certain area of the sea. The basic notion is that the closed area could act as a nursery ground which could safely supply the fishable area through migration or transport of larvae. Lauck, Clark, Mangel, and Munro (Chapter 29) argue for marine reserves on the basis of the inherent and irreducible uncertainties involved in managing fisheries. They provide an analysis which shows that marine reserves are the equivalent of bet hedging or portfolio diversification when faced with uncertainty. Hannesson (Chapter 30) produces a descriptive model which shows how an otherwise open access fishery will respond to marine reserves. The key variables are the percentage of the total area that is closed to fishing and the effective speed at which the fish migrate. While not taking the bet hedging question into account, he shows that much of the potential gains in stock size will be lost due to migration and there is no potential for economic gains because of increasing costs. He shows that the percentage of the area that will have to be closed in order to get any kind of stock enhancement will be very large indeed.

Beyond the Schaefer Model

The basic Schaefer population dynamics model assumes that recruitment, growth, and natural mortality can all be lumped together in a single function to show how stock growth is a function of stock size. While this is a very simplified model, it does capture the essence of stock reproduction and it has the added advantage that it allows for analytical solutions for models of open access and optimal utilization. The Beverton–Holt model which considers age structure, individual growth, and natural mortality separately is a more realistic biological model. A fish stock is divided into year classes called cohorts, the number of which is determined by the maximum age a fish can achieve. Each year a certain number of fish are recruited into the first year class. Over time, these individuals will grow to some maximum size and will be subject to natural mortality. The time path of biomass for the cohort will be a single peaked curve. Initially the increase in individual size will have more influence than will natural mortality, and biomass will increase, but eventually a point will be reached where the reverse is true and biomass will decrease to zero when the last fish dies. At any point in time, the biomass of the total fish stock will depend upon the total number of cohorts and the recruitment pattern over the previous years.

The analytics of economically optimal utilization are much more difficult in this case and Clark, Edwards, and Friedlaender (Chapter 31) were the first to analyze it. Looking first at the optimal utilization of a single year class, they show that there is an optimal time period over which the year class should be fished. Assuming that the biomass gets large enough for harvesting to be profitable, fishing should start at that point where rate of growth in the net value of the biomass (measured in terms of its market value minus harvesting cost) is equal to the interest rate. Fishing should continue until the biomass is reduced through fishing and natural mortality such that the net value of a unit of harvest is zero. They go on to analyze the more complex cases of conglomerate and sequential year classes under the assumption of constant recruitment and they describe the optimal harvest policy for both an over-exploited and a virgin stock. It is de monstrated that if mesh restrictions are not possible, pulse fishing may be a second best option. See also Hannesson (1975, 1986).

Sanchirico and Wilen (Chapter 32) present a model which takes the spatial dynamics of stocks and boats into account. The standard Schaefer analysis assumes that fish are uniformly distributed over the fishing grounds. Gordon did lay some groundwork for this analysis by considering the exploitation of two stocks in two different areas by the same fleet. Their analysis goes further and considers how a stock of fish can inhabit a broad area with concentrations in various patches. The size of the population in a patch will vary with reproduction but also, because of migration, between patches which may be related to the relative density of the stock in the two areas. When commercial fishing is considered in this case, it is necessary to take into account the relative profitability of fishing the different patches as well as the cost of moving from patch to patch. Fishing can affect stock size directly and indirectly by changing density and hence migration patterns. The fishery will tend to an equilibrium where the return net of transportation costs is equal across all areas.

Schooling Species

Another type of fishery which merits attention is schooling species, such as herring and sardines, which tend to gather in groups rather than disperse evenly across the fishing grounds. As a result, effective stock density will not decline with stock size. Therefore, if the schools can be easily located, the marginal productivity of fishing effort will not decrease with stock size. Such stocks are more vulnerable to extinction because fishing may be profitable 'until the last school is harvested.' Bjørndal and Conrad (Chapter 33) analyze the North Sea herring, a schooling fishery, and provide an empirical demonstration of the Smith model of an open access fishery by estimating equations for the rate of change in stock and fleet. The analysis is useful for its depiction of an actual open access time path of a fishery, but in addition, it shows the special problem of possible extinction facing a schooling species. As a contrast, Bjørndal (Chapter 34) studies the same fishery with respect to dynamically optimal utilization and estimates the required equations to derive the optimal stock size and he discusses the optimal trajectory for obtaining it, taking the special attributes of schooling species into account.

The Share System

While much of the early analysis of fisheries focused on the return to labour, the fact that returns to labour are based on a share system where trip revenues are divided among the boat owner, the captain and the crew are almost never considered in the context of open access and optimal utilization. Anderson (Chapter 35) presents such an analysis. He shows that in order to achieve efficient production in terms of input selection, the share system should allow for the sharing of revenues net of *all* variable costs. This will provide incentives to consider the cost of fishing and particularly the contribution of any one input to revenues relative to its purchase price. He also demonstrates that in a deterministic world, there will be incentives for the market to determine a share rate that will guarantee that crew members receive their opportunity wage and that boat owners receive returns equal to the annual opportunity cost of the boat. The analysis goes on to show how imperfections in the market which determines the share rate can affect the open access equilibrium and the desired effects of different regulation strategies. See also Sutinen (1979) and Ferris and Plourde (1982).

Recreational Fisheries

While much of the literature on optimal exploitation focuses on commercial fisheries, the basic results apply to recreational fisheries as well. Recreational fishers do not operate for profit but for the satisfaction or utility that is achieved by participation. Utility is a function of both quantity (the number of trips) and quality (catch per day or average size of catch, and so on). Since the quality variables are a function of stock size and sometimes the number of other participants, each participant can affect the utility of the others both in the short and long run. This is strictly analogous to the way commercial participants affect profits of other participants in the short and long run. Open access utilization will not lead to a time path of participation levels and stock size that will maximize the NPV of utility.

McConnell and Sutinen (Chapter 36) present a model where the quality variable is a function of stock size only, which is analogous to having a linear short run production function in the commercial fisheries. The only externalities will be in the long term through changes in stock size. Using this model they compare the difference between open access and optimal dynamic utilization. They also analyze open access and optimal utilization of the joint utilization of a single stock by commercial and recreational fishers. An interesting result is that optimal utilization will not necessarily result in equal percentage reductions in commercial and recreational effort.

Anderson (Chapter 37) expands the analysis by allowing the number of participants to be a variable, distinguishing between catch and landings so as to consider catch and release programs, considering heterogeneous participants, and using a quality variable which is a function of current participation as well as stock size. The latter assumption is analogous to a nonlinear short term production function in the commercial sector; there will be short and long run interactions.[7] Again the difference between open access and dynamic optimal utilization are demonstrated. Although optimal utilization will require a reduction in total fishing mortality, the analysis shows that this can be obtained by restricting the number of participants, the number of trips per participant, and landings per trip. It is not possible to generalize about how these reductions should be made; it will depend upon the nature of the utility functions. The response of individuals to various sorts of commonly used regulations is also analyzed.

Analysis of Management Agencies

In order to consider the NPV of fisheries utilization, it is necessary to consider all costs including those associated with fisheries management. Ideally this would include the costs of instituting and operating the government agencies which perform the required biological research, and determine and implement specific regulations.[8] If these costs vary with fisheries output (as opposed to being fixed annual expenditures), they will affect the optimal utilization path. Sutinen and Andersen (Chapter 38) consider this issue by analyzing optimal utilization with costly and imperfect enforcement. They study the profit maximizing behaviour of individual participants in a regulated fishery. Using the analysis of Becker, the decision to conform to regulations will, in part, depend upon the gains from cheating and the expected penalties. The standard results apply (higher fines and higher probability of being caught will increase compliance) but in addition it can be shown that increases in stock size may reduce compliance because of the increased returns from cheating. They also show that when enforcement costs depend upon stock size and the actual amount of harvest, the dynamically optimal stock size is not as large as it would be with costless enforcement. This is because with the enforcement cost, the total cost of investing in the stock is increased and so the amount of investment should be reduced.

Anderson and Lee (Chapter 39) take a different slant on this analysis by looking more closely at exactly what government agencies can do to achieve compliance. Agencies do not control output directly. Rather they can choose the governing instrument (taxes, quotas, and so on), the enforcement procedures and penalty structure, and the operation level of each. The effectiveness and the cost of enforcement will depend upon the combinations chosen. Also, participants may use costly resources to avoid detection. For example, they may off-load illegal catches at distant ports where the chance of being checked is lower. Not only will this mean that catch will be higher than desired but the overall cost of the fishery will increase. They provide an analysis of optimal enforcement policy. The results demonstrate the importance of the interrelationship among the three policy control variables. In addition it stresses the importance of efficacy versus cost. For example, while many of the traditional regulations may increase the real cost of producing effort, some of them may be relatively inexpensive to enforce, and effective in reducing catch. Overall, they may be the best choice.

Homans and Wilen (Chapter 40) take the analysis of management agencies one step further by making regulation an endogenous variable. They use a positive approach to try to shed light on why regulated fisheries operate the way they do as opposed to the traditional normative approach which focuses on optimal utilization. In their view, regulators are goal oriented and choose the type and level of regulation in order to achieve those goals. They model this in the case of TAC regulation by assuming the annual TAC is set according to a linear function of stock size: specifically, annual TAC $= -c + dX$ when that value is positive. The goals implicit in this policy are that stock size should never be below c/d and the desired stock size is where the TAC line intersects the Schaefer growth curve. See their Figure 2. If the stock is below the minimum acceptable level, the TAC will be zero. If stock is above the minimum acceptable level but below the desired level, TAC will be positive but less than growth and the stock will grow. If stock is above the desired level, the TAC will be greater than growth and stock will decrease. Basically they add another dimension to the Smith model. In any period, fishermen will try to maximize profits subject to the given set of regulations. Over time, the fleet will increase or decrease depending upon current returns, the stock will increase or decrease according to the difference between catch and stock growth, and the regulations will change depending upon the success of meeting the regulation goals. A true equilibrium will be achieved when fleet size, stock size, and regulation level do not change.

Applications

The final set of essays demonstrate how the theory can be applied in real world cases. They are important for the descriptions of the required data, the methods used, and the modifications to the basic theory which are necessary, in some cases, in order to perform the analysis. They also demonstrate how and where data can be obtained and how data availability can affect the types of work that can be accomplished. Other application essays are listed in their references.

Agnello and Donnelly (Chapter 41) verify the basic result of rent dissipation in open access fisheries, showing that property rights in oyster fisheries do in general make a significant difference in average labour productivity. Bjørndal (Chapter 42) uses duality theory to estimate a fisheries production function, and using a population dynamics model that captures reality better than the simple Schaefer model, is able to explain the workings of Clark's marginal stock effect. He also provides estimates of the optimal stock size. Kellogg, Easley, and Johnson (Chapter 43) look at the problem of the optimal utilization of a cohort, and estimate the equations necessary to specify the optimal time to open the North Carolina Bay Scallop fishery. Conrad (Chapter 44) provides the analysis to estimate an optimal time path of harvest in the Pacific whiting fishery using an 'approximately optimal rule'. Weninger (Chapter 45) estimates the efficiency gains from implementing an ITQ regime. He stresses that the rate at which these gains are achieved will depend upon the rate at which more efficient vessels can be constructed and that rate at which the quota is transferred to more efficient users. See Weninger and Just (1997) and Anderson (2000). He also provides a theoretical analysis of whether there will be specialized or diversified vessels when there is an ITQ program for more than one species. Thunberg, Helser, and Mayo (Chapter 46) provide an empirical bioeconomic simulation of the Atlantic Silver hake fishery that allows for an analysis of the selection pattern of age at first capture.

Notes

1 Jens Warming, a Danish Economist, covered many of the same issues in 1911. See the English translation of his article by Peder Andersen (1983).
2 The early analysis used the term common property, but open access is perhaps the more appropriate term. See Ciriacy-Wantrup and Bishop (1975).
3 The reader is warned that although the economic and biological equations are similar in most papers on fisheries economics, the notation is often different.
4 A more common form of the logistic equation is $rX(1-X/M)$ where r is the intrinsic growth rate of the stock. They are the same if $k_1 = r/M$.
5 See the Clark and Munro paper for a discussion of the possibility of a unique equilibrium.

6 The following discussion is based on their analysis of quasi-malleable capital.
7 McConnell and Sutinen also touch on catch and release and different classes of participants.
8 To be precise, it should also include any research costs undertaken by participants, but not funding for 'pure' research on fisheries biology.

Bibliography

Abdullah, Nik Mustapha Raja, Kuperan, K. and Pomeroy, R.S. (1998), 'Transaction Costs and Fisheries Co-Management', *Marine Resource Economics*, **13**(2), pp. 103–14.

Agnello, Richard J. and Donnelley, Lawrence P. (1975a), 'Prices and Property Rights in the Fisheries', *Southern Economic Journal*, **42**(2), pp. 253–62.

Agnello, Richard J. and Donnelley, Lawrence P. (1975b), 'Property Rights and Efficiency in the Oyster Industry', *The Journal of Law and Economics*, **18**, pp. 521–33.

Agnello, Richard J. and Donnelley, Lawrence P. (1984), 'Regulation and the Structure of Property Rights: The Case of the U.S. Oyster Industry', *Research in Law and Economics*, **6**, pp. 267–81.

Alvarez, Luis H.R. (1998), 'Optimal Harvesting Under Stochastic Fluctuations and Critical Depensation', *Mathematical Biosciences*, **152**, pp. 63–85.

Amundsen, Eirik S., Bjørndal, Trond and Conrad, Jon M. (1995), 'Open Access Harvesting of the Northeast Atlantic Minke Whale', *Environmental and Resource Economics*, **6**, pp. 167–85.

Andersen, Peder (1982), 'Commercial Fisheries Under Price Uncertainty', *Journal of Environmental Economics and Management*, **9**, pp. 11–28.

Andersen, Peder (1983), '"On Rent of Fishing Grounds": A Translation of Jens Warming's 1911 Article, with an Introduction', *History of Political Economy*, **15**(3), pp. 391–6.

Andersen, Peder and Sutinen, Jon G. (1984), 'Stochastic Bioeconomics: A Review of Basic Methods and Results', *Marine Resource Economics*, **1**(2), pp. 117–36.

Anderson, Eric E. (1986), 'Taxes vs. Quotas for Regulating Fisheries Under Uncertainty: A Hybrid Discrete-Time Continuous-Time Model', *Marine Resource Economics*, **3**(3), pp. 183–208.

Anderson, James L. (1985a), 'Market Interactions Between Aquaculture and the Common-Property Commercial Fishery', *Marine Resource Economics*, **2**(1), pp. 1–23.

Anderson, James L. (1985b), 'Private Aquaculture and Commercial Fisheries: Bioeconomics of Salmon Ranching', *Journal of Environmental Economics and Management*, **12**, pp. 353–70.

Anderson, Lee G. (1973), 'Optimum Economic Yield of a Fishery Given a Variable Price of Output', *Journal of the Fisheries Research Board of Canada*, **30**, pp. 509–18.

Anderson, Lee G. (1975a), 'Analysis of Open Access Commercial Exploitation and Maximum Economic Yield in Biologically and Technologically Interdependent Fisheries', *Journal of the Fisheries Research Board of Canada*, **32**, pp. 1825–42.

Anderson, Lee G. (1975b), 'Optimum Economic Yield in an Internationally Utilized Fishery', *Fishery Bulletin*, **70**(1), pp 51–66.

Anderson, Lee G. (1976), 'The Relationship between Firm and Fishery in Common Property Fisheries', *Land Economics*, **52**(1), pp.180–91.

Anderson, Lee G. (1980), 'The Necessary Components of Economic Surplus in Fisheries Economics', *Canadian Journal of Fisheries and Aquatic Sciences*, **37**(5), pp. 858–70.

Anderson, Lee G. (1982a), 'Optimal Utilization of Fisheries with Increasing Costs of Effort', *Canadian Journal of Fisheries and Aquatic Sciences*, **39**(2), pp. 211–14.

Anderson, Lee G. (1982b), 'The Share System in Open-Access and Optimally Regulated Fisheries', *Land Economics*, **58**(4), pp 435–49.

Anderson, Lee G. (1983), 'The Demand Curve for Recreational Fishing with an Application to Stock Enhancement Activities', *Land Economics*, **59**(3), pp. 279–86.

Anderson, Lee G. (1985), 'Potential Economic Benefits from Gear Restrictions and License Limitations', *Land Economics*, **61**(4), pp. 409–18.

Anderson, Lee G. (1986), *The Economics of Fisheries Management*, Johns Hopkins University Press, Baltimore, p. 265.

Anderson, Lee G. (1987), 'A Management Agency Perspective of the Economics of Fisheries Regulation', *Marine Resource Economics*, **4**(2), pp. 123–31.

Anderson, Lee G. (1989a), 'Enforcement Issues in Selecting Fisheries Management Policy', *Journal of Marine Resource Economics*, **6**, pp. 261–77.

Anderson, Lee G. (1989b), 'Optimal Intra- and Intersessional Harvesting Strategies When Price Varies with Individual Size', *Marine Resource Economics*, **6**(2), pp. 145–62.

Anderson, Lee G. (1991a), 'Efficient Policies to Maintain Total Allowable Catches in ITQ Fisheries with At-Sea Processing', *Land Economics*, **67**(2), pp. 141–57.

Anderson, Lee G. (1991b), 'A Note on Market Power in ITQ Fisheries', *Journal of Environmental Economics and Management*, **21**(3), pp. 291–6.

Anderson, Lee G. (1993), 'Toward a Complete Economic Theory of the Utilization and Management of Recreational Fisheries', *Journal of Environmental Economics and Management*, **24**(3), pp. 272–95.

Anderson, Lee G. (1994), 'An Economic Analysis of Highgrading in ITQ Fisheries Regulation Programs', *Marine Resource Economics*, **9**(3), pp. 209–26.

Anderson, Lee G. (1995), 'Privatizing Open Access Fisheries: Individual Transferable Quotas', *The Handbook of Environmental Economics*, D.W. Bromley, Oxford, UK, Blackwell, pp. 453–74.

Anderson, Lee G. (1999a), 'ITQs in Share System Fisheries: Implications for Efficiency, Distribution, and Tax Policy' in Ragnar Arnason and Hannes H. Gissurarson (eds) *Individual Transferable Quotas in Theory and Practice*, University of Iceland Press, Reykjavik, pp. 169–84.

Anderson, Lee G. (1999b), 'The Microeconomics of Vessel Behavior: A Detailed Short-run Analysis of the Effects of Regulation', *Marine Resource Economics*, **14**(2), pp. 29–50.

Anderson, Lee G. (2000), 'The Effects of ITQ Implementation: A Dynamic Approach', *Natural Resource Modeling*, **13**(4), pp. 1–36.

Anderson, Lee and Lee, Dwight (1986), 'Optimal Governing Instrument, Operation Level, and Enforcement in Natural Resource Regulation: The Case of the Fishery', *American Journal of Agricultural Economics*, **68**(3), pp. 678–90.

Andrews, Elizabeth J. and Wilen, James E. (1988), 'Angler Response to Success in the California Salmon Sportfishery: Evidence and Management Implications', *Marine Resource Economics*, **5**(2), pp. 125–38.

Androkovich, Robert A. and Stollery, Kenneth R. (1989), 'Regulation of Stochastic Fisheries: A Comparison of Alternative Methods in the Pacific Halibut Fishery', *Marine Resource Economics*, **6**(2), pp. 109–22.

Androkovich, Robert A. and Stollery, Kenneth R. (1991), 'Tax Versus Quota Regulation: A Stochastic Model of the Fishery', *American Journal of Agricultural Economics*, May, pp. 300–8.

Androkovich, Robert A. and Stollery, Kenneth R. (1994), 'A Stochastic Dynamic Programming Model of Bycatch Control in Fisheries', *Marine Resource Economics*, **9**(1), pp. 19–30.

Annala, J.H. (1996), 'New Zealand's ITQ System: Have the First Eight Years Been a Success or a Failure?', *Reviews in Fish Biology and Fisheries*, **6**(1), pp. 43–62.

Apostle, R., McCay, B.J. *et al.* (1997), 'The Politcal Construction of an ITQ Management System: The Mobile Gear ITQ Experiment in the Scotia Fundy Region of Canada', *Social Implications of Quota Systems in Fisheries*, G. Pálsson and G. Pétursdóttir, Copenhagen, Nordic Council of Ministers, pp. 27–49.

Armstrong, C.W. (1994), 'Co-operative Solutions in a Transboundary Fishery: The Russian–Norwegian Co-Management of the Arcto-Norwegian Cod Stock', *Marine Resource Economics*, **9**, pp. 329–51.

Arnason, Ragnar (1990), 'Minimum Information Management in Fisheries', *Canadian Journal of Economics*, **23**(3), pp. 630–53.

Arnason, Ragnar (1993), 'The Icelandic Individual Transferable Quota System: A Descriptive Account', *Marine Resource Economics*, **8**(3), pp. 201–18.

Arnason, Ragnar (1994), 'On Catch Discarding in Fisheries', *Marine Resource Economics*, **9**(3), pp. 189–207.

Arnason, Ragnar (1996), 'On the Individual Transferable Quota Fisheries Management System in Iceland', *Reviews in Fish Biology and Fisheries*, **6**(1), pp. 63–90.

Arnason, Ragnar (1998), 'Ecological Fisheries Management Using Individual Transferable Quotas', *Ecological Applications*, **8**(1) supplement, pp. 1151–9.

Bach, G.L. and Saunders, Phillip (1965), 'Economic Education: Aspirations and Achievements', *The American Economic Review*, **55**, pp. 329–56.

Barbier, E. and Strand, I. (1998), 'Valuing Mangrove-Fishery Linkages: A Case Study of Mexican Shrimp in Campeche, Mexico', *Environmental and Resource Economics*, **12**(2), pp. 151–66.

Batkin, K.M. (1996), 'New Zealand's Quota Management System: A Solution for the United States' Federal Fisheries Management Crisis?', *Natural Resources Journal*, **36**(4), pp. 855–80.

Battaglene, T. and Campbell, D. (1991), 'Survey Reveals Shark Fishermen are Struggling', *Australian Fisheries*, **50**(5), pp. 12–15.

Becker, Gary S. (1968), 'Crime and Punishment: An Economic Approach', *Journal of Political Economy*, **76**, pp. 169–217.

Beddington, John R. and Cooke, J.G. (1982), 'Harvesting from a Prey-Predator Complex', *Ecological Modelling*, **14**, pp. 155–77.

Beddington, John R. and May, Robert M. (1977), 'Harvesting Natural Populations in a Randomly Fluctuating Environment', *Science*, **19**, pp. 463–5.

Beddington, John R. and May, Robert M. (1982), 'The Harvesting of Interacting Species in a Natural Ecosystem', *Scientific American*, **247**, pp. 62–9.

Beddington, John R., Watts, C.M.K. and Wright, W.D.C. (1975), 'Optimal Cropping of Self-Reproducible Natural Resources', *Econometrica*, **43**(4), pp. 789–802.

Bell, Frederick W. (1968), 'The Pope and the Price of Fish', *American Economic Review*, **50**, pp. 1346–50.

Bell, Frederick W. (1972a), 'On Models of Commercial Fishing: A Defense of the Traditional Literature', *Journ. Pol. Econ.*, **80**, pp. 761–8.

Bell, Frederick W. (1972b), 'Technological Externalities and Common-Property Resources: An Empirical Study of the U.S. Northern Lobster Fishery', *Journal of Political Economy*, **80**(1), pp. 148–58.

Bell, Frederick W. (1978), *Food From The Sea: The Economics and Politics of Ocean Fisheries*, Boulder: Westview Press, 380 pp.

Bell, Frederick W. (1986), 'Competition from Fish Farming in Influencing Rent Dissipation: The Crawfish Fishery', *American Journal of Agricultural Economics*, February, pp. 95–101.

Bell, F.W., Carlsson, E.W. and Fullenbaum, R.F. (1971), 'Economics of Production from Natural Resources: Comment', *Am. Econ. Rev.*, **51**, pp. 483–7.

Benford, Frank A. (1995), 'A Model of the North Carolina Bay Scallop Fishery with Endogenous Fishing Effort and Entry', *Natural Resources Modeling*, **9**(3), pp. 197–228.

Berck, Peter (1979), 'Open Access and Extinction', *Econometrica*, **47**(4), pp. 877–82.

Berck, Peter (1981), 'Optimal Management of Renewable Resources with Growing Demand and Stock Externalities', *Journal of Environmental Economics and Management*, **8**, pp. 105–17.

Bishop, Richard C. (1973), 'Limitation of Entry in the United States Fishing Industry: An Economic Appraisal of a Proposed Policy', *Land Economics*, **49**(4), pp. 381–90.

Bishop, Richard C. and Samples, Karl C. (1980), 'Sport and Commerical Fishing Conflicts: A Theoretical Analysis', *Journal of Environmental Economics and Management*, **7**(3), pp. 220–33.

Bjørndal, Trond (1987), 'Production Economics and Optimal Stock Size in a North Atlantic Fishery', *Scandinavian Journal of Economics*, **89**(2), pp. 145–64.

Bjørndal, Trond (1988), 'The Optimal Management of North Sea Herring', *Journal of Environmental Economics and Management*, **15**, pp. 9–29.

Bjørndal, Trond (1989), 'Production in a Schooling Fishery: The Case of the North Sea Herring Fishery', *Land Economics*, **65**(1), pp. 49–56.

Bjørndal, Trond and Conrad, Jon M. (1987a), 'Capitol Dynamics in the North Sea Herring Fishery', *Marine Resource Economics*, **4**(1), pp. 63–74.

Bjørndal, Trond and Conrad, Jon M. (1987b), 'The Dynamics of an Open Access Fishery', *Canadian Journal of Economics*, **20**(1), pp. 74–85.

Bjørndal, Trond and Conrad, Jon M. (1993), 'Stock Size, Harvesting Costs, and the Potential for Extinction: The Case of Sealing', *Land Economics*, **69**(2), pp. 156–67.

Bjørndal, Trond and Conrad, Jon M. (1998), 'A Report on the Norwegian Minke Whale Hunt', *Marine Policy*, **22**(2), pp. 161–74.

Bjørndal, T. and Gordon, D.V. (1989), 'Price Response and Optimal Vessel Size in a Multi-Output Fishery', in *Rights Based Fishing* (eds P.A. Neher, R. Arnason and N. Mollett), Dordrecht: Kluwer Academic Publishers.

Bjørndal, Trond and Gordon, Daniel V. (1993), 'The Opportunity Cost of Capital and Optimal Vessel Size in Norwegian Fishing', *Land Economics*, **69**(1), pp. 98–107.

Bjørndal, Trond and Munro, Gordon (1997), 'The Economics of Fisheries Management: A Survey', manuscript.

Bjørndal, Trond and Salvanes, Kjell G. (1995), 'Gains from Deregulation? An Empirical Test for Efficiency Gains in Norwegian Fish Farming', *Journal of Agricultural Economics*, **46**(1), pp. 113–26.

Black, Neal D. (1997), 'Balancing the Advantages of Individual Transferable Quotas Against Their Redistributive Effects: The Case of Alliance Against IFQs v. Brown', *Georgetown International Law Review*, **9**(3), pp. 727–46.

Bockstael, Nancy E. and Opaluch, James J. (1983), 'Discrete Modeling of Supply Response Under Uncertainty: The Case of the Fishery', *Journal of Environmental Economics and Management*, **10**(2), pp. 125–37.

Bockstael, N.E., Strand, I.E. and Lipton, D.W. (1992), 'The Effect of Common Property on Optimal Generic Advertising', *Marine Resource Economics*, **7**, pp. 189–208.

Boyce, John R. (1992), 'ITQs and Production Externalities in a Fishery', *Natural Resource Modeling*, **6**(4), pp. 385–408.

Boyce, John R. (1995), 'Optimal Capital Accumulation in a Fishery: A Nonlinear Irreversible Investment Model', *Journal of Environmental Economics and Management*, **28**(3), pp. 324–39.

Boyce, John R. (1996), 'An Economic Analysis of the Fisheries Bycatch Problem', *Journal of Environmental Economics and Management*, **31**(3), pp. 314–36.

Boyd, R.O. and Dewees, C.M. (1992), 'Putting Theory into Practice: Individual Transferable Quotas in New Zealand's Fisheries', *Society and Natural Resources*, **5**(2), pp. 179–98.

Brandt, S.J. (1994), 'Effects of Limited Access Management on Substitutable Resources: A Case Study of the Surf Clam and Ocean Quahog Fishery', *Journal of Environmental Systems*, **23**(1), pp. 21–49.

Brown, Gardner (1974), 'An Optimal program for Managing Common Property Resources with Congestion Externalities', *Journal of Political Economy*, **82**(1), pp. 163–73.

Brown, Gardner and Roughgarden, Jonathan (1997), 'A Metapopulation Model with Private Property and a Common Pool', *Ecological Economics*, **22**(1), pp. 65–71.

Brubaker, E. (1996), 'The Ecological Implications of Establishing Property Rights in Atlantic Fisheries', in B.L. Crowley (ed.), *Taking Ownership: Property Rights and Fishery Management on the Atlantic Coast*, Halifax, Nova Scotia: Atlantic Institute for Market Studies, pp. 221–51.

Burt, O. and Cummings, R. (1969), 'The Economics of Production from Natural Resources: Note', *Am. Econ. Rev.*, **59**, pp. 985–90.

Burt, O. and Cummings, R. (1970), 'Production and Investment in Natural Resources Industries', *Am. Econ. Rev.*, **60**, pp. 576–90.

Burt, Oscar R. and Cummings, Ronald G. (1977), 'Natural Resource Management, the Steady State, and Approximately Optimal Decision Rules', *Land Economics*, **53**(1), pp. 1–22.

Buss, A., Strand, I. and Kirkley, J. (1996), 'Backward-Bending Labor Supply in Fisheries', *Journal of Environmental Economics and Management*, **31**(3), pp. 352–67.

Caddy, John F. and Seijo, Juan Carlos (1998), 'Application of a Spatial Model to Explore Rotating Harvest Strategies for Sedentary Species', in *Proc. of the North Pacific Symposium on Invertebrate Stock Accessment and Management*, (eds) G.S. Jamieson and A. Campbell, Can. Spec. Publ. Fish. Aquat. Sci., **125**, pp. 359–65.

Campbell, D., Battaglene, T. and Shafron, W. (1992), 'The Economics of Resource Conservation in the Southern Shark Fishery', *Australian Journal of Marine and Freshwater Research*, **43**(2), pp. 251–62.

Campbell, D., Brown, D. and Battatglene, T. (2000), 'Individual Transferable Catch Quotas: Austral-
ian Experience in the Southern Bluefin Tuna Fishery', *Marine Policy*, **24**(2), pp. 109–17.
Campbell, H.F. (1989), 'Fishery Buy-Back Programmes and Economic Welfare', *Australian Journal
of Agricultural Economics*, **33**(1), pp. 20–31.
Campbell, H.F. (1991), 'Estimating the Elasticity of Substitution Between Restricted and Unrestricted
Inputs in a Regulated Fishery: A Probit Approach', *Journal of Environmental Economics and
Management*, **20**(3), pp. 262–74.
Campbell, H.F. and Lindner, R.K. (1989), 'A Note on Optimal Effort in the Maldavian Tuna Fishery',
Marine Resource Economics, **6**(2), pp. 173–6.
Campbell, H.F. and Lindner, R.K. (1990), 'The Production of Fishing Effort and the Economic
Performance of Licence Limitation Programs', *Land Economics*, **66**(1), pp. 56–66.
Campbell, H.F. and Nicholl, R.B. (1994), 'Can Purse Seiners Target Yellowfin Tuna?', *Land Econom-
ics*, **70**(3), pp. 345–53.
Campbell, H.F., Hand, A.J. and Smith, A.D.M. (1993), 'A Bioeconomic Model for Management of
Orange Roughy Stocks', *Marine Resource Economics*, **8**(2), pp. 155–72.
Casey, Keith E., Dewees, Christopher M., Turris, Bruce R. and Wilen, James E. (1995), 'The Effects
of Individual Vessel Quotas in the British Columbia Halibut Fishery', *Marine Resource Economics*,
10(3), pp. 211–30.
Chambers, R.G. and Strand, I.E. (1985), 'Estimating Parameters of a Renewable Resource Model
Without Stock Data', *Marine Resource Economics*, **2**(3), pp. 263–74.
Charles, Anthony T. (1983a), 'Optimal Fisheries Investment: Comparative Dynamics for a Determin-
istic Seasonal Fishery', *Canadian Journal of Fisheries and Aquatic Sciences*, **40**, pp. 2069–79.
Charles, Anthony T. (1983b), 'Optimal Fisheries Investment Under Uncertainty', *Canadian Journal
of Fisheries and Aquatic Sciences*, **40**, pp. 2080–91.
Charles, Anthony T. (1985), 'Nonlinear Costs and Optimal Fleet Capacity in Deterministic and
Stochastic Fisheries', *Mathematical Biosciences*, **73**, pp. 271–99.
Charles, Anthony T. (1988), 'Fishery Socioeconomics: A Survey', *Land Economics*, **64**(3), pp. 276–
95.
Charles, Anthony T. (1989), 'Bio-Socio-Economic Fishery Models: Labour Dynamics and
Multiobjective Management', *Canadian Journal of Fisheries and Aquatic Sciences*, **46**(8), pp.
1313–22.
Charles, A.T. (1992), 'Fisheries Conflicts: A Unified Framework', *Marine Policy*, **16**(5), pp. 376–93.
Charles, Anthony T. and Munro, Gordon R. (1985), 'Irreversible Investment and Optimal Fisheries
Management: A Stochastic Analysis', *Marine Resource Economics*, **1**(3), pp. 247–64.
Charles, Anthony T. and Reed, William J. (1985), 'A Bioeconomic Analysis of Sequential Fisheries:
Competition, Coexistence, and Optimal Harvest Allocation Between Inshore and Offshore Fleets',
Canadian Journal of Fisheries and Aquatic Sciences, **42**(5), pp. 952–62.
Chen, Kuo-Shung and Lin, Chau-Jy (1981), 'Analysis of Modified Model for Commerical Fishing
with Possible Extinctive Fishery Resources', *Journal of Environmental Economics and Manage-
ment*, **8**, pp. 151–5.
Cheng, Hsiang-tai and Townsend, Ralph (1993), 'Potential Impact of Seasonal Closures in the U.S.
Lobster Fishery', *Marine Resource Economics*, **8**(2), pp. 101–17.
Cheng, Kuo-Shung, Lin, Chau-Li and Wang, Ar-Young (1981), 'Analysis of Modified Model for
Commercial Fishing with Possible Extinctive Fishery Resources', *Journal of Environmental Eco-
nomics and Management*, **8**, pp. 151–5.
Chiarella, Carl, Kemp, Murray C., Van Long, Ngo and Okuguchi, Koji (1984), 'On the Economics of
International Fisheries', *International Economic Review*, **25**(1), pp. 85–92.
Christy, Francis T., Jr. (1974), 'Fisherman Quotas: A Tentative Suggestion for Domestic Manage-
ment', *Occasional Paper No. 19*, Law of the Sea Institute, University of Rhode Island.
Christy, Francis T., Jr. (1975), 'Propery Rights in the World Ocean', *Natural Resources Journal*, **15**,
pp. 695–712.
Christy, F.T. (1996), 'The Death Rattle of Open Access and the Advent of Property Rights Regimes in
Fisheries', *Marine Resource Economics*, **11**(4), pp. 287–304.

Christy, F.T., Jr. and Scott, A.D. (1965), 'The Common Wealth in Ocean Fisheries', *Resources for the Future*, Baltimore: Johns Hopkins Press, p. 265.

Ciriacy-Wantrup, S.V. and Bishop, Richard C. (1975), '"Common Property" as a Concept in Natural Resources Policy', *Natural Resource Journal*, **15**(4), pp. 713–27.

Clark, Colin W. (1973a), 'The Economics of Overexploitation', *Science*, **181**, pp. 630–4.

Clark, Colin W. (1973b), 'Profit Maximization and the Extinction of Animal Species', *The Journal of Political Economy*, **81**(4), pp. 950–61.

Clark, Colin W. (1980), 'Towards a Predictive Model for the Economic Regulation of Commercial Fisheries', *Canadian Journal of Fisheries and Aquatic Sciences*, **37**(7), pp. 1111–29.

Clark, Colin W. (1985), *Bioeconomic Modelling and Fisheries Management*, Wiley-Interscience, New York, 291 pp.

Clark, Colin W. (1990), *Mathematical Bioeconomics: The Optimal Management of Renewable Resources*, John Wiley & Sons, Inc., New York, 386 pp.

Clark, Colin W., Clarke, Frank H. and Munro, Gordon R. (1979), 'The Optimal Exploitation of Renewable Resource Stocks: Problems of Irreversible Investment', *Econometrica*, **47**(1), pp. 25–47.

Clark, Colin W., Edwards, Gordon and Friedlaender, Michael (1973), 'Beverton-Holt Model of a Commercial Fishery: Optimal Dynamics', *Journal of the Fisheries Research Board of Canada*, **30**, pp. 1629–40.

Clark, Colin W. and Kirkwood, Geoffrey P. (1986), 'On Uncertain Renewable Resource Stocks: Optimal Harvest Policies and the Value of Stock Surveys', *Journal of Environmental Economics and Management*, **13**(3), pp. 235–44.

Clark, Colin W. and Mangel, Marc (1979), 'Aggregation and Fishery Dynamics: A Theoretical Study of Schooling and the Purse Seine Tuna Fisheries', *Fishery Bulletin*, **77**(2), pp. 317–37.

Clark, Colin W. and Munro, Gordon R. (1975), 'The Economics of Fishing and Modern Capital Theory: A Simplified Approach', *Journal of Environmental Economics and Management*, **2**, pp. 96–106.

Clark, Colin W. and Munro, Gordon R. (1978), 'Renewable Resource Management and Extinction', *Journal of Environmental Economics and Management*, **5**, pp. 198–205.

Clark, Colin W. and Munro, Gordon R. (1979), 'Fisheries and the Processing Sector: Some Implications for Management Policy', *The Bell Journal of Economics*, **7**, pp. 603–16.

Clark, I.N. (1993), 'Individual Transferable Quotas: The New Zealand Experience', *Marine Policy*, **17**, pp. 340–2.

Clarke, F.H. and Munro, G.R. (1987), 'Coastal States, Distant Water Fishing Nations and Extended Jurisdiction: A Principal-Agent Analysis', *Natural Resource Modeling*, **2**(1), pp. 81–107.

Clarke, Raymond P., Yoshimoto, Stacey S. and Pooley, Samuel G. (1992), 'A Bioeconomic Analysis of the Northwestern Hawaiian Islands Lobster Fishery', *Marine Resource Economics*, **7**, pp. 115–40.

Comitini, Salvatore and Huang, David S. (1967), 'A Study of Production and Factor Shared in the Halibut Fishing Industry', *The Journal of Political Economy*, **75**(4), pp. 366–71.

Conklin, James and Kolberg, William (1994), 'Chaos for the Halibut?', *Marine Resource Economics*, **9**(2), pp. 159–82.

Conrad, Jon M. (1982), 'Management of a Multiple Cohort Fishery: The Hard Clam in Great South Bay', *American Journal of Agricultural Economics*, August, 1982, pp. 463–74.

Conrad, Jon M. (1989), 'Bioeconomics and the Bowhead Whale', *Journal of Political Economy*, **97**(4), pp. 974–87.

Conrad, Jon M. (1992), 'A Bioeconomic Model of the Pacific Whiting', *Bulletin of Mathematical Biology*, **54**(2,3), pp. 219–39.

Conrad, Jon M. (1995), 'Bioeconomic Models of the Fishery', in Daniel W. Bromley (ed.) *The Handbook of Environmental Economics*, Blackwell Publishers: Cambridge, MA.

Conrad, Jon M. and Adu-Asamoah, Richard (1986), 'Single and Multispecies Systems: The Case of Tuna in the Eastern Tropical Atlantic', *Journal of Environmental Economics and Management*, **13**, pp. 50–68.

Conrad, Jon M. and Bjørndal, Trond (1991), 'A Bioeconomic Model of the Harp Seal in the North-west Atlantic', *Land Economics*, **67**(2), pp. 158–71.

Conrad, Jon M. and Clark, Colin W. (1987), *Natural Resource Economics: Notes and Problems*, New York: University of Cambridge, 213 pp.

Copes, Parzival (1970), 'The Backward-Bending Supply Curve of the Fishing Industry', *Scottish Journal of Political Economy*, **17**(1), pp. 69–77.

Copes, Parzival (1972), 'Factor Rents, Sole Ownership and the Optimum Level of Fisheries Exploitation', *The Manchester School of Economics and Social Studies*, **41**(2), pp. 145–63.

Copes, Parzival (1986), 'A Critical Review of the Individual Quotas: A Device in Fisheries Management', *Land Economics*, **62**(3), pp. 278–91.

Copes, Parzival (1997), 'Social Impacts of Fisheries Management Regimes Based on Individual Transferable Quotas', in G. Pálsson and G. Pétursdóttir (eds), *Social Implications of Quota Systems in Fisheries*, Copenhagen: Nordic Council of Ministers, pp. 61–90.

Cropper, M.L., Lee, Dwight R. and Pannu, Sukhraj Singh (1979), 'The Optimal Extinction of a Natural Resource', *Journal of Environmental Economics and Management*, **6**, pp. 341–49.

Cropper, M.L. (1988), 'A Note on the Extinction of Renewable Resources', *Journal of Environmental Economics and Management*, **15**, pp. 64–70.

Crothers, S. (1988), 'Individual Transferable Quotas: The New Zealand Experience', *Fisheries*, **13**(1), pp. 10–12.

Crutchfield, James A. (1956), 'Common Property Resources and Factor Allocation', *Canadian Journal of Economics and Political Science*, **22**, pp. 292–300.

Crutchfield, James A. (1961), 'An Economic Evaluation of Alternative Methods of Fishery Regulation', *Journal of Law and Economics*, **4**, pp. 131–43.

Crutchfield, James A. (1962), 'Valuation of Fishery Resources', *Land Economics*, **38**, pp. 145–54.

Crutchfield, James A. (1979), 'Economic and Social Implications of the Main Policy Alternatives for Controlling Fishing Effort', *Journal of the Fisheries Research Board of Canada*, **36**, pp. 742–52.

Crutchfield, J.A. and Pontecorvo, G. (1969), 'The Pacific Salmon Fisheries: A Study in Irrational Conservation', *Resources for the Future*, Baltimore: Johns Hopkins Press.

Crutchfield, J.A. and Zellner, A. (1962), 'Economic Aspects of the Pacific Halibut Fishery', *Fish. Ind. Research*, **1**(1), U.S. Gov't Printing Office, Washington, D.C.

Crutchfield, Stephen R. (1983), 'A Bioeconomic Model of an International Fishery', *Journal of Environmental Economics and Management*, **10**(4), pp. 310–28.

Crutchfield, Stephen R. and Gates, John M. (1985), 'The Impact of Extended Fisheries Jurisdiction on the New England Otter Trawl Fleet', *Marine Resource Economics*, **2**(2), pp. 153–74.

Cummings, Ronald G. and Burt, Oscar R. (1969), 'The Economics of Production from Natural Resources: Note', *The American Economic Review*, **59**, pp. 985–90.

Davis, A. (1996), 'Barbed Wire and Bandwagons: A Comment on ITQ Fisheries Management', *Reviews in Fish Biology and Fisheries*, **6**, pp. 97–108.

Deacon, Robert T. (1989), 'An Empirical Model of Fishery Dynamics', *Journal of Environmental Economics and Management*, **16**(2), pp. 167–83.

De Meza, David and Gould, J.R. (1985), 'Free Access vs. Private Ownership: A Comparison', *Journal of Economic Theory*, **36**, pp. 387–91.

Dewees, C.M. (1989), 'Assessment of the Implementation of Individual Transferable Quotas in New Zealand's Inshore Fishery', *North American Journal of Fisheries Management*, **9**, pp. 131–9.

Diaz-de-Leon, Antonio J. and Seijo, Juan Carlos (1992), 'A Multi-Criteria Non-Linear Optimization Model for the Control and Management of a Tropical Fishery', *Marine Research Economics*, **7**(2), pp. 23–40.

Dnes, Antony W. (1985), 'Rent Seeking Behavior and Open Access Fishing', *Scottish Journal of Political Economy*, **32**(2), pp. 159–70.

Doll, John P. (1988), 'Traditional Economic Models of Fishing Vessels: A Review with Discussion', *Marine Resource Economics*, **5**(2), pp. 99–123.

Dow, James P., Jr. (1993), 'Dynamic Regulation of Fisheries: The Case of the Bowhead Whale', *Marine Resource Economics*, **8**(2), pp. 145–54.

Duarte, C.C. (1992), Targeted Versus Nontargeted Multispecies Fishing', *Environmental and Resource Economics*, **2**, pp. 259–81.

Dudley, Norman and Waugh, Geoffrey (1980), 'Exploitation of a Single-Cohort Fishery Under Risk: A Simulation-Optimization Approach', *Journal of Environmental Economics and Management*, **7**, pp. 234–55.

Dupont, Diane (1990), 'Rent Dissipation in Restricted Access Fisheries', *Journal of Environmental Economics and Management*, **19**(1), pp. 26–44.

Dupont, Diane P. (1991), 'Testing for Input Substitution in a Regulated Fishery', *American Journal of Agricultural Economics*, **73**, pp. 155–64.

Dupont, Diane P. (1993), 'Price Uncertainty, Expectations Formation and Fisher's Location Choices', *Marine Resource Economics*, **8**(3), pp. 219–47.

Dupont, Diane P. and Phipps, Shelley A. (1991), 'Distributional Consequences of Fisheries Regulations', *Canadian Journal of Economics*, **24**(1), pp. 206–20.

Eales, James and Wilen, James E. (1986), 'An Examination of Fishing Location Choice in the Pink Shrimp Fishery', *Marine Resource Economics*, **2**(4), pp. 331–52.

Easley, J.E., Jr. (1992), 'Selected Issues in Modeling Allocation of Fishery Harvests', *Marine Resource Economics*, **7**(2), pp. 41–56.

Edwards, Steven F. and Murawski, Steven A. (1993), 'Potential Economic Benefits from Efficient Harvest of New England Groundfish', *North American Journal of Fisheries Management*, **13**, pp. 437–49.

Emerson, William and Anderson, James (1989), 'A Spatial Allocation Model for the New England Fisheries', *Marine Resource Economics*, **6**(2), pp. 123–44.

Feeny, David, Hanna, Susan and McEvoy, Arthur F. (1996), 'Questioning the "Tragedy of the Commons" Model of Fisheries', *Land Economics*, **72**(2), pp. 187–205.

Ferris, J.S. and Plourde, C.G. (1982), 'Labour Mobility, Seasonal Unemployment Insurance, and the Newfoundland Inshore Fishery', *Canadian Journal of Economics*, **15**(3), pp. 426–41.

Fischer, Ronald D. and Mirman, Leonard J. (1992), 'Strategic Dynamic Interactions: Fish Wars', *Journal of Economic Dynamic Control*, **16**, pp. 267–87.

Fischer, Ronald D. and Mirman, Leonard J. (1996), 'The Compleat Fish Wars: Biological and Dynamic Interactions', *Journal of Environmental Economics and Management*, **30**(1), pp. 34–42.

Fisher, Anthony C., Hanneman, W. Michael and Keeler, Andrew G. (1991), 'Integrating Fishery and Water Resources Management: A Biological Model of a California Salmon Fishery', *Journal of Environmental Economics and Management*, **20**(3), pp. 234–61.

Flaaten, Ola (1983), 'The Optimal Harvesting of a Natural Resource with Seasonal Growth', *Canadian Journal of Economics*, **16**(3), pp. 447–62.

Flaaten, Ola (1991), 'Bioeconomics of Sustainable Harvest of Competing Species', *Journal of Environmental Economics and Management*, **20**, pp. 163–80.

Flaaten, Ola, Heen, Knut and Salvanes, Kjell G. (1995), 'The Invisible Resource Rent in Limited Entry and Quota Managed Fisheries: The Case of Norwegian Purse Seine Fisheries', *Marine Resource Economics*, **10**(4), pp. 341–56.

Flam, S.D. and Storoy, S. (1982), 'Capacity Reduction on Norwegian Industrial Fisheries', *Canadian Journal of Fisheries and Aquatic Sciences*, **39**(9), pp. 1314–17.

Fletcher, Jerald J., Howitt, Richard E. and Johnston, Warren E. (1987), 'Management of Multipurpose, Heterogeneous Fishing Fleets Under Uncertainty', *Marine Resource Economics*, **4**(4), pp. 249–70.

Frick, Harold C. (1957), 'The Optimum Level of Fisheries Exploitation', *J. Fish. Res. Bd. Canada*, **14**(5), pp. 683–6.

Furlong, William J. (1991), 'The Deterrent Effect of Regulatory Enforcement in the Fishery', *Land Economics*, **67**(1), pp. 116–29.

Gallastegui, Carmen (1983), 'An Economic Analysis of Sardine Fishing in the Gulf of Valencia (Spain)', *Journal of Environmental Economics and Management*, **10**, pp. 138–50.

Garza-Gil, M.D. (1998), 'ITQ Systems in Multifleet Fisheries: An Application for Iberoatlantic Hake', *Environmental and Resource Economics*, **11**(1), pp. 79–92.

Gates, J.M. (1974), 'Demand Price, Fish Size and the Price of Fish', *Canadian Journal of Agricultural Economics*, **22**(3), pp. 1–12.

Gautam, Amy Buss, Strand, Ivar and Kirkley, James (1996), 'Leisure/Labor Tradeoffs: The Backward-Bending Labor Supply in Fisheries', *Journal of Environmental Economics and Management*, **31**(3), pp. 352–67.

Gauvin, John R., Ward, John M. and Burgess, Edward E. (1994), 'Description and Evaluation of the Wreckfish (*Polyprion Americanus*) Fishery Under Individual Transferable Quotas', *Journal of Marine Resource Economics*, **5**, pp. 1–20.

Geen, Gerry, and Nayar, Mark (1988), 'Individual Transferable Quotas in the Sourthern Bluefin Tuna Fishery: An Economic Appraisal', *Marine Resource Economics*, **5**(4), pp. 365–88.

Gillis, D.M., Peterman, R.M. and Pikitch, E.K. (1995), 'Implications of Trip Regulations for High-Grading: A Model of the Behavior of Fishermen', *Canadian Journal of Fisheries and Aquatic Sciences*, **52**, pp. 402–15.

Ginter, J.J.C. (1995), 'The Alaska Community Development Quota Fisheries Management Program', *Ocean and Coastal Management*, **28**(1–3), pp. 147–63.

Gordon, H. Scott (1953), 'An Economic Approach to the Optimum Utilization of Fisheries Resources', *Journal of the Fisheries Research Board of Canada*, **10**(7), pp. 442–57.

Gordon, H. Scott (1954), 'The Economic Theory of a Common-Property Resource: The Fishery', *Journal of Political Economy*, **62**(2), pp. 124–42.

Gould, J.R. (1972a), 'Externalities, Factor Proportions and the Level of Exploitation of Free Access Resources', *Economica*, **39**, pp. 383–402.

Gould, J.R. (1972b), 'Extinction of a Fishery by Commercial Exploitation: A Note', *Journal of Political Economy*, **80**, pp. 1031–38.

Grafton, R. Quentin (1992), 'Rent Capture in an ITQ Fishery', *Canadian Journal of Fisheries and Aquatic Sciences*, **49**(3), pp. 497–503.

Grafton, R. Quentin (1994), 'A Note on Uncertainty and Rent Capture in an ITQ Fishery', *Journal of Environmental Economics and Management*, **27**, pp. 286–94.

Grafton, R. Quentin (1995), 'Rent Capture in a Rights-Based Fishery', *Journal of Environmental Economics and Management*, **28**(1), pp. 48–67.

Grafton, R. Quentin (1996a), 'Individual Transferable Quotas and Canada's Atlantic Fisheries', in D.V. Gordon and G.R. Munro (eds), *Fisheries and Uncertainty: A Precautionary Approach to Resource Management*, Calgary: University of Calgary Press, pp. 129–54.

Grafton, R. Quentin (1996b), 'Individual Transferable Quotas: Theory and Practice', *Reviews in Fish Biology and Fisheries*, **6**(1), pp. 5–20.

Grafton, R.Q., Squires, D. *et al.* (1996), 'Private Property Rights and Crises in World Fisheries: Turning the Tide', *Contemporary Economic Policy*, **14**(4), pp. 90–9.

Griffen, Wade L. and Berattie, Bruce R. (1978), 'Economic Impact of Mexico's 200-Mile Offshore Fishing Zone on the United States Gulf of Mexico Shrimp Fishery', *Land Economics*, **54**(1), pp. 27–37.

Griffen, Wade, Hendrickson, Holly, Oliver, Chris, Matlock, Gary, Bryan, C.E., Reichers, Robin and Clark, Jerry (1993), 'An Economic Analysis of Texas Shrimp Season Closures', *Marine Fisheries Review*, **54**(3), pp. 21–8.

Hanemann, W.M. and Strand, I.E. (1993), 'Natural Resource Damage Assessment: Economic Implications for Fisheries Management', *American Journal of Agricultural Economics*, **75**(5), pp. 1188–94.

Hannesson, Rögnvaldur (1974), 'Relation Between Reproductive Potential and Sustained Yield of Fisheries', *Journ. Fish. Res. Bd. Canada*, **31**, pp. 359–62.

Hannesson, Rögnvaldur (1975), 'Fishery Dynamics: A North Atlantic Cod Fishery', *Canadian Journal of Economics*, **8**(2), pp. 151–73.

Hannesson, Rögnvaldur (1982), 'A Note on Socially Optimal versus Monopolistic Exploitation of a Renewable Resource', *Journal of Economics*, **43**(1), pp. 63–70.

Hannesson, Rögnvaldur (1983a), 'Bioeconomic Production Function in Fisheries: Theoretical and Empirical Analysis', *Canadian Journal of Fisheries and Aquatic Science*, **40**, pp. 968–82.

Hannesson, Rögnvaldur (1983b), 'Optimal Harvesting of Ecologically Interdependent Fish Species', *Journal of Environmental Economics and Management*, **10**, pp. 329–45.

Hannesson, Rögnvaldur (1985), 'The Effects of a Fishermen's Monoply in the Market for Inprocessed Fish', *Marine Resource Economics*, **2**(1), pp. 75–85.

Hannesson, Rögnvaldur (1986a), 'The Effect of Discount Rate on the Optimal Exploitation of Renewable Resources', *Marine Resource Economics*, **3**(4), pp. 319–30.

Hannesson, Rögnvaldur (1986b), 'Inefficiency Through Government Regulations: The Case of Norway's Fishery Policy', *Marine Resource Economics*, **2**(2), pp. 115–42.

Hannesson, Rögnvaldur (1986c), 'Optimal Thinning of a Year-Class with a Density-Dependent Growth', *Canadian Journal of Fisheries and Aquatic Science*, **43**, pp. 889–92.

Hannesson, Rögnvaldur (1987), 'Optimal Catch Capacity and Fishing Effort in Deterministic and Stochastic Fishery Models', *Fisheries Research*, **5**, pp. 1–21.

Hannesson, Rögnvaldur (1989), 'Fixed or Variable Catch Quotas? The Importance of Population Dynamics and Stock Dependent Costs', *Marine Resource Economics*, **5**(4), pp. 415–32.

Hannesson, Rögnvaldur (1991), 'How to Set Catch Quotas: Constant Effort or Constant Catch?', *Journal of Environmental Economics and Management*, **20**, pp. 71–91.

Hannesson, Rögnvaldur (1993a), *Bioeconomic Analysis of Fisheries*, Cambridge, MA: Blackwell Scientific Publications, 138 pp.

Hannesson, Rögnvaldur (1993b), 'Fishing Capacity and Harvest Rules', *Marine Resource Economics*, **8**, pp. 133–43.

Hannesson, Rögnvaldur (1994a), 'International Transfer of Excess Allowable Catches', *Land Economics*, **70**(3), pp. 330–44.

Hannesson, Rögnvaldur (1994b), 'Optimum Fishing Capacity and International Transfer of Excess Allowable Catch', *Land Economics*, **70**(3), pp. 330–44.

Hannesson, Rögnvaldur (1996), 'Sequential Fishing: Cooperative and Non-Cooperative Equilibria', *Natural Resource Modeling*, **9**(1), pp. 51–9.

Hannesson, Rögnvaldur (1997), 'Fishing as a Supergame', *Journal of Environmental Economics and Management*, **32**(3), pp. 309–22.

Hannesson, Rögnvaldur (1998), 'Marine Reserves: What Do They Accomplish', *Marine Resource Economics*, **13**(3), pp, 159–70.

Hannesson, Rögnvaldur and Steinshamn, Stein Ivar (1991), 'How to Set Catch Quotas: Constant Effort or Constant Catch', *Journal of Environmental Economics and Management*, **20**(1), pp. 71–91.

Hardin, G. (1968), 'The Tragedy of the Commons', *Science*, **162**, pp. 1143–248.

Henderson, J.V. and Tugwell, M. (1979), 'Exploitation of the Lobster Fishery: Some Empirical Results', *Journal of Environmental Economics and Management*, **6**, pp. 287–96.

Hendrickson, H.M. and Griffin, W.L. (1993), 'An Analysis of Management Policies for Reducing Shrimp By-catch in the Gulf of Mexico', *North American Journal of Fisheries Management*, **13**, pp. 686–97.

Hermann, M. (1996), 'Estimating the Induced Price Increase for Canadian Pacific Halibut with the Introduction of the Individual Vessel Quota Program', *Canadian Journal of Agricultural Economics*, **44**(2), July 1996, pp. 151–64.

Herrick, Samuel F., Jr. and Strand, Ivar (1994), 'Application of Benefit-Cost Analysis to Fisheries Allocation Decisions: The Case of Alaska Walleye Pollock and Pacific Cod', *North American Journal of Fisheries Management*, **14**(4), pp. 726–41.

Herrick, S., Strand, I., Squires, D., Miller, M., Lipton, D., Walden, J. and Freese, S. (1994), 'Fundamental Problems Using Benefit Cost Analysis in Fisheries', *North American Journal of Fisheries Management*, **14**, pp. 726–41.

Hoagland, Porter and Jin, Di (1997), 'A Model of By-catch Involving a Passive Use Stock', *Marine Resource Economics*, **12**(1), pp. 11–28.

Holland, Daniel S. and Brazee, Richard J. (1996), 'Marine Reserves for Fisheries Management', *Marine Resource Economics*, **11**(2): pp. 157–71.

Homans, Frances R. and Wilen, James E. (1997), 'A Model of Regulated Open Acess Resource Use', *Journal of Environmental Economics and Management*, **32**(1), pp. 1–21.

Huppert, D.D. (1979), 'Implications of Multipurpose Fleets and Mixed Stocks for Control Policies', *Journal of the Fisheries Research Board of Canada*, **36**, pp. 845–54.

Huppert, Daniel D. and Squires, Dale (1986), 'Potential Economic Benefits and Optimum Fleet Size in the Pacific Coast Trawl Fleet', *Marine Resource Economics*, **34**(4), pp. 297–318.

Huppert, Daniel D., Ellis, Gregory M. and Noble, Benjamin (1996), 'Do Permit Prices Reflect the Discontinued Value of Fishing? Evidence from Alaska's Commercial Salmon Fisheries', *Canadian Journal of Fisheries and Aquatic Sciences*, **53**(4), pp. 761–68.

Johnson, Barry L., Milliman, Scott R., Bishop, Richard C. and Kitchell, James F. (1992), 'Evaluating Fishery Rehabilitation under Uncertainty: A Bioeconomic Analysis of Quota Management', *North American Journal of Fisheries Management*, **12**, pp. 703–20.

Johnson, Ronald N. (1995), 'Implications of Taxing Quota Value in an Individual Transferable Quota Fishery', *Marine Resource Economics*, **10**(4), pp. 327–40.

Johnson, Ronald N. and Libecap, Gary D. (1982), 'Contracting Problems and Regulation: The Case of the Fishery', *The American Economic Review*, **72**(5), pp. 1005–22.

Johnston, Robert J. and Sutinen, Jon G. (1996), 'Uncertain Biomass Shift and Collapse: Implications for Harvest Policy in the Fishery', *Land Economics*, **72**(4), pp. 500–18.

Kahn, James R. (1987), 'Measuring the Economic Damages Associated with Terrestrial Pollution of Marine Ecosystems', *Marine Resource Economics*, **4**(3), pp. 193–209.

Kahn, James R. and Kemp, W. Michael (1985), 'Economic Losses Associated with the Degradation of an Ecosystem: The Case of Submerged Aquatic Vegetation in Chesapeake Bay', *Journal of Environmental Economics and Management*, **12**, pp. 246–63.

Kahn, James R. and Rockel, Mark (1988), 'Measuring the Economic Effects of Brown Tides', *Journal of Shellfish Research*, **7**(4), pp. 677–82.

Kaitala, Viejo and Munro, Gordon R. (1993), 'The Management of High Seas Fisheries', *Marine Resource Economics*, **8**(4), pp. 313–29.

Kamien, Morton, Levhari, David and Mirman, Leonard J. (1985), 'Dynamic Model of Fishing: The Relationship to Conjectural Variations', *Journal of Environmental Economics and Management*, **12**(4), pp. 308–21.

Karpoff, Jonathan M. (1984), 'Insights from the Markets for Limited Entry Permits in Alaska', *Canadian Journal of Fisheries and Aquatic Sciences*, **41**(8), pp. 1160–6.

Karpoff, Jonathan M. (1985a), 'Non-Pecuniary Benefits in Commercial Fishing: Empirical Findings from the Alaska Salmon Fisheries', *Economic Inquiry*, **23**, pp. 159–74.

Karpoff, Jonathan M. (1985b), 'Time, Capital Intensity, and the Cost of Fishing Effort', *Western Journal of Agricultural Economics*, **10**(2), pp. 254–8.

Karpoff, Jonathan M. (1987), 'Suboptimal Controls in Common Resource Management: The Case of the Fishery', *Journal of Political Economy*, **95**(1), pp. 179–94.

Katz, Eliakim and Smith, J. Barry (1988), 'Rent-Seeking and Optimal Regulation in Replenishable Resource Industries', *Public Choice*, **59**, pp. 25–36.

Kellogg, Robert L., Easley, J.E. Jr. and Johnson, Thomas (1988), 'Optimal Timing of Harvest for the North Carolina Bay Scallop Fishery', *American Agricultural Economics Association*, **70**, pp. 50–62.

Kennedy, John O.S. (1987), 'A Computable Game Theoretic Approach to Modeling Competitive Fishing', *Marine Resource Economics*, **4**, pp. 1–14.

Kennedy, John O.S. (1992), 'Optimal Annual Changes from Multicohort Fish Stocks: The Case of the Western Mackeral', *Marine Resource Economics*, **7**(3), pp. 95–114.

Kennedy, John O.S. and Watkins, James W. (1986), 'Time-Dependent Quotas for the Southern Bluefin Tuna Fishery', *Marine Resource Economics*, **2**(4), pp. 293–314.

Kirkley, James E. and Strand, Ivar E. (1988), 'The Technology and Management of Multi-species Fisheries', *Applied Economics*, **20**, pp. 1279–92.

Kirkley, J.E., Squires, D. and Strand, I. (1995), 'Assessing Technical Efficiency in Commercial Fisheries: An Application to the Mid-Atlantic Sea Scallop Fishery', *American Journal of Agricultural Economics*, **77**(3), pp. 686–97.

Kirkley, J., Squires, D. and Strand, I. (1998), 'Characterizing Managerial Skill and Technical Efficiency in a Fishery', *Journal of Productivity Analysis*, **9**, pp. 145–60.

Koenig, Evan F. (1984a), 'Controlling Stock Externalities in a Common Property Fishery Subject to Uncertainty', *Journal of Environmental Economics and Management*, **11**(2), pp. 124–38.

Koenig, Evan F. (1984b), 'Fisheries Regulation Under Uncertainty: A Dynamic Analysis', *Marine Resource Economics*, **1**(2), pp. 193–208.

Kolbert, William C. (1993), 'Quick and Easy Optimal Approach Paths for Nonlinear Natural Resource Models', *American Journal of Agricultural Economics*, **75**, pp. 685–95.

Kuperan, K. and Abdullah, Nik Mustapha Raja (1994), 'Coastal Small-Scale Fisheries and Co-Management', *Marine Policy*, **18**(4), pp. 306–13.

Kuperan, K. and Sutinen, Jon G. (1998), 'Blue Water Crime, Deterrence, Legitimacy and Compliance in Fisheries', *Law and Society Review*, **32**(2), pp. 309–37.

Kuronuma, Yoshihiro and Tisdell, Clement A. (1994), 'Economics of Antarctic Minke Whale Catches: Sustainability and Welfare Considerations', *Marine Resource Economics*, **9**(2), pp. 141–58.

Larkin, Sherry L. and Sylvia, Gilbert (1999), 'Intrinsic Fish Characteristics and Intraseason Production Efficiency: A Management-Level Bioeconomic Analysis of a Commercial Fishery', *American Journal of Agricultural Economics*, **81**(1), pp. 29–43.

Larson, D.M., House, B.W. and Terry, J.M. (1996), 'Toward Efficient Bycatch Management in Multispecies Fisheries: A Nonparametric Approach', *Marine Resource Economics*, **11**, pp. 181–201.

Larson, Douglas M., House, Brett W. and Terry, Joseph M. (1998), 'Bycatch Control in Multipurpose Fisheries', *American Journal of Agricultural Economics*, **80**(4), pp. 778–92.

Lauck, Tim, Clark, Colin W., Mangel, Marc and Munro, Gordon R. (1998), 'Implementing the Precautionary Principle in Fisheries Management Through Marine Reserves', *Ecological Applications*, **8**(1) supplement, pp. s72–s78.

Layman, R. Craig, Boyce, John R. and Criddle, Keith R. (1996), 'Economic Valuation of the Chinook Salmon Sport Fishery of the Gulkana River, Alaska, under Current and Alternative Management Plans', *Land Economics*, **72**(1), pp. 113–28.

Lepiz, Luis Guillermo and Sutinen, Jon G. (1985), 'Surveillance and Enforcement Operations in the Costa Rican Tuna Fishery', *Marine Policy*, **10**(5), pp. 310–21.

Leung, Anthony and Wang, Ar-Young (1976), 'Analysis of Models for Commercial Fishing: Mathematical and Economical Aspects', *Econometrica*, **44**(2), pp. 295–303.

Levhari, David and Michener, Ron (1981), 'Dynamic Programming Models of Fishing: Competition', *The American Economic Review*, **71**(4), pp. 649–61.

Levhari, David and Mirman, Leonard J. (1980), 'The Great Fish War: An Example Using a Dynamic Cournot-Nash Solution', *The Bell Journal of Economics*, **11**(1), pp. 322–34.

Lewis, Tracy R. (1981), 'Exploitation of a Renewable Resource Under Uncertainty', *Canadian Journal of Economics*, **14**(3), pp. 422–39.

Lewis, T.R. and Schmalensee, R. (1977), 'Nonconvexity and the Optimal Exhaustion of Renewable Resources', *International Economic Review*, **18**, pp. 535–52.

Lindner, R.K., Campbell, H.F. and Bevin, G.F. (1992), 'Rent Generation During the Transition to a Managed Fishery: The Case of the New Zealand ITQ', *Marine Resource Economics*, **7**(4), pp. 229–48.

Lipton, Douglas W. and Strand, Ivar E., Jr. (1989), 'The Effect of Common Property on the Optimal Structure of the Fishing Industry', *Journal of Environmental Economics and Management*, **16**, pp. 45–51.

Lipton, Douglas W. and Strand, Ivar E. (1992), 'Effect of Stock Size and Regulations on Fishing Industry Cost and Structure: The Surf Clam Industry', *American Journal of Agricultural Economics*, February, pp. 197–208.

McCay, B.J., Apostle, R. *et al.* (1995), 'Individual Transferable Quotas (ITQs) in Canadian and U.S. Fisheries', *Ocean and Coastal Management*, **28**(1–3), pp. 85–116.

McCay, B.J., Apostle, R. *et al.* (1998), 'Individual Transferable Quotas, Comanagement, and Community: Lessons from Nova Scotia', *Fisheries*, **23**(4), pp. 20–3.

McConnell, Kenneth E. (1995), 'The Economics of Outdoor Recreation', in Allen V. Kneese and James L. Sweeney (eds), *Handbook of Natural Resource and Energy Economics, II*, Amsterdam: North Holland.

McConnell, Kenneth E. and Strand, Ivar E. (1989), 'Benefits from Commercial Fisheries When Demand and Supply Depend on Water Quality', *Journal of Environmental Economics and Management*, **17**(3), pp. 284–92.

McConnell, Kenneth E. and Sutinen, John G. (1979), 'Bioeconomic Models of Marine Recreational Fishing', *Journal of Environmental Economics and Management*, **6**, pp. 127–39.

McDonald, A. David and Hanf, Claus-Hennig (1992), 'Bio-economic Stability of the North Sea Shrimp Stock with Endogeneous Fishing Effort', *Journal of Environmental Economics and Management*, **22**, pp. 38–56.

McKelvey, Robert (1982), 'The Fishery in a Fluctuating Environment: Coexistence of Specialist and Generalist Fishing Vessels in a Multipurpose Fleet', *Journal of Environmental Economics and Management*, **10**, pp. 287–309.

McKelvey, Robert (1986), 'Economic Regulation of Targeting Behavior in Multispecies Fishery', *Natural Resource Modelling*, **1**, pp. 171–89.

MacMillan, Douglas C. and Ferrier, Robert C. (1994), 'A Bioeconomic Model for Estimating the Benefits of Acid Rain Abatement to Salmon Fishing: A Case Study in South West Scotland', *Journal of Environmental Planning and Management*, **37**(2), pp. 131–44.

McRae, James J. (1978), 'Optimal and Competitive Use of Replenishable Natural Resources by Open Economics', *Journal of International Economics*, **8**, pp. 29–54.

Mangel, Marc and Clark, Colin W. (1983), 'Uncertainty, Search, and Information in Fisheries', *J. Cons. Int. Explor. Mer.*, **41**, pp. 93–103.

Marasco, R.J. and Terry, J.M. (1982), 'Controlling Incidental Catch: An Economic Analysis of Six Management Options', *Marine Policy*, **8**, pp. 131–9.

Markusen, James R. (1976), 'Production and Trade from International Common Property Resources', *Canadian Journal of Economics*, **9**(2), May, pp. 309–19.

Marvin, K.A. (1992), 'Protecting Common Property Resources Through the Marketplace: Individual Transferable Quotas for Surf Clams and Ocean Quahogs', *Vermont Law Review*, **16**(Spring), pp. 1127–68.

Matthiasson, Thorolful (1992), 'Principles for Distribution of Rent from a "Commons"', *Marine Policy*, **16**(3), pp. 210–31.

Matthiasson, Thorolful (1996), 'Why Fishing Fleets Tend to be "Too Big"', *Marine Resource Economics*, **11**(3), pp. 173–9.

Matthiasson, Thorolful (1997), 'Consequences of Local Government Involvement in the Icelandic ITQ Market', *Marine Resource Economics*, **12**(2), pp. 107–26.

Matulich, S.C., Mittelhammer, R.C. *et al.* (1996), 'Toward a More Complete Model of Individual Transferable Fishing Quotas: Implications of Incorporating the Processing Sector', *Journal of Environmental Economics and Management*, **31**(1), pp. 112–28.

May, Robert, Beddington, John R., Clark, Colin W., Holt, Sidney J. and Laws, Richard M. (1979), 'Management of Multispecies Fisheries', *Science*, **205**, pp. 267–77.

Megrey, Bernard A. and Wespestad, Vidar G. (1990), 'Alaskan Groundfish Resources: 10 Years of Management Under the Magnuson Fishery Conservation and Management Act', *North American Journal of Fisheries Management*, **10**(2), pp. 125–43.

Mendelssohn, Roy (1978), 'Optimal Harvesting Strategies for Stochastic Single-Species, Multiage Class Models', *Mathematical Biosciences*, **41**, pp. 159–74.

Mendelssohn, Roy (1982), 'Discount Factors and Risk Aversion in Managing Random Fish Populations', *Canadian Journal of Fisheries and Aquatic Sciences*, **39**(9), pp. 1252–7.

Merrifield, J. (1999), 'Implementation Issues: The Political Economy of Efficient Fishing', *Ecological Economics*, **30**(1), pp. 5–12.

Mesterton-Gibbsons, Michael (1987), 'On the Optimal Policy for Combined Harvesting of Independent Species', *Natural Resource Modeling*, **2**, pp. 109–34.

Mesterton-Gibbsons, Michael (1988), 'On the Optimal Policy for Combining Harvesting of Predator and Prey', *Natural Resource Modeling*, **3**(1), pp. 3–43.

Milliken, W.J. (1994), 'Individual Transferable Fishing Quotas and Antitrust Law', *Ocean and Coastal Law Journal*, **1**, pp. 35–58.

Milliman, Scott R. (1986), 'Optimal Fishery Management in the Presence of Illegal Activity', *Journal of Environmental Economics and Management*, **13**, pp. 363–81.

Milliman, Scott R. and Johnson, Barry L. (1992), 'The Bioeconomics of Resource Rehabilitation: A Commercial-Sport Analysis for a Great Lakes Fishery', *Land Economics*, **68**(2), pp. 191–210.

Mirman, Leonard J. and Spulber, Daniel F. (1985), 'Fishery Regulation with Harvest Uncertainty', *International Economic Review*, **26**(3), pp. 731–45.

Moloney, David G. and Pearse, Peter H. (1979), 'Quantitative Rights as an Instrument for Regulating Commercial Fisheries', *J. Fish. Res. Board*, **36**, pp. 859–66.

Morey, Edward R. (1980), 'Fishery Economics: An Introduction and Review', *Natural Resources Journal*, October, 20, pp. 827–51.

Morey, Edward R. (1986), 'A Generalized Harvest Function for Fishing: Allocating Effort Among Common Property Cod Stocks', *Journal of Environmental Economics and Management*, **13**, pp. 30–49.

Moxnes, E. (1998a), 'Not Only the Tragedy of the Commons, Misperceptions of Bioeconomics', *Management Science*, **44**(9), pp. 1234–48.

Moxnes, E. (1998b), 'Overexploitation of Renewable Resources: The Role of Misperceptions', *Journal of Economic Behavior and Organization*, **37**(1), pp. 107–27.

Munro, Gordon R. (1979), 'The Optimal Management of Transboundary Renewable Resources', *Canadian Journal of Economics*, **12**, pp. 355–76.

Munro, Gordon R. (1982), 'Fisheries, Extended Jurisdiction and the Economics of Common Property Resources', *Canadian Journal of Economics*, **15**(3), pp. 405–25.

Munro, Gordon R. (1986), 'The Management of Shared Fishery Resources Under Extended Jurisdiction', *Marine Resource Economics*, **3**(4), pp. 271–96.

Munro, Gordon R. (1990), 'The Optimal Management of Transboundary Fisheries: Game Theoretic Considerations', *Natural Resource Modeling*, **4**(4), pp. 403–25.

Munro, Gordon R. (1992), 'Mathematical Bioeconomics and the Evolution of Modern Fisheries Economics', *Bulletin of Mathematical Biology*, **54**(2/3), pp. 163–84.

Munro, Gordon and Scott, Anthony D. (1985), 'The Economics of Fisheries Management', in Allen V. Kneese and James L. Sweeney (eds), *Handbook of Natural Resource and Energy Economics, II*, Amsterdam: North Holland.

Neher, Philip A. (1974), 'Notes on the Volterra-Quadratic Fishery', *Journal of Economic Theory*, **8**, pp. 39–49.

Neher, P.A., Arnason, R. and Mollett, N. (1989), *Rights Based Fishing*, Dordrecht: Kluwer Academic Publishers.

Newbery, David M. (1975), 'Congestion and Over-Exploitation of Free Access Resources', *Economica*, August, pp. 243–60.

O'Boyle, R., Annand, C. *et al.* (1994), 'Individual Quotas in the Scotian Shelf Groundfishery off Nova Scotia, Canada', in K. Gimbel (ed.), *Limiting Access to Marine Fisheries: Keeping the Focus on Conservation*, Washington, DC: Center for Marine Conservation and World Wildlife Fund, pp. 152–68.

Opsomer, Jean-Didier and Conrad, Jon M. (1994), 'An Open-Access Analysis of the Northern Anchovy Fishery', *Journal of Environmental Economics and Management*, **27**(1), pp. 21–37.

Organisation for Economic Co-operation and Development (1993), *The Use of Individual Quotas in Fisheries Management*, Paris, France, 221 pp.

Organisation for Economic Co-operation and Development (1997), *Towards Sustainable Fisheries: Economic Aspects of the Management of Living Marine Resources*, Paris, France, 268 pp.

Pearse, P.H. (1991), *Building on Progress: Fisheries Policy Development in New Zealand*, Wellington, NZ: Ministry of Fisheries.

Pearse, Peter H. and Walter, Carl J. (1992), 'Harvesting Regulation Under Quota Management Systems for Ocean Fisheries: Decision Making in the Face of Natural Variability, Weak Information, Risks and Conflicting Incentives', *Marine Policy*, **16**(3), pp. 167–82.

Placenti, V., Rizzo, G. and Spagnolo, M. (1992), 'A Bio-Economic Model for the Optimization of a Multi-Species, Multi-Gear Fishery: The Italian Case', *Marine Resource Economics*, **7**(4), pp. 275–95.

Plourde, C.G. (1970), 'A Simple Model of Replenishable Resource Use', *American Economic Review*, **60**, pp. 518–22.

Plourde, C.G. (1971), 'Exploitation of Common Property Replenishable Natural Resources', *Western Economic Journal*, **9**(3), pp. 256–66.

Plourde, C.G. (1979), 'Diagrammatic Representations of the Exploitation of Replenishable Natural Resources: Dynamic Iterations', *Journal of Environmental Economics and Management*, **6**, pp. 119–26.

Plourde, C.G. and Smith, J. Barry (1989), 'Crop Sharing in the Fishery and Industry Equilibrium', *Marine Resource Economics*, **6**(3), pp. 179–93.

Quiggin, John (1992), 'How to Set Catch Quotas: A Note on the Superiority of Constant Effort Rules', *Journal of Environmental Economics and Management*, **22**(2), pp. 199–203.

Quirk, J.P. and Smith, V.L. (1970), 'Dynamic Economic Models of Fishing', in A.D. Scott (ed.), *Economics of Fisheries Management: A Symposium*, Vancouver, Canada: Institute of Animal Resource Ecology, University of British Columbia, pp. 3–32.

Radomski, P.J. (1999), 'Commerical Overfishing and Property Rights', *Fisheries*, **24**(6), pp. 22–29.

Ragozin, David L. and Brown, Gardner Jr. (1985), 'Harvest Policies and Nonmarket Valuation in a Predator-Prey System', *Journal of Environmental Economics and Management*, **12**, pp. 155–68.

Reed, W. (1974), 'A Stochastic Model for the Economic Management of a Renewable Resource', *Mathematical Biosciences*, **22**, pp. 313–37.

Reed, W. (1979), 'Optimal Escapement Levels in Stochastic and Deterministic Harvesting Models', *Journal of Environmental Economics and Management*, **6**, pp. 350–63.

Reed, William J. (1989), 'Optimal Investment in the Protection of a Vulnerable Biological Resource', *Natural Resource Modeling*, **3**(4), pp. 463–80.

Richardson, Edward J. and Gates, John M. (1986), 'Economic Benefits of American Lobster Fishery Management Regulations', *Marine Resource Economics*, **2**(4), pp. 353–82.

Rosenman, Robert E. (1986), 'The Optimal Tax for Maximum Economic Yield: Fishery Regulation Under Rational Expectations', *Journal of Environmental Economics and Management*, **13**(4), pp. 348–62.

Rosenman, Robert E. (1987), 'Structural Modeling of Expectations and Optimization in a Fishery', *Natural Resource Modeling*, **2**(2), pp. 245–58.

Rosenman, Robert E. (1991), 'Impacts of Recreational Fishing on the Commercial Sector: An Empirical Analysis of Atlantic Mackerel', *Natural Resource Modeling*, **5**(2), pp. 239–57.

Rosenman, Robert E. and Whiteman, Charles H. (1987), 'Fishery Regulation Under Rational Expectations and Costly Dynamic Adjustment', *Natural Resource Modeling*, **1**(2), pp. 297–320.

Salvanes, K.G. and Squires, D. (1995), 'Transferable Quotas, Enforcement Costs and Typical Firms: An Empirical Application to the Norwegian Trawler Fleet', *Environmental and Resource Economics*, **6**(1), pp. 1–21.

Salvanes, Kjell G. and Steen, Frode (1994), 'Testing for Relative Performance Between Seasons in a Fishery', *Land Economics*, **70**(4), pp. 431–47.

Sampson, David B. (1992), 'Fishing Technology and Fleet Dynamics: Predictions from a Bioeconomic Model', *Marine Resource Economics*, **7**(1), pp. 37–58.

Sampson, D.B. (1994), 'Fishing Tactics in a Two-Species Fisheries Model: The Bioeconomics of Bycatch and Discarding', *Canadian Journal of Fisheries and Aquatic Sciences*, **51**, pp. 2688–94.

Sanchirico, James N. and Wilen, James E. (1999), 'Bioeconomics of Spatial Exploitation in a Patchy Environment', *Journal of Environmental Economics and Management*, **37**(2), pp. 129–50.

Sandal, Leif K. and Steinshamn, Stein I. (1997a), 'A Feedback Model for the Optimal Management of Renewable Natural Resource Capital Stocks', *Canadian Journal of Fisheries and Aquatic Sciences*, **54**, pp. 2475–82.

Sandal, Leif K. and Steinshamn, Stein I. (1997b), 'Optimal Steady States and the Effect of Discounting', *Marine Resource Economics*, **12**(2), pp. 95–106.

Santopietro, George D. and Shabman, Leonard A. (1992), 'Can Privatization Be Inefficient?: The Case of the Chesapeake Bay Oyster Fishery', *Journal of Economic Issues*, **26**(2), pp. 407–18.

Sathiendrakumar, R. and Tisdell, C.A. (1987), 'Optimal Economic Fishery Effort in the Maldivian Tuna Fishery: An Appropriate Model', *Marine Resource Economics*, **4**(1), pp. 15–44.

Schaefer, Milner B. (1954), 'Some Aspects of the Dynamics of Populations Important to the Management of the Commercial Marine Fisheries', *Bulletin of the Inter-American Tropical Tuna Commission*, **2**, pp. 27–56.

Schaefer, Milner B. (1957), 'Some Considerations of the Population Dynamics and Economics in Relation to the Management of the Commercial Marine Fisheries', *Journal of the Fisheries Research Board of Canada*, **14**, pp. 669–81.

Schaefer, Milner B. (1959), 'Biological and Economic Aspects of the Management of the Commercial Marine Fisheries', *Transactions of the American Fisheries Society*, **88**, pp. 100–4.

Schellberg, Thomas (1993), 'The Problem of Nonmalleable Capital Revisited: A Study of the Pacific Halibut Fishery', *Natural Resource Modeling*, **7**(3), pp. 245–76.

Schworm, William E. (1983), 'Monopsonistic Control of a Common Property Renewable Resource', *Canadian Journal of Economics*, **16**(2), pp. 275–87.

Scott, Anthony (1955), 'The Fishery: The Objectives of Sole Ownership', *The Journal of Political Economy*, **63**(2), pp. 116–24.

Scott, Anthony (1979), 'Development of Economic Theory on Fisheries Regulation', *Journal of Fisheries Resources Board of Canada*, **36**, pp. 725–41.

Scott, Anthony (1986), 'Catch Quotas and Shares in the Fishstock as Property Rights', in E. Miles, R. Pealy, and R. Stokes (eds), *Natural Resource Economics and Policy Applications*, Seattle: University of Washington Press.

Scott, Anthony D. (1988), 'Development of Property in the Fishery', *Marine Resource Economics*, **5**(4), pp. 289–311.

Scott, Anthony (1993), 'Obstacles to Fishery Self Government', *Marine Resource Economics*, **8**(3), pp. 187–99.

Seijo, Juan Carlos (1993), 'Individual Transferable Grounds in a Community Managed Artisanal Fishery', *Marine Resource Economics*, **8**, pp. 78–81.

Sharp, Basil M.H. (1997), 'From Regulated Access to Transferable Harvesting Rights: Policy Insights from New Zealand', *Marine Policy*, **21**(6), pp. 501–17.

Sissenwine, M.P. and Mace, P.M. (1992), 'ITQs in New Zealand: The Era of Fixed Quota in Perpetuity', *Fisheries Bulletin*, **90**, pp. 147–60.

Smith, Courtland L. (1990), 'Resource Scarcity and Inequality in the Distribution of Catch', *North American Journal of Fisheries Management*, **10**, pp. 269–78.

Smith, C.L. and Hanna, S.S. (1990), 'Measuring Fleet Capacity and Capacity Utilization', *Canadian Journal of Fisheries and Aquatic Sciences*, **47**(11), pp. 2085–91.

Smith, I.R. (1981), 'Improving Fishing Incomes When Resources are Overfished', *Marine Policy*, **5**(1), pp. 17–22.

Smith, J. Barry (1980), 'Replenishable Resource Management Under Uncertainty: A Reexamination of the U.S. Northern Lobster Fishery', *Journal of Environmental Economics and Management*, **7**, pp. 209–19.

Smith, J. Barry (1985), 'A Discrete Model of Replenishable Resource Management Under Uncertainty', *Marine Resource Economics*, **1**(2), pp. 283–308.

Smith, J. Barry (1986), 'Stochastic Steady-State Replenishable Resource Management Policies', *Marine Resource Economics*, **3**(2), pp. 155–68.

Smith, Vernon L. (1968), 'Economics of Production from Natural Resources', *The American Economic Review*, **58**(3), pp. 409–31.

Smith, Vernon L. (1969), 'On Models of Commercial Fishing', *Journal of Political Economy*, **77**(2), pp. 181–98. See also comment by F.W. Bell, E.W. Carlson and R.F. Fullenbaum in Vol. **80** (1972) of the same journal pp. 761–8, and a reply by Smith pp. 776–8.

Smith, Vernon L. (1971), 'Economics of Production from Natural Resources: Reply', *Am. Econ. Rev.*, **61**, pp. 488–91.

Smith, Vernon L. (1972), 'On Models of Commercial Fishing: The Traditional Literature Needs No Defenders', *Journ. Pol. Econ.*, **80**, pp. 776–8.

Smith, Vernon L. (1974), 'General Equilibrium with a Replenishable Natural Resource', *Review of Economic Studies:* Symposium on the Economics of Exhaustible Resources, pp. 105–15.

Southey, Clive (1972), 'Policy Prescriptions in Bionomic Models: The Case of the Fishery', *Journal of Political Economy*, **80**(4), pp. 769–75.

Spulber, Daniel F. (1985), 'The Multicohort Fishery Under Uncertainty', *Marine Resource Economics*, **1**(2), pp. 265–82.

Squires, Dale (1987), 'Fishing Effort: Its Testing, Specification, and Internal Structure in Fisheries Economics and Management', *Journal of Environmental Economics and Management*, **14**, pp. 268–82.

Squires, Dale and Kirkley, J. (1991), 'Production Quota in Multiproduct Pacific Fisheries', *Journal of Environmental Economics and Management*, **21**(2), pp. 109–26.

Squires, Dale, Alauddin, Mohammad and Kirkley, James (1994), 'Individual Transferable Quota Markets and Investment Decisions in the Fixed Gear Sablefish Industry', *Journal of Environmental Economics and Management*, **27**(2), pp. 185–204.

Squires, D., Kirkley, J. *et al.* (1995), 'Individual Transferable Quotas as a Fisheries Management Tool', *Reviews in Fisheries Science*, **3**(2), pp. 141–69.

Squires, D., Cunningham, S. *et al.* (1998), 'Individual Transferable Quotas in Multispecies Fisheries', *Marine Policy*, **22**(2), pp. 135–59.

Stefanou, Spiro E. and Wilen, James E. (1992), 'License Values in Restricted Access Fisheries', *Bulletin of Mathematical Biology*, **54**(2/3), pp. 209–18.

Stollery, Kenneth (1984), 'Optimal versus Unregulated Industry Behavior in a Beverton-Holt Multicohort Fishery Model', *Canadian Journal of Fisheries and Aquatic Sciences*, **41**, pp. 446–50.

Stollery, Kenneth R. (1986a), 'Monopsony Processing in an Open-Access Fishery', *Marine Resource Economics*, **3**(4), pp. 331–52.

Stollery, Kenneth (1986b), 'A Short-Run Model of Capital Stuffing in the Pacific Halibut Fishery', *Marine Resource Economics*, **3**(2), pp. 137–54.

Strand, I.E. and Cessine, Robert (1979), 'An Analysis of Surf Clam Production Using an Exhaustible Resource Model', *Journal of the Northeast Agricultural Economics Council*, **7**(2), pp. 99–103.

Sutinen, Jon G. (1979), 'Fishermen's Remuneration Systems and Implications for Fisheries Development', *Scottish Journal of Political Economy*, **26**, pp. 147–62.

Sutinen, Jon G. and Andersen, Peder (1985), 'The Economics of Fisheries Law Enforcement', *Land Economics*, **61**(4), pp. 387–97.

Sutinen, Jon G. and Kuperan, K. (1999), 'A Socioeconomic Theory of Regulatory Compliance in Fisheries', *International Journal of Social Economics*, **26**(1/2/3), pp. 174–93.

Sutinen, Jon G., Rieser, Alison and Gauvin, John R. (1990), 'Measuring and Explaining Noncompliance in Federally Managed Fisheries', *Ocean Development and International Law*, **21**, pp. 335–72.

Swallow, Stephen K. (1994), 'Intraseason Harvest Regulation for Fish and Wildlife Recreation: An Application to Fishery Policy', *American Journal of Agricultural Economics*, **76**, pp. 924–35.

Sylvia, Gilbert and Enriquez, Roberto R. (1994), 'Multiobjective Bioeconomic Analysis: An Application to the Pacific Whiting Fishery', *Marine Resource Economics*, **9**(4), pp. 311–28.

Terkla, David G. and Doeringer, Peter B. (1988), 'Widespread Labor Stickiness in the New England Offshore Fishery Industry: Implications for Adjustment and Regulation', *Land Economics*, **64**(1), pp. 73–82.

Terrebonne, R. Peter (1995), 'Property Rights and Entreprenurial Income in Commercial Fisheries', *Journal of Environmental Economics and Management*, **28**, pp. 68–82.

Thunberg, E.M., Helser, T.E. and Mayo, R.K. (1998), 'Bioeconomic Analysis of Alternative Selection Patterns in the United States Atlantic Silver Hake Fishery', *Marine Resource Economics*, **13**(1), pp. 51–74.

Thurman, Walter N. and Easley, J.E., Jr. (1992), 'Valuing Changes in Commercial Fishery Harvest: A General Equilibrium Derived Demand Analysis', *Journal of Environmental Economics and Management* **22**(3), pp. 226–40.

Townsend, Ralph E. (1986), 'A Critique of Models of the American Lobster Fishery', *Journal of Environmental Economics and Management*, **13**, pp. 227–91.

Townsend, Ralph E. (1990), 'Entry Restrictions in the Fishery: A Survey of the Evidence', *Land Economics*, **66**(4), pp. 361–78.

Townsend, Ralph E. (1992), 'A Fractional Licensing Program for Fisheries', *Land Economics*, May, **68**(2), pp. 185–90.

Townsend, Ralph E. (1995a), 'Fisheries Self-Governance: Corporate or Cooperative Structures', *Marine Policy*, **19**(1), pp. 39–45.

Townsend, Ralph E. (1995b), 'Transferable Dynamic Stock Rights', *Marine Policy*, **19**(2), pp. 153–8.

Townsend, R.E. (1998), 'Beyond ITQs: Property Rights as a Management Tool', *Fisheries Research*, **37**(1–3), pp. 203–10.

Townsend, Ralph E. and Pooley, Samuel G. (1995), 'Fractional Licenses: An Alternative to License Buy-Backs', *Land Economics*, **71**(1), pp. 141–3.

Tu, Pierre N.V. and Wilman, Elizabeth A. (1992), 'A Generalized Predator-Prey Model: Uncertainty and Management', *Journal of Environmental Economics and Management*, **23**, pp. 123–38.

Turner, Matthew A. (1996), 'Value-based ITQs', *Marine Resource Economics*, **11**(2), pp. 59–69.

Turner, Matthew A. (1997), 'Quota-induced Discarding in Heterogeneous Fisheries', *Journal of Environmental Economics and Management*, **33**(2), pp. 186–95.

Turvey, Ralph (1964), 'Optimization and Suboptimization in Fishery Regulation', *The American Economic Review*, **54**, pp. 64–76.

Van Meir, Lawrence W. (1969), 'An Economic Evaluation of Alternative Management Systems for Commercial Fisheries', *Transactions of the American Fisheries Society*, **2**, pp. 347–50.

Vestergaard, Niels (1996), 'Discard Behavior, Highgrading and Regulation: The Case of the Greenland Shrimp Fishery', *Marine Resource Economics*, **11**, pp. 247–66.

Vislie, Jon (1987), 'On the Optimal Management of Transboundary Renewable Resources: A Comment on Munro's Paper', *Canadian Journal of Economics*, **20**(4), pp. 870–75.

Walker, James M., Gardner, Roy and Ostrom, Elinor (1990), 'Rent Dissipation in a Limited-Access Common-Pool Resource: Experimental Evidence', *Journal of Environmental Economics and Management*, **19**(3), pp. 203–11.

Wallace, S.W. and Bekke, K. (1986), 'Optimal Fleet Size When National Quotas can be Traded', *Marine Resource Economics*, **2**(4), pp. 315–29.

Wang, Ar-Young and Cheng, Kuo-Shung (1978), 'Dynamic Analysis of Commercial Fishing Model', *Journal of Environmental Economics and Management*, **5**(2), pp. 113–27.

Wang, S.D. (1995), 'The Surf Clam ITQ Management: An Evaluation', *Marine Resource Economics*, **10**(1), pp. 93–8.

Ward, John M. (1994), 'The Bioeconomic Implications of a Bycatch Reduction Device as a Stock Conservation Management Measure', *Marine Resource Economics*, **9**(3), pp. 227–40.

Ward, John M. and Sutinen, Jon G. (1994), 'Vessel Entry-Exit Behavior in the Gulf of Mexico Shrimp Fishery', *American Journal of Agricultural Economics*, **76**, pp. 916–23.

Warming, Jens (1911), 'Om Grundrentaf Fiskegrunde', *Nationalokonomisk Tidskrift*, **49**, pp. 499–505. Translated by Peder Anderson as 'On Rent of Fishing Grounds', *History of Political Economy*, **15**(3), pp. 391–6.

Waters, James R. (1991), 'Restricted Access vs. Open Access Methods of Management: Toward More Effective Regulation', *Marine Fisheries Review*, **53**(3), pp. 1–10.

Waters, James R., Easley, J.E., Jr. and Danielson, Leon E. (1980), 'Economic Trade-Offs and the North Carolina Shrimp Fishery', *American Journal of Agricultural Economics*, **62**(1), pp. 125–9.

Waugh, Geoffrey (1984), *Fisheries Management: Theoretical Developments and Contemporary Applications*, Boulder, CO: Westview Press, p. 245.

Weitzman, Martin L. (1974), 'Free Access vs. Private Ownership as Alternative Systems for Managing Common Property', *Journal of Economic Theory*, **8**, pp. 225–34.

Weninger, Q. (1998), 'Assessing Efficiency Gains from Individual Transferable Quotas: An Application to the Mid-Atlantic Surf Clam and Ocean Quahog Fishery', *American Journal of Agricultural Economics*, **80**(4), pp. 750–64.

Weninger, Q. and Just, R.E. (1997), 'An Analysis of Transition from Limited Entry to Transferable Quotas: Non-Marshallin Principles for Fisheries Management', *Natural Resource Modeling*, **10**, pp.53–83.

Wilen, James E. (1985), 'Towards a Theory of the Regulated Fishery', *Marine Resource Economics*, **1**(4), pp. 369–88.

Wilen, James E. (1988), 'Limited Entry Licensing: A Retrospective Assessment', *Marine Resource Economics*, **5**(4), pp. 313–24.

Wilen, James E. (1995), 'Bioeconomics of Renewable Resource Use', in Allen V. Kneese and James L. Sweeney (eds), *Handbook of Natural Resource and Energy Economics, I*, Amsterdam: North Holland.

Wilen, James and Brown, Gardner (1986), 'Optimal Recovery Paths for Perturbations of Trophic Level Bioeconomic Systems', *Journal of Environmental Economics and Management*, **13**, pp. 225–34.

Wilson, James A. (1980), 'Adaptation to Uncertainty and Small Numbers Exchange: The New England Fresh Fish Market', *The Bell Journal of Economics*, **11**(2), pp. 491–504.

Wilson, James A. (1982), 'The Economical Management of Multispecies', *Land Economics*, **58**(4), pp. 417–34.

Wilson, James A. (1990), 'Fishing for Knowledge', *Land Economics*, **66**(1), pp. 12–29.

Wilson, James A., Acheson, James M., Metcalfe, Mark and Kleban, Peter (1994), 'Chaos, Complexity and Community Management of Fisheries', *Marine Policy*, **18**(4), pp. 291–305.

Yohe, Gary W. (1984), 'Regulation Under Uncertainty: An Intuitive Survey and Application to Fisheries', *Marine Resource Economics*, **1**(2), pp. 171–92.

Young, M.D. (1995), 'The Design of Fishing-Right Systems: The New South Wales Experience', *Ocean and Coastal Management*, **28**(1–3), pp. 45–61.

Part I
The Basics

[1]

THE ECONOMIC THEORY OF A COMMON-PROPERTY RESOURCE: THE FISHERY[1]

H. SCOTT GORDON

Carleton College, Ottawa, Ontario

I. INTRODUCTION

THE chief aim of this paper is to examine the economic theory of natural resource utilization as it pertains to the fishing industry. It will appear, I hope, that most of the problems associated with the words "conservation" or "depletion" or "overexploitation" in the fishery are, in reality, manifestations of the fact that the natural resources of the sea yield no economic rent. Fishery resources are unusual in the fact of their common-property nature; but they are not unique, and similar problems are encountered in other cases of common-property resource industries, such as petroleum production, hunting and trapping, etc. Although the theory presented in the following pages is worked out in terms of the fishing industry, it is, I believe, applicable generally to all cases where natural resources are owned in common and exploited under conditions of individualistic competition.

II. BIOLOGICAL FACTORS AND THEORIES

The great bulk of the research that has been done on the primary production phase of the fishing industry has so far been in the field of biology. Owing to the lack of theoretical economic research,[2] biologists have been forced to extend the scope of their own thought into the economic sphere and in some cases have penetrated quite deeply, despite the lack of the analytical tools of economic theory.[3] Many others, who have paid no specific attention to the economic aspects of the problem have nevertheless recognized that the ultimate question is not the ecology of life in the sea as such, but man's use of these resources for his own (economic) purposes. Dr. Martin D. Burkenroad, for example, began a recent article on fishery management with a section on "Fishery Management as Political Economy," saying that "the Management of fisheries is intended for the benefit of man, not fish; therefore effect of management upon fishstocks cannot be regarded as beneficial *per se*."[4] The

---

[1] I want to express my indebtedness to the Canadian Department of Fisheries for assistance and co-operation in making this study; also to Professor M. C. Urquhart, of Queen's University, Kingston, Ontario, for mathematical assistance with the last section of the paper and to the Economists' Summer Study Group at Queen's for affording opportunity for research and discussion.

[2] The single exception that I know is G. M. Gerhardsen, "Production Economics in Fisheries," *Revista de economia* (Lisbon), March, 1952.

[3] Especially remarkable efforts in this sense are Robert A. Nesbit, "Fishery Management" ("U.S. Fish and Wildlife Service, Special Scientific Reports," No. 18 [Chicago, 1943]) (mimeographed), and Harden F. Taylor, *Survey of Marine Fisheries of North Carolina* (Chapel Hill, 1951); also R. J. H. Beverton, "Some Observations on the Principles of Fishery Regulation," *Journal du conseil permanent international pour l'exploration de la mer* (Copenhagen), Vol. XIX, No. 1 (May, 1953); and M. D. Burkenroad, "Some Principles of Marine Fishery Biology," *Publications of the Institute of Marine Science* (University of Texas), Vol. II, No. 1 (September, 1951).

[4] "Theory and Practice of Marine Fishery Management," *Journal du conseil permanent international pour l'exploration de la mer*, Vol. XVIII, No. 3 (January, 1953).

great Russian marine biology theorist, T. I. Baranoff, referred to his work as "bionomics" or "bio-economics," although he made little explicit reference to economic factors.[5] In the same way, A. G. Huntsman, reporting in 1944 on the work of the Fisheries Research Board of Canada, defined the problem of fisheries depletion in economic terms: "Where the take in proportion to the effort fails to yield a satisfactory living to the fisherman";[6] and a later paper by the same author contains, as an incidental statement, the essence of the economic optimum solution without, apparently, any recognition of its significance.[7] Upon the occasion of its fiftieth anniversary in 1952, the International Council for the Exploration of the Sea published a *Rapport Jubilaire*, consisting of a series of papers summarizing progress in various fields of fisheries research. The paper by Michael Graham on "Overfishing and Optimum Fishing," by its emphatic recognition of the economic criterion, would lead one to think that the economic aspects of the question had been extensively examined during the last half-century. But such is not the case. Virtually no specific research into the economics of fishery resource utilization has been undertaken. The present state

of knowledge is that a great deal is known about the biology of the various commercial species but little about the economic characteristics of the fishing industry.

The most vivid thread that runs through the biological literature is the effort to determine the effect of fishing on the stock of fish in the sea. This discussion has had a very distinct practical orientation, being part of the effort to design regulative policies of a "conservation" nature. To the layman the problem appears to be dominated by a few facts of overriding importance. The first of these is the prodigious reproductive potential of most fish species. The adult female cod, for example, lays millions of eggs at each spawn. The egg that hatches and ultimately reaches maturity is the great exception rather than the rule. The various herrings (Clupeidae) are the most plentiful of the commercial species, accounting for close to half the world's total catch, as well as providing food for many other sea species. Yet herring are among the smallest spawners, laying a mere hundred thousand eggs a season, which, themselves, are eaten in large quantity by other species. Even in inclosed waters the survival and reproductive powers of fish appear to be very great. In 1939 the Fisheries Research Board of Canada deliberately tried to kill all the fish in one small lake by poisoning the water. Two years later more than ninety thousand fish were found in the lake, including only about six hundred old enough to have escaped the poisoning.

The picture one gets of life in the sea is one of constant predation of one species on another, each species living on a narrow margin of food supply. It reminds the economist of the Malthusian law of population; for, unlike man, the

---

[5] Two of Baranoff's most important papers— "On the Question of the Biological Basis of Fisheries" (1918) and "On the Question of the Dynamics of the Fishing Industry" (1925)—have been translated by W. E. Ricker, now of the Fisheries Research Board of Canada (Nanaimo, B.C.), and issued in mimeographed form.

[6] "Fishery Depletion," *Science*, XCIX (1944), 534.

[7] "The highest take is not necessarily the best. The take should be increased only as long as the extra cost is offset by the added revenue from sales" (A. G. Huntsman, "Research on Use and Increase of Fish Stocks," *Proceedings of the United Nations Scientific Conference on the Conservation and Utilization of Resources* [Lake Success, 1949]).

fish has no power to alter the conditions of his environment and consequently cannot progress. In fact, Malthus and his law are frequently mentioned in the biological literature. One's first reaction is to declare that environmental factors are so much more important than commercial fishing that man has no effect on the population of the sea at all. One of the continuing investigations made by fisheries biologists is the determination of the age distribution of catches. This is possible because fish continue to grow in size with age, and seasonal changes are reflected in certain hard parts of their bodies in much the same manner as one finds growth-rings in a tree. The study of these age distributions shows that commercial catches are heavily affected by good and bad brood years. A good brood year, one favorable to the hatching of eggs and the survival of fry, has its effect on future catches, and one can discern the dominating importance of that brood year in the commercial catches of succeeding years.[8] Large broods, however, do not appear to depend on large numbers of adult spawners, and this lends support to the belief that the fish population is entirely unaffected by the activity of man.

There is, however, important evidence to the contrary. World Wars I and II, during which fishing was sharply curtailed in European waters, were followed by indications of a significant growth in fish populations. Fish-marking experiments, of which there have been a great number, indicate that fishing is a major cause of fish mortality in developed fisheries. The introduction of restrictive laws has often been followed by an increase in fish populations, although the evidence on this point is capable of other interpretations which will be noted later.

General opinion among fisheries biologists appears to have had something of a cyclical pattern. During the latter part of the last century, the Scottish fisheries biologist, W. C. MacIntosh,[9] and the great Darwinian, T. H. Huxley, argued strongly against all restrictive measures on the basis of the inexhaustible nature of the fishery resources of the sea. As Huxley put it in 1883: "The cod fishery, the herring fishery, the pilchard fishery, the mackerel fishery, and probably all the great sea fisheries, are inexhaustible: that is to say that nothing we do seriously affects the number of fish. And any attempt to regulate these fisheries seems consequently, from the nature of the case, to be useless."[10] As a matter of fact, there was at this time relatively little restriction of fishing in European waters. Following the Royal Commission of 1866, England had repealed a host of restrictive laws. The development of steam-powered trawling in the 1880's, which enormously increased man's predatory capacity, and the marked improvement of the trawl method in 1923 turned the pendulum, and throughout the interwar years discussion centered on the problem of "overfishing" and "depletion." This was accompanied by a considerable growth of restrictive regula-

---

[8] One example of a very general phenomenon: 1904 was such a successful brood year for Norwegian herrings that the 1904 year class continued to outweigh all others in importance in the catch from 1907 through to 1919. The 1904 class was some thirty times as numerous as other year classes during the period (Johan Hjort, "Fluctuations in the Great Fisheries of Northern Europe," *Rapports et procès-verbaux, Conseil permanent international pour l'exploration de la mer*, Vol. XX [1914]; see also E. S. Russell, *The Overfishing Problem* [Cambridge, 1942], p. 57).

[9] See his *Resources of the Sea* published in 1899.

[10] Quoted in M. Graham, *The Fish Gate* (London, 1943), p. 111; see also T. H. Huxley, "The Herring," *Nature* (London), 1881.

THEORY OF A COMMON-PROPERTY RESOURCE   127

tions.[11] Only recently has the pendulum begun to reverse again, and there has lately been expressed in biological quarters a high degree of skepticism concerning the efficacy of restrictive measures, and the Huxleyian faith in the inexhaustibility of the sea has once again begun to find advocates. In 1951 Dr. Harden F. Taylor summarized the overall position of world fisheries in the following words:

> Such statistics of world fisheries as are available suggest that while particular species have fluctuated in abundance, the *yield of the sea fisheries as a whole or of any considerable region has not only been sustained, but has generally increased with increasing human populations*, and there is as yet no sign that they will not continue to do so. No single species so far as we know has ever become extinct, and no regional fishery in the world has ever been exhausted.[12]

In formulating governmental policy, biologists appear to have had a hard struggle (not always successful) to avoid oversimplification of the problem. One of the crudest arguments to have had some support is known as the "propagation theory," associated with the name of the English biologist, E. W. L. Holt.[13] Holt advanced the proposition that legal size limits should be established at a level that would permit every individual of the species in question to spawn at least once. This suggestion was effectively demolished by the age-distribution studies whose results have been noted above. Moreover, some fisheries, such as the "sardine" fishery of the Canadian Atlantic Coast, are specifically for *immature* fish. The history of this particular fishery shows no evidence whatever that

the landings have been in any degree reduced by the practice of taking very large quantities of fish of prespawning age year after year.

The state of uncertainty in biological quarters around the turn of the century is perhaps indicated by the fact that Holt's propagation theory was advanced concurrently with its diametric opposite: "the thinning theory" of the Danish biologist, C. G. J. Petersen.[14] The latter argued that the fish may be too plentiful for the available food and that thinning out the young by fishing would enable the remainder to grow more rapidly. Petersen supported his theory with the results of transplanting experiments which showed that the fish transplanted to a new habitat frequently grew much more rapidly than before. But this is equivalent to arguing that the reason why rabbits multiplied so rapidly when introduced to Australia is because there were no rabbits already there with which they had to compete for food. Such an explanation would neglect all the other elements of importance in a natural ecology. In point of fact, in so far as food alone is concerned, thinning a cod population, say by half, would not double the food supply of the remaining individuals; for there are other species, perhaps not commercially valuable, that use the same food as the cod.

Dr. Burkenroad's comment, quoted earlier, that the purpose of practical policy is the benefit of man, not fish, was not gratuitous, for the argument has at times been advanced that commercial fishing should crop the resource in such a way as to leave the stocks of fish in the sea completely unchanged. Baranoff was largely responsible for destroying this

[11] See H. Scott Gordon, "The Trawler Question in the United Kingdom and Canada," *Dalhousie Review*, summer, 1951.

[12] Taylor, *op. cit.*, p. 314 (Dr. Taylor's italics).

[13] See E. W. L. Holt, "An Examination of the Grimsby Trawl Fishery," *Journal of the Marine Biological Association* (Plymouth), 1895.

[14] See C. G. J. Petersen, "What Is Overfishing?" *Journal of the Marine Biological Association* (Plymouth), 1900–1903.

approach, showing most elegantly that a commercial fishery cannot fail to diminish the fish stock. His general conclusion is worth quoting, for it states clearly not only his own position but the error of earlier thinking:

> As we see, a picture is obtained which diverges radically from the hypothesis which has been favoured almost down to the present time, namely that the natural reserve of fish is an inviolable capital, of which the fishing industry must use only the interest, not touching the capital at all. Our theory says, on the contrary, that a fishery and a natural reserve of fish are incompatible, and that the exploitable stock of fish is a changeable quantity, which depends on the intensity of the fishery. The more fish we take from a body of water, the smaller is the basic stock remaining in it; and the less fish we take, the greater is the basic stock, approximating to the natural stock when the fishery approaches zero. Such is the nature of the matter.[15]

The general conception of a fisheries ecology would appear to make such a conclusion inevitable. If a species were in ecological equilibrium before the commencement of commercial fishing, man's intrusion would have the same effect as any other predator; and that can only mean that the species population would reach a new equilibrium at a lower level of abundance, the divergence of the new equilibrium from the old depending on the degree of man's predatory effort and effectiveness.

The term "fisheries management" has been much in vogue in recent years, being taken to express a more subtle approach to the fisheries problem than the older terms "depletion" and "conservation." Briefly, it focuses attention on the quantity of fish caught, taking as the human objective of commercial fishing the derivation of the largest sustainable

catch. This approach is often hailed in the biological literature as the "new theory" or the "modern formulation" of the fisheries problem.[16] Its limitations, however, are very serious, and, indeed, the new approach comes very little closer to treating the fisheries problem as one of human utilization of natural resources than did the older, more primitive, theories. Focusing attention on the maximization of the catch neglects entirely the inputs of other factors of production which are used up in fishing and must be accounted for as costs. There are many references to such ultimate economic considerations in the biological literature but no analytical integration of the economic factors. In fact, the very conception of a *net economic yield* has scarcely made any appearance at all. On the whole, biologists tend to treat the fisherman as an exogenous element in their analytical model, and the behavior of fishermen is not made into an integrated element of a general and systematic "bionomic" theory. In the case of the fishing industry the large numbers of fishermen permit valid behavioristic generalization of their activities along the lines of the standard economic theory of production. The following section attempts to apply that theory to the fishing industry and to demonstrate that the "overfishing problem" has its roots in the economic organization of the industry.

III. ECONOMIC THEORY OF THE FISHERY

In the analysis which follows, the theory of optimum utilization of fishery re-

---

[15] T. I. Baranoff, "On the Question of the Dynamics of the Fishing Industry," p. 5 (mimeographed).

[16] See, e.g., R. E. Foerster, "Prospects for Managing Our Fisheries," *Bulletin of the Bingham Oceanographic Collection* (New Haven), May, 1948; E. S. Russell, "Some Theoretical Considerations on the Overfishing Problem," *Journal du conseil permanent international pour l'exploration de la mer*, 1931, and *The Overfishing Problem*, Lecture IV.

THEORY OF A COMMON-PROPERTY RESOURCE   129

sources and the reasons for its frustration in practice are developed for a typical demersal fish. Demersal, or bottom-dwelling fishes, such as cod, haddock, and similar species and the various flat-fishes, are relatively nonmigratory in character. They live and feed on shallow continental shelves where the continual mixing of cold water maintains the availability of those nutrient salts which form the fundamental basis of marine-food chains. The various feeding grounds are separated by deep-water channels which constitute barriers to the movement of these species; and in some cases the fish of different banks can be differentiated morphologically, having varying numbers of vertebrae or some such distinguishing characteristic. The significance of this fact is that each fishing ground can be treated as unique, in the same sense as can a piece of land, possessing, at the very least, one characteristic not shared by any other piece: that is, location.

(Other species, such as herring, mackerel, and similar pelagic or surface dwellers, migrate over very large distances, and it is necessary to treat the resource of an entire geographic region as one. The conclusions arrived at below are applicable to such fisheries, but the method of analysis employed is not formally applicable. The same is true of species that migrate to and from fresh water and the lake fishes proper.)

We can define the optimum degree of utilization of any particular fishing ground as that which maximizes the net economic yield, the difference between total cost, on the one hand, and total receipts (or total value production), on the other.[17] Total cost and total production

can each be expressed as a function of the degree of fishing intensity or, as the biologists put it, "fishing effort," so that a simple maximization solution is possible. Total cost will be a linear function of fishing effort, if we assume no fishing-induced effects on factor prices, which is reasonable for any particular regional fishery.

The production function—the relationship between fishing effort and total value produced—requires some special attention. If we were to follow the usual presentation of economic theory, we should argue that this function would be positive but, after a point, would rise at a diminishing rate because of the law of diminishing returns. This would not mean that the fish population has been reduced, for the law refers only to the *proportions* of factors to one another, and a fixed fish population, together with an increasing intensity of effort, would be assumed to show the typical sigmoid pattern of yield. However, in what follows it will be assumed that the law of diminishing returns in this pure sense is inoperative in the fishing industry. (The reasons will be advanced at a later point in this paper.) We shall assume that, as fishing effort expands, the catch of fish increases at a diminishing rate but that it does so because of the effect of catch upon the fish population.[18] So far as the argument of the next few pages is concerned, all that is formally necessary is to assume that, as fishing intensity increases, catch will grow at a diminishing rate. Whether this reflects the pure law of diminishing returns or the reduction

---

[17] Expressed in these terms, this appears to be the monopoly maximum, but it coincides with the social optimum under the conditions employed in the analysis, as will be indicated below.

[18] Throughout this paper the conception of fish population that is employed is one of *weight* rather than *numbers*. A good deal of the biological theory has been an effort to combine growth factors and numbers factors into weight sums. The following analysis will neglect the fact that, for some species, fish of different sizes bring different unit prices.

H. SCOTT GORDON

of population by fishing, or both, is of no particular importance. The point at issue will, however, take on more significance in Section IV and will be examined there.

Our analysis can be simplified if we retain the ordinary production function instead of converting it to cost curves, as is usually done in the theory of the firm. Let us further assume that the functional relationship between average production (production-per-unit-of-fishing-effort) and the quantity of fishing effort is uniformly linear. This does not distort the

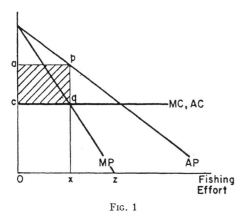

FIG. 1

results unduly, and it permits the analysis to be presented more simply and in graphic terms that are already quite familiar.

In Figure 1 the optimum intensity of utilization of a particular fishing ground is shown. The curves AP and MP represent, respectively, the average productivity and marginal productivity of fishing effort. The relationship between them is the same as that between average revenue and marginal revenue in imperfect competition theory, and MP bisects any horizontal between the ordinate and AP. Since the costs of fishing supplies, etc., are assumed to be unaffected by the amount of fishing effort, marginal cost and average cost are identical and

constant, as shown by the curve MC, AC.[19] These costs are assumed to include an opportunity income for the fishermen, the income that could be earned in other comparable employments. Then Ox is the optimum intensity of effort on this fishing ground, and the resource will, at this level of exploitation, provide the maximum net economic yield indicated by the shaded area $apqc$. The maximum sustained physical yield that the biologists speak of will be attained when marginal productivity of fishing effort is zero, at Oz of fishing intensity in the chart shown. Thus, as one might expect, the optimum economic fishing intensity is less than that which would produce the maximum sustained physical yield.

The area $apqc$ in Figure 1 can be regarded as the rent yielded by the fishery resource. Under the given conditions, Ox is the best rate of exploitation for the fishing ground in question, and the rent reflects the productivity of that ground, not any artificial market limitation. The rent here corresponds to the extra productivity yielded in agriculture by soils of better quality or location than those on the margin of cultivation, which may produce an opportunity income but no more. In short, Figure 1 shows the determination of the intensive margin of utilization on an intramarginal fishing ground.

We now come to the point that is of greatest theoretical importance in understanding the primary production phase of the fishing industry and in distinguishing it from agriculture. In the sea fish-

[19] Throughout this analysis, fixed costs are neglected. The general conclusions reached would not be appreciably altered, I think, by their inclusion, though the presentation would be greatly complicated. Moreover, in the fishing industry the most substantial portion of fixed cost—wharves, harbors, etc.—is borne by government and does not enter into the cost calculations of the operators.

eries the natural resource is not private property; hence the rent it may yield is not capable of being appropriated by anyone. The individual fisherman has no legal title to a section of ocean bottom. Each fisherman is more or less free to fish wherever he pleases. The result is a pattern of competition among fishermen which culminates in the dissipation of the rent of the intramarginal grounds. This can be most clearly seen through an analysis of the relationship between the

fishermen are free to fish on whichever ground they please, it is clear that this is not an equilibrium allocation of fishing effort in the sense of connoting stability. A fisherman starting from port and deciding whether to go to ground *1* or *2* does not care for *marginal* productivity but for *average* productivity, for it is the latter that indicates where the greater total yield may be obtained. If fishing effort were allocated in the optimum fashion, as shown in Figure 2, with *Ox* on

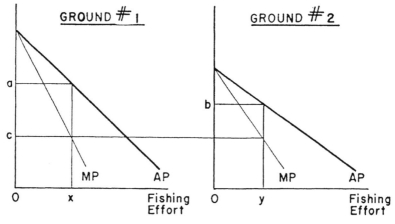

Fig. 2

intensive margin and the extensive margin of resource exploitation in fisheries.

In Figure 2, two fishing grounds of different fertility (or location) are shown. Any given amount of fishing effort devoted to ground *2* will yield a smaller total (and therefore average) product than if devoted to *1*. The maximization problem is now a question of the allocation of fishing effort between grounds *1* and *2*. The optimum is, of course, where the marginal productivities are equal on both grounds. In Figure 2, fishing effort of *Ox* on *1* and *Oy* on *2* would maximize the total net yield of *Ox* + *Oy* effort if marginal cost were equal to *Oc*. But if under such circumstances the individual

1, and *Oy* on *2*, this would be a disequilibrium situation. Each fisherman could expect to get an average catch of *Oa* on *1* but only *Ob* on *2*. Therefore, fishermen would shift from *2* to *1*. Stable equilibrium would not be reached until the average productivity of both grounds was equal. If we now imagine a continuous gradation of fishing grounds, the extensive margin would be on that ground which yielded nothing more than outlaid costs plus opportunity income—in short, the one on which average productivity and average cost were equal. But, since average cost is the same for all grounds and the average productivity of all grounds is also brought to equality by the free and competitive nature of fishing, this means that the intramarginal grounds also yield no rent. It is entirely possible that some grounds would be exploited at a level of *negative* marginal productivity. What happens is that the rent which the intramarginal grounds are capable of yielding is dissipated through misallocation of fishing effort.

This is why fishermen are not wealthy, despite the fact that the fishery resources of the sea are the richest and most indestructible available to man. By and large, the only fisherman who becomes rich is one who makes a lucky catch or one who participates in a fishery that is put under a form of social control that turns the open resource into property rights.

Up to this point, the remuneration of fishermen has been accounted for as an opportunity-cost income comparable to earnings attainable in other industries. In point of fact, fishermen typically earn less than most others, even in much less hazardous occupations or in those requiring less skill. There is no effective reason why the competition among fishermen described above must stop at the point where opportunity incomes are yielded. It may be and is in many cases carried much further. Two factors prevent an equilibration of fishermen's incomes with those of other members of society. The first is the great immobility of fishermen. Living often in isolated communities, with little knowledge of conditions or opportunities elsewhere; educationally and often romantically tied to the sea; and lacking the savings necessary to provide a "stake," the fisherman is one of the least mobile of occupational groups. But, second, there is in the spirit of every fisherman the hope of the "lucky catch." As those who know fishermen well have often testified, they

are gamblers and incurably optimistic. As a consequence, they will work for less than the going wage.[20]

The theory advanced above is substantiated by important developments in the fishing industry. For example, practically all control measures have, in the past, been designed by biologists, with sole attention paid to the production side of the problem and none to the cost side. The result has been a wide-open door for the frustration of the purposes of such measures. The Pacific halibut fishery, for example, is often hailed as a great achievement in modern fisheries management. Under international agreement between the United States and Canada, a fixed-catch limit was established during the early thirties. Since then, catch-per-unit-effort indexes, as usually interpreted, show a significant rise in the fish population. W. F. Thompson, the pioneer of the Pacific halibut management program, noted recently that "it has often been said that the halibut regulation presents the only definite case of sustained improvement of an overfished deep-sea fishery. This, I believe, is true and the fact should lend special importance ·to the principles which have been deliberately used to obtain this improvement."[21] Actually, careful study of the statistics indicates that the estimated recovery of halibut stocks could not have been due principally to the control measures, for the average catch was, in fact, greater during the recovery years than during the years of

---

[20] "The gambling instinct of the men makes many of them work for less remuneration than they would accept as a weekly wage, because there is always the possibility of a good catch and a financial windfall" (Graham, *op. cit.*, p. 86).

[21] W. F. Thompson, "Condition of Stocks of Halibut in the Pacific," *Journal du conseil permanent international pour l'exploration de la mer*, Vol. XVIII, No. 2 (August, 1952).

decline. The total amount of fish taken was only a small fraction of the estimated population reduction for the years prior to regulation.[22] Natural factors seem to be mainly responsible for the observed change in population, and the institution of control regulations almost a coincidence. Such coincidences are not uncommon in the history of fisheries policy, but they may be easily explained. If a long-term cyclical fluctuation is taking place in a commercially valuable species, controls will likely be instituted when fishing yields have fallen very low and the clamor of fishermen is great; but it is then, of course, that stocks are about due to recover in any case. The "success" of conservation measures may be due fully as much to the sociological foundations of public policy as to the policy's effect on the fish. Indeed, Burkenroad argues that biological statistics in general may be called into question on these grounds. Governments sponsor biological research when the catches are disappointing. If there are long-term cyclical fluctuations in fish populations, as some think, it is hardly to be wondered why biologists frequently discover that the sea is being depleted, only to change their collective opinion a decade or so later.

Quite aside from the *biological* argument on the Pacific halibut case, there is no clear-cut evidence that halibut fishermen were made relatively more prosperous by the control measures. Whether or not the recovery of the halibut stocks was due to natural factors or to the catch limit, the potential net yield this could have meant has been dissipated through a rise in fishing costs. Since the method of control was to halt fishing when the limit had been reached, this created a

great incentive on the part of each fisherman to get the fish before his competitors. During the last twenty years, fishermen have invested in more, larger, and faster boats in a competitive race for fish. In 1933 the fishing season was more than six months long. In 1952 it took just twenty-six days to catch the legal limit in the area from Willapa Harbor to Cape Spencer, and sixty days in the Alaska region. What has been happening is a rise in the average cost of fishing effort, allowing no gap between average production and average cost to appear, and hence no rent.[23]

Essentially the same phenomenon is observable in the Canadian Atlantic Coast lobster-conservation program. The method of control here is by seasonal closure. The result has been a steady growth in the number of lobster traps set

---

[22] See M. D. Burkenroad, "Fluctuations in Abundance of Pacific Halibut," *Bulletin of the Bingham Oceanographic Collection*, May, 1948.

[23] The economic significance of the reduction in season length which followed upon the catch limitation imposed in the Pacific halibut fishery has not been fully appreciated. E.g., Michael Graham said in summary of the program in 1943: "The result has been that it now takes only five months to catch the quantity of halibut that formerly needed nine. This, *of course*, has meant profit, where there was none before" (*op. cit.*, p. 156; my italics). Yet, even when biologists have grasped the economic import of the halibut program and its results, they appear reluctant to declare against it. E.g., W. E. Ricker: "This method of regulation does not necessarily make for more profitable fishing and certainly puts no effective brake on waste of effort, since an unlimited number of boats is free to join the fleet and compete during the short period that fishing is open. However, the stock is protected, and yield approximates to a maximum if quotas are wisely set; as biologists, perhaps we are not required to think any further. Some claim that any mixing into the economics of the matter might prejudice the desirable biological consequences of regulation by quotas" ("Production and Utilization of Fish Population," in a Symposium on Dynamics of Production in Aquatic Populations, Ecological Society of America, *Ecological Monographs*, XVI [October, 1946], 385). What such "desirable biological consequences" might be, is hard to conceive. Since the regulatory policies are made by man, surely it is necessary they be evaluated in terms of human, not piscatorial, objectives.

134   H. SCOTT GORDON

by each fisherman. Virtually all available lobsters are now caught each year within the season, but at much greater cost in gear and supplies. At a fairly conservative estimate, the same quantity of lobsters could be caught with half the present number of traps. In a few places the fishermen have banded together into a local monopoly, preventing entry and controlling their own operations. By this means, the amount of fishing gear has been greatly reduced and incomes considerably improved.

That the plight of fishermen and the inefficiency of fisheries production stems from the common-property nature of the resources of the sea is further corroborated by the fact that one finds similar patterns of exploitation and similar problems in other cases of open resources. Perhaps the most obvious is hunting and trapping. Unlike fishes, the biotic potential of land animals is low enough for the species to be destroyed. Uncontrolled hunting means that animals will be killed for any short-range human reason, great or small: for food or simply for fun. Thus the buffalo of the western plains was destroyed to satisfy the most trivial desires of the white man, against which the long-term food needs of the aboriginal population counted as nothing. Even in the most civilized communities, conservation authorities have discovered that a bag-limit *per man* is necessary if complete destruction is to be avoided.

The results of anthropological investigation of modes of land tenure among primitive peoples render some further support to this thesis. In accordance with an evolutionary concept of cultural comparison, the older anthropological study was prone to regard resource tenure in common, with unrestricted exploitation, as a "lower" stage of development comparative with private and group

property rights. However, more complete annals of primitive cultures reveal common tenure to be quite rare, even in hunting and gathering societies. Property rights in some form predominate by far, and, most important, their existence may be easily explained in terms of the necessity for orderly exploitation and conservation of the resource. Environmental conditions make necessary some vehicle which will prevent the resources of the community at large from being destroyed by excessive exploitation. Private or group land tenure accomplishes this end in an easily understandable fashion.[24] Significantly, land tenure is found to be "common" only in those cases where the hunting resource is migratory over such large areas that it cannot be regarded as husbandable by the society. In cases of group tenure where the numbers of the group are large, there is still the necessity of co-ordinating the practices of exploitation, in agricultural, as well as in hunting or gathering, economies. Thus, for example, Malinowski reported that among the Trobriand Islanders one of the fundamental principles of land tenure is the co-ordination of the productive activities of the gardeners by the person possessing magical leadership in the group.[25] Speaking generally, we may say that stable primitive cultures appear to have discovered the dangers of common-property tenure and to have de-

[24] See Frank G. Speck, "Land Ownership among Hunting Peoples in Primitive America and the World's Marginal Areas," *Proceedings of the 22nd International Congress of Americanists* (Rome, 1926), II, 323–32.

[25] B. Malinowski, *Coral Gardens and Their Magic*, Vol. I, chaps. xi and xii. Malinowski sees this as further evidence of the importance of magic in the culture rather than as a means of co-ordinating productive activity; but his discussion of the practice makes it clear that the latter is, to use Malinowski's own concept, the "function" of the institution of magical leadership, at least in this connection.

veloped measures to protect their resources. Or, if a more Darwinian explanation be preferred, we may say that only those primitive cultures have survived which succeeded in developing such institutions.

Another case, from a very different industry, is that of petroleum production. Although the individual petroleum producer may acquire undisputed lease or ownership of the particular plot of land upon which his well is drilled, he shares, in most cases, a common pool of oil with other drillers. There is, consequently, set up the same kind of competitive race as is found in the fishing industry, with attending overexpansion of productive facilities and gross wastage of the resource. In the United States, efforts to regulate a chaotic situation in oil production began as early as 1915. Production practices, number of wells, and even output quotas were set by governmental authority; but it was not until the federal "Hot Oil" Act of 1935 and the development of interstate agreements that the final loophole (bootlegging) was closed through regulation of interstate commerce in oil.

Perhaps the most interesting similar case is the use of common pasture in the medieval manorial economy. Where the ownership of animals was private but the resource on which they fed was common (and limited), it was necessary to regulate the use of common pasture in order to prevent each man from competing and conflicting with his neighbors in an effort to utilize more of the pasture for his own animals. Thus the manor developed its elaborate rules regulating the use of the common pasture, or "stinting" the common: limitations on the number of animals, hours of pasturing, etc., designed to prevent the abuses of excessive individualistic competition.[26]

There appears, then, to be some truth in the conservative dictum that everybody's property is nobody's property. Wealth that is free for all is valued by none because he who is foolhardy enough to wait for its proper time of use will only find that it has been taken by another. The blade of grass that the manorial cowherd leaves behind is valueless to him, for tomorrow it may be eaten by another's animal; the oil left under the earth is valueless to the driller, for another may legally take it; the fish in the sea are valueless to the fisherman, because there is no assurance that they will be there for him tomorrow if they are left behind today. A factor of production that is valued at nothing in the business calculations of its users will yield nothing in income. Common-property natural resources are free goods for the individual and scarce goods for society. Under unregulated private exploitation, they can yield no rent; that can be accomplished only by methods which make them private property or public (government) property, in either case subject to a unified directing power.

IV. THE BIONOMIC EQUILIBRIUM OF THE FISHING INDUSTRY

The work of biological theory in the fishing industry is, basically, an effort to delineate the ecological system in which a particular fish population is found. In the main, the species that have been extensively studied are those which are subject to commercial exploitation. This is due not only to the fact that funds are forthcoming for such research but also because the activity of commercial fishing vessels provides the largest body of data upon which the biologist may work.

[26] See P. Vinogradoff, *The Growth of the Manor* [London, 1905], chap. iv; E. Lipson, *The Economic History of England* [London, 1949], I, 72.

Despite this, however, the ecosystem of the fisheries biologist is typically one that excludes man. Or, rather, man is regarded as an exogenous factor, having influence on the biological ecosystem through his removal of fish from the sea, but the activities of man are themselves not regarded as behaviorized or determined by the other elements of a system of mutual interdependence. The large number of independent fishermen who exploit fish populations of commercial importance makes it possible to treat man as a behavior element in a larger, "bionomic," ecology, if we can find the rules which relate his behavior to the other elements of the system. Similarly, in their treatment of the principles of fisheries management, biologists have overlooked essential elements of the problem by setting maximum physical landings as the objective of management, thereby neglecting the economic factor of input cost.

An analysis of the bionomic equilibrium of the fishing industry may, then, be approached in terms of two problems. The first is to explain the nature of the equilibrium of the industry as it occurs in the state of uncontrolled or unmanaged exploitation of a common-property resource. The second is to indicate the nature of a socially optimum manner of exploitation, which is, presumably, what governmental management policy aims to achieve or promote. These two problems will be discussed in the remaining pages.

In the preceding section it was shown that the equilibrium condition of uncontrolled exploitation is such that the net yield (total value landings *minus* total cost) is zero. The "bionomic ecosystem" of the fishing industry, as we might call it, can then be expressed in terms of four variables and four equations. Let P rep-

resent the population of the particular fish species on the particular fishing bank in question; L the total quantity taken or "landed" by man, measured in value terms; E the intensity of fishing or the quantity of "fishing effort" expended; and C the total cost of making such effort. The system, then, is as follows:

$$P = P(L), \tag{1}$$

$$L = L(P, E), \tag{2}$$

$$C = C(E), \tag{3}$$

$$C = L. \tag{4}$$

Equation (4) is the equilibrium condition of an uncontrolled fishery.

The functional relations stated in equations (1), (2), and (3) may be graphically presented as shown in Figure 3. Segment *1* shows the fish population as a simple negative function of landings. In segment *2* a map of landings functions is drawn. Thus, for example, if population were P_3, effort of Oe would produce Ol of fish. For each given level of population, a larger fishing effort will result in larger landings. Each population contour is, then, a production function for a given population level. The linearity of these contours indicates that the law of diminishing returns is not operative, nor are any landings-induced price effects assumed to affect the value landings graphed on the vertical axis. These assumptions are made in order to produce the simplest determinate solution; yet each is reasonable in itself. The assumption of a fixed product price is reasonable, since our analysis deals with one fishing ground, not the fishery as a whole. The cost function represented in equation (3) and graphed in segment *3* of Figure 3 is not really necessary to the determination, but its inclusion makes the matter somewhat clearer. Fixed prices of input

THEORY OF A COMMON-PROPERTY RESOURCE   137

factors—"fishing effort"—is assumed, which is reasonable again on the assumption that a small part of the total fishery is being analyzed.

Starting with the first segment, we see that a postulated catch of Ol connotes an equilibrium population in the biological ecosystem of Op. Suppose this population to be represented by the contour P_3 of segment 2. Then, given P_3, Oe is the effort required to catch the postulated landings Ol. This quantity of effort involves a total cost of Oc, as shown in segment 3 of the graph. In full bionomic

found. If the case were represented by C and L_1, the fishery would contract to zero; if by C and L_2, it would undergo an infinite expansion. Stable equilibrium requires that either the cost or the landings function be nonlinear. This condition is fulfilled by the assumption that population is reduced by fishing (eq. [1] above). The equilibrium is therefore as shown in Figure 5. Now Oe represents a fully stable equilibrium intensity of fishing.

The analysis of the conditions of stable equilibrium raises some points of general theoretical interest. In the foregoing we

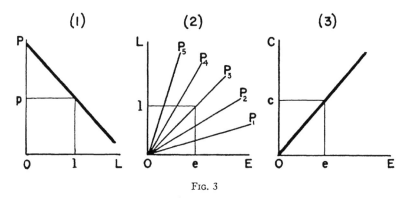

Fig. 3

equilibrium, $C = L$, and if the particular values Oc and Ol shown are not equal, other quantities of all four variables, L, P, E, and C, are required, involving movements of these variables through the functional system shown. The operative movement is, of course, in fishing effort, E. It is the equilibrating variable in the system.

The equilibrium equality of landings (L) and cost (C), however, must be a position of stability, and $L = C$ is a necessary, though not in itself sufficient, condition for stability in the ecosystem. This is shown by Figure 4. If effort-cost and effort-landings functions were both linear, no stable equilibrium could be

have assumed that stability results from the effect of fishing on the fish population. In the standard analysis of economic theory, we should have employed the law of diminishing returns to produce a landings function of the necessary shape. Market factors might also have been so employed; a larger supply of fish, forthcoming from greater fishing effort, would reduce unit price and thereby produce a landings function with the necessary negative second derivative. Similarly, greater fishing intensity might raise the unit costs of factors, producing a cost function with a positive second derivative. Any one of these three— population effects, law of diminishing re-

138   H. SCOTT GORDON

turns, or market effects—is alone sufficient to produce stable equilibrium in the ecosystem.

As to the law of diminishing returns, it has not been accepted per se by fisheries biologists. It is, in fact, a principle that becomes quite slippery when one applies it to the case of fisheries. Indicative of this is the fact that Alfred Marshall, in whose *Principles* one can find extremely little formal error, misinterprets the application of the law of dimin-

estingly enough, his various criticisms of the indexes were generally accepted, with the significant exception of this one point. More recently, A. G. Huntsman warned his colleagues in fisheries biology that "[there] may be a decrease in the take-per-unit-of-effort without any decrease in the total take or in the fish population. . . . This may mean that there has been an increase in fishermen rather than a decrease in fish."[29] While these statements run in terms of average

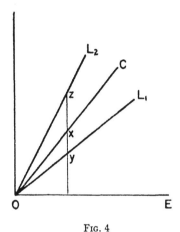

Fig. 4

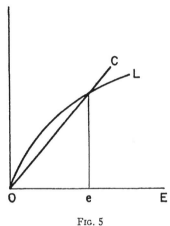

Fig. 5

ishing returns to the fishing industry, arguing, in effect, that the law exerts its influence through the reducing effect of fishing on the fish population.[27] There have been some interesting expressions of the law or, rather, its essential varying-proportions-of-factors aspect, in the biological literature. H. M. Kyle, a German biologist, included it in 1928 among a number of reasons why catch-per-unit-of-fishing-effort indexes are not adequate measures of population change.[28] Inter-

rather than marginal yield, their underlying reasoning clearly appears to be that of the law of diminishing returns. The point has had little influence in biological circles, however, and when, two years ago, I advanced it, as Kyle and Huntsman had done, in criticism of the standard biological method of estimating population change, it received pretty short shrift.

[27] See H. Scott Gordon, "On a Misinterpretation of the Law of Diminishing Returns in Alfred Marshall's *Principles*," *Canadian Journal of Economics and Political Science*, February, 1952.

[28] "Die Statistik der Seefischerei Nordeuropas," *Handbuch der Seefischerei Nordeuropas* (Stuttgart, 1928).

[29] A. G. Huntsman, "Fishing and Assessing Populations," *Bulletin of the Bingham Oceanographic Collection* (New Haven), May, 1948.

In point of fact, the law of diminishing returns is much more difficult to sustain in the case of fisheries than in agriculture or industry. The "proof" one finds in standard theory is not empirical, although the results of empirical experiments in agriculture are frequently adduced as subsidiary corroboration. The main weight of the law, however, rests on a *reductio ad absurdum*. One can easily demonstrate that, were it not for the law of diminishing returns, all the world's food could be grown on one acre of land. Reality is markedly different, and it is because the law serves to render this reality intelligible to the logical mind, or, as we might say, "explains" it, that it occupies such a firm place in the body of economic theory. In fisheries, however, the pattern of reality can easily be explained on other grounds. In the case at least of developed demersal fisheries, it cannot be denied that the fish population is reduced by fishing, and this relationship serves perfectly well to explain why an infinitely expansible production is not possible from a fixed fishing area. The other basis on which the law of diminishing returns is usually advanced in economic theory is the prima facie plausibility of the principle as such; but here, again, it is hard to grasp any similar reasoning in fisheries. In the typical agricultural illustration, for example, we may argue that the fourth harrowing or the fourth weeding, say, has a lower marginal productivity than the third. Such an assertion brings ready acceptance because it concerns a process with a zero productive limit. It is apparent that, ultimately, the land would be completely broken up or the weeds completely eliminated if harrowing or weeding were done in ever larger amounts. The law of diminishing returns signifies simply that

such a zero limit is *gradually approached*, all of which appears to be quite acceptable on prima facie grounds. There is nothing comparable to this in fisheries at all, for there is no "cultivation" in the same sense of the term, except, of course, in such cases as oyster culture or pond rearing of fish, which are much more akin to farming than to typical sea fisheries.

In the biological literature the point has, I think, been well thought through, though the discussion does not revolve around the "law of diminishing returns" by that name. It is related rather to the fisheries biologist's problem of the interpretation of catch-per-unit-of-fishing-effort statistics. The essence of the law is usually eliminated by the assumption that there is no "competition" among units of fishing gear—that is, that the ratio of gear to fishing area and/or fish population is small. In some cases, corrections have been made by the use of the compound-interest formula where some competition among gear units is considered to exist.[30] Such corrections, however, appear to be based on the idea of an increasing catch-population ratio rather than an increasing effort-population ratio. The latter would be as the law of diminishing returns would have it; the idea lying behind the former is that the total population in existence represents the maximum that can be caught, and, since this maximum would be gradually approached, the ratio of catch to population has some bearing on the efficiency of fishing gear. It is, then, just an aspect of the population-reduction effect. Similarly, it has been pointed out that, since fish are recruited into the

[30] See, e.g., W. F. Thompson and F. H. Bell, *Biological Statistics of the Pacific Halibut Fishery, No. 2: Effect of Changes in Intensity upon Total Yield and Yield per Unit of Gear: Report of the International Fisheries Commission* (Seattle, 1934).

140 H. SCOTT GORDON

catchable stock in a seasonal fashion, one can expect the catch-per-unit-effort to fall as the fishing season progresses, at least in those fisheries where a substantial proportion of the stock is taken annually. Seasonal averaging is therefore necessary in using the catch-effort sta-

the fishery, nor is there any prima facie ground for its acceptance.

Let us now consider the exploitation of a fishing ground under unified control, in which case the equilibrium condition is the maximization of net financial yield, $L - C$.

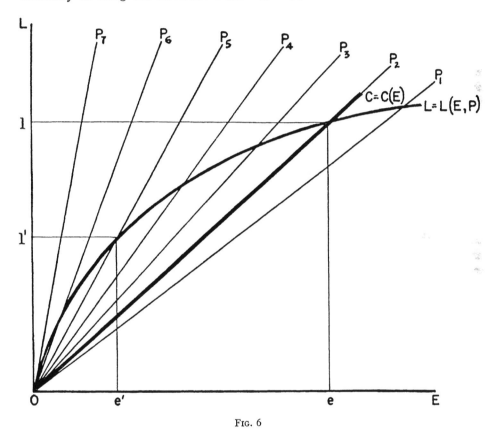

Fig. 6

tistics as population indexes from year to year. This again is a population-reduction effect, not the law of diminishing returns. In general, there seems to be no reason for departing from the approach of the fisheries biologist on this point. The law of diminishing returns is not necessary to explain the conditions of stable equilibrium in a static model of

The map of population contours graphed in segment 2 of Figure 3 may be superimposed upon the total-landings and total-cost functions graphed in Figure 5. The result is as shown in Figure 6. In the system of interrelationships we have to consider, population changes affect, and are in turn affected by, the amount of fish landed. The map of population contours does not include this roundabout effect that a population change has upon itself. The curve labeled L, however, is a landings function which accounts for the fact that larger landings reduce the population, and this is why it is shown to have a steadily diminishing slope. We may regard the landings function as moving progressively to lower population contours P_7, P_6, P_5, etc., as total landings increase in magnitude. As a consequence, while each population contour represents many hypothetical combinations of E, L, and P, only one such combination on each is actually compatible in this system of interrelationships. This combination is the point on any contour where that contour is met by the landings function L. Thus the curve labeled L may be regarded as tracing out a series of combinations of E, L, and P which are compatible with one another in the system.

The total-cost function may be drawn as shown, with total cost, C, measured in terms of landings, which the vertical axis represents.[31] This is a linear function of effort as shown. The optimum intensity of fishing effort is that which maximizes $L - C$. This is the monopoly solution; but, since we are considering only a single fishing ground, no price effects are introduced, and the social optimum coincides with maximum monopoly revenue. In this case we are maximizing the yield of a natural resource, not a privileged position, as in standard monopoly theory. The rent here is a social surplus yielded by the resource, not in any part due to artificial scarcity, as is monopoly profit or rent.

If the optimum fishing intensity is that which maximizes $L - C$, this is seen to

be the position where the slope of the landings function equals the slope of the cost function in Figure 6. Thus the optimum fishing intensity is Oe' of fishing effort. This will yield Ol' of landings, and the species population will be in continuing stable equilibrium at a level indicated by P_5.

The equilibrium resulting from uncontrolled competitive fishing, where the rent is dissipated, can also be seen in Figure 6. This, being where $C = L$, is at Oe of effort and Ol of landings, and at a stable population level of P_2. As can be clearly seen, the uncontrolled equilibrium means a higher expenditure of effort, higher fish landings, and a lower continuing fish population than the optimum equilibrium.

Algebraically, the bionomic ecosystem may be set out in terms of the optimum solution as follows. The species population in equilibrium is a linear function of the amount of fish taken from the sea:

$$P = a - bL . \qquad (1)$$

In this function, a may be described as the "natural population" of the species— the equilibrium level it would attain if not commercially fished. All natural factors, such as water temperatures, food supplies, natural predators, etc., which affect the population are, for the purposes of the system analyzed, locked up in a. The magnitude of a is the vertical intercept of the population function graphed in segment I of Figure 3. The slope of this function is b, which may be described as the "depletion coefficient," since it indicates the effect of catch on population. The landings function is such that no landings are forthcoming with either zero effort or zero population; therefore,

$$L = cEP . \qquad (2)$$

---

[31] More correctly, perhaps, C and L are both measured in money terms.

The parameter c in this equation is the technical coefficient of production or, as we may call it simply, the "production coefficient." Total cost is a function of the amount of fishing effort.

$$C = qE .$$

The optimum condition is that the total net receipts must be maximized, that is,

$$L - C \text{ to be maximized} .$$

Since q has been assumed constant and equal to unity (i.e., effort is counted in "dollars-worth" units), we may write $L - E$ to be maximized. Let this be represented by R:

$$R = L - E , \qquad (3)$$

$$\frac{dR}{dE} = 0 . \qquad (4)$$

The four numbered equations constitute the system when in optimality equilibrium. In order to find this optimum, the landings junction (2) may be rewritten, with the aid of equation (1), as:

$$L = cE (a - bL) .$$

From this we have at once

$$L (1 + cEb) = cEa ,$$

$$L = \frac{caE}{1 + cbE} .$$

To find the optimum intensity of effort, we have, from equation (3):

$$\frac{dR}{dE} = \frac{dL}{dE} - \frac{dE}{dE}$$

$$= \frac{(1 + cbE)(ca) - caE(cb)}{(1 + cbE)^2} - 1,$$

$$= \frac{ca}{(1 + cbE)^2} - 1 ;$$

for a maximum, this must be set equal to zero; hence,

$$ca = (1 + cbE)^2 ,$$

$$1 + cbE = \pm \sqrt{ca} ,$$

$$E = \frac{-1 \pm \sqrt{ca}}{cb} .$$

For positive E,

$$E = \frac{\sqrt{ca} - 1}{cb} .$$

This result indicates that the effect on optimum effort of a change in the production coefficient is uncertain, a rise in c calling for a rise in E in some cases and a fall in E in others, depending on the magnitude of the change in c. The effects of changes in the natural population and depletion coefficient are, however, clear, a rise (fall) in a calling for a rise (fall) in E, while a rise (fall) in b means a fall (rise) in E.

[2]

THE FISHERY: THE OBJECTIVES OF SOLE OWNERSHIP[1]

ANTHONY SCOTT

University of British Columbia

> The rights of property, as such, have not been venerated by those master minds who have built up economic science; but the authority of the science has been wrongly assumed by some who have pushed the claims of vested rights to extreme and antisocial uses. It may be well therefore to note that the tendency of careful economic study is to base the rights of private property not on any abstract principle, but on the observation that in the past they have been inseparable from economic progress. . . .—ALFRED MARSHALL, *Principles of Economics* (8th ed.), p. 48.

IT IS *a* commonplace to observe that for natural resources—as for other types of wealth—"everybody's property is nobody's property." No one will take the trouble to husband and maintain a resource unless he has a reasonable certainty of receiving some portion of the product of his management; that is, unless he has some property right in the yield. Yet the mere existence of the institution of private property is not sufficient to insure the efficient management of natural resources; the property must be allocated on a *scale* sufficient to insure that one management has complete control of the asset. In this paper, for example, I shall show that private property in fishing boats is not a sufficient condition for efficiency; sole ownership of the fishery is also necessary. Some assets, such as oil fields, fisheries, and watersheds, occur on an immense scale, and it is a very real problem to know whether the efficiency gained from unified management provides a social gain sufficient to offset the possible dangers of the creation of some immense sole-ownership organization (such as a co-operative, a government board, a private corporation, or an international authority).

This paper continues the discussion of the economics of private and common property undertaken in "The Economic Theory of a Common-Property Resource: The Fishery," by H. Scott Gordon, which appeared in the *Journal of Political Economy* for April, 1954 (pp. 124–42). Gordon's contribution was a most stimulating, original, and important study of the advantages of sole ownership, which seem practically to have escaped theoretical discussion since Marshall's time.[2] While the economics of the farm and the forest are continually under revision, the earlier economists' insistence[3] that efficiency in production depends upon scarce wealth

---

[1] In writing this paper, I have had interesting and helpful discussions with Messrs. D. C. Corbett, Stuart Jamieson, W. J. Anderson, O. R. Reischer, and E. E. Snyder.

[2] See Marshall, *Principles*, pp. 166–67 and 369–72; see also Gordon's article in the *Canadian Journal of Economics and Political Science*, February, 1952.

[3] See J. S. Mill, *Principles* (5th ed.; New York, 1897), Book II, chap. ii, sec. 5, and the citation to Sismondi's *Étude sur l'économie politique* (n.d.) in the footnote.

THE FISHERY: THE OBJECTIVES OF SOLE OWNERSHIP 117

being "appropriated" has been relegated to the introductory chapters of principles textbooks and is scarcely considered in them again. This is all the more reason for welcoming Gordon's bringing some "political economy" back into economics.

In this paper I wish to compare the use of a fishery by competing fishermen with the mode of management that would be most profitable to a "sole owner" of the same fishery. In particular, I wish to show that *long-run* considerations of efficiency suggest that sole ownership is a much superior regime to competition but that in the *short run* in the ordinary case there is little difference between the efficiency of common and of private property.[4]

I do not wish to dispute in any way the facts about fisheries presented by Gordon (except those about opportunity costs) but to welcome them and to use them myself. Particularly interesting was the point made about the ignorance among fisheries biologists as to whether recent conservation measures have ever produced changes in demersal (deep-water, sea-bottom) fish populations. It is remarkable that many popular books advocating the conservation of natural resources, after describing the undoubted success of conservation in some pelagic (surface) and in-shore fisheries, then—perhaps unintentionally—give the reader to understand that the same conservation might be achieved by controlling demersal

fisheries, an assertion that is apparently as yet unverified.

I

In many ways the central part of Gordon's paper is Section IV, "The Bionomic Equilibrium of the Fishing Industry." In this section he sets out to suggest the nature of the equilibrium of this common-property industry as it occurs in the state of uncontrolled or unmanaged exploitation. Later, he undertakes to indicate the nature of a socially optimum manner of exploitation, which is presumably what governmental management policy seeks to achieve or promote.[5] In a subsequent section of the present paper I shall discuss this optimum, but I should like first to recapitulate his exposition of the equilibrium of a given fishery under conditions of competitive exploitation by individual fishermen.

Gordon first argues that, because there is no sole owner to capture for himself whatever gain there may be from using the fishery conservatively, it will pay every fisherman to enter the industry so long as he can earn something above his cash expenses plus his opportunity costs. As long as fishermen do this, the tendency will be for exploitation to continue beyond the point where the marginal product of fishing effort equals its marginal cost, to the point where the average product of effort just covers the marginal cost of effort (where, in fact, every fisherman just covers his opportunity costs, and average cost is equal to price). There tends, it is argued, to be no "surplus" earned in the industry—the dollar value of the catch exactly equals

---

[4] "Sole ownership" is not monopoly but merely complete appropriation of all of a natural resource in a particular location. Putting a resource into sole ownership is sometimes called making a resource "specific" to one owner (see my forthcoming *Natural Resources: The Economics of Conservation* [Toronto: University of Toronto Press]).

[5] These two aims have been paraphrased from Gordon, *op. cit.*, p. 136.

the dollar cost of landing the catch.

Under such a regime, the argument continues, the possible equilibrium of effort, fish population, and income may be described in terms of a system of four variables, which may be drawn in four figures. These diagrams can be conveniently studied in one four-quadrant diagram (Fig. 1).

If we start with the two upper quadrants, we see that output (or landings) L depends upon both the size of the input (or effort) E and the size of capital stock, or population, P. The size

of the capital stock, or population, is in turn itself assumed to be dependent upon the effort.[6] In the southeast quadrant is a cost function, C, showing constant marginal costs of effort, and in the southwest quadrant is a simple transition function of 45° which confronts the cost of various kinds of effort with the revenue from various sizes of landing. The condition of equilibrium is that the total costs must equal the total

[6] Although Gordon admits that this is not always the case, the analysis holds only when population *is* affected by effort.

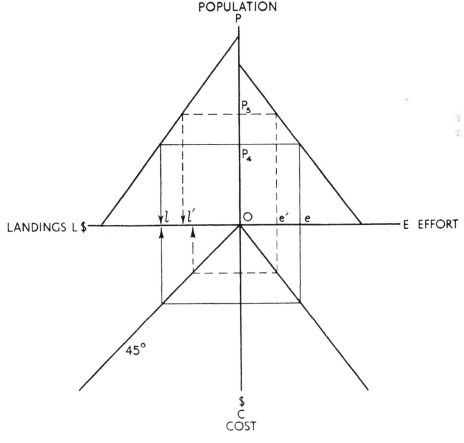

FIG. 1

THE FISHERY: THE OBJECTIVES OF SOLE OWNERSHIP   119

revenues, *Ol*. In diagrammatic terms there must be an inscribed rectangle in the four quadrants. Hence the dotted lines do not represent an equilibrium, but the solid lines do.

Assume that, in equilibrium, a certain total "effort," *E* (that is, of fishermen-plus-equipment), is being expended in the fishery. This effort, expended annually, is compatible with a certain size of fish population, shown on the *P*-axis, and there is an equilibrium annual catch that can be achieved by this effort with this population, shown on the *L*-axis. But, as was argued above, the fishery will not be in equilibrium unless the cost of expending that effort (shown on the *C*-axis) is exactly equal to the revenue, *L*. When *L* and *C* coincide, the fishery is in equilibrium.

A condition for achieving a positive, stable equilibrium is that there must be one and only one rectangle that will fit into the four quadrants. There may be no unique equilibrium possible, given the three functions, unless one of them is curvilinear, or (as here) unless the *E*- and *L*-functions have different intercepts on the *P*-axis.[7]

It is also possible to show the suggested equilibrium position of a sole owner. The dotted lines may now be used to represent this equilibrium: the difference at *l'* between the cost and the landings revenue is assumed to have been maximized. Gordon's description of the equilibrium applies to this diagram:

The optimum intensity of fishing effort is that which maximizes *L* — *C*. This is the

monopoly solution; but, since we are considering only a single fishing ground, no price effects are introduced, and the social optimum coincides with maximum monopoly revenue. In this case we are maximizing the yield of a natural resource, not a privileged position, as in standard monopoly theory. The rent here is a social surplus yielded by the resource, not in any part due to artificial scarcity, as is monopoly profit or rent.

If the optimum fishing intensity is that which maximizes *L* — *C*, . . . the optimum fishing intensity is *Oe'* of fishing effort. This will yield *Ol'* of landings, and the species population will be in continuous stable equilibrium at a level indicated by P_5.[8]

II

Gordon calls these two situations "positions of equilibrium." It is true that in the first, the competitive exploitation of the fishery, the equilibrium indicates values of *C, P,* and *E* that are compatible with one another, though we are not told just how the fisherman arrives at this equilibrium. It should be noted that it is quite possible that a competitive fishery in the process of production—which might be described diagrammatically as a trial-and-error procedure of searching for the inscribed rectangle—might easily miss an unstable equilibrium and gravitate rapidly between the alternate extremes of zero output and zero population. Furthermore, in those fisheries there may be no stable equilibrium short of zero landings and full natural population. It is not unlikely that fished-out lakes —which are, after all, demersal fisheries in the *economic* sense of Gordon's paper—would be examples of this impossibility of finding a competitive equilibrium position.

"Would be" because the fundamental assumption in Gordon's paper (apart from the admittedly unknown

[7] For those readers who would like to identify this diagram with the algebraic system given in Gordon's paper, we can say that $P = a - bL$; $P = a/bCE + 1$; $C = qE$; $C = L$ (Gordon, *op. cit.*, pp. 141–42).

[8] *Ibid.*, p. 141.

nature of the biological relationship between landings and demersal populations) is that there are in fishing no diminishing returns and hence no increasing costs and no incentive to stop operations short of the equality of total costs and landings. Surely this fundamental assumption is incorrect; surely *in the short run* (with population and equipment fixed) each fishing boat will experience increasing costs as it attempts to increase its landings.

Gordon's analysis, which I have followed in Figure 1, relies upon the depletion of the population to produce a species of "diminishing returns" effect that will explain, with price given, why the competitive fishery does not expand indefinitely. But this explanation applies only to the long run and cannot hold within a single season, when the fish population is one of the fixed inputs. In the short run, fishermen do not expand their catch indefinitely because they *do* experience increasing costs in attempting to increase their landings. Gordon depends upon the omnibus variable "effort" to cover the changeable combinations of men, boats, and other equipment used by individual fishermen. But, if we look through this omnibus variable, we see that in fact the short-run situation in a fishery exploited by competing fishermen will be very like the standard situation in pure competion. The supply curve of this fishery (with the price given by the world market situation) will be made up by the addition of the relevant portions of the supply curves of the individual fishermen. These curves will slope upward because, with fixed equipment and a fixed number of boats, there will be some number of landings per boat which has a least cost; if the crew is worked long hours, or the boat is kept running without time for maintenance or repair, the cost per landing will begin to rise. Each boat will increase its landings until its supply price (marginal cost) is equal to the going price. The "surplus" that might be captured in this situation is the usual quasi-rent, available to each boat by operating at the point where marginal costs are equal to marginal revenue.

Now (if we continue to consider only short-run decisions), would a sole owner select a different rate of output than that which was determined under competition? There are two possible situations here: (1) the sole owner may take over an existing competitive fishery, boats, canneries, and crews. (2) The sole owner may reorganize the fishery in the most efficient way; this is not the *same* short-run situation but an alternative situation.

1. If the sole owner were taking over *for a season only* a fishery that had been equipped in the manner suitable for operation by competing fishermen, he would operate it in exactly the same way as they had, that is, at the output for which the marginal cost of fishing equaled the price of the product. There is, however, one qualification of this assertion. If it were the case that competing fishermen were so numerous that boats got in each other's way, then the sole owner would rationally lay off some of the boats (and perhaps canneries and collecting boats) for the season. In this way he could reduce the external diseconomies of fishing. But, apart from this qualification (which is really a matter of the long run), the sole owner and competitive fisherman would in the short run oper-

THE FISHERY: THE OBJECTIVES OF SOLE OWNERSHIP   121

ate the fleet identically, so that marginal cost equaled price and so that the marginal product of labor equaled the price of labor.

2. However, if a sole owner expected to have permanent tenure, then even in the short run his organization of the fishery would probably be quite different from that of small competing fishermen. For instance, it has been suggested that on the West Coast the sole owner of a salmon fishery would rely more on traps than on vessels; doubtless economically similar techniques are known in the demersal fisheries. Not only would a sole owner prevent the wasteful interference of competing fishermen with each other, but he would also design his fleet and his transport and packing facilities so as to take advantage of the economies of integration and scale. When he had worked this out, assuming that he was in competition with the owners of other fisheries, he would still tend to operate where short-run marginal cost equaled price. Whether this rule would result in his using more or less variable factors, and whether his catch would be larger or smaller, it is impossible to guess a priori. There is no reason to believe that it would be significantly different, although the productivity of all inputs would almost certainly be higher, since the sole owner has the choice of a wider range of techniques.[9]

Hence, we can say that, as a general rule, the mere fact of sole ownership does not bring about a significant

[9] Another external diseconomy arises from the shortage of really skilled labor. A sole owner could either plan to economize on the use of labor by adopting labor-saving techniques (which, to an extent, is a method also open to competing boats) or to act as a monopsonist in the purchase of local labor services, or both.

change in the exploitation of the fishery in the short run. Both the sole owner and the competing fisherman will operate at an output which is theoretically similar (in its equality of marginal cost and marginal revenue) to that in other industries. Only if there is an opportunity for adopting alternative fishing techniques that reduce the investment necessary for a given output is there an argument in favor of sole ownership. Some efficient techniques may be profitable only on the assumption that a very large fishing operation can adopt them. But why cannot there be large-scale efficient operations under competition among fishermen? Perhaps because of the danger of diminishing the population or of the fear of its diminution; but these are long-run dangers which I shall discuss below.

While I am on the short-run part of the argument, it is relevant to comment on the subject of the cost of the variable factors. These can be divided into cash costs and opportunity costs of the fishermen. The smaller the opportunity costs—perhaps because of the immobility of the fishermen—the greater the use of factors in fishing, regardless of whether the industry is competitive or typified by sole ownership. The low opportunity costs do not provide a basic explanation of the inefficiency of competitive exploitation of fisheries; it is the inability to control the size of the fish population in the long run which does that. Hence even in areas where relevant opportunity costs are high, as they are in the West Coast industry, we find more men and more rigs employed than would be employed in a "monopolized" fishery. The price system, when it works well, does not depend only upon high opportu-

122   ANTHONY SCOTT

nity costs to draw factors into the most
productive employment. It also relies
on employers dispensing with factors
that are not needed; and our subject
here is really the alleged failure of
competitive fisheries to do this. Low
opportunity costs are not relevant to
the immediate problem. Where Gordon
brings in the low opportunity costs

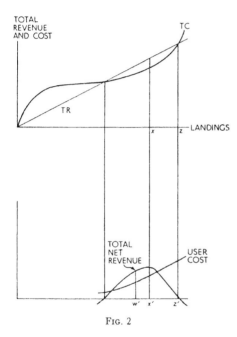

Fig. 2

of the industry, he drags in a red
herring.[10]

III

But it is when we come to the long
period that we see where the four-
variable analysis fails as a description
of the sole owner's optimum. What is
now needed is an indication of the best
use of the factors of production and
of the fishery over time. It is not to be
concluded, for example, that the ra-

[10] *Op. cit.*, p. 132.

tional owner would even wish to find
an "equilibrium" size for the fish population. His most profitable action might
be instead to deplete the fishery, gradually, over time; or, alternatively, to
build it up over time. *As long as the
user of a fishery is sure that he will
have property rights over the fishery
for a series of periods in the future*,
he can plan the use of the fishery in
such a way as to maximize the present
value (future net returns discounted
to the present) of his enterprise. From
the social point of view it can be said
that he will bring about the "best" use
of the fishery and of all other factors
invested in it over future periods by
thus allocating outputs and outlays
over time in accordance with the current rate of discount.

I can best illustrate the nature of
this decision-making process by using
the following diagrams (Fig. 2).[11] The
first diagram indicates the total costs
and total revenues of a fishery in a
period. The total revenue curve is
shown as linear to indicate that the
output (landings) of this one fishery
has no effect on the price at which it
may be sold. The total cost curve is
shaped to suggest diminishing marginal
returns (increasing costs) in the short
run.

Under competitive fishing conditions,
or if the fishery is taken into sole ownership, the tendency is to maximize
net returns from the fishery by producing x, where TC is parallel to TR
(that is, where marginal cost equals
price). This holds only in the short
run.[12] However, if the catch today has
an influence on the population and so

[11] See my "Notes on User Cost," *Economic
Journal*, June, 1953, p. 372.

[12] The output z is the competitive equilibrium
(no-profit) output suggested by Gordon.

THE FISHERY: THE OBJECTIVES OF SOLE OWNERSHIP   123

on the catch tomorrow, the sole owner will wish not only to maximize current returns but also to arrange for the optimum series of landings through the ensuing future periods. He will, in fact, wish to maximize the present value of his property. This he will do by investigating the effect of his marginal current output on the present value (or sum of the discounted net returns of all future periods) and by fixing current output where marginal current net revenue is equal to marginal user cost. In such a position, since a unit of output is produced only if its addition to current net revenue exceeds its cost in diminished present value, the sole owner succeeds in keeping the future returns from the fishery as high as possible while maximizing current income. This position is to be found at w' where the total net revenue curve is parallel to the user-cost curve. The total net revenue curve is derived directly from the TC and TR curves and shows the difference between them. The user-cost curve shows the effect of succeeding units of current output on the "present value" of the enterprise. The greater the rate of interest (or the personal rate of time preference of the owner), the lower the valuation put on landings in the remote future, and the lower the user cost.

If increased output tends to diminish the population and so to reduce the net revenues that could be earned in other periods had output been restrained today, the user-cost curve will slope upward, marginal user cost will equal marginal net revenue at less than the maximum total net revenue, and sole ownership will result in a still greater reduction of desired output than would be the case if short-run considerations

only were at stake. This slope of the UC curve is presumed to describe the situation in those fisheries that are exhaustible. Pelagic fisheries such as salmon and seal might be suggested.

If, on the other hand, increased output should tend to increase the population and the net revenues to be earned in future periods, the user-cost curve would slope downward. This, it has been said, is true up to a point of some fisheries: effort today not only produces a catch today but also improves conditions for increase of the fishery. But this too may be truer of pelagic than of demersal fisheries. In these special circumstances the current rate of output of a sole owner would be somewhat larger than that which yielded the maximum current return, and perhaps even more than that of Gordon's competing fishermen.

If landings have no effect on population (or, more precisely, on future landings), there is no user cost, no user-cost curve, and the output x is most profitable even when long-run effects are taken into consideration. This seems to be the thoroughgoing "demersal" case.

IV

What has been left out of this picture of the sole owner's planning for the long run?

In the first place, following Gordon, I have not given much attention to the nature of the fixed equipment used in the industry. It may not be possible to replan the duration of the fishery (that is, the size of the population) from period to period with complete freedom because the type of equipment needed to bring in small, unusually large, or postponed landings may not be available. In actual practice it is necessary to plan the "scale" of a private fishery at the outset, to fit into the complementary provision of canneries, transport, etc. Once this "scale" has been established, it is not easy to change the general range of output per season if user cost changes. The problem here could be illustrated diagrammatically by showing a variety of short-run cost curves corresponding to the choice among "scales" of fishing industry. The net revenue curves so established would have to be confronted with a series of user-cost curves corresponding to alternative future long-run scales of the industry.[13]

In the second place, I have assumed that the sole owner is not the monopolist of his product. If he were a monopolist and could influence the price by his output, he would be confronted by a nonlinear total revenue curve in Fig. 2, and it is conceivable that his landings per period would be even smaller than those of a corresponding solely owned fishery competing with many other fisheries. Also, if he were a monopolist, he would be able to influence the future price and even the trend of taste and demand for his output. There are so many possible consequences of this power that it is impossible to generalize about them. One

[13] Professor Donald Carlisle in his recent "The Economics of a Fund Resource: Mining," *American Economic Review*, September, 1954, p. 609, has suggested a three-dimensional cost surface, with rate of mining on one axis, level (or scale) of the mine on another, and costs on the vertical axis. Instead of the usual U-shaped cost curve, a saucer-shaped surface is developed. It is not easy to see how the "present value optimum," which corresponds to our long-run equality of marginal user cost and marginal net revenue, is determined —its position is merely drawn into the diagram, apparently arbitrarily. However, the diagram is a useful reminder that scale and current output cannot be dissociated; each must be constantly reappraised.

important possibility, however, is that the uncertainty surrounding the price and sales in each period would be somewhat less than the uncertainty borne by competitive producers. Such monopolists might well increase the investment in the scale of the fishery. This, in turn, would tend to keep the fish population high and to promote a sustained yield rather than the gradually decreasing yield that is a likely outcome of sole ownership.

Finally, I have asserted that the equilibrium of the sole owner who maximized the present value of the fishery would correspond more closely to the social optimum than would the competitive equilibrium. This is true only if the other enterprises in the economy are run by purely competitive businessmen who attempt to maximize their profits and to maximize the present value of their enterprises in terms of the market rate of discount. If these assumptions are satisfied by equalizing marginal user cost and marginal net return, we have a situation where the marginal productivity of each factor is the same wherever it is used, and where the allocation of production over time corresponds to the rate of discount determined by the marginal rate of time preference of the community. In such circumstances, given the endowment of the economy with factors and resources, and given the tastes of savers and consumers, it would not be possible to increase the value of output of any product without reducing some other output by a greater amount. In this sense, the social optimum in both the long run and the short run would demand that common-property resources be allocated to maximizing owners, associations, co-operatives, or governments.

[3]

Some Considerations of Population Dynamics and Economics in Relation to the Management of the Commercial Marine Fisheries[1]

By Milner B. Schaefer

Inter-American Tropical Tuna Commission,
Scripps Institution of Oceanography, La Jolla, California

INTRODUCTION

FISHING is one of man's oldest occupations; some of the sea fisheries pre-date recorded history. So long, however, as men relied on oar and sail to reach the fishing grounds, and on simple, hand-operated gear to catch the fish, the intensity of fishing on the high seas remained low, so that the amount of the catch had apparently little effect on the magnitude of the fish stocks. There were great variations in the harvests, of course, but these were due to fluctuations in the fish populations quite independent of the amount of fishing. A fisherman's success depended on uncontrollable natural factors, and was not much affected by whether the number of fishermen was many or few.

With the industrialization of sea fishing in the latter part of the last century, bringing steam and later diesel power to the vessels, and bringing new and more efficient types of fishing gear, and machinery to handle it, the sea fisheries near northern Europe and in some other parts of the world began to show signs of diminishing return per unit of fishing effort. Near the turn of the century, there was considerable controversy as to whether or not the amount of a given kind of fish which man is able to take from the sea is sufficient to have any noticeable effect on the supply. This matter was discussed at some length, for example, by McIntosh (1899), Garstang (1900) and others. Alfred Marshall was preparing the first edition of his famous *Principles of Economics* (1890) at the time when the industrialization of the British trawl fishery was proceeding rapidly and this controversy was going on. It was, therefore, yet a moot question whether his Law of Diminishing Returns applied to the sea fisheries (Marshall, 8th edition, 1938, p. 166).

Subsequent history of the North Sea demersal species, the haddock of Iceland and the Northwest Atlantic, the Pacific halibut, and of numerous other fisheries, leaves little room to doubt that a modern commercial fishery can so affect the stock of fish in the sea that the return per unit of fishing effort is thereby diminished, and can even become so intense that the *total* harvest is also reduced. It may be noted here, although the matter will be developed in more detail later, that the law of diminishing returns as applied by Marshall and others to agriculture is somewhat different than the application to the sea fisheries. As originally developed, the law holds that the increased application of other factors of

---

[1]Received for publication January 14, 1957.

J. FISH. RES. BD. CANADA, 14(5), pp. 669–681, 1957.
Printed in Canada.

670

production to the land results in a decreased *rate* of return, but so long as the fundamental fertility of the land is not reduced, the *total* return would not diminish, but would increase, at a falling rate, toward some upper limit determined by the fertility of the land, the rainfall, amount of solar radiation, etc. (Ricardo's "original and indestructible powers of the soil".)

Experience having shown that the stock of commercial sizes of a sea-fish species, and the annual harvest obtainable from that stock, is related to the amount of fishing effort applied, there arises the important question of how the amount of fishing should be managed in order to provide the greatest benefits to mankind. This, of course, is a socio-economic problem which will have unique solutions only if it can be specified what situation among possible alternatives is to be regarded as most beneficial. Fundamental to rational consideration of the matter, however, is knowledge respecting what are the possibilities, which must depend on the dynamic relationships between amount of fishing and amount and yield of the fish stocks, and corollary economic implications.

Attempts to systematize some of the significant biological and economic facts bearing on this problem have been made by a number of persons in recent years, among which may be cited Russel (1931), Graham (1935, 1939, 1953), Beverton (1953), Gordon (1953, 1954), Burkenroad (1951), and Schaefer (1954a, b). The main result has been a considerable advance in our understanding of the biological and economic principles involved. There is, however, a fairly large degree of confusion, resulting from the biologists' inadequate consideration of economic principles, and, in part, from economists' failure to fully consider the properties of a self-renewing natural resource, the rate of renewal of which is dependent on the magnitude of the stock of the resource, which importantly distinguish such a resource from other classes of natural resources.

It seems worthwhile, therefore, to consider together some significant aspects both of the population dynamics of commercial fish stocks and of the economics of commercial fishing in order to arrive at a rational basis of considering the social problem of fisheries management.

In considering the bionomic properties of a fishery we shall be concerned with the "long run" relationships. That is, we are interested in the average annual harvests that will be *sustained* by the fish population indefinitely at different levels of fishing effort, the monetary value of the harvests, and the monetary cost of the fishing effort. This approach has been admirably and carefully applied by Gordon (1954), but some further consideration appears to be necessary, because (1) he has not made the necessary distinction between the self-regulating, density dependent fish resources and other common-property resources of a different nature, (2) the mathematical model in Section IV of his paper is not consistent with the assumptions in Section III (and is not quite in accord with some dynamic properties of fish populations), and (3) he has (p. 129) defined the optimum degree of utilization of any particular fish stock as that which maximizes the net economic yield, the difference between total cost, on the one hand, and total receipts (or total value production), on the other. This is one possible choice, of course, but it is not immediately obvious that it is the social optimum;

indeed other possibilities have been explicitly chosen both for particular fisheries[2] and as a general objective of fishery management.[3]

DYNAMICS OF RENEWABLE NATURAL RESOURCES

The natural resources upon which mankind depends are of two classes. In one class are those resources, such, for example, as fossil fuels, mineral deposits, and elements capable of yielding energy by the conversion of mass to energy by nuclear processes, which are non-renewable, or for which the rate of renewal is so slow that it is infinitesimal compared to the rate of use. To this class of resources the concept of indefinitely sustainable yield cannot be applied. Man can use these resources rapidly or slowly, efficiently or inefficiently, but the total quantity on this planet is limited and is subject to being completely used up in finite time. In the second category are those resources which are constantly renewed, and are, therefore, capable of yielding sustained production indefinitely.

The different nature of renewable and non-renewable resources was clearly distinguished by Marshall (p. 166–167). Among the renewable resources there are, again, two types with fundamental difference, those for which the rate of renewal is dependent on the amount of the resources which is left unharvested to perpetuate itself, and those where such dependence does not exist, or is negligible.

NON-SELF-REGULATING RESOURCES

For the second type, noted in the preceding paragraph, which we may call *non-self-regulating resources*, the amount which is available to be used each year is determined by natural phenomena other than the magnitude of the resource itself, and the amount which is used during a given year has no effect on the amount which will be available for use during the next year. The renewal of resources of this type is, consequently, independent of the rate of use.

The concept of "land" as one of the factors of production in classical economic science is apparently based on the idealization of this type of resource. The concept is summarized by Marshall (p. 144):

"While man has no power of creating matter, he creates utilities by putting things into a useful form; and the utilities made by him can be increased in supply if there is an increased demand for them: they have a supply price. But there are other utilities over the supply of which he has no control; *they are given as a fixed quantity by nature* and have therefore no supply price. The term "land" has been extended by economists so as to include the permanent sources of these utilities; whether they are found in land, as the term is commonly used, or in the seas and rivers, in sunshine and rain, in winds and waterfalls."

---

[2]The Convention between Canada and the United States of America for the Preservation of the Halibut Fishery of the Northern Pacific Ocean and Bering Sea, for example, provided for regulations "designed to develop the stocks of halibut in the Convention waters to those levels which will permit maximum sustained yield".

[3]The International Technical Conference on the Conservation of the Living Resources of the Sea, 16 April to 10 May, 1955 (Rome), attended by representatives of 45 nations, decided "The principal objective of conservation of the living resources of the seas is to obtain the optimum sustainable yield so as to secure a maximum supply of food and other marine products".

672

One example of this type of resource is hydro power. The amount of power which can be obtained annually from a river system depends on the rainfall, the topography of the land, and certain other physical factors (Massé and Rousselier, 1951). The amount which is used is always less than the amount potentially available, because of economic factors, but the amount used in a year has no effect, in general, on the amount which can be obtained in subsequent years.

A more familiar example, which has been the basis of much classical economic theory, is the use of agricultural land to produce cultivated crops from planted seed. As we have noted above, the land is considered to be a fixed factor of production, the amount of which can neither be increased nor diminished by man's actions. Since the quantity of this factor is fixed, the application of increasingly large quantities of other factors of production (labour and capital) to the land results in a decreasing return per unit of these other factors. So long as the inherent natural properties of the land are not destroyed, increasing effort will give increasing return, but the return per "dose" of capital and labour decreases. In other terms, production as a function of effort (number of "doses" of capital and labour) is a monotonically increasing function with negative second derivative, approaching some upper limiting value asymptotically. This may be illustrated by Fig. 1.

It should be noted that, for this type of resources, there is no maximum in the production function, since $\frac{dL}{dE} > 0$ for all finite E. In this respect such resources differ importantly from the self-regulating resources, the supply of which *is* influenced by man's action.

SELF-REGULATING RESOURCES

Populations of sea fish belong to a different type of natural resources, for which the annual rate of renewal of the resource is a function both of the physical environment, which is presumably constant, on the average, over the long run, and of the magnitude of the standing crop, or population, of the resource, which is diminished by the rate of harvesting. Some other populations of organisms are also of this same type, such as populations of wild, fur-bearing animals, forests which depend on self-seeding for renewal, herds of range cattle, range grasses (forage), and insect populations. In this last instance, man is usually interested in minimizing the population rather than in obtaining a high yield, but the same basic principles of population dynamics are involved. I shall, in the following, discuss primarily sea-fish populations, but the analogies for other similar resources should be obvious.

An outstanding common characteristic of populations of fishes, and other natural populations of organisms, is that they tend to remain in dynamic balance, neither falling to zero nor increasing without limit. Over any reasonably long period of time, losses from the population must be balanced by accessions to the population. When, however, the percentage rate of loss is increased, decreasing the size of the population, from whatever cause, the percentage rate of renewal must increase also, so that the population again comes into balance. This has been demonstrated in some detail by a number of biologists, from both theoretical and experimental viewpoints, including Nicholson (1954,a,b), Ricker (1954), Lack (1954), and others. From the standpoint of the fish population, the harvesting by man is simply an additional source of mortality by predation, which is met by a compensatory increase in the rate of population renewal, so that the population again comes into balance at some lower population level at which the increased rate of growth of the population balances the harvest taken by man. For each size of population, there is a certain *rate of natural increase*, which is, under average environmental conditions, some single valued function of population size. In mathematical notation

$$\frac{dP}{dt} = f(P) \tag{1}$$

The catch, or landings, L, during a year is some function of the size of population and the amount of the other factors of production, which we collectively term "fishing effort", E.

$$L = \phi(P,E) \tag{2}$$

In the equilibrium state, which we are here discussing, the catch is exactly equal to the rate of natural increase. This has been called the *equilibrium catch* by Schaefer (1954a). This equilibrium catch is the long-term annual production of the fishery for a given level of population (and effort).

It immediately follows, of course, from (1) and (2) that under equilibrium conditions population size is some function of fishing effort

$$P = \Psi(E) \tag{3}$$

Data from experimental animal populations and from the commercial fisheries (Büchmann 1938, Graham 1935, 1939, Schaefer 1954a) indicate that $f(P)$ is a single valued positive function, falling to zero at $P = O$ and at $P = M$, the maximum population which the environment will support under average conditions, with no fishing, and having a maximum at some intermediate value of P. It further appears that a reasonably good first approximation is the quadratic

$$f(P) = k_1 P(M\text{-}P) \tag{4}$$

where k_1 and M are constants.

It also appears that, to a good degree of approximation,

$$L = k_2 EP \tag{5}$$

where k_2 is a constant so that, under equilibrium conditions,

$$k_2 EP = k_1 P(M\text{-}P) \tag{6}$$

and, consequently

$$P = M - \frac{k_2}{k_1} E \tag{7}$$

That is, for equilibrium conditions population size is a linear function of fishing effort; and, from (5) and (7)

$$L = k_2 E \left(M - \frac{k_2}{k_1} E \right) \tag{8}$$

It is to be noted that this mathematical model is identical with that derived by Gordon (1954) in section III of his paper and illustrated in his Fig. 1. It does *not* correspond to the model in section IV of his paper, where he assumed the population size to be a linear function of equilibrium catch, which is not in agreement with his earlier model, and is not in accordance with the experimental

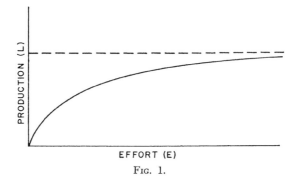

EFFORT (E)

Fig. 1.

and observational data, because a necessary consequence of his last model is that equilibrium catch increases continually with fishing effort, having no maximum value at some intermediate level of fishing effort.

The production function, in terms of total weight of catch, of a fishery is the equation (8), shown graphically in Fig. 2.

For some fisheries, where the fish below the minimum commercial size constitute an important part of the breeding stock, the production may not fall to zero with increasing effort applied to the commercial stock, but will asymptotically approach some limiting value of production greater than zero, as shown, for example, by Beverton, *op. cit.* Over the range of values of effort which will be encountered in practice in such fisheries, however, my equation (8) still seems to provide a fairly good approximation.

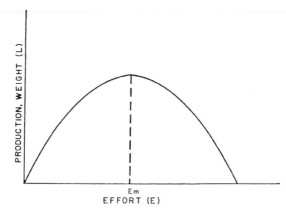

Em

EFFORT (E)

Fig. 2.

If the price of fish remains constant, i.e. if the demand is perfectly elastic, this is also the production function in terms of value. If, however, there is some decrease of unit prices with quantity marketed, the production function in terms of value, V, will have a somewhat different form.

The form of V as a function of E will depend on the demand-price function, but it is to be noted that so long as the price elasticity of demand remains greater than unity, the production function in terms of value will have its maximum at the same level of effort as the production function in terms of quantity (Fig. 3, Curve A). If the elasticity of demand becomes less than unity for some levels of production which the fishery is able to reach, there may be two maxima in the production function in terms of value (Fig. 3, Curve B). The zero points of the two functions will be identical under any circumstances, because zero catch is of zero value whatever the unit price.

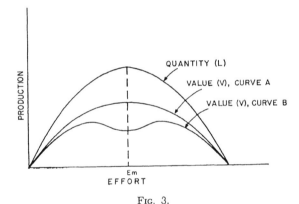

FIG. 3.

For most individual stocks of sea fish, the catch is a rather small share of the total production of all fish with which it competes in the market, and, therefore, it seems reasonable to assume that for the products of a particular fish stock the elasticity of demand is large. Indeed, we should not go far wrong in assuming, as has Gordon (1954), that the price does not vary with landings. In the economic model, analysed below, I shall consider only this case.

From comparison of Fig. 1 and 2, the difference in the law of diminishing returns for a non-self-regulating resource, and a self-regulating, density de-pendent resource, such as a sea fish population, may be readily appreciated. In the former case as other factors of production are added, i.e. as the effort is increased, the rate of increase of production diminishes, but the gross produc-tion increases monotonically toward some limiting upper value as an asymptote. In the case of the fish resource, the rate of increase of production diminishes continually with increasing effort also, but, in addition, after a certain level of effort is reached, the gross production falls off as well. In the extreme case

676

the resource, and production from it, could be reduced to zero, although this does not often happen in practice due to the cost of effort, which will be discussed below.

ECONOMIC MODEL OF THE FISHERY

To complete the economic model of the fishery, in order to investigate some of its properties, we need to consider, in addition to the foregoing, the cost of the fishing effort. The unit costs of the factors of production in a fishery are largely determined by the general economy, so we should not be making any very serious departure from reality, if we assume with Gordon (1954) that the cost of the fishing effort, C, is directly proportional to the amount of effort.

$$C = aE \qquad (9)$$

where a is the cost of a unit of effort.

If we consider the simple case where the unit value of the catch is constant, and equal to β, we then have with the aid of (9) and (8) (combining some constants in 8 for simplicity of notation), for a complete model

$$
\begin{aligned}
L &= aE\,(b - E) \\
V &= \beta L \\
C &= aE
\end{aligned} \qquad (10)
$$

We may now investigate some of the properties of this model.

As we have noted previously, the production function is a second degree polynomial, having a maximum at some intermediate value of fishing effort. This is shown as curve V in Fig. 4, where we have designated the amount of fishing

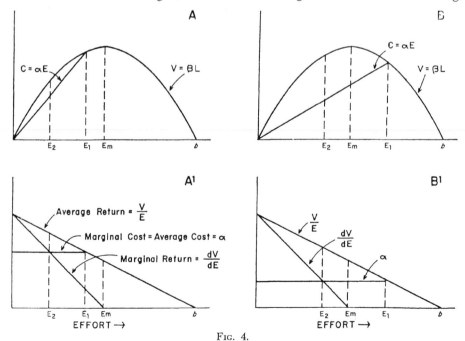

Fig. 4.

effort corresponding to maximum sustainable production by E_m. In this same figure, we show the cost function as the line C. We have in Fig. 4A and Fig. 4B illustrated two cases, first where a is so large that the amount of fishing effort (E_1), for which total cost equals total product, is less than E_m; and second where a is so small that E_1 is greater than E_m.

The necessary condition for maximum production is, of course,

$$\frac{dV}{dE} = a\beta(b - 2E) = 0$$

whence

$$E_m = \frac{b}{2} \tag{11}$$

From the first two equations of (10)

$$V = a\beta E(b - E)$$

The average return is then

$$\frac{V}{E} = a\beta(b - E) \tag{12}$$

and the marginal return is then

$$\frac{dV}{dE} = a\beta(b - 2E) \tag{13}$$

while marginal cost and average cost are equal:

$$\frac{C}{E} = a \tag{14}$$

and

$$\frac{dC}{dE} = a \tag{15}$$

These functions are illustrated in Fig. 4, A[1] and B[1], for the same two values of a as in Fig. 4, A and B.

It may be seen that marginal return falls to zero at $E = b/2$, which is also the amount of effort E_m for which total product is maximum.

The *net economic yield* is the difference between total cost and total product

$$Y = C - V = aE - \beta aE(b - E) \tag{16}$$

and will be maximum for that value of effort where

$$\frac{dY}{dE} = a - a\beta(b - 2E) = 0$$

or

$$a = a\beta(b - 2E) \tag{17}$$

$$E_2 = \frac{1}{2}\left(b - \frac{a}{a\beta}\right) \tag{18}$$

This is also, of course, the value of E for which the marginal cost equals the marginal return. It is labelled E_2 in Fig. 4.

In a fishery which is a common property resource, where anyone who wishes to do so is free to enter, new operators will be attracted to come into the fishery so long as the average cost is less than the average return (cost, of course, including all the costs of factors of production, including interest on capital investments and the normal entrepreneurs' fee), so that in the unrestricted common-property fishery the effort will grow until it reaches E_1, where average cost equals average return and the net economic yield is zero.

For average cost equal to average return

$$a = a\beta(b - E)$$

so that

$$E_1 = b - \frac{a}{a\beta} \tag{19}$$

If unit cost is high relative to unit price of product (Fig. 4A, A^1), E_1 may be at a level of effort below that corresponding to maximum sustainable yield (E_m). In this case no increase in total yield can be obtained from restrictions on fishing effort. Any increase in yield must involve increasing unit prices or decreasing unit costs by some artificial means, such as price supports, subsidies, etc.

If, on the other hand, unit price is sufficiently high relative to unit cost (Fig. 4B, B^1), E_1 may be at a level of effort higher than the level E_m where maximum yield is obtainable. In this case, the yield can be increased by restricting the amount of fishing effort.

The maximum net economic yield will be obtained by restricting the fishing effort to E_2. From (11), (18), and (19) it may be seen that this is always (1) at a lower level of fishing intensity than that at which maximum total catch is obtained (E_m) and, (2) at half the intensity (E_1) at which the unrestricted common-property fishery will arrive. Since E_2 is always less than E_1 and E_m, maximizing the net economic yield must always entail some sacrifice of total production.

If it is desirable to maximize the net economic yield, this may be accomplished, as has been advocated by Gordon (1953, 1954), by making the fishery a property right, under unified control, because in this event that intensity of fishing will result where marginal cost equals marginal return, which corresponds exactly, as has been shown above, to the condition of maximum net economic yield which, in this event, would be the rent from the property.

Under some circumstances, it may be desirable to reduce fishing effort below E_1 but yet not reduce it to E_2. This, for example, is the case where, as in Fig. 4B, the uncontrolled common-property fishery will become so intense that the total catch is less than the maximum sustainable catch, and it is desired to maximize the sustainable catch. This may be accomplished by restricting the effort to E_m by suitable measures. If these regulations are so selected that the cost per unit of fishing effort is not increased, there will be some net economic yield also, measured by the difference V–C at E_m. Regulations in practice are often used such that the unit cost is increased at the same time the effort is restricted, so that the net economic yield is partly or completely dissipated.

SOME SOCIAL AND POLITICAL CONSIDERATIONS

Two of the possible benefits from fishery management are increased total production of food and other marine products, and increased net economic return to the fishermen. It may be seen that these are to some degree mutually exclusive, however. In the case of Fig. 4A, no increase in production can be obtained by curtailing effort, and an increase in net economic yield must entail sacrifice of some production. In the case of Fig. 4B, it is possible to increase production and at the same time obtain some net economic return by curtailing fishing effort, until it reaches the level E_m. Below E_m, however, again increased net economic return inevitably implies decreased production.

Gordon (1953) has strongly advocated managing a fishery to maximize the net economic return on the theory (p. 443) that, ". . . with every productive enterprise—the measure of its own contribution to human economic welfare is determined by its *net* output, after the costs of the factors necessary to that output's production have been deducted". He does, however, qualify this by noting ". . . the economic optimum is not necessarily the *human* optimum. Under certain circumstances we may well prefer to have an economically 'inefficient' fishery if the other effects of organizing the fishery along economically optimum lines are politically difficult or socially undesirable."

One of the important social considerations is the increasing requirement of protein food to feed the world's rapidly growing human population. The UNSCUR Conference in 1949 spent a great deal of time considering current food shortages, the prospective requirements in the near future, and means by which they can be satisfied. Broadley (1950), for example, pointed out that even before the Second World War food standards in many countries fell short of a minimum caloric diet, and fell even further short of minimum requirements of proteins, vitamins, and minerals. The forecast for 1960 indicated a very large increase in the requirements. It was indicated by other persons at the conference that a large increase in production from agriculture could be obtained by bringing land now sub-marginal into production, and by increasing the efficiency of agriculture. This, however, will quite obviously require great efforts both nationally and internationally, without much regard to the obtaining of a net economic yield from the increase of production. The existence of agricultural subsidies, direct or indirect, in many countries indicates the necessity of sacrificing some economic yield in order to obtain the production needed to feed the human population.

There is also some doubt whether the supply of proteins required can be produced on the land. Le Gall (1951) states, with respect to the world requirements of proteins: "The present annual needs are approximately 50 million tons; 75 million tons will be required by 1960, and the land alone will not be able to provide them."

It would seem, therefore, that there is adequate reason to give first priority to maximizing the yield of the sea fisheries. This choice has been the basis of fishery management, in general, in the United States, and has, as noted by Graham (1956), been explicit in all the recent international conventions, in the New World. He also stated, that in the Old World there has not, as yet, been made any explicit choice among the possible qualities of the fishery. The conclusions of the recent United Nations Conference on the Conservation of the Living Resources of the Sea (see footnote on page 671) indicates, however, that all the nations in attendance are giving high priority to this alternative.

For the fisheries of the high seas, which are exploited by several nations jointly, there also arises the difficulty of applying any common economic criterion. The level of exploitation giving maximum sustainable production is a property of the fish population, and is the same regardless of who catches the fish. The maximum net economic yield, on the other hand, depends not only on the dynamics of the fish population, but also on the value of the catch and the costs of making the catch. Since the value of a given kind of fish varies from nation to nation, and the cost of a unit of fishing effort also varies from nation to nation, it is difficult to see how any common agreement as to the level of fishing effort corresponding to maximum net economic yield could be arrived at in an international, high-seas fishery. For this reason, even aside from the question of the social desirability, the economic yield appears, at this time at least, to be a quite impossible basis of agreement among nations for the joint management of the high-seas fisheries.

In a fishery such that, without regulation, the fishing intensity can exceed the level where the sustainable production is the maximum (Fig. 4B), the production can be maximized by curtailment of the fishing intensity. Such curtailment can be accomplished in a variety of ways, which leads to further problems of a social nature. If, in such a fishery, the fishing intensity is limited to E_m, the maximum average sustainable catch will be obtained. Some net economic return is also possible (the difference V–C at E_m), provided that the cost per unit of fishing effort is not increased. This, however, involves, in one way or another, establishing at least a limited property right in the fishery, because, if it remains open to exploitation by all-comers, with only the total annual effort being limited (as, for example, by the establishment only of annual total catch limits, as in the North Pacific Halibut fishery) there is a tendency for additional participants to enter the fishery, decreasing the length of the season, or the amount gear fished per vessel, with a consequent increase in unit costs, dissipating some or all of the economic yield. It has been argued by Nesbit (1943) and others that the net economic yield should be preserved by a preferential licensing system, or other means of limiting the number of persons engaged in the fishery. This, however, comes into conflict with another possible benefit from fishery management, the providing of maximum opportunity for employment in the fishery. Many social and political considerations enter into the choice involved here. It is not my purpose to attempt even to review these considerations, but merely to point out that free access to the fishery by all citizens and the obtaining of the possible economic yield are mutually exclusive.

LITERATURE CITED

BEVERTON, R. J. N. 1953. Some observations on the principles of fishery regulation. *J. Cons. Expl. Mer*, 19(1): 56–68.

BROADLEY, HERBERT. 1950. Critical shortages of food. *United Nations Conference on the Conservation and Utilization of Resources*, Vol. 1 (plenary meetings), pp. 30–34. Rome.

BÜCHMANN, A. 1938. Ueber den Höchstertrag der Fischerei und die Gesetze organischem Wachstums. *Ber. der Deutschen Wissenschaft. Komm. f. Meeresforschung*, N.F., 9: 16–48.

BURKENROAD, M. 1951. Some principles of marine fishery biology. *Publ. Inst. Mar. Sci. (Texas)*, 2(1): 177–212.

GARSTANG, W. 1900. The impoverishment of the sea. *J. Mar. Biol. Assoc. U.K.*, N.S., 6(1): 1–69.

GORDON, H. S. 1953. An economic approach to the optimum utilization of fishery resources. *J. Fish. Res. Bd. Canada*, 10(7): 442–457.

——— 1954. The economic theory of a common property resource: the fishery. *J. Political Economy*, 62(2): 124–142.

GRAHAM, M. 1935. Modern theory of exploiting a fishery, and applications to North Sea trawling. *J. Cons. Expl. Mer*, 10(3): 263–274.

——— 1939. The sigmoid curve and the overfishing problem. Rapp. et Proc.-Verb., Cons. Expl. Mer, 110: 15–20.

——— 1953. Overfishing and optimum fishing. *Ibid.*, 132: 72–78.

——— 1956. Concepts of conservation. *Papers, Int. Tech. Conf. on Conservation of the Living Resources of the Sea*, pp. 1–13.

LACK, D. 1954. The natural regulation of animal numbers. Clarendon Press, Oxford, 343 pp.

LE GALL, JEAN. 1951. The present world problem of sea fisheries. *United Nations Conference on the Conservation and Utilization of Resources*, Vol. 7 (*Wildlife and fish resources*), pp. 11–13.

MARSHALL, ALFRED. 1938. Principles of economics. 8th edition, MacMillan Co., New York, 871 pp.

MASSÉ, P., AND M. ROUSSELIER. 1951. Hydro power and conservation of power resources. *United Nations Conference on the Conservation and Utilization of Resources*, Vol. 4 (*water resources*), pp. 430–432.

MCINTOSH, W. C. 1899. The resources of the sea. J. C. Cloy and Sons, London, 248 pp.

NESBIT, R. A. 1943. Biological and economic problems of fishery management. *U.S. Fish and Wildlife Service, Special Sci. Rept.*, No. 18, pp. 23–53.

NICHOLSON, A. J. 1954a. An outline of the dynamics of animal populations. *Australian J. Zool.*, 2(1): 9–65.

——— 1954b. Compensatory reactions of populations to stresses and their evolutionary significance. *Ibid.*, 2(1): 1–8.

RICKER, W. E. 1954. Stock and recruitment. *J. Fish. Res. Bd. Canada*, 11: 559–623.

RUSSEL, E. S. 1931. Some theoretical consideration on the "overfishing" problem. *J. Cons. Expl. Mer*, 6(1): 3–20.

SCHAEFER, M. B. 1954a. Some aspects of the dynamics of populations important to the management of the commercial marine fisheries. *Bull. Inter-American Tropical Tuna Comm.*, 1(2): 25–26.

——— 1954b. Fisheries dynamics and the concept of maximum equilibrium catch. *Proc. Gulf and Caribbean Fisheries Institute*, 6th Annual Session, pp. 53–64.

[4]

JOURNAL OF ENVIRONMENTAL ECONOMICS AND MANAGEMENT 2, 92–106 (1975)

The Economics of Fishing and Modern Capital Theory: A Simplified Approach [1]

COLIN W. CLARK AND GORDON R. MUNRO

Departments of Mathematics and Economics, The University of British Columbia, Vancouver, Canada V6T 1W5

Received February 18, 1975

While the link between fisheries economics and capital theory has long been recognized, fisheries economics has, until the last few years, developed largely along nondynamic lines. The purpose of this article is to demonstrate that, with the aid of optimal control theory, fisheries economics can without difficulty be cast in a capital–theoretic framework yielding results that are both general and readily comprehensible.

We commence by developing a dynamic linear autonomous model. The static version of the fisheries economics model is seen to be the equivalent of a special case of the dynamic autonomous model. The model is then extended, first by making it nonautonomous and second, nonlinear. Problems arising therefrom, such as multiple equilibria, are considered.

1. INTRODUCTION

It has been recognized, virtually from the time of its inception, that fisheries economics, like other aspects of resource economics, should ideally be cast in capital–theoretic terms. The fish population, or biomass, can be viewed as a capital stock in that, like "conventional" or man-made capital, it is capable of yielding a sustainable consumption flow through time. As with "conventional" capital, today's consumption decision, by its impact upon the stock level, will have implications for future consumption options. The resource management problem thus becomes one of selecting an optimal consumption flow through time, which in turn implies selecting an optimal stock level as a function of time.

In a pioneering and much cited paper, Scott [34] attempted to cast the problem of the management of a fishery resource as a problem in capital theory. The attempt was followed by Crutchfield and Zellner's [18] formulation of the problem in terms of a dynamic mathematical model. In spite of these works, however, the received theory of fisheries economics, founded by Gordon [21], continued to be formulated largely in static terms.[2] Indeed, one finds the static analysis being employed right up to the present day.[3] Reasons for the retreat to nondynamic analysis are not difficult to discover. While being warned that explicit consideration of time ought properly to be brought into the analysis, the reader was also advised that this could be an extraor-

---

[1] The authors express their gratitude to Professors A. D. Scott and H. F. Campbell for their helpful criticisms of and comments on earlier drafts of this article.

[2] See, for example, [1, 5, 7, 11, 16, 17, 25, 36, 37, 39].

[3] Christy [10].

dinarily complex, if not impossible undertaking.[4] It is perhaps reasonable to argue that the problem lay, not with the attempt to apply capital theory to fisheries economics, but rather with capital theory itself, which as Dorfman [20] has argued, suffered from an inadequacy of mathematical instruments.

Since the work of Ramsey [31], it has been clearly recognized that capital theory is in essence a problem in the calculus of variations. It was also recognized, however, that, in its classical formulation, the techniques provided by calculus of variations were inadequate to the task [20]. The extensions of calculus of variations provided by optimal control theory [6, 29] eliminated the inadequacies of the classical techniques to a large extent. Economists were quick to appreciate the implications for capital theory; indeed, Dorfman goes so far as to argue that modern capital theory traces its origins to the development of optimal control theory.[5] It seemed only a matter of time before the techniques of optimal control theory would be brought to bear upon fisheries economics.

Several attempts in this direction have by now been made.[6] This paper, with the aid of a simple linear model, summarizes most of the major results achieved so far, but does so in a manner such that the links with capital theory are made transparent. The paper then explores two sets of problems that have yet to be properly dealt with in the fisheries economics literature. The first concerns the optimal approach to the equilibrium stock, i.e., the optimal "investment" policy. The second set of problems arises from the relaxation of the highly restrictive assumption of autonomy (i.e., the assumption that the parameters are independent of time). The paper then concludes with the examination of the complexities that can arise when the assumption of linearity is relaxed.

2. THE BASIC MODEL

We commence with a simple dynamic model used widely in fisheries economics (e.g., [12, 18, 27]) that is usually associated with the name of Schaefer [32]. The model rests upon the Pearl-Verhulst, or logistic, equation of population dynamics.

Let $x = x(t)$ represent the biomass at time t. Corresponding to each level of biomass, there exists (according to Schafer) a certain natural rate of increase, $F(x)$:

$$dx/dt = F(x). \tag{2.1}$$

Equation (2.1) can be viewed as the net recruitment function or as the "natural" production function.[7] It is assumed that[8]

$$F(x) > 0 \quad \text{for} \quad 0 < x < K, \qquad F(0) = F(K) = 0, \qquad F''(x) < 0, \tag{2.2}$$

where K denotes the carrying capacity of the environment, i.e., $\lim_{t \to \infty} x(t) = K$.

---

[4] See [18, Appendix I]. Turvey [39, p. 75], on the other hand, argued rather curiously (and incorrectly in our belief) that even if one did make the analysis dynamic, no new interesting results would be forthcoming.

[5] Dorfman [20, p. 817].

[6] See [8, 12, 13, 19, 26, 27, 28, 30, 38].

[7] The "natural" production function can also be expressed as $\dot{x} = G(x, z)$, where z denotes the input of the aquatic environment. The input z is normally assumed to be constant; thus $\dot{x} = G(x, z)$ can be reduced to $\dot{x} = F(x)$. The fixity of z, of course, explains the diminishing returns to which x is assumed to be subject [$F''(x) < 0$]; cf. [32].

[8] Conditions (2.2) are automatically satisfied for the case $F(x) = rx(1 - x/K)$, which is the standard Pearl-Verhulst logistic model. Virtually all of our analysis is valid, however, under the less restrictive hypothesis (2.2).

When harvesting is introduced, Eq. (2.1) is altered to

$$dx/dt = F(x) - h(t), \qquad (2.3)$$

where $h(t) \geq 0$ represents the harvest rate, assumed to be equal to the consumption rate, and where dx/dt can be interpreted as the rate of investment (positive or negative[9]) to the stock (biomass).

Society's basic resource-management problem is that of determining the optimal consumption/harvest time path with the object of maximizing social utility (welfare). From Eq. (2.3) it is clear that this is the equivalent of determining the optimal stock-level time path.

There is, of course, the complication to be faced that the biological constraint (2.3) is accompanied by a harvesting cost constraint. The harvesting cost function is dependent upon an effort cost function and a "harvest" production function. We shall assume, in keeping with many standard fisheries models (e.g., the model of Crutchfield and Zellner [18]), that

$$C_E = aE, \qquad (2.4)$$

where C_E is total effort cost, E is effort, and a is a constant.[10] We also assume that

$$h(t) = bE^\alpha x^\beta, \qquad (2.5)$$

where b, α, and β are constants. It is further assumed that α is equal to 1.[11] The implications of (2.4) and (2.5) are that harvesting costs are linear in harvesting but are a decreasing function of biomass x (so long as $\beta > 0$).[12]

Given these assumptions, the complication introduced by positive harvesting costs does not alter the basic nature of society's optimization problem. It remains, in essence, the selection of an optimal consumption-flow/stock-level time path. Indeed, it will be demonstrated that for most of the cases encountered in this paper, the one major consequence of positive harvesting costs will be to introduce an effect directly analogous to the "wealth effect" encountered in modern capital theory.

In addition to the above assumptions, we abstract from all second-best considerations, and assume that the price of fish adequately measures the marginal social benefit (gross) derived from the consumption of the fish, and also that the demand for fish is infinitely elastic.[13] The problem can thus be viewed in terms of rent maximization as in the received theory.[14]

In the model the fundamental differential equation or state equation is (2.3):

$$dx/dt = F(x) - h(t), \quad x(0) = x_0;$$

[9] It may be worth stressing that in contrast to the standard or typical model in capital theory, disinvestment is not only allowed for in the fisheries model, but plays a critical role.

[10] That is, the supply function of effort is infinitely elastic.

[11] We do not assume that β is constrainted to equal 1, but only that $\beta \geq 0$.

[12] The assumption that harvesting costs are a decreasing function of the biomass is almost universal in the literature, although exceptions to this rule can be found in the literature. See, for example, [36]. The assumption that costs are, or can be, linear in harvesting is employed by Schaefer and by those who have used this model. This assumption implies that $\partial h / \partial E$ is independent of E, an assumption that seems very restrictive, but one that is widely used by fisheries biologists. The reader is referred to [21, pp. 138–140], who which gives a strong defense of the use of the assumption.

[13] Although this assumption appears to be highly restrictive, it is reasonable when applied to fisheries where the harvested fish are sold in large markets supplied by many other fisheries.

[14] The theory as expounded by Gordon [21], Christy and Scott [11], and others.

FISHERIES AND CAPITAL THEORY   95

the variable $x = x(t) \geq 0$ is the state variable and $h = h(t)$ is the control variable.[15] The initial population $x(0)$ is assumed to be known. The control $h(t)$ is assumed subject to the constraints

$$0 \leq h(t) \leq h_{max}, \tag{2.6}$$

where h_{max} may in general be a given function, $h_{max} = h_{max}[t; x(t)]$. The constraint h_{max} may be viewed as being determined by the fishing industry's capacity to harvest at any point in time. The mathematical implications of this constraint are described further in the Appendix.

The object is to maximize the present value of rent derived from fishing. Given the assumptions of constant price and costs linear in harvesting, the objective functional can be expressed as

$$PV = \int_0^\infty e^{-\delta t}\{p - c[x(t)]\}h(t)dt, \tag{2.7}$$

where δ is the instantaneous social rate of discount, p the price, and $c(x)$ the unit cost of harvesting.

Given that the objective functional is linear in the control variable, $h(t)$, we face a linear optimal control problem. The problem is to determine the optimal control $h(t) = h^*(t)$, $t \geq 0$, and the corresponding optimal population $x(t) = x^*(t)$, $t \geq 0$, subject to the state equation (2.3) and the control constraints (2.6), such that the objective functional (2.7) assumes a maximum value. The problem is straightforward and easily solved via the maximum principle.

The Hamiltonian of our problem is

$$H = e^{-\delta t}[\{p - c(x)\}h(t) + \psi(t)\{F(x) - h(t)\}]$$
$$= \sigma(t)h(t) + e^{-\delta t}\psi(t)F(x), \tag{2.8}$$

where $\sigma(t)$, the switching function, is given by

$$\sigma(t) = e^{-\delta t}[p - c(x) - \psi(t)] \tag{2.9}$$

and where $\psi(t)$ is the adjoint or costate variable.

The standard procedure for solving linear optimal control problems proceeds as follows (see Appendix). First, one determines the so-called singular solution, which arises when

$$\sigma(t) \equiv 0. \tag{2.10}$$

We shall see that in the model so far developed this gives rise to an equilibrium solution[16] $x^* = $ constant. The question of the optimal approach to this equilibrium solution will be discussed at a later point.

By a routine calculation (see Appendix), Eq. (2.10) leads to the equation for the singular solution x^*:

$$(1/\delta)[(d/dx^*)\{(p - c(x^*))F(x^*)\}] = p - c(x^*). \tag{2.11}$$

---

[15] While we have chosen to use $h(t)$ as the control variable, we could have used effort $E(t)$. The results would have been identical, but the notation more cumbersome.

[16] In general, the equilibrium solution x^* may not be uniquely determined. For the commonly used logistic model, where $f(x) = rx(1 - x/K)$ and $c(x) = c/x$, however, a unique solution $x^* > 0$ exists provided only that $p > c/K$, as is easily seen from Eq. (2.11). In the discussion following, we shall assume uniqueness of x^*.

This equation does not involve time t explicitly. Hence, as asserted above, the solution x^* is constant, a steady-state solution.

Equation (2.11) can be interpreted without difficulty. The l.h.s. is the present value of the marginal sustainable rent, $(d/dx^*)\{[p - c(x^*)]F(x^*)\}$, afforded by the marginal increment to the stock. The r.h.s. is the marginal rent enjoyed from *current* harvesting. On the one hand, the l.h.s. of (2.11) can be interpreted as an expression of marginal user cost [33], in that it shows the present cost of capturing the marginal increment of fish, a cost that has to be weighed against the marginal gain from current capture. On the other hand, the l.h.s. and r.h.s. of the equation can be viewed as the imputed demand price and the supply price, respectively, of the capital "asset" at time t.[17]

A more transparent form of (2.11) is obtained by carrying out the differentiation on the l.h.s. and then multiplying through by $\delta/(p - c(x^*))$:

$$F'(x)^* - [c'(x^*)F(x^*)/(p - c(x^*))] = \delta. \tag{2.12}$$

This equation is recognizable from capital theory as a modified golden-rule equilibrium equation, being modified both by the discount rate and by what we shall refer to as the marginal stock effect. The l.h.s. of (2.12) is the "own rate of interest," i.e., the instantaneous marginal sustainable rent divided by the supply price of the asset. Thus, (2.12) states simply that the optimal stock is the one at which the own rate of interest of the stock is equal to the social rate of discount. The own rate of interest consists of two components: $F'(x^*)$, the instantaneous marginal physical product of the capital, and $- c'(x^*)F(x^*)/(p - c(x^*))$, the marginal stock effect.

The marginal stock effect is analogous to the "wealth effect" to be found in modern capital theory. As defined by Kurz [23], a wealth effect means that the objective functional is sensitive, not only to the consumption flow, but to the capital stock as well [23, p. 352]. In the fisheries model, the objective functional is sensitive to the stock level, because the size of the stock influences harvesting costs. The term "wealth effect" is inappropriate in this context; hence, we replace it with "stock effect."

The numerator of the marginal stock effect term in (2.12) is simply the partial derivative of total harvesting costs with respect to x^*. The denominator indicates that the marginal harvesting cost gain/loss has to be adjusted by the supply price of capital. Caeteris paribus, the higher the supply price, the smaller (in absolute terms) the marginal stock effect.

Two points implied by the modified golden-rule equation require emphasis. The first is that in this linear model, harvesting costs influence the stock-level optimization *only* through the stock effect. If harvesting costs are insensitive to the size of the biomass,[18] they become irrelevant to the optimization process, so long as $c(K) < p$. The second point is that the two correctives in (2.12) are pulling in opposite directions. One cannot determine a priori whether or not the rational social manager would

---

[17] It is usual in the literature to refer to the adjoint variable as the "imputed price," or more properly, as the "imputed demand price of capital" [35]. The l.h.s. of (2.11) is identical to the adjoint variable along the singular path. We know that achieving the optimal capital stock involves maximizing the Hamiltonian with respect to the control variable; i.e., $\partial H/\partial h = \sigma = 0$. This implies that $\psi(t) = p - c(x)$. We know that along the singular path $p - c(x) = (1/\delta)\{[p - c(x)]F'(x) - c'(x)F(x)\}$. Thus the l.h.s. of (2.11) can be seen as the adjoint variable along the singular path.

In using the term "supply price" here, we are using an essentially Marshallian/Keynesian definition; namely, the amount that must be paid to obtain the additional increment to the stock. In the context of this model the amount that must be "paid" is the current rent forgone at the margin [22, p. 135].

[18] It seems unlikely that harvesting costs will not be affected by the biomass size. However, Smith [36, p. 413] suggests that this may in fact be the case for certain species.

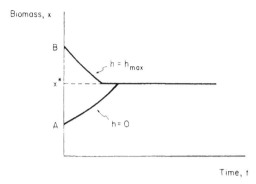

FIGURE 1

engage in biological overfishing. This will depend in the first instance upon the relative sizes of the two correctives. If one lets

$$R \equiv -c'(x_{MSY})F(x_{MSY})/(p - c(x_{MSY})),$$

it can be stated that

$$x^* \begin{cases} < x_{MSY} & \text{if } \delta > R \\ = x_{MSY} & \text{if } \delta = R \\ > x_{MSY} & \text{if } \delta < R. \end{cases} \qquad (2.13)$$

One must also recognize that a necessary condition for biological overfishing, or "depletion," is that $p > c(x_{MSY})$. Clearly, if $p < c(x_{MSY})$, no rational social manager would engage in biological overfishing even if $\delta = \infty$.

The received theory cast in static terms can easily be compared with the dynamic model developed in this section if it is realized that the static model is the equivalent of a dynamic model with $\delta = 0$.[19] If we return to Eq. (2.11), we can see that if $\delta = 0$, (2.11) reduces to

$$(d/dx^*)\{(p - c(x^*))F(x^*)\} = 0. \qquad (2.14)$$

This equation states that the optimal biomass level is the level that maximizes the sustainable level of rent, one of the basic conclusions of the received theory. The dynamic model simply modifies the received theory by allowing for a discount rate greater than zero.

We turn now to the nature of the optimal approach to the steady-state solution, x^*. As clarified in the Appendix, it turns out that (as a result of the linearity of the present model and provided that x^* is unique[20]) the optimal approach is a so-called bang-bang approach:

$$h^*(t) = h_{max} \quad \text{whenever} \quad x(t) > x^*$$
$$= 0 \qquad \text{whenever} \quad x(t) < x^*. \qquad (2.15)$$

This in turn implies that maximum disinvestment is optimal whenever $x(t) > x^*$, and that maximum investment is optimal whenever $x(t) < x^*$. The resulting optimal population level $x(t)$ is indicated in Fig. 1.

[19] Clark [12].
[20] See Footnote 16.

The rationale of this strategy is straightforward. If unit harvest costs do not vary with the harvest rate, and if one commences at $t = 0$ at point A, the object would be to disinvest at maximum rate until reaching x^*, when disinvestment would abruptly cease. On the other hand, if one commences at B, one would want to invest at maximum rate until reaching x^*.

Once x^* is reached, fishing should proceed on a sustained-yield basis (at least until a parameter shift occurs). Fishing on other than a sustained-yield basis would imply further investment or disinvestment, and thus an adjustment of the stock to a non-optimal level.

Although the rationale for such a strategy is fairly obvious, given the linear dependence of both revenue and costs on the harvest rate h, the mathematical justification may not be so transparent. Indeed, as soon as the linearity hypothesis is dropped (see Section 3), the bang-bang adjustment phase is no longer optimal. These matters are discussed in detail in the Appendix.

3. NONAUTONOMOUS MODELS

The model in the previous section rested upon highly restrictive assumptions of linearity and autonomy. However, the model can quite easily be extended to make it either nonautonomous or nonlinear. In this section we make the model nonautonomous while retaining the linearity assumptions. Nonlinearity assumptions are introduced in the following section. The model is made nonautonomous by introducing continuous parameter shifts through time. It is unreasonable, after all, to assume that prices will remain constant through time or that cost functions will not shift. Demand shifts through time are bound to occur; technology changes influencing costs are equally likely to occur.

We shall confine the analysis to price and harvesting cost shifts. It should be pointed out, however, that the analysis could easily be applied to the continuous shifts in other parameters such as the discount rate.

While revenue and harvesting costs continue to be assumed linear in harvesting, it will be assumed that both price and the harvesting-cost function will be subject to known continuous shifts over the time range $t = 0$ to $t = \infty$; i.e., the future time paths of price and costs are fully known. Price can now be expressed as $p(t)$. Unit costs of harvesting $c(x, t)$ will now be expressed as

$$c(x, t) = \phi(t)c(x(t)), \tag{3.1}$$

where $\phi(t) \geq 0$ is a variable coefficient that permits us to account for shifts in the cost function.

The objective functional can now be expressed as

$$PV = \int_0^\infty e^{-\delta t}[p(t) - \phi(t)c(x(t))]h(t)dt. \tag{3.2}$$

The present value of rent is to be maximized subject to the usual conditions. A routine calculation as before (see Appendix) leads easily to the equation for the singular solution, $x(t) = x^*(t)$:

$$F'(x^*) - \frac{\phi(t)c'(x^*)F(x^*)}{p(t) - \phi(t)c(x^*)} = \delta - \frac{\dot{p}(t) - \dot{\phi}(t)c(x^*)}{p(t) - \phi(t)c(x^*)}. \tag{3.3}$$

FISHERIES AND CAPITAL THEORY   99

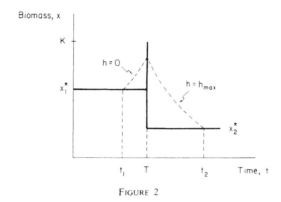

FIGURE 2

If we can assume that there exists a unique solution x^* at each point t, then $x^*(t)$ can be seen as tracing out an optimal time path for the stock or biomass level.

The effect of incorporating continuous shifts in price and the harvesting-cost function is to introduce an additional corrective to the modified golden-rule equation (2.12), a corrective that can be interpreted as the instantaneous percentage change in the supply price of capital. The effect of the new corrective upon the process of stock-level optimization can be seen as follows. Let it be supposed that the supply price of capital is expected to increase, the expected increase being due either to an expected increase in the price of fish or to an expected exogeneous decrease in harvesting costs, or to both. From Eq. (2.12) it can be seen that upward shifts in the supply price of capital lead to lower optimal stock levels. The effect of *anticipation* of the immediate increase in supply price, however, will be to increase the optimal stock level at time t. The rationale is obvious enough. Anticipation of greater benefits from fishing tomorrow will cause a reduction in harvesting today.

The new modified golden rule (3.3) is "myopic" [2, 3] in the sense that the decision rule is independent of both the past and the future, except to the extent that one must anticipate the immediate change in the capital supply price. Thus, the information demands imposed by the rule are extremely limited. In spite of the fact that prices and costs may be fluctuating steadily over time, the only information required for determining the optimal stock, $x^*(t)$, is the marginal product of x at time t, the price of fish and harvesting costs at time t plus the instantaneous rates of change of $p(t)$ and $c(x, t)$.

The myopic rule holds so long as the constraints upon the control variable do not become binding, which means that the adjustments in x^* demanded by supply price changes are not so great as to drive $h(t)$ to $h = 0$ to $h = h_{max}$. If in fact the constraints do become binding, then the stock level $x(t)$ is temporarily forced off the singular path $x^*(t)$. We are then faced with what Arrow [3] refers to as a "blocked interval." Consequently, the myopic rule must be modified and the optimization problem becomes more difficult, but not insoluble.

Consider for example the effect of a large discrete increase in the supply price of the capital occurring at time T. The supply price is assumed to be constant before T and after T (Fig. 2.) The singular path follows the solid line.[21] Ideally, x should be in-

---

[21] Note that the singular path has a "spike" at $t = T$; this corresponds to the occurrence of the term $\dot{p}(T) = +\infty$ in Eq. (3.3), resulting from the discrete jump in $p(t)$ at that point.

creased to the biological maximum, $x = K$, the instant before T [as $(d/dt)(p - \phi c(x))$ goes to infinity] and then should immediately be reduced to x_2^* at time T. This is clearly impossible, as h cannot be reduced below $h = 0$ nor increased above $h = h_{max}$. In other words, the constraints upon the control variable become binding. If $x(T)$ is to be larger than x_1^*, then at some point of time $t_1 < T$, harvesting must be reduced in order to permit the stock level to increase. The time period $t_1 \le t \le T$, then, is a blocked interval, in the sense that we are forced temporarily off the singular path. At time T, the stock cannot be instantaneously reduced to x_2^*. The best that can be done is to harvest at the maximum rate until x_2^* is reached at some time $t_2 > T$. The time period $T \le t \le t_2$ then constitutes a second blocked interval.

The problem, then, is to select the optimal value of t_1, i.e., the point in time at which harvesting is reduced to permit x to rise above x_1^*. Once t_1 is selected, $x(T)$ is automatically determined, as is t_2. It is a consequence of the maximum principle (see Appendix) that the optimal harvest rates during the blocked intervals are $h(t) = 0$ and $h(t) = h_{max}$, respectively.

Conceptually, the nature of this optimization problem is straightforward. The later t_1 is, the shorter the period of sacrifice during which society forgoes harvesting will be; e.g., if $t_1 = T$, there will be no period of sacrifice. On the other hand, the later t_1 is, the smaller $x(T)$ will be, and the smaller the benefit enjoyed from harvesting at the higher supply price.

Let the present value to be enjoyed from harvesting the fish now be expressed as

$$PV = \int_0^{t_1} e^{-\delta t}(p_1 - \phi_1 c(x_1^*))F(x_1^*)dt$$

$$+ \int_T^{t_2} e^{-\delta t}(p_2 - \phi_2 c(x, t))h_{max}dt + \int_{t_2}^{\infty} e^{-\delta t}(p_2 - \phi_2 c(x_2^*))F(x_2^*)dt. \quad (3.4)$$

Differentiating PV with respect to t_1 and setting $\partial PV/\partial t_1 = 0$, we have

$$e^{-\delta t_1}[p_1 - \phi_1 c(x_1^*)]F(x_1^*)$$

$$= \{e^{-\delta t_2}[p_2 - \phi_2 c(x_2^*)](F(x_2^*) - h_{max})\}\frac{dt_2}{dt_1} + \int_T^{t_2} e^{-\delta t}\frac{\partial \phi_2 c[x(t, t_1)]h_{max}}{\partial t_1}dt. \quad (3.5)$$

This apparently complex equation can be interpreted simply as stating that the optimal t_1 will be that at which the marginal benefit derived from extending t_1 (l.h.s.) is equal to the marginal cost from so doing (r.h.s.).

Finally, it should be clear that when one is faced with blocked intervals the information demands could become extensive. The supply price change occurring at T will have to be anticipated in advance of T, possibly far in advance.

4. NONLINEAR MODELS

We turn now to a consideration of nonlinear models, while at the same time restoring the assumption of autonomy. Nonlinearity is introduced, first by relaxing the assumption that the demand function for fish has a price elasticity equal to infinity, and second by permitting effort costs to be nonlinear in effort, which in turn implies that harvesting costs are nonlinear in harvesting. It shall be assumed that $\partial^2 C_E/\partial E^2 > 0$, thus implying that $\partial^2 C_h/\partial h^2 > 0$, where C_E and C_h denote total effort costs and total harvesting costs, respectively.

FISHERIES AND CAPITAL THEORY   101

Copes [16] demonstrates in a lucid fashion that once one relaxes the assumption that the demand for fish is perfectly elastic and relaxes the assumption that effort costs are linear in effort, one can no longer express the object of social utility maximization solely in terms of maximization of resource rent, i.e., total revenue derived from the sale of the fish minus harvesting costs. In effect, one now has to take into account consumers' surplus and producers' surplus as well.[22]

Given our earlier assumption that the price of the fish adequately represents the marginal social benefit (gross) derived from consuming harvested fish, we represent the gross social benefit derived from a given rate of harvest, h, as $U(h)$, where

$$U(h) = \int_0^h p(h)dh.$$

We assume that $U'(h) > 0$ and $U''(h) < 0$. The object is to maximize the present value of the *net* social benefit derived from harvesting fish through time. Assuming no divergence between private and social costs of effort, the objective functional can be expressed as

$$PV = \int_0^\infty e^{-\delta t}[U(h) - c(x, h)h]dt, \tag{4.1}$$

where $c(x, h)$ denotes unit harvesting costs as a function of x and h. As the objective functional (4.1) is a nonlinear function of the control variable, h, the model itself is now nonlinear.

The Hamiltonian of the above nonlinear optimal control problem is

$$H = e^{-\delta t}\{U(h) - c(x, h)h + \psi(t)(F(x) - h)\}. \tag{4.2}$$

From the maximum principle in the nonlinear case (see Appendix),[23] we obtain the equation for equilibrium solutions (i.e., with $h = F(x^*)$):

$$F'(x^*) - \frac{[\partial c(x^*, F(x^*))/\partial x^*] \cdot F(x^*)}{p(F(x^*)) - [c(x^*, F(x^*)) + [\partial c(x^*, F(x^*))/\partial h] \cdot F(x^*)]} = \delta. \tag{4.3}$$

The marginal stock-effect term now looks somewhat formidable, but can be interpreted in the same way as before. The numerator is the partial derivative of total harvesting costs with respect to the biomass, while the denominator is a more complex version of the supply price of capital. The expression $[c(x^*, F(x^*)) + [\partial c(x^*, F(x^*))/\partial h] \cdot F(x^*)]$ is the partial derivative of total harvesting costs with respect to the harvest rate.

An interesting feature of Eq. (4.3) is that it may give rise to multiple equilibria. It has long been recognized that multiple equilibria could emerge in the case of competitive, unregulated fisheries when the demand function had a price elasticity less than infinity.[24] It appeared, however, that one could confidently assume there would be a unique optimal solution for the socially managed fishery.[25] Equation (4.3) indicates that, in the context of a dynamic model, the confidence is unwarranted.

---

[22] We assume that the harvesting and consumption of the fish are internal to the economy in question; e.g., there are no exports of fish to foreign consumers.

[23] Our assumptions imply that the integrand in Eq. (4.1) is a concave function of the control variable h, so that the maximum principle is relevant.

[24] For example, [1, 11, 16].

[25] Anderson [1].

102   CLARK AND MUNRO

If there are three equilibria,[26] i.e., an unstable equilibrium bounded by two stable equilibria, no serious problem exists, so long as the initial position is given [14]. Suppose that the equilibrium stocks are x_1^*, x_2^*, and x_3^*, where $x_1^* < x_2^* < x_3^*$. The stock level x_2^*, the unstable equilibrium, constitutes a "watershed"[27] in the sense that if $x(0) < x_2^*$, the optimal equilibrium stock will be x_1^*, whereas if $x(0) > x_2^*$ the optimal equilibrium stock will be x_3^*. It is conceivable, however, that one might encounter more than three equilibria, in which case selecting an optimum optimorum could prove to be extremely difficult, if not impossible.

Next we observe that in the nonlinear model, the optimal approach to equilibrium (even where a unique equilibrium exists) will differ from that encountered in the linear model. The "bang-bang" approach of the linear case will be replaced by an asymptotic approach. The decision rule to be applied along the approach path can be expressed as

$$F'(x) - \frac{[\partial c(x, h)/\partial x] \cdot h}{p(h) - [c(x, h) + \partial c(x, h)/\partial h) \cdot h]} + \frac{\dot{\psi}}{\psi} = \delta, \qquad (4.4)$$

where ψ, it will be recalled, is the demand price of the resource. As one approaches the equilibrium stock level, ψ will be subject to continuous change. Thus, capital gains (losses) will be continuously generated, which must be accounted for in the decision rule. When the equilibrium stock, x^*, is reached, $\dot{\psi}$ will equal zero and Eq. (4.4) reduces to Eq. (4.3).

After having discussed nonautonomous and nonlinear models individually, it would seem advisable to discuss models that are both nonautonomous and nonlinear. However, we shall not do so, because such models present complexities that would carry us far beyond the scope of this paper. Further comments on nonautonomous, nonlinear models and the difficulties they pose will be found in the Appendix.

5. CONCLUSION

As has been recognized from its inception, the economics of fishing, like other branches of natural resource economics, should ideally be cast in capital–theoretic terms. The fact that what we have termed the received theory was cast in nondynamic terms was as much as anything a reflection on the inadequacies of capital theory. With the advent of optimal control theory, capital theory became transformed into a powerful and flexible tool of analysis. This in turn has led to various attempts to reformulate the economic theory of fishing in dynamic terms. The purpose of this paper has been to attempt to explore the relationships between the economics of fishing and modern capital theory in a systematic and rigorous manner, but to do so in such a way that the reader does not lose sight of the economics by becoming enmeshed in unnecessarily complex mathematical formulations.

The study commences with a simple linear autonomous model of optimal fishery management. Here the results are particularly straightforward. An optimal stationary equilibrium exists, determined by a generalized "modified golden rule." The optimal management policy that emerges is that of following the "bang-bang" feedback

---

[26] Cases can arise in which Eq. (4.3) possesses an even number of solutions, but except under pathological circumstances $x = 0$ will then also become a stable equilibrium, so that the number of equilibria remains odd. For example, if (4.3) has no solutions, then $x = 0$ becomes a stable equilibrium, and optimal harvesting may lead to the extinction of the fishery; see [4, 12, 13].

[27] Leviatan and Samuelson [4].

control law: Adjust the stock level toward the stationary equilibrium as rapidly as possible.

The model is then extended in two directions by relaxing in turn the assumptions of autonomy and linearity. While the basic results obtained can be readily interpreted, the simplicity of the linear autonomous theory is soon lost with the advent of such complications as blocked intervals and multiple equilibria. However, the presence of such difficulties should evoke no surprise. The complexities arising from nonautonomous and nonlinear models are, after all, major sources of uncertainty and controversy in present-day capital theory.

The models developed in this article have been confined entirely to fishing. It should be clear, however, that the analysis could, mutatis mutandis, be extended to other areas of renewable resource management.[28]

APPENDIX

As linear optimal control problems arise only infrequently in economics, it may be of some benefit to the reader to have the techniques of the linear theory outlined and contrasted with the more familiar techniques of nonlinear optimal control theory. Further details can be found in the work of Bryson and Ho [9]. The discussion is based upon the Pontrjagin maximum principle [29].

In the general case (linear or nonlinear), we begin with a state equation

$$dx/dt = F(x; t; u), \quad 0 \le t \le T, \tag{A1}$$

and an objective functional

$$J = \int_0^T G(x, t; u)dt, \tag{A2}$$

which is to be maximized by appropriate choice of the control $u(t)$, subject to (A1). The maximum principle is formulated in terms of the Hamiltonian expression[29]

$$H(x, t; u, \lambda) = G(x, t; u) + \lambda(u) + \lambda(t)F(x, t; u), \tag{A3}$$

where $\lambda(t)$, the adjoint variable, is to be determined.

In the nonlinear (classical) case, the maximum principle asserts the following two equations (plus appropriate transversality conditions) as necessary conditions for optimality,

$$\partial H/\partial u = 0, \tag{A4}$$

$$\partial H/\partial x = -d\lambda/dt. \tag{A5}$$

Since H is nonlinear in u, Eq. (A4) can in principle be solved (by virtue of the implicit function theorem) for u in terms of x and λ. Substituting this solution into (A1) and (A5) then yields a coupled system of two differential equations, determining the optimal trajectories $(x(t), \lambda(t))$. If the original problem (A1), (A2) is autonomous, the same will be true of the differential equation (A5), so that the well-developed theory of plane autonomous systems can be utilized. The typical problem in capital theory possesses a unique solution (x^*, λ^*), which turns out to be a saddle point. Hence,

---

[28] See [15].

[29] This formulation assumes "normality" of the given optimal control problem; cf. [9]. Also, the adjoint variable is now expressed as $\lambda(t)$, rather than as $e^{-\delta t}\lambda(t)$ as previously.

optimal trajectories can be seen to possess the "catenary turnpike" property (in finite time–horizon problems) (Fig. 3).

Next consider the linear case, in which

$$F(x, t, u) = F_1(x, t)u + F_2(x, t),$$
$$G(x, t, u) = G_1(x, t)u + G_2(x, t).$$

Thus, the Hamiltonian is also linear:

$$H(x, t; u, \lambda) = (G_1 + \lambda F_1)u + (G_2 + \lambda F_2). \tag{A6}$$

Let $\sigma(t)$ denote the coefficient of u in this expression:

$$\sigma(t) = G_1(x(t), t) + \lambda(t)F_1(x(t), t). \tag{A7}$$

Since $\partial H/\partial u$ does not contain u, the approach used in the nonlinear case is unsuccessful. Rather, one requires the generalized (Pontrjagin) version of (A4), namely,

$$u(t) \text{ maximizes } H(x(t), t; u; \lambda(t)) \text{ for all } t, \tag{A8}$$

where the maximization is taken over u belonging to a predetermined control set, e.g.,

$$a \leq u \leq b. \tag{A9}$$

Here a and b may depend explicitly on t and $x(t)$.

Clearly, at any time t we must have $u(t) = a$ or b ("bang-bang control"), or else $\sigma(t) = 0$. Of particular interest is the case, called "singular control," in which

$$\sigma(t) \equiv 0 \tag{A10}$$

over an open time interval (t_1, t_2). A standard algorithm for solving linear problems now proceeds as follows.

First we determine the singular solution, using (A10) and its derived form

$$\frac{d\sigma}{dt} = \left\{ \frac{\partial G_1}{\partial x} + \frac{\partial F_1}{\partial x} \right\} \frac{dx}{dt} + \frac{\partial G_1}{\partial t} + \lambda \frac{\partial F_1}{\partial t} + F_1 \frac{d\lambda}{dt} \equiv 0. \tag{A11}$$

Substituting from Eqs. (A1) and (A5) we then obtain a single equation for the singular path $x(t) = x^*(t)$, since $u(t)$ drops out at this stage. We assume that $x^*(t)$ is uniquely determined (the contrary case is interesting but difficult to handle).

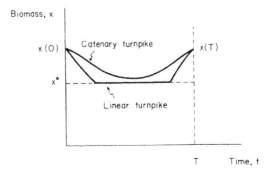

FIGURE 3

Next, if $x^*(0) \neq x_0$, the given initial value, then we must utilize a bang-bang adjustment control to drive the state variable $x(t)$ to the singular path.[30] A similar terminal adjustment phase may also be required. The resulting optimal path is illustrated in Fig. 3. Finally, it may happen that the control constraints (A9) prevent $x(t)$ from following the singular path $x^*(r)$; giving rise to a so-called "blocked interval," as encountered in Section 3 above.

In this paper we have made use of the fact that nonautonomous linear control problems can often be solved quite easily by using the above algorithm. On the other hand, nonautonomous nonlinear control problems are generally much more difficult, for the good reason that nonautonomous nonlinear differential equations are difficult. Specialized methods, usually based on numerical computation, are required for each particular type of problem. Many additional complexities can arise.

[30] For the sake of uniformity in handling both linear and nonlinear control problems, we have chosen to discuss both on the basis of the mathematically profound maximum principle, omitting numerous technical details. It happens, however, that the linear case (in one state dimension) can be treated rigorously on a much more elementary level: a single application of Green's theorem in the plane immediately establishes optimality of the bang-bang singular solution described here.

REFERENCES

1. L. G. Anderson, Optimum economic yield of a fishery given a variable price of output, *J. Fish. Res. Board Canada* **30**, 509–518 (1973).
2. K. J. Arrow, Optimal capital policy, the cost of capital, and myopic decision rules, *Ann. Inst. Statist. Math.* **16**, 21–30 (1964).
3. K. J. Arrow, Optimal capital policy with irreversible investment, *in* "Value Capital and Growth, Papers in Honour of Sir John Hicks" (J. N. Wolfe, Ed.), pp. 1–20, Edinburgh University Press, Edinburgh (1968).
4. J. R. Beddington, C. M. K. Watts, and W. D. C. Wright, Optimal cropping of self-reproducible natural resources, *Econometrica* **43**, 789–802 (1975).
5. F. W. Bell, Technological externalities and common-property resources: An empirical study of the U. S. Northern lobster fishery, *J. Polit. Econ.* **80**, 148–158 (1972).
6. R. Bellman, "Dynamic Programming," Princeton University Press, Princeton, N. J. (1957).
7. P. G. Bradley, Seasonal models of the fishing industry, *in* "Economics of Fisheries Management: A Symposium" (A. D. Scott, Ed.), pp. 33–44, University of British Columbia, Institute for Animal Resource Ecology, Vancouver (1970).
8. G. Brown, An optimal program for managing common property resources with congestion externalities, *J. Polit. Econ.* **82**, 163–174 (1974).
9. A. E. Bryson and Y. C. Ho, "Applied Optimal Control," Blaidsell, Waltham, Mass. (1969).
10. F. T. Christy, "Alternative Arrangements for Marine Fisheries: An overview," Resources for the Future, Washington, D. C. (1973).
11. F. T. Christy and A. D. Scott, "The Common Wealth in Ocean Fisheries," John Hopkins Press, Baltimore, Md. (1965).
12. C. W. Clark, The economics of overexploitation, *Science* **181**, 630–634 (1973).
13. C. W. Clark, Profit maximization and the extinction of animal species, *J. Polit. Econ.* **81**, 950–961 (1973).
14. C. W. Clark, Supply and demand relationships in fisheries economics, *in* "Proceedings of the IRIA Symposium on Control Theory" (Paris) (1974).
15. C. W. Clark and J. de Pree, A general linear model for the optimal exploitation of renewable resources, to appear.
16. P. Copes, The backwards-bending supply curve of the fishing industry, *Scott. J. Polit. Econ.* **17**, 69–77 (1970).
17. P. Copes, Factor rents, sole ownership and the optimum level of fisheries exploitation, *Manchester School Econ. Soc. Studies* **40**, 145–164 (1972).
18. J. A. Crutchfield and A. Zellner, Economic aspects of the Pacific halibut fishery, *Fish. Ind. Res.* **1**, No. 1 (1962). U. S. Department of the Interior, Washington, D. C.

19. R. G. Cummings and O. R. Burt, The economics of production from natural resources. A Note, *Amer. Econ. Rev.* **59**, 985–990 (1969).

20. R. Dorfman, An economic interpretation of optimal control theory, *Amer. Econ. Rev.* **59**, 817–831 (1969).

21. H. Gordon, The economic theory of a common-property resource: The fishery, *J. Polit. Econ.* **62**, 124–142 (1954).

22. J. M. Keynes, "The General Theory of Employment, Interest and Money," Harcourt, Brace, New York (1936).

23. M. Kurz, Optimal economic growth and wealth effects. *Intern. Econ. Rev.* **9**, 348–357 (1968).

24. N. Leviatan and P. A. Samuelson, Notes on turnpikes: Stable and unstable, *J. Econ. Theory* **1**, 454–475 (1969).

25. H. S. Mohring, The costs of inefficient fishery regulation: A partial study of the Pacific Coast halibut industry, unpublished.

26. P. A. Neher, Notes on the Volterra-quadratic fishery, *J. Econ. Theory* **6**, 39–49 (1974).

27. C. G. Plourde, A simple model of replenishable natural resource exploitation, *Amer. Econ. Rev.* **60**, 518–522 (1970).

28. C. G. Plourde, Exploitation of common-property replenishable resources, *West. Econ. J.* **9**, 256–266 (1971).

29. L. S. Pontrjagin, V. S. Boltjanskii, R. V. Gamkrelidze and E. F. Mishchenko, "The Mathematical Theory of Optimal Processes," Wiley, New York (1962).

30. J. P. Quirk and V. L. Smith, Dynamic economic models of fishing, *in* "Economics of Fisheries Management: A Symposium" (A. D. Scott, Ed.), pp. 1–32, University of British Columbia, Institute for Animal Resource Ecology, Vancouver (1970).

31. F. P. Ramsey, Mathematical theory of saving, *Econ. J.* **38**, 543–559 (1928).

32. M. B. Schaefer, Some considerations of population dynamics and economics in relation to the management of marine fisheries, *J. Fish. Res. Board Canada* **14**, 669–681 (1957).

33. A. D. Scott, Notes on user cost, *Econ. J.* **63**, 368–384 (1953).

34. A. D. Scott, The fishery: The objectives of sole ownership, *J. Polit. Econ.* **63**, 116–124 (1955).

35. K. Shell, Applications of Pontrjagin's maximum principle to economics, *in* "Mathematical Systems Theory in Economics I" (H. W. Kuhn and G. P. Szego, Eds), pp. 241–292, Lecture Notes in Operations Research and Mathematical Economics, Vol. 2, Springer–Verlag, Berlin (1962).

36. V. L. Smith, Economics of production from natural resources, *Amer. Econ. Rev.* **58**, 409–431 (1968).

37. V. L. Smith, On models of commercial fishing, *J. Polit. Econ.* **77**, 181–198 (1969).

38. M. Spence, Blue whales and applied control theory, Technical Report No. 108, Stanford University, Institute for Mathematical Studies in the Social Sciences (1973).

39. R. Turvey, Optimization and suboptimization in fishery regulation, *Amer. Econ. Rev.* **54**, 64–76 (1964).

[5]

Econometrica, Vol. 47, No. 1 (January, 1979)

THE OPTIMAL EXPLOITATION OF RENEWABLE RESOURCE STOCKS:
PROBLEMS OF IRREVERSIBLE INVESTMENT[1]

By COLIN W. CLARK, FRANK H. CLARKE, and GORDON R. MUNRO

This paper studies the effects of irreversibility of capital investment upon optimal exploitation policies for renewable resource stocks. It is demonstrated that although the long-term optimal sustained yield is not affected by the assumption of irreversibility (except in extreme cases), the short-term dynamic behavior of an optimal policy may depend significantly upon the assumption. It is suggested that the results may have profound implications for problems of rehabilitation of overexploited fisheries and other renewable resource stocks.

1. INTRODUCTION

THIS PAPER IS CONCERNED with the problem of "non-malleability" of capital and the implications thereof for the optimal exploitation of renewable resource stocks over time. We use the term "non-malleability" to refer to the existence of constraints upon the disinvestment of capital assets utilized in exploiting the resource stock (cf. Arrow [1], Arrow and Kurz [2]).

Previous studies of renewable resource economics have assumed, either explicitly [3] or implicitly [8], that capital stocks were perfectly malleable. It is easy to see that this implies that the variable representing the capital stock can be eliminated from the analysis. This is no longer possible when capital is assumed to be non-malleable, however, with the result that the corresponding optimization problem becomes considerably more complex, since it necessarily involves a minimum of two state variables. It will be shown that in fact the non-malleability assumption has a significant influence on the form of optimal exploitation policies. This has long been suspected on intuitive grounds; for example, numerous discussions of the practical problems of fishery management created by the non-malleability of capital (both physical and human) can be found in the literature (e.g. [5, p. 222; 19]).

In Section 2 we describe the model to be used as the basis for our investigations. This model, associated with the names of Gordon [11] and Schaefer [17], has often been used in the study of commercial fisheries [8, 10]. In spite of its somewhat specialized nature, we are confident that the results will remain qualitatively valid for a wide choice of alternative models of renewable resource exploitation.

In Section 3 we review briefly the case in which capital is perfectly malleable. It will be verified that in this case the capital stock variable can indeed be eliminated from the analysis and that the capital input can be treated as a flow. The model thus reduces to the single-state-variable model studied earlier [8, 10].

[1] This research has been supported in part by the National Research Council of Canada (Grant numbers A-3990 and A-9082), and also by the Donner Canadian Foundation as part of a project on "Canada and the International Management of the Oceans" sponsored by the Institute of International Relations, University of British Columbia, Canada, along with the University of B.C. Programme in Natural Resource Economics, financed by the Canada Council. The suggestions of the referees are gratefully acknowledged.

In Sections 4–6 we introduce three alternative assumptions of non-malleability of capital. In Section 4 we deal with the case of perfectly non-malleable capital in which the depreciation rate is equal to zero and in which the capital has a negligible scrap value. In Section 5 we continue to assume that the scrap value is zero, but allow for a positive depreciation rate. In this case we say that capital is "quasi-malleable." In Section 6 we allow for a positive unit scrap value, which is significantly below the replacement price.

The material presented in Sections 4–6 is of a descriptive nature. A rigorous analytic proof of all our results appears in Section 7. Our proof is based upon a method due to Carathéodory [4], which does not seem to have been employed previously in economic analysis.

Inasmuch as our model involves two stock variables (resource stock, capital stock) and two control variables (harvest rate, investment rate), it is structurally similar to a number of two-sector models of economic growth that have appeared in the literature [12, 14]. Such models are known to give rise to relatively complex optimal trajectories, frequently involving multiple switches of the control variables. Our model is no exception to this rule. On the other hand, our model is much less abstract than these growth models. It is an eight-parameter model, each parameter representing a measurable biological or economic variable. Because of the concrete nature of the model, the bio-economic reasons underlying the form of optimal exploitation policies become quite transparent, particularly since we are able to express the solution explicitly in synthesized (feedback) form: see Figures 1–3.

2. THE GENERAL MODEL

For the sake of explicitness we shall henceforth restrict our attention to a bio-economic model of the commercial fishery under sole ownership. The biological basis of the model is the general production model developed by Schaefer [17] and Pella and Tomlinson [15]; the economic basis stems from the work of Gordon [11], Crutchfield and Zellner [10], and others (see Clark and Munro [8]).

The population dynamics of the fishery resource is modeled by the equation

$$(2.1) \qquad \frac{dx}{dt} = F(x) - qEx, \quad x(0) = x^0,$$

where $x(t)$ is population biomass at time t, $F(x)$ is the natural growth function, q is the catchability coefficient (constant), and $E(t)$ is fishing effort at time t. Regarding the natural, or biological, growth function $F(x)$, we shall assume that

$$(2.2) \qquad F(0) = F(\bar{x}) = 0, \quad F(x) > 0, \quad F''(x) < 0 \quad \text{for} \quad 0 < x < \bar{x}.$$

In equation (2.1) the rate of harvest $h(t)$ is of the form

$$(2.3) \qquad h(t) = qE(t)x(t);$$

this particular form of the Cobb-Douglas production function is the traditional harvest production function employed in fishery models.

The variables $x(t)$ and $E(t)$ of equation (2.1) are subject to the constraints

(2.4) $x(t) \geqslant 0$

and

(2.5) $0 \leqslant E(t) \leqslant E_{\max} = K(t)$

where $E_{\max}$ is maximum effort capacity and $K(t)$ is the amount of capital invested in the fishery at time t. We shall think of $K(t)$ as representing the number of "standardized" fishing vessels available to the fishery. Equation (2.5) then asserts that the maximum effort capacity equals the number of vessels available, and that the actual level of effort employed at any time cannot exceed $E_{\max}$.

Possible adjustments to the level of capital are modeled by the equation

(2.6) $\dfrac{dK}{dt} = I - \gamma K, \quad K(0) = K^0,$

where $I(t)$ is gross investment rate at time t (expressed in physical terms) and γ is the rate of depreciation (constant). The fish biomass $x(t)$ and capital stock $K(t)$ are subject to the constraints

(2.7) $x(t) \geqslant 0; \quad K(t) \geqslant 0.$

The assumption of non-malleability is embodied in the following constraint:

(2.8) $0 \leqslant I(t) \leqslant +\infty.$

The case $I(t) = +\infty$ allows for instantaneous jump increases in the level of capital. Admitting this possibility simplifies the analysis and lets us concentrate on the phenomenon of non-malleability of capital.

As our objective function we employ the discounted net cash flow:

(2.9) $J = \displaystyle\int_0^\infty e^{-\delta t}\{ph(t) - cE(t) - \pi I(t)\}\, dt$

where δ is the instantaneous rate of discount (constant), p is price of landed fish (constant), c is operating cost per unit effort (constant), and π is price (purchase or replacement) of capital (constant). All the parameters of our model, viz q, γ, δ, p, c, and π, are taken as given constants, and all are assumed to be positive, except for the depreciation rate γ, which we merely assume to be $\geqslant 0$. We shall refer to $\gamma = 0$ as the case of perfect non-malleability, and to $\gamma > 0$ as the case of quasi-malleability of capital.

The problem we face is that of determining the optimal effort and investment policies $E(t)$, $I(t)$, leading to the maximization of the objective (2.9). In the next three sections we describe the solution to this problem, and discuss a variety of policy implications of the model. The rigorous justification of this solution is provided in Section 7.

In Section 6 we shall discuss an alternative model, in which disinvestment is unconstrained, but in which unwanted capital can be sold only as scrap. Let

$$\pi_s = \text{unit scrap value of capital (constant)};$$

we shall assume that

(2.10) $0 < \pi_s < \pi.$

For this model we suppose that gross investment is unrestricted:

(2.11) $-\infty \leq I(t) \leq +\infty.$

However, the objective functional is now replaced by

(2.12) $J = \displaystyle\int_0^\infty e^{-\delta t}\{ph(t) - cE(t) - \phi(I(t))\}\, dt$

where

(2.13) $\phi(I) = \begin{cases} \pi I & \text{if } I > 0, \\ \pi_s I & \text{if } I < 0. \end{cases}$

The corresponding control problem is as before.

3. PERFECTLY MALLEABLE CAPITAL

To set the stage for the more difficult problems generated by non-malleability of capital we review briefly the case in which capital is perfectly malleable, in the sense that

(3.1) $-\infty \leq I(t) \leq +\infty$

and that $\pi = \pi_s$. The relevant objective functional is (2.9).

Under this assumption it is clear that for an optimal policy there will never be excess harvesting capacity, i.e., we will always have

(3.2) $K(t) = E(t)$

since any unused capacity can immediately be disposed of at the purchase price π. Consequently $I(t) = \dot{K} + \gamma K = \dot{E} + \gamma E$ and (2.9) becomes, after integration by parts,

$$J = \int_0^\infty e^{-\delta t}\{ph - cE - \pi(\delta + \gamma)K\}\, dt + \pi K^0$$

(3.3)

$$= \int_0^\infty e^{-\delta t}\{pqx - c_{\text{total}}\}E\, dt + \text{constant}$$

where

(3.4) $c_{\text{total}} = c + (\delta + \gamma)\pi$

represents the total cost per unit of fishing effort. In this expression the term c

denotes unit operating cost, while $(\delta + \gamma)\pi$ can be viewed as the unit "rental" cost of capital.

Thus when capital is assumed to be perfectly malleable, the stock variable K can be eliminated entirely from the model (as pointed out earlier by Beddington, Watts, and Wright [3]; see also Hadley and Kemp [12, Chapter 6]). The problem reduces to the maximization of expression (3.3) subject to the biomass equation (2.1) and to constraints (2.4) and (2.5) on $x(t)$ and $E(t)$. But in fact this is precisely the problem of the received dynamic model of the commercial fishery [8, 10]. We see therefore that the received theory assumes, whether explicitly or implicitly, that capital is perfectly malleable. The consequences of relaxing this assumption will become clear in the sequel.

The maximization problem for equation (3.3) subject to (2.1) has a particularly simple solution [6, 8]. Namely, there exists an optimal biomass level $x = x^*$, determined by the "modified golden rule":

$$(3.5) \qquad F'(x^*) - \frac{c'_{\text{total}}(x^*)F(x^*)}{p - c_{\text{total}}(x^*)} = \delta$$

where $c_{\text{total}}(x)$ denotes unit harvesting costs:

$$(3.6) \qquad c_{\text{total}}(x) = \frac{c_{\text{total}}}{qx}.$$

The left-hand side of equation (3.5) is simply the own rate of interest of the resource biomass x^*; the second term is referred to as the marginal stock effect [8]. As an alternative, equation (3.5) can be written in the form

$$(3.7) \qquad \frac{1}{\delta}\left[\frac{d}{dx^*}\{(p - c_{\text{total}}(x^*))F(x^*)\}\right] = p - c_{\text{total}}(x^*)$$

where the left-hand side can be interpreted as the shadow price or imputed demand price of the biomass and the right-hand side as the supply price of the biomass (see [8, pp. 95–96]).

The optimal approach to x^* from a non-optimal initial biomass level $x^0 \neq x^*$ is the "most rapid" [18] or "bang-bang" [8] approach:

$$(3.8) \qquad E(t) = \begin{cases} E_{\max} & \text{whenever} \quad x(t) > x^*, \\ E_{\min} & \text{whenever} \quad x(t) < x^*, \end{cases}$$

where in our present model $E_{\min} = 0$ and $E_{\max}$ is an ad hoc upper bound.

It is interesting to note in passing the rather extreme policy implications of (3.8). Suppose, for example, that the fishery being subject to an optimal management policy has hitherto been an open-access fishery with the consequence that $x(0) \ll x^*$. According to (3.8) the appropriate policy would be the drastic one of shutting the fishery down entirely until the biomass has recovered to the level x^*. (If for political or other reasons not considered here, it is practically impossible to shut the fishery down entirely, then condition (3.8) prescribes the reduction of E to $E_{\min} > 0$ until the biomass grows to x^*.)

Finally we consider for the sake of future reference the (limiting) case in which capital is "free" in the sense that $\pi = \pi_s = 0$. In this case the optimal biomass level becomes $x = \tilde{x}$, where $\tilde{x}$ is determined from the equation

(3.9) $F'(\tilde{x}) - \dfrac{c'(\tilde{x})F(\tilde{x})}{p - c(\tilde{x})} = \delta$

or

(3.10) $\dfrac{1}{\delta} \cdot \dfrac{d}{d\tilde{x}}\{(p - c(\tilde{x}))F(\tilde{x})\} = p - c(\tilde{x})$

where

(3.11) $c(x) = \dfrac{c}{qx}.$

Thus operating costs alone are relevant to the determination of $\tilde{x}$. Assuming that x^* and $\tilde{x}$ are uniquely determined by these equations we have

(3.12) $\tilde{x} < x^*.$

The role of the two biomass levels $\tilde{x}$ and x^* will become clear in the following discussions.

4. NON-MALLEABLE CAPITAL

We turn now to the main problem, that of delineating the optimal harvest and investment policies under the assumption that capital is non-malleable; see equation (2.8). As will become apparent, the form of the solution is complicated significantly by this assumption. In this and the next two sections we present a description of the optimal solution and a discussion of some of its implications, without attempting to give a proof of optimality. The proof is discussed in detail, however, in Section 7.

We commence with the easiest of the non-malleable capital cases, that in which capital is perfectly non-malleable. In this section, therefore, we shall assume that

(4.1) $\gamma = 0.$

A. *The Unexploited Fishery*

First we consider the case of an initially unexploited fishery resource:

(4.2) $x^0 = \tilde{x}$ and $K^0 = 0.$

We also assume that $\tilde{x} > x^*$, where x^* is defined by equation (3.5), for otherwise no development of the fishery proves to be worthwhile.

Let us refer to Figure 1, which constitutes a feedback control diagram (in the $x - K$ state space) for the optimal harvest and investment policy, in the case of perfectly non-malleable capital ($\gamma = 0$; $I \geqslant 0$). The optimal values of E and I are given in this figure as functions of the current state variables $x = x(t)$ and

$K = K(t)$. The arrows in Figure 1 show the time motions (trajectories) of an optimally controlled biomass-capital system.

Notice that there are three classes of optimal policy, which are utilized, respectively, in the three subregions R_1, R_2 (shaded region), R_3, viz.

$$\text{in } R_1: \quad E = I = 0,$$

(4.3) $$\text{in } R_2: \quad E = E_{max} = K, \quad I = 0,$$

$$\text{in } R_3: \quad I = +\infty.$$

In addition there are certain equilibrium positions, indicated by the heavy curve BCD in the figure. The biomass levels $\tilde{x}$ and x^* (and the corresponding levels of capital $\tilde{K} = F(\tilde{x})/q\tilde{x}$ and $K^* = F(x^*)/qx^*$) play an important role in the solution.

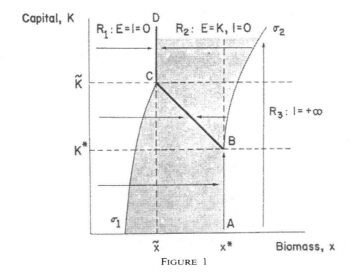

FIGURE 1

The initial point (4.2) lies in region R_3. Consequently the optimal policy requires a once-and-for-all jump increase in the level of capital, up to the level K specified by the "switching curve" σ_2. By assumption, this investment is completed instantaneously at time $t = 0$. Note that $K > K^*$, so that the optimal level of capital exceeds the long-run optimum level that would prevail under conditions of perfect malleability (but not necessarily the short-run optimum).

Once the investment occurs at $t = 0$, capital costs cease to be relevant. The optimal biomass level thus becomes the "free capital" level, $\tilde{x}$. If the optimal capital stock is $K > \tilde{K}$, we set $E(t) = K$, reducing the biomass stock, $x(t)$, as rapidly as possible, until reaching the optimal stock level $\tilde{x}$. Upon reaching $\tilde{x}$, we then set $E = \tilde{K}$ and henceforth harvest $h = F(\tilde{x})$ on a sustained-yield basis. In a sense the fishery is then "overcapitalized," in that the excess capital $K - \tilde{K}$ is redundant.

The overcapitalization, however, is only *ex post* and not *ex ante*, since the initial level of capital K is assumed to have been optimally determined.

Whether in fact $K > \check{K}$, given that $x^0 = \bar{x}$, depends on the parameters of the model. If $K < \check{K}$, then the optimal biomass level $\bar{x}$ is not attainable at the effort rate $E = K$. The optimal policy calls for setting $E = K$ for all $t > 0$, and an equilibrium level lying between $\bar{x}$ and x^* (see Figure 1) will thus be approached. In a sense we are "trapped" in that it neither pays to expand the capital stock sufficiently to make $\bar{x}$ a feasible equilibrium biomass level, nor does it pay to build up the biomass level to x^*.

How is the switching curve σ_2, which determines the optimal level of capitalization (given that $x^0 > x^*$), determined? Consider a point (x, K) in the state space, with $K > K^*$ and such that (x, K) lies to the right of the line BCD. Let $S(x, K)$ denote the "return function," starting at time $t = 0$ at the point (x, K), and using the policy just described, i.e., $I = 0$ for all $t \geq 0$, and $E = K$ as long as $x > \bar{x}$, otherwise $E = \check{K}$. Thus $S(x, K)$ is simply the present-value integral

$$S(x, K) = \int_0^\infty e^{-\delta t}\{pqx - c\}E(t)\, dt,$$

where $E(t)$ is as above, and $x(t)$ is the corresponding biomass, given by equation (2.1). The switching curve σ_2 is then determined by the equation

(4.4) $\dfrac{\partial S(x, K)}{\partial K} = \pi.$

Equation (4.4) is simply the usual Keynesian investment rule.

B. *The Overexploited Fishery*

Consider now the case of an initially overexploited fishery:

(4.5) $x^0 < \bar{x}.$

Clearly a policy of recovery, or rehabilitation, of the fish stock is indicated in these circumstances. The only questions to be answered are the extent of the rehabilitation and the desired speed of recovery. Not surprisingly in light of the previous discussion the answers will depend upon the initial level of capital, K^0.

Let it be supposed first that $K^0 > \check{K}$. The cost of initial investment is a bygone and is thus irrelevant to future decisions. Moreover, the capital stock K^0 is abundant in the sense that there is more than sufficient capital to permit us to harvest $F(\bar{x})$ on a sustained basis. Hence the optimal harvest policy is to set $E = 0$ until the biomass has grown to $\bar{x}$ and then to set $E = \check{K}$, i.e., to harvest $F(\bar{x})$ on a sustained basis. From our discussion in Section 3 we recognize that at $x = \bar{x}$ the demand price of the biomass is equal to its supply price, given that the only relevant effort costs are operating costs.

Next let it be supposed that $K^* < K^0 < \check{K}$. (To consider a practical example, the fishery may hitherto have been subject to uncontrolled international exploitation, but is now encompassed by a coastal state's exclusive economic zone. Internal political pressures compel the coastal state to exclude all foreign vessels. The remaining domestic fleet constitutes K^0.) The initial stock of capital is no longer abundant, because it is insufficient to harvest $F(\tilde{x})$ on a sustained basis. The capital stock can be built up to $\tilde{K}$, but only at a cost of π per unit. Yet, as will be proven in Section 7, at every point to the left of x^* the demand price of capital, which we have characterized as $\partial S(x, K)/\partial K$, will be less than the supply price. Hence investment is non-optimal, so that we will be prevented from harvesting on a sustained basis at $\tilde{x}$. We will thus be confronted with an enforced "conservationist" policy in which the biomass will necessarily rise above $\tilde{x}$ to an equilibrium level lying in the "trap" between $\tilde{x}$ and x^*. Moreover, as a consequence of the coming enforced conservation policy, it will no longer be optimal to refrain from harvesting until the biomass has grown to the level $\tilde{x}$. Rather, the optimal policy calls for switching from zero harvesting to maximum harvesting as $x(t)$ crosses the switching curve σ_1 (see Figure 1). This premature switching phenomenon is a common occurrence in linear optimal control problems.

Finally in the case that $K^0 < K^*$ we see that the biomass level eventually recovers to the level $x = x^*$. At this moment it becomes optimal to (suddenly) increase capital to the level K^*, thus establishing a long-run equilibrium at the point (x^*, K^*).

The switching curve σ_1 can be determined by the same method used above for σ_2. Specifically, for an initial position (x, K) to the left of line ABC, and below $K = \tilde{K}$, let $S(x, K)$ denote the return function corresponding to the policy $I = 0$, $E = K$ as long as x remains below x^*, together with an impulse jump in K to K^* if and when $x(t) = x^*$. The switching curve σ_1 is then given by the solution of the equation

$$(4.6) \qquad \frac{\partial S(x, K)}{\partial x} = p - \frac{c}{qx},$$

where the left-hand side and right-hand side represent the demand price (or shadow price) and the supply price of the biomass, respectively. The reason for premature switching thus becomes transparent. If we commence at a point to the left of σ_1 with $K < \tilde{K}$, it will pay to increase the biomass, but not to the same extent that it would if capital were abundant.

It is instructive to compare the above rehabilitation policy with the corresponding policy in the case of perfect malleability of capital, in which it is optimal to shut the fishery down entirely until the biomass has grown to x^*. If harvesting capital is perfectly non-malleable, the optimal policy is radically altered. It will then never pay to refrain from harvesting once the biomass reaches the lower level $\tilde{x}$. If $K < \tilde{K}$, it will be optimal to commence harvesting at the maximum rate even before $x(t)$ reaches $\tilde{x}$. The ultimate optimal equilibrium biomass level depends upon the initial stock of capital K^0.

If capital depreciates at a positive rate, these conclusions must be altered. Let us now turn to this more interesting case.

5. QUASI-MALLEABLE CAPITAL

In this section we consider the case

(5.1) $\gamma > 0$;

we continue to impose the constraint of non-negative gross investment, $I \geq 0$. Figure 2 is the feedback optimal control diagram for this case. The similarity with Figure 1 is apparent, but there are also significant differences.

As before, the state plane is divided into three control regions R_1–R_3, and the control law (4.3) again applies. The biomass level $\tilde{x}$ is the same as before, but x^* (which is determined by the total cost $c_{\text{total}} = c + \pi(\delta + \gamma)$) has moved to the right. The most important new feature of Figure 2, however, concerns the dynamic behavior of the system. Because of the depreciation of capital, all trajectories

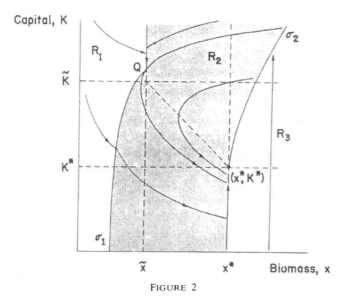

FIGURE 2

(outside of R_3) now move downwards with the passage of time. It follows that, in contrast to the perfectly non-malleable case, the system now tends to a uniquely determined long-run equilibrium, at (x^*, K^*). For this reason, it is appropriate to refer to x^* as the *long-run optimum* biomass level, and to $\tilde{x}$ as a (possible) *short-run optimum* biomass level.

The description and interpretation of the optimal harvest and investment policies from any initial position is now straightforward. Two aspects of the optimal policy bear emphasizing, however. First note that the switching curve σ_1 now meets the line $x = \tilde{x}$ at the point Q above $K = \check{K}$. The implication of this is as follows. Consider the system in short-run equilibrium with $x = \tilde{x}$ and (x, K) lying above Q. Although the existing stock of capital is abundant, the abundance is strictly temporary, because of depreciation. Anticipation of the coming "short-age" of capital—in the sense that the singular solution $x(t) = \tilde{x}$ will eventually become non-feasible—leads to a switch in the harvesting policy at the point Q. The switch is to a policy of harvesting at maximum rate $(E = K)$, *even though this switch will cause $x(t)$ to fall temporarily below the short-run optimum $\tilde{x}$.*

The second and more important aspect concerns the general problem of restoration of an overexploited fishery. Although the optimal restoration policy ultimately leads to a long-run equilibrium at $x = x^*$ and $K = K^*$, and although this equilibrium is the same as for the case of perfectly malleable capital, the restoration policies in the two cases are notably different. In the malleable case, the optimal restoration policy requires a complete moratorium. In the case that capital is quasi-malleable, however, the optimal approach to x^* is far more gradual. Indeed, after a certain point it will be optimal to harvest at the maximum rate with the existing stock of capital. In other words, the disruptive consequences of a fishing moratorium can only be considered optimal in the case that fishing vessels have viable alternative uses, except that a brief moratorium may be optimal if the fish stock is very severely depleted (i.e., to the left of σ_1).

6. A MARKET FOR SCRAP

We turn next to the alternative model specified by equations (2.10)–(2.13). Whereas the original model was linear in both control variables E, I, this alternative model displays a minor but significant nonlinearity, inasmuch as the function $\phi(I)$ consists of two linear segments of unequal slope π_s and π, respectively. As shown elsewhere [7] in a simpler setting, this nonlinearity gives rise to a new type of optimal control ("corner control"), which persists for time intervals during which $I(t) \equiv 0$.

Let x_s denote the solution to the equation

$$(6.1) \qquad F'(x) - \frac{c_s'(x)F(x)}{p - c_s(x)} = \delta$$

where

$$(6.2) \qquad c_s(x) = \frac{c + (\delta + \gamma)\pi_s}{x}.$$

Thus x_s represents the optimal biomass level for a model of perfectly malleable capital with price (purchase price *and* unit scrap value) equal to π_s. Since $0 < \pi_s < \pi$ we have

$$(6.3) \qquad \tilde{x} < x_s < x^*.$$

The feedback control diagram for this model is given in Figure 3. This diagram is almost identical to that of Figure 2 except for the presence of an additional "overlay" disinvestment region R_4, bounded by the line $x = x_s$ and a new switching curve σ_3. (Depending on parameter values, σ_3 may cross σ_1 as shown, or

the two curves may fail to intersect.) The optimal harvest and investment policies
are also identical with those of the previous model, except for (a) initial positions
(x, K) lying within R_4 and (b) initial positions that lead to trajectories that
penetrate R_4, i.e., that hit the line $x = x_s$ above the point Q_s.

In case (a) the optimal polic_/ requires immediate disinvestment, down to the
level given by σ_3. Following this initial disinvestment, the policy given in Section 5
is optimal. In case (b), an impulse disinvestment must be undertaken, to the level
Q_s, at the instant when $x(t)$ reaches x_s.

The optimal disinvestment policy is thus essentially symmetric to the optimal
investment policy of region R_3. This of course reflects the close symmetry of the
present model with respect to investment and disinvestment. (Region R_2, by the
way, is an instance of "corner" control mentioned above.)

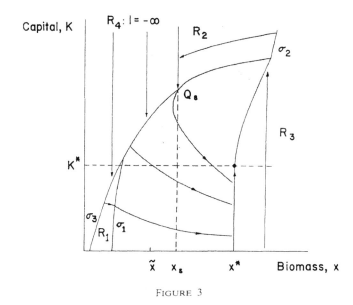

FIGURE 3

The switching curve σ_3 is determined from the equation $\partial S/\partial K = \pi_s$.

The rationale behind this solution is quite straightforward. At any given time,
the decision must be taken whether to disinvest, to invest, or to do neither. Just as
investment is indicated when there are "too few" vessels and "too many" fish, so is
disinvestment indicated only when these conditions are reversed. Like investment, disinvestment turns out to be a once-and-for-all decision (there is an
exception regarding investment, when $x(t)$ returns to x^*). If the decision is taken
not to disinvest, then the optimal policy goes ahead *exactly as if the disinvestment
opportunity did not exist.* Disinvestment may occur later, namely if the trajectory
later hits $x = x_s$ above Q_s.

Let us also note that, because of the disinvestment opportunity, there is now no short-run optimal equilibrium biomass level. (In control-theoretic language, x_s is not a singular solution for this model, although $\tilde{x}$ is such for the previous model.) From an arbitrary initial position the system ultimately converges to the same long-run equilibrium solution (x^*, K^*) as in the case of quasi-malleable capital. The proof of optimality for this model is outlined in the next section.

7. PROOF OF OPTIMALITY

In this section we shall prove that the policy described above is indeed optimal for the problem at hand. The method we shall employ is one of verification rather than deduction. The solution was arrived at by the study of necessary conditions, a knowledge of the solution when K is constant, and a certain amount of trial and error. However, we shall not give a detailed account of this procedure; instead we simply prove optimality. A word about the necessary conditions: it is not possible to apply the Pontryagin maximum principle to the problem, since this principle is not valid when the permissible values of the controls (E, I) depend on the current value of the state variables (x, K), as they do here $(E \le K)$. We used instead the more general formulation of "differential inclusions" and some recent results (see [9] for a complete discussion) to first observe the existence of the two distinguished ("singular") values $\tilde{x}$ and x^* (see Section 3 above).

We remark in passing that the methods developed here should prove useful in other control-theoretic problems for which the maximum principle may prove inappropriate. For example, an exhaustible resource optimization model similar to our model has been discussed by Puu [16]; our method can be used to complete and correct the proposed solution for this model. A recent model of capital accumulation and durable goods production (Kamien and Schwartz [13]) also has a similar structure—and similar properties of optimal policies—to our model; it seems likely that our approach could be applied to models of this kind.

The verification technique we shall employ would be recognized by the expert as an adaptation of a classical approach in the calculus of variations sometimes labeled "the royal road of Carathéodory" (see, for example, [4]). It has the advantage of being simple-minded and elementary. Of course, it depends upon knowing the answer in the first place! The basic idea is the following: let a harvesting-investment policy be specified from any initial value (x, K), and let the resulting net discounted return (see (2.9)) be denoted $S(x, K)$. Suppose that for all values of (x, K), for all E in $[0, K]$ and $I \ge 0$, the following inequality holds:

$$(7.1) \quad \delta S(x, K) + E[qxS_x(x, K) - pqx + c] - F(x)S_x(x, K) + \gamma K S_K(x, K) + I(\pi - S_K(x, K)) \ge 0$$

where S_x and S_K denote partial derivatives. Under these conditions, we claim that the given policy is optimal. To see this, let any other control policy $E(t)$, $I(t)$ be given, with I finite. Then, if (x_0, K_0) is the starting point,

$$\int_0^\infty e^{-\delta t}\{(pqx-c)E - \pi I\}\, dt - S(x_0, K_0)$$

$$= \int_0^\infty e^{-\delta t}\{(pqx-c)E - \pi I\}\, dt + \int_0^\infty \frac{d}{dt}\{e^{-\delta t}S(x(t), K(t))\}\, dt$$

$$= \int_0^\infty e^{-\delta t}\{(pqx-c)E - \pi I - \delta S(x, K) + S_x \dot{x} + S_K \dot{K}\}\, dt$$

$$= -\int_0^\infty e^{-\delta t}\{\delta S + E[qxS_x - pqx + c] - F(x)S_x + \gamma KS_K + I(\pi - S_K)\}\, dt$$

$$\leqslant 0 \quad \text{(since the integrand is always nonnegative)}.$$

But this says that the return from any other policy does not exceed the return from our given policy (it suffices to know this for policies with finite investment rates to conclude optimality).

In summary, one can verify the optimality of a policy by producing a function S with the properties mentioned above. We shall now do just this for the policy described in the preceding sections. There remains also the task of precisely defining the switching curves, and there will be an added complication due to the fact that S will sometimes fail to be differentiable along these curves, but the underlying idea remains unchanged.

We assume henceforth that δ, $\gamma > 0$ (the case $\gamma = 0$ is simpler and can be treated with minor modifications). The constraint $I \geqslant 0$ is assumed. We also set $q = \pi = 1$, which merely amounts to a scaling of the variables E and K, but simplifies the notation. We assume that F is twice continuously differentiable and satisfies $F'' < 0$ in the interval $(0, \bar{x})$, in which F is positive. This says that F is strictly concave, and assures among other things that $K = F(x)/x$ defines K as a strictly decreasing function of x.

Now let (see (3.5), (3.9))

$$\tilde{\phi}(x) = \delta - F'(x) + c'(x)F(x)/[p - c(x)],$$

$$\phi^*(x) = \delta - F'(x) + c'_{\text{total}}(x)F(x)/[p - c_{\text{total}}(x)].$$

We assume the following: the equation $\tilde{\phi}(x) = 0$ has a unique solution $\tilde{x}$ in the interval $(x_\infty, \bar{x})$; for $x > \tilde{x}$ we have $\tilde{\phi}(x) > 0$, and for $x < \tilde{x}$, $\tilde{\phi}(x) < 0$; $p\tilde{x} - c > 0$. These conditions are easily seen to hold if the marginal stock effect (last term on the right-hand side) is a decreasing function of x. We make similar assumptions regarding ϕ^*, denoting the solution to $\phi^*(x) = 0$ by x^*, which is necessarily greater than $\tilde{x}$. We let $\tilde{K} = F(\tilde{x})/\tilde{x}$, $K^* = F(x^*)/x^*$.

Let us remark before proceeding to the systematic construction of the optimal policy and corresponding return that all our hypotheses are satisfied (for a suitable range of parameters) if F is the familiar logistic growth function, i.e., if

$$F(x) = rx(\bar{x} - x).$$

RENEWABLE RESOURCE STOCKS 39

Finally we note the interpretation of (2.9) when $I = \infty$ is allowed: set $t_0 = 0$, and suppose jumps in the value of K occur at times $t_0, t_1, t_2, \ldots$. If we denote the values immediately before and after these jumps by $K(t_i-)$ and $K(t_i+)$, respectively, and if a finite investment rate $I(t)$ is employed between the jumps, then (2.9) is given by

$$(7.2) \qquad \int_0^\infty e^{-\delta t}\{(px - c)E - I\}\,dt - \sum_{i=0}^\infty e^{-\delta t_i}[K(t_i+) - K(t_i-)].$$

We let C_1 be the locus of points (x_0, K_0) in the part of the (x, K)-plane $x \geq x^*$, $K \geq K^*$ such that the trajectory $x(t), K(t)$ originating from (x_0, K_0) with $E(t) = K(t), I(t) = 0$ passes through (x^*, K^*) (see Figure 4). It follows that for any (x, K) such that $0 < x < x^*, K > 0$, or such that $x \geq x^*$ and (x, K) lies above C_1, the policy $E = K, I = 0$ will result after finite time in arriving at $x(t) = x^*, 0 < K(t) \leq K^*$ (a sample trajectory is indicated in Figure 4). Let this arrival time be denoted $\tau(x_0, K_0)$, and let K be increased to K^* at time τ; for $t \geq \tau$ we remain at (x^*, K^*) by setting $E = E^*, I = \gamma K^*$. We denote the net discounted return from this policy $S(x, K)$.

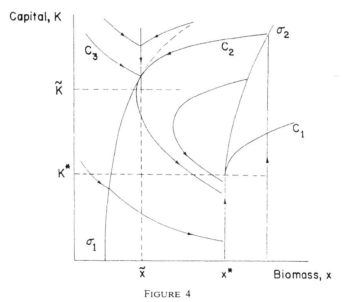

FIGURE 4

LEMMA 1: *In the interior of its domain of definition, S is given by*

$$S(x, K) = K + \int_0^\infty e^{-\delta t}\{px(t) - c - \delta - \gamma\}K(t)\,dt;$$

$S(x, K)$ *is twice continuously differentiable and satisfies*

$$\delta S + (Kx - F)S_x + \gamma K S_K - K(px - c) = 0.$$

40   C. W. CLARK, F. H. CLARKE, AND G. R. MUNRO

PROOF: Using the notation established for (7.2), we have by integration by parts:

$$\int_0^\infty e^{-\delta t} I \, dt = \int_0^\infty e^{-\delta t} (\dot{K} + \gamma K) \, dt$$

$$= \int_0^\infty e^{-\delta t} (\delta + \gamma) K(t) \, dt + \sum_{i \geq 1} e^{-\delta t_i} K(t_i -)$$

$$- \sum_{i \geq 1} e^{-\delta t_{i-1}} K(t_{i-1} +).$$

When this is substituted into (7.2), along with $E(t) = K(t)$, we obtain the stated expression (all the discrete summation terms cancel except $e^{-\delta t_0} K(t_0-) = K$).

It is a consequence of the implicit function theorem that τ is a twice continuously differentiable function of (x, K), and it is also known that $x(t)$ and $K(t)$ are (for each $t \leq \tau$) twice continuously differentiable functions of their initial value (x, K). When we make use of the preceding formula to write

$$S(x, K) = K + \int_0^\tau e^{-\delta t} \{px - c - \delta - \gamma\} K \, dt + e^{-\delta \tau} \{px^* - c - \delta - \gamma\} K^* / \delta,$$

the twice continuous differentiability of S becomes apparent.

If we proceed to observe the identity

$$e^{-\delta t} S(x(t), K(t)) = \int_0^t e^{-\delta s} \{px - c\} K(s) \, ds + S(x, K)$$

(which holds for $t < \tau$), differentiate with respect to t, and then set $t = 0$, we obtain the required partial differential equation.   Q.E.D.

We shall show presently that the equations

$$S_x = p - c/x,$$

$$S_K = 1,$$

define two curves σ_1 and σ_2, respectively, located essentially as indicated in Figure 4.

LEMMA 2: $S_K < 1$ for $x < x^*$, and for points (x^*, K) with $K > K^*$.

PROOF: If in the formula for S derived in the preceding lemma we substitute $K(t) = F(x(t)) - \dot{x}(t)$, we see that

$$S(x, K) = K + \int e^{-\delta t} \{p - (c + \delta + \gamma)/x\} F(x) \, dt$$
$$- e^{-\delta t} \{p - (c + \delta + \gamma)/x\} \, dx,$$

where the line integral is taken in the (t, x)-plane along the curve $x = x(t)$, $t \geq 0$. If we now increase K to $K + \Delta$, the graph of the resulting $x(t)$ will lie below that of

RENEWABLE RESOURCE STOCKS 41

the original $x(t)$ for all t, and using Green's theorem we obtain

$$S(x, K + \Delta) - S(x, K) = \Delta + \iint e^{-\delta t}\{p - (c + \delta + \gamma)/x\}\phi^*(x)\, dt\, dx,$$

where the double integral is taken over the (compact) region in the (t, x)-plane between the graphs. Since $\phi^*(x) < 0$ in that region, we obtain immediately $S_K(x, K) \leq 1$. An elementary estimate shows that the double integral is bounded above by a term $-\varepsilon\Delta$ $(\varepsilon > 0)$, so that we have in fact $S_K < 1$. *Q.E.D.*

LEMMA 3: $S_x < p - c/x$ *for $x \geq \tilde{x}$, $K \leq \hat{K}$.*

PROOF: Let the trajectory originating from (x, K) be denoted as usual $x(t)$, $K(t)$, and let the trajectory originating from $(x + \Delta, K)(\Delta > 0)$ be denoted $x_1(t)$, $K_1(t)$. If we proceed as in the previous lemma, keeping account of the line segment from x to $x + \Delta$ in the (t, x)-plane which together with the graphs of $x(t)$ and $x_1(t)$ comprise our closed curve, we obtain

$$S(x, K) - S(x + \Delta, K) = \oint e^{-\delta t}\{p - c/x\}F(x)\, dt - e^{-\delta t}\{p - c/x\}\, dx$$

$$+ \int_0^\infty e^{-\delta t}(\delta + \gamma)(K_1(t) - K(t))\, dt$$

$$+ \int_x^{x+\Delta} -(p - c/x)\, dx$$

$$\geq \iint e^{-\delta t}\{p - c/x\}\tilde{\phi}(x)\, dt\, dx - \int_x^{x+\Delta} (p - c/x)\, dx,$$

since $K_1(t) \geq K(t)$.
   This implies

$$-S_x \geq -(p - c/x),$$

since $\tilde{\phi} > 0$ in the region in question. Strict inequality follows as in Lemma 2.
   Q.E.D.

From Lemma 3 and the facts that S_x is bounded (as is easily shown) and $p - c/x$ tends to $-\infty$ as x decreases to 0, we deduce that the locus of points satisfying $S_x = p - c/x$ defines a curve σ_1 which lies between the lines $x = 0$ and $x = \tilde{x}$ for $K \leq \hat{K}$. We shall prove later (see Lemma 4, Corollary) that σ_1 crosses the line $x = \tilde{x}$ at one point $(\tilde{x}, \bar{K})$ with $\bar{K} > \hat{K}$.

LEMMA 4: *Along σ_1, we have*

$$(F(x) - Kx)(S_{xx} - c/x^2) - \gamma K S_{xK} = (p - c/x)\tilde{\phi}(x),$$

and any trajectory employing $E = K$, $I = 0$ meets σ_1 at most once for $x < \tilde{x}$.

PROOF: The above equation follows immediately upon differentiating the equation of Lemma 1 with respect to x and using the equality $S_x = p - c/x$. Since

$\tilde{\phi}(x)<0$ when $x<\tilde{x}$, the above equality implies that whenever any trajectory as stated meets σ_1, we have at that point

$$\frac{d}{dt}[S_x(x(t), K(t))-p+c/x(t)]<0.$$

Thus the quantity in brackets can only vanish once along the trajectory.   Q.E.D.

COROLLARY: σ_1 *crosses the line* $x = \tilde{x}$ *at a unique point* $(\tilde{x}, \bar{K})$ *with* $\bar{K} > \tilde{K}$; *when* $Kx > F(x)$ *the graph of* σ_1 *is that of a function* $K = g(x)$.

PROOF: Let σ_1 cross the graph of $K = F(x)/x$ at $x = x_1$. It follows from Lemma 4 that S_{xK} is positive at that point. If one recalls that the normal vector to σ_1 (in the direction of increasing S_x) is $[S_{xx} - c/x^2, S_{xK}]$, we see from Lemma 4 that for $x > x_1$, σ_1 defines a function $K = g(x)$, since its tangent vector can never be vertical.

It follows that g cannot grow to ∞, but must have a finite value $g(\tilde{x}) = \bar{K} > \tilde{K}$ at $\tilde{x}$ (otherwise some trajectory $x(t)$, $K(t)$ would meet σ_1 twice).

We let C_2 be the locus of points $x(t)$, $K(t)$ (for $x(t) \geq \tilde{x}$) lying on the trajectory which has $E = K, I = 0$ and which passes through $(\tilde{x}, \bar{K})$. The next result follows from Lemma 4 as did the above Corollary.

LEMMA 5: C_2 *lies below* σ_1 *(for* $x \geq \tilde{x}$*)*.

Let C_3 (see Figure 4) be the locus of points $x(t)$, $K(t)$ (for $x(t) \leq \tilde{x}$) lying on the trajectory which has $E = 0, I = 0$ and which passes through $(\tilde{x}, \bar{K})$. We define the policy above C_3 as follows: employ $E = 0 = I$ until $x(t) = \tilde{x}$, then use $E = F(\tilde{x})/\tilde{x}$ and $I = 0$ until $K = \bar{K}$, then proceed according to the earlier policy. The resulting return is $S(x, K)$ on the region in question.

LEMMA 6: *Above* C_3, S *is twice continuously differentiable,* $S_K \leq 1$, $S_x \geq p - c/x$, *and* S *satisfies*

$$\delta S - F(x)S_x + \gamma K S_K = 0.$$

PROOF: The smoothness of S and the equation that it satisfies follow as in Lemma 1. The inequality involving S_x may be proven by the method of Lemma 3, leaving only the verification that $S_K \leq 1$.

For each (x, K) in the region in question, let $T(x, K)$ denote the time t at which the trajectory $x(t)$, $K(t)$ beginning at x, K and using $E = 0$, $I = 0$ arrives at $x(t) = \tilde{x}$. Then we have

$$S(x, K) = e^{-\delta T}S(\tilde{x}, K e^{-\gamma T}).$$

Consequently we find (note $\partial T/\partial K = 0$)

$$S_K(x, K) = e^{-\delta T}S_K(\tilde{x}, K e^{-\gamma T})e^{-\gamma T}.$$

Since it is easy to see that $S_K \leq 1$ when $x = \tilde{x}$, we conclude that $S_K(x, K) \leq 1$ for all x, K. Q.E.D.

Our next redefinition occurs in the region bounded by $x = 0$, σ_1 and C_3. There we employ $E = 0$, $I = 0$ until we reach σ_1, and after that we switch to $E = K$, $I = 0$ and proceed as before. Note that Lemma 4 assures that this policy is well defined. The proof of the following (which parallels that of Lemma 6) is omitted.

LEMMA 7: *In the above region, S is twice continuously differentiable, $S_K \leq 1$, $S_x \geq p - c/x$, and*

$$\delta S - F(x)S_x + \gamma K S_K = 0.$$

We next study the situation in the region lying to the right of $x = x^*$.

LEMMA 8: *Along C_1 we have*

$$S(x, K) - [K - F(x)/x] > (px - c - \gamma)(F(x)/x)/\delta.$$

PROOF: We first observe that the left-hand side is the return obtained from starting at $(x, F(x)/x)$ if we immediately increase the number of boats to K and use our stated policy from then on, while the right-hand side is the return obtained from using $E = F(x)/x$, $I = \gamma K$ and thus staying at (x, K). Proceeding as in Lemma 2, we may express the difference between these two returns as

$$\int\int e^{-\delta t}\{p - (c + \delta + \gamma)/x\}\phi^*(x)\, dt\, dx,$$

which is positive in the region in question since $x > x^*$. Q.E.D.

LEMMA 9: *Along C_1 $(x > x^*)$ we have $S_K > 1$.*

PROOF: It is possible to use classic theorems in differential equations to calculate $\partial x(t)/\partial K_0$, where $x(t)$ is the value at time t of the solution of our differential equation (with $E = K$, $I = 0$) and K_0 is the initial value of K. We remark only that at $t = 0$, this partial derivative is known to be zero. If we differentiate (with respect to K_0) the expression for S given in Lemma 1, it then follows that we obtain an expression which along C_1 is of the form

$$1 + \int_0^\tau f(t)\, dt,$$

where the integrand is recognizably positive for sufficiently small t. It follows that $S_K > 1$ when (x_0, K_0) lies on C_1, $x > x^*$, and τ is sufficiently small (i.e., when we are close to (x^*, K^*)).

Now let σ_2 be the curve $S_K = 1$. From Lemma 2 and the above, it follows that σ_2 lies strictly between C_1 and the line $x = x^*$, at least in a neighborhood of (x^*, K^*).

If we differentiate the equation of Lemma 1 with respect to K, set $S_K = 1$ and use the inequality of Lemma 8 (which also holds along σ_2 because $S_K \geq 1$ between σ_2 and C_1) we obtain

$$S_{Kx}(F(x) - Kx) - S_{KK}\gamma K < 0$$

along σ_2. This shows that no trajectory can intersect σ_2 twice; in particular we conclude that σ_2 is strictly above C_1 for $x > x^*$. $\qquad$ Q.E.D.

As noted above, we have shown:

COROLLARY: *Any trajectory using $E = K$, $I = 0$ meets σ_2 at most once.*

Let us observe in passing that Lemma 9 shows that an optimal policy always results in temporary "overcapacity" under the circumstances that $x_0 > x^*$.

There are now two cases that present themselves, depending on whether σ_2 intersects C_2 at a point P having $x < \tilde{x}$, or whether σ_2 lies below C_2 in the region $x^* < x < \tilde{x}$. We shall discuss the latter case, where it suffices to discuss the definition of S below σ_2 and above C_2. (In the former case the redefinition of S above C_2 and the left of P necessitates a redefinition of the switching curve σ_2 above C_2.)

Above C_2, we redefine our policy, and hence S, as follows: employ $E = K$, $I = 0$ until $x = \tilde{x}$, then proceed from that point as previously defined (preceding Lemma 6). The resulting net discounted return is $S(x, K)$ for that region. Below σ_2, we immediately increase the number of boats to the value placing us on σ_2, and then we proceed as per our previously adopted policy. Note that by the preceding Corollary, we do not encounter σ_2 again.

LEMMA 10: *Above C_2, we have $S_x \leq p - c/x$, $S_K \leq 1$, and $\delta S + (Kx - F)S_x + \gamma K S_K - K(px - c) = 0$. Below σ_2, we have $\delta S + (Kx - F)S_x + \gamma K S_K - K(px - c) \geq 0$, $S_x < p - c/x$ and $S_K = 1$.*

PROOF: The first two inequalities may be proven by much the same arguments as in Lemmas 2 and 3, and the equation also follows as before. To prove the latter set of inequalities, we use the fact that σ_2 defines a function $K = h(x)$ for $x > x^*$. (This follows from the inequality $S_{KK} < 0$, which says that the marginal value of boats decreases as the number of boats increases.) Thus below σ_2, S is given by

$$S(x, K) = S(x, h(x)) - (h(x) - K),$$

and we see immediately $S_x < p - c/x$, $S_K = 1$. The remaining inequality is then seen to be equivalent to

$$\delta(S(x, h(x)) - h(x) + K) + (Kx - F)S_x(x, h(x)) + \gamma K - K(px - c) \geq 0,$$

which we now establish. The derivative of the left-hand side with respect to K is $\delta + \gamma + c - px + xS_x(x, h(x))$. The inequality of Lemma 8 (which holds also along σ_2 as noted previously) and the equation of Lemma 1 imply that this last term is nonpositive. Thus it suffices to prove the required inequality when $K = h(x)$. But in that case it is already known (in fact, equality holds by Lemma 1). *Q.E.D.*

We have now defined a function S everywhere in the (x, K)-plane, and a corresponding policy of harvest and investment for which S is the return. We have seen that S is a smooth function except possibly along a finite number of curves (where its method of definition changes), and S is everywhere continuous. If we consider the optimality argument following (7.1), we see that it is unimpaired as long as the trajectories $x(t)$, $K(t)$, being otherwise arbitrary, are such that $S(x(t), K(t))$ is differentiable with respect to t except for a finite number of points t. But the trajectories $x(t)$, $K(t)$ which only cross the curves where S is non-differentiable a finite number of times (and never travel along these curves) are dense in the space of all admissible trajectories (i.e., any trajectory can be approximated to any degree of closeness by one of this kind).

Consequently it suffices to know that our stated policy yields a return as good as any of these special trajectories; that is, it suffices to know that S satisfies (7.1) wherever it is differentiable. Thus the following concludes the proof:

LEMMA 11: *S satisfies* (7.1) *at all points not on the curves* $\sigma_1, \sigma_2, C_1, C_2, C_3$ *or the lines* $x = \tilde{x}, x = x^*$.

PROOF: In the region $x > x^*$ below σ_2, this is a consequence of Lemma 10. A perusal of all other cases will show that S always satisfies the equation

$$\delta S + K \min \{0, xS_x - px + c\} - FS_x + \gamma KS_K = 0$$

and that S_K is always less than or equal to 1. A moment's thought suffices to see that this implies (7.1). *Q.E.D.*

Finally we shall consider the alternative model discussed in Section 6. For this case, equation (7.1) must be modified by replacing the term $I(\pi - S_K)$ by

$$\psi(I) - IS_K = \begin{cases} I(\pi - S_K) & \text{for} \quad I > 0, \\ I(\pi_s - S_K) & \text{for} \quad I < 0. \end{cases}$$

To establish this modified inequality, we first use the equation

$$\frac{\partial S}{\partial K} = \pi_s$$

to define a new switching curve σ_3, the geometrical properties of which are established as in Lemma 4 above. The return function $S(x, K)$ is then redefined in region R_4 and also at all points (x, K) from which trajectories ultimately penetrate R_4.

We must then re-establish all of the inequalities proved above, and also verify that for all (x, K)

$$\pi_s \leqslant \frac{\partial S}{\partial K} \leqslant \pi.$$

46   C. W. CLARK, F. H. CLARKE, AND G. R. MUNRO

However, the present function $S(x, K)$ is identical to the previous function $S(x, K)$ except for R_4 and points influenced by R_4. Hence it is only in these regions that further verifications are required. These additional verifications are sufficiently similar to those already discussed that we can safely leave them to the reader.

8. SUMMARY AND CONCLUSIONS

This paper has investigated the implications of restricted malleability of capital for the optimal exploitation of renewable resource stocks. While the study has been carried out on the basis of a specific model of the commercial fishing industry, we believe that the qualitative nature of our results will prove to be robust.

Under the non-malleability assumption the dynamics of the optimally controlled fishery can be described in terms of short-run versus long-run behavior. Over the long run (unless capital is perfectly non-malleable) the fishery reaches an equilibrium state corresponding to "optimum sustained yield," for which the relevant cost function incorporates the full cost of fishing, i.e., operating plus capital costs. Following the initial development of the fishery, however, there is a short-run phase during which capital is excessive (from the long-run viewpoint), and only operating costs are relevant to the management decision. The development of the fishery thus follows a complex pattern of expansion, "overcapacity," and gradual contraction via depreciation, leading ultimately to the OSY equilibrium. We emphasize again that this pattern is an optimal one under the assumptions of our model.

In deriving these results we have been forced to adopt several simplifying assumptions. Perhaps the most serious of these lies in the autonomous nature of our model. Practically speaking, variability of economic parameters over time is more the rule than the exception in renewable resource industries. We make no attempt to analyze the effects of such variations here (the malleable case has been discussed in [8]). Some information can be gleaned from a comparative dynamics approach, i.e., by studying the sensitivity of the solution to the parameters of the model. For example, it is easy to verify that the purchase price of capital π has no effect on the short-run equilibrium $\tilde{x}$, but affects x^* positively and also affects the switching curve σ_2 in a negative sense. Thus higher capital costs have no effect on "bygones," but decrease the optimal (ex ante) level of capitalization, and result in lower levels of exploitation over the long run. The effects of varying other parameters are also easily worked out.

Finally, the policy implications of our study are sufficiently clear from a qualitative viewpoint. On the one hand, the analysis supports the accepted belief that excessive capitalization is likely to occur during the initial development of a common-property resource, although a certain degree of overcapitalization is now shown to be generally acceptable. On the other hand, the analysis shows that extreme policies of stock rehabilitation (e.g., fishing moratoria), may be unwarranted unless the stock has become very severely depleted. The less transferable

RENEWABLE RESOURCE STOCKS 47

are capital assets, the more important this latter consideration becomes. (Along these lines, it is clear that non-transferability of labor would have similar implications.) The application of these findings to explicit resource-management problems will require additional research.

University of British Columbia

Manuscript received April, 1977; revision received November, 1977

REFERENCES

[1] ARROW, K. J.: "Optimal Capital Policy with Irreversible Investment," in *Value, Capital, and Growth: Papers in Honour of Sir John Hicks*, ed. by J. N. Wolfe. Edinburgh: Edinburgh University Press, 1968, pp. 1–20.

[2] ARROW, K. J., AND M. KURZ: *Public Investment, the Rate of Return, and Optimal Fiscal Policy.* Baltimore: Johns Hopkins Press, 1970.

[3] BEDDINGTON, J. R., C. M. K. WATTS, AND W. D. C. WRIGHT: "Optimal Cropping of Self-Reproducible Natural Resources," *Econometrica*, 43 (1975), 789–802.

[4] CARATHÉODORY, C.: *Calculus of Variations and Partial Differential Equations of the First Order* (English translation). San Francisco: Holden-Day, 1966.

[5] CHRISTY, F. T., JR., AND A. D. SCOTT: *The Common Wealth in Ocean Fisheries.* Baltimore: Johns Hopkins University Press, 1965.

[6] CLARK, C. W.: *Mathematical Bioeconomics: The Optimal Management of Renewable Resources.* New York: Wiley-Interscience, 1976.

[7] ———: "Optimal Capital Policy and Scrap Value," to appear.

[8] CLARK, C. W., AND G. R. MUNRO: "The Economics of Fishing and Modern Capital Theory: A Simplified Approach," *Journal of Environmental Economics and Management*, 2 (1975), 92–106.

[9] CLARKE, F. H.: "Necessary Conditions for a General Control Problem," in *Calculus of Variations and Control Theory*, ed. by D. L. Russell. New York: Academic Press, 1976.

[10] CRUTCHFIELD, J. A., AND A. ZELLNER: "Economic Aspects of the Pacific Halibut Fishery," Fishery Industrial Research, Vol. 1, No. 1. Washington: U.S. Department of the Interior, 1962.

[11] GORDON, H. S.: "The Economic Theory of a Common-Property Resource: the Fishery," *Journal of Political Economy*, 62 (1954), 124–142.

[12] HADLEY, G., AND M. C. KEMP: *Variational Methods in Economics.* Amsterdam: North-Holland, 1971.

[13] KAMIEN, M. I., AND N. L. SCHWARTZ: "Optimal Capital Accumulation and Durable Goods Production," *Zeitschrift für Nationalökonomie*, 37 (1977), 25–43.

[14] NAGATANI, K., AND P. A. NEHER: "On Adjustment Dynamics—An Exercise in Traverse," in *Equilibrium and Disequilibrium in Economic Theory*, ed. by G. Schwödiauer. Dordrecht, Holland: D. Reidel, 1977, pp. 369–396.

[15] PELLA, J. J., AND P. K. TOMLINSON: "A Generalized Stock Production Model," *Bulletin of the Inter-American Tropical Tuna Commission*, 13 (1969), 421–496.

[16] PUU, T.: "On the Profitability of Exhausting Natural Resources," *Journal of Environmental Economics and Management*, 4 (1977), 185–199.

[17] SCHAEFER, M. B.: "Some Aspects of the Dynamics of Populations Important to the Management of the Commercial Marine Fisheries," *Bulletin of the Inter-American Tropical Tuna Commission*, 1 (1954), 25–56.

[18] SPENCE, M., AND D. STARRETT: "Most Rapid Approach Paths in Accumulation Problems," *International Economic Review*, 16 (1975), 388–403.

[19] UNITED NATIONS, FOOD AND AGRICULTURAL ORGANIZATION: *Economic Aspects of Fisheries Regulations.* Rome: F.A.O. Fisheries Report, No. 5, 1962.

Part II
Refinements

Demand, Supply, and Inputs

[6]

OPTIMIZATION AND SUBOPTIMIZATION IN FISHERY REGULATION

By Ralph Turvey*

The purpose of this article is to show that fishery regulation is one of those spheres of economic policy where what is the best thing to do depends on what can be done. This is usually illustrated by the analogy that, if one wants to climb as high as possible but cannot climb all the way up the highest mountain, the best thing to do may be to walk in the opposite direction and climb to the top of a lower one. If the *optimum optimorum* is to be reached (the highest mountain scaled), then regulation must extend not only to the scale but also to the mode of operation.

That this rather general point has an important application in fisheries management became apparent to me during a conference in Ottawa of economists, biologists, and administrators organized by the Food and Agricultural Organization in 1961. The proceedings of this conference have now been published [8]. The two general survey papers by Professor Scott and Dr. Dickie at the beginning of this volume afford an admirable survey of the whole subject, including a review of the literature. I shall therefore not attempt to summarize their contributions, and what follows is a self-contained argument which may, however, incidentally serve to introduce readers to a fascinating subject where economists and biologists are clearly complements and only talk as though they were substitutes when they misunderstand one another.

What follows is a static (i.e., steady state) analysis of a single-trawl fishery where there is only one fish stock, fished from ports which supply a common market and which are equidistant from the fishing ground. It is assumed that the port market is competitive and that there are no restrictions on entry into the fishery. These assumptions serve to make the exposition reasonably simple; their removal would not destroy the essential argument of the paper.

* The author is reader in economics at the London School of Economics. He is much indebted to R. J. H. Beverton, L. M. Dickie, B. B. Parrish, F. A. Popper, and A. D. Scott for their constructive comments on an earlier draft of this paper.

I. *Catch*

The determinants of the weight of catch, under these assumptions, are twofold,[1] given constant natural population parameters. The first is the rate of capture (the fishing mortality rate) of the fish liable to capture, which is proportionate to the amount of fishing effort. This is defined as the product of the number of hours' fishing (or some other index of fishing time) and of fishing power, where the fishing power of any particular ship is defined by reference to a standard vessel by comparing their catches when fishing at the same time and place. In practice, the fishing power of trawlers is broadly related to the gross tonnages, so that fishing effort in a given time interval may be measured as the average gross tonnage of the fishing fleet multiplied by the total number of hours' fishing. The relevance of fishing effort is simply that, if it is decreased, the stock of fish will grow and the average age, weight, and size of the fish will increase, making fishing easier, i.e., increasing catch per unit of fishing effort.

The second determinant of the weight of catch is the size (age) at which the fish become liable to capture, i.e., are "recruited" to the fishery. In a trawl fishery, which is the example we take here, this is regulated by the mesh size of the cod-ends of the trawls used. This determines the minimum size of fish caught; the larger the mesh, the older, larger, and heavier will be the fish liable to capture. Thus, other things remaining the same, an increase in mesh size will initially reduce the catch but may ultimately raise it by increasing the stock. Here, since the analysis is static, it is only this ultimate effect which concerns us.

Biologists have formulated a number of models which provide a function relating the weight of catch to the two variables, mesh size and fishing effort. These models deal with the rates of natural mortality, growth, and recruitment, each of which is a function of the size of the stock and its age distribution. As in economics, metrics is several steps behind theory, however, so that not all the parameters of the more complicated models can at present be estimated, owing to lack of data. Things which are known to be relevant, such as the spatial distribution of fishing effort, just have to be ignored. If we confine ourselves to the operational models used for most fishery assessments, the simplified biological interrelationships that have been introduced are the following: (a) natural mortality a function of age; (b) growth of the fish, by weight and length, a function of its age; (c) recruitment to the fish stock, i.e., the number of fish reaching catchable size, treated as exogenous; and (d) fishing mortality rate proportional to fishing effort.

[1] The standard work on all this is by R. J. H. Beverton and S. J. Holt [1]. On the measurement of "fishing effort" see also J. A. Gulland [3]. A useful survey article, of which Dr. Parrish has kindly lent me a translation, is by G. Hempel and D. Sahrhage [4].

The steady-state catch in weight from a fish stock is then determined by the interaction between the rates of these natural processes and the fishing mortality rate. The relationship between weight of catch and age of recruitment (mesh size) and fishing mortality rate (fishing effort), estimated from a simple, dynamic model in which the rates of the natural processes are assumed constant, takes the form shown in Figure 1, where fishing effort is measured horizontally and weight of catch is measured vertically. Each yield curve shows catch as a function of fishing effort for a given age of recruitment to the fished stock. (The dotted line will be explained later.)

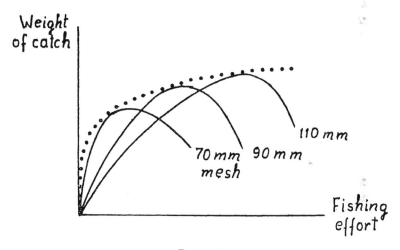

FIGURE 1

II. *Costs*

The study of fishing costs is a complicated matter. To illustrate, a large part of fishermen's remuneration is a share in profits rather than a wage.[2] Such matters are not relevant to the argument of the present paper, however, so I shall exclude them by simply assuming that over the relevant range of catch, the total cost of fishing effort rises more than proportionately to the amount of fishing effort because the minimum earnings necessary to attract and retain labor and capital at the margin rise as more of these resources are employed in fishing. Total cost is defined to include the "rents" of the intramarginal factors, i.e., the excess of their actual earnings over the minimum required to retain them in the industry. This is the only aspect of costs that we require in the present argument, so I need not discuss whether few ships fish

[2] See Zouteweij, "Fishermen's Remuneration" in [7]. Economies of scale in shore installations constitute another example.

many hours or vice versa, since fishing effort as defined above is the independent variable of the cost function. What is significant in this formulation is the implicit assumption that mesh size has no noticeable effect upon the costs of fishing effort. This is probably true within the range of mesh size that is relevant in practice.

III. *Revenue*

The price of fish, given income levels and the prices of other foods, depends not only upon the total weight of the catch but also upon its size distribution and upon its freshness. Here we shall assume that there is some given minimum marketable size of fish; otherwise I neglect these points.[3] Thus I postulate a given function relating total revenue to weight of catch.

IV. *Steady-State Bionomic Equilibrium*

Let us postulate that: (1) any fluctuations in the natural processes are small and have no trend; and (2) there is free entry of resources to the fishery. Then, given all the various functions discussed above, the long-run equilibrium position can be deduced.

On the economic side, the equilibrium condition is that total revenue equals total cost including the rent of intramarginal factors. We put it this way rather than in terms of average revenue and long-run marginal cost because average revenue is related to weight of catch and marginal cost to fishing effort. In order to relate average revenue to marginal cost, we would have to convert catch into effort, and write the condition as:

Price × Catch per unit effort = Marginal cost of fishing effort excluding rent

i.e., Average revenue × Catch per unit effort = Average cost including rent

which gives us:

Price × Catch = Average cost including rent × No. of units of effort

i.e., Total revenue = Total cost including rent.

If total revenue exceeds total costs, resources not in the industry will find that it is worth while to move into it, while in the reverse case some resources at the margin will be earning less than is required to retain them in the industry, and they will leave it.

Each fisherman will want to maximize the marketable value of his catch at any given level of costs. This means, on our assumptions, that

---

[3] The way in which price depends upon size and freshness can, of course, be observed but it would probably be extremely difficult to estimate how the price pattern would change following alterations in these variables. Note that freshness is related to length of stay at sea, which in turn is related to the size of ship and hence affects costs.

he will wish to maximize the weight of his catch of fish above the minimum marketable size for any given level of costs. Thus for the present argument we will assume that he will choose a mesh which limits his catch to fish above that size. If he used a smaller mesh, he would have to throw part of his catch back into the sea, which would involve unnecessary trouble.

We can now describe the equilibrium position. The minimum marketable size of fish determines mesh size. This, in conjunction with fishing effort, determines the weight of catch. This, in turn, determines total revenue which must equal the total costs of the amount of fishing involved. All this is shown in Figure 2 whose N.E. quadrant corresponds to Figure 1 and where the equality of total revenue and total costs is shown by using a 45° line in the S.W. quadrant.

As Figure 2 is drawn, either a general fall in cost or a rise in the demand for fish will cause an increase in the equilibrium catch. But if fishing effort has been carried beyond the point of maximum yield this standard conclusion does not follow, as Figure 3 shows. Here the dashed line shows one equilibrium position and the dotted line shows another; the latter corresponds to a higher level of demand, represented by the dotted total revenue curve. In this case the higher level

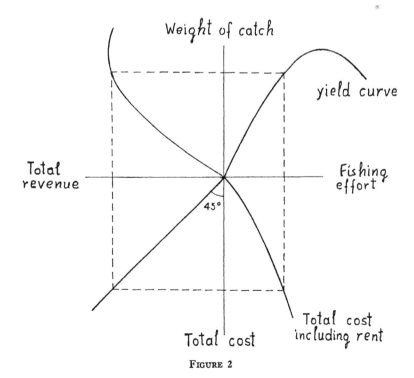

FIGURE 2

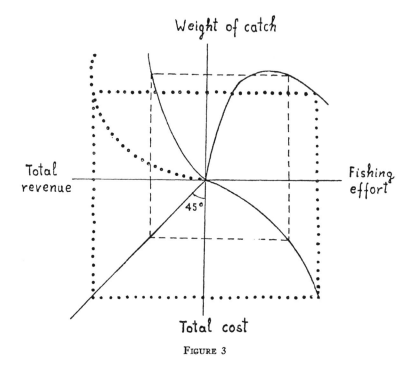

FIGURE 3

of demand means a greater employment of resources in the industry but a smaller total product!

V. *External Diseconomies*

Such a peculiar result indicates that there is something special about the industry. There is indeed. But from what has been said it should be apparent that two problems are involved, not one. The first is that while the catch of the individual fisherman is proportionate to his own fishing effort, the same is not true of all fishermen together; i.e., the yield curves for all fishermen in Figure 1 are not straight lines through the origin.[4] Each fisherman imposes an external diseconomy upon his brethren; the marginal private product of his fishing effort exceeds the marginal social product.

The second problem involves mesh size. By catching small fish, fishermen are reducing the number of large fish to be caught later. If an individual fisherman were to raise his mesh size he would lose by increasing the number of hauls necessary to achieve any given weight of catch. Yet in the long run, his use of a larger mesh may lower the costs

---

[4] They are straight lines in an underdeveloped fishery where natural mortality is infinite relative to fishing mortality.

of all fishermen together and, if all of them used larger meshes, all would benefit. Here again, social and private product diverge.

VI. *The Optimum Optimorum*

When external economies are involved both in the level of fishing effort and in the choice of mesh size, it is clear that to achieve the optimum resource allocation requires regulation of *both* these variables.

Let us assume for the present that the conditions for optimal resource allocation are fulfilled in the rest of the economy, so that no problems of "second best" arise. Let us further assume, first, that the effect of changes in the price of fish upon the distribution of real income between fish consumers and others is unimportant and, second, that a level of earnings in the industry which is equal (at the margin) to the earnings .those resources could obtain elsewhere is socially acceptable. Finally, let us consider only those cases where the fishery is a very small part of the economy.

Under these conditions, the *optimum optimorum* is reached when G is maximized, G being the excess of the value of the catch to consumers over the value to them of the alternative goods and services sacrificed by devoting resources to fishing. Now the value of the catch to consumers is the maximum they would pay rather than go without it, i.e., what they do pay (total revenue) *plus* consumer surplus: $TR + S$, the area under the demand curve. The value of goods and services sacrificed is equal to the contribution to production that the resources used in fishing would make if they were not so used and this, on our assumptions, is what they could earn elsewhere. It is therefore measured by the total costs of the fishery *less* the rents of the intramarginal resources, $TC - R$. Thus the *optimum optimorum* is to be reached by maximizing:

$$G = (TR + S) - (TC - R).$$

A necessary, but insufficient, condition for maximization is that mesh size be such as to maximize the catch for the actual level of fishing effort. In terms of Figure 1 this means that the fishery must be "on" the dotted envelope curve, known as the "eumetric yield curve" (Beverton and Holt [1]). It is a property of this curve that, unlike the individual yield curves for given mesh sizes, it is asymptotic to the horizontal, thus rising throughout its length.

Figure 4 shows the determination of the optimum. The level of fishing effort, OF, must be so chosen as to maximize G. This is achieved when the slope of the $TR + S$ curve (i.e., price per unit catch) multiplied by the slope of the eumetric yield curve (i.e., marginal social yield of fishing effort) equals the slope of $TC - R$ (i.e., the marginal cost of

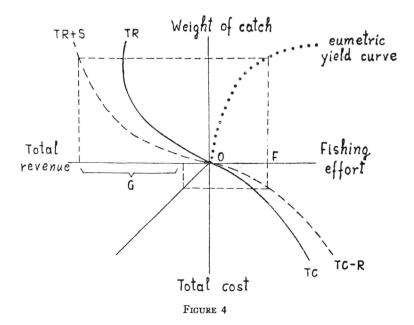

FIGURE 4

fishing effort). When this is the case, i.e., when:

> Price × Marginal social yield = Marginal cost of fishing effort excluding rent

> i.e., Price × Marginal social yield = Average cost of fishing effort including rent

it must be true that:

> Price × Average yield > Average cost of fishing effort including rent

(since the average yield of fishing effort *exceeds* the marginal social yield),[5] which means that total revenue exceeds total costs including rent. $G = S + R + (TR - TC)$, the maximum gain, is the sum of consumers' surplus, producers' surplus (the rents of intramarginal labor and capital), and the rent of that other scarce resource, the fish stock.[6]

In an unregulated fishery, where resource allocation is nonoptimal, free entry means that total revenue will equal total costs so that no rent of the fish stock is achieved. Under these circumstances, the total gain from the fishery $G = S + R$.[7] Whether a shift from this situation to the *optimum optimorum* will raise or lower the total catch (and hence

[5] I.e., the eumetric curve is concave from below.

[6] Cf. H. S. Gordon [2]. Note that if capital and fishermen of equal skill are both in infinitely elastic supply and if the fishery meets a very small part of the total demand, R and S respectively will be zero. Nevertheless, the *optimum optimorum* still requires regulation of both effort and mesh size.

[7] Which will equal zero in the special case of the previous footnote!

S) and whether it will raise or lower the amount of fishing effort (and hence R) cannot be determined a priori.

VII. *Suboptimization of Mesh Size*

In order to show this it is convenient to consider the regulation of mesh size and the regulation of fishing effort separately and to show that in some circumstances these may have opposite effects. It is of great importance to consider the suboptima that can be reached by regulating only one or the other of these variables since, in practice, political and administrative considerations frequently make one or the other type of regulation impossible to introduce. This is the case, for example, with the regulation of international trawl fisheries, where agreement upon mesh size is much easier to attain than agreement about the share of fishing effort (and hence catch) to be apportioned to each nation.

When mesh size alone can be regulated, it is obvious that the fishery ought to move on to the eumetric curve, since this is the boundary of the production function. At the original level of fishing effort and costs this will raise the weight of catch but may either lower total revenue (elasticity of demand less than unity) or raise it (elasticity greater than unity). In the first case resources will leave the industry, so that fishing effort is reduced; conversely in the second case. Even in the first case, however, catch will be greater in the new suboptimum equilibrium than it was in the initial nonoptimum equilibrium. If it were not greater, total revenue would be higher or the same as before while fishing effort and total costs would be lower, so that total revenue would exceed total costs—a state of affairs incompatible with equilibrium when entry of resources is unrestricted.

Suboptimization with respect to mesh size thus involves maximizing:

$$G = S + R$$

subject to the constraint $TR = TC$. Moving to the suboptimum, equilibrium must raise S, since catch rises, but may either raise or lower R. If R falls, the rise in S must exceed the fall.[8] But if there were constant costs (R always zero) and an infinitely elastic demand (S always zero), there would be no point in regulating mesh size.

VIII. *Suboptimization of Fishing Effort*

While, as we have just seen, suboptimizing mesh size must raise catch but may either raise or lower fishing effort and costs, suboptimizing

[8] Because TR and TC fall equally (by assumption); R must fall less than TC (when the supply curves of factors are rising); S must rise more than TR falls (since $TR + S$ rises with movement down any downward-sloping demand curve).

fishing effort will always lower fishing effort and costs but may either raise or lower catch. The analysis of this case follows the same lines as that of the *optimum optimorum* (see Figure 4), save that the yield curve is not the eumetric curve but a constant-mesh yield curve. Now such a curve, as shown in Figure 1, has a maximum. If the initial non-optimum equilibrium is to the left of this maximum, suboptimizing with respect to fishing effort will *reduce* both fishing effort and catch. But if the initial equilibrium is to the right of this maximum, so that the marginal social product of fishing effort is actually *negative*, sub-optimization will increase the catch. North Sea plaice and haddock are examples of trawl fisheries where fishing effort is to the right of the maximum for the mesh size in use. The classic case of this is not a trawl fishery but is said to be the Pacific halibut fishery.

How can fishing effort be regulated? The answer is surely to charge fishermen a rent for fishing, so that they economize in the use of the scarce resource. This rent would take the form of a charge per unit of weight of catch proportionate to the difference between marginal private yield (catch per unit of effort) and marginal social yield (the slope of the yield curve). The effect of this would be to raise the marginal private cost of catching fish to equal the marginal social cost.

An alternative method would be to impose quantitative restriction upon the amount of fishing effort. In practice, such restriction would produce an inferior suboptimum, unless it were so all-embracing as to amount to the institution of a sole-owner of the fishery. The reason is that any restriction must be framed in terms of a limit upon one or more of the resources used in the industry. But to ration the supply of any resource is to invite the substitution of other resources for it. Thus if the number of boats were limited, larger engines would be installed, bigger crews carried, and more resources devoted to securing a rapid turn-around in port. None of this would be efficient, in terms of resource allocation; it would simply increase the total costs of all alternative levels of fishing effort. Furthermore, it would concentrate the rent to be had from fishing economically, i.e., the gain from suboptimizing fishing effort, in the hands of the owners of the boats. In some circumstances this would be less desirable from the point of view of income distribution than a charge that placed the rent in the hands of the government which could dispose of it in a variety of ways. Thus a tax on catch is generally necessary for suboptimization. To reach the *optimum optimorum*, however, such a tax is necessary but not sufficient.

There is a great deal more that can be said about the effects of restricting entry to a fishery in ways that lead to economic inefficiency, but I shall not pursue the matter here. My purpose is merely to assert the fundamental principle that either mesh regulation or the control of fishing effort is better than nothing but that regulation of both is still better.

IX. *Second Best*

If the assumptions listed at the beginning of the section on the *optimum optimorum* are not met, allowance must be made for this fact in deciding policy. I do not believe that the distributional effects of changes in fish prices are worth bothering about, but there might well be circumstances in which regulation adversely affected the earnings of fishermen to an undesirable extent. Where this resulted from the creation of an excess of total revenue over total costs, the government would acquire more than sufficient funds to compensate the displaced fishermen. Where mesh regulation alone was involved, however, no such funds would be available, and the problem acquires important political elements.

Leaving these distributional questions aside, the recommendation to optimize or, if that is impossible, to suboptimize depends crucially upon the assumptions that the costs of fishing effort do reflect the value of resources in alternative uses and that the price of fish does measure the marginal contribution of fish to consumer welfare. The economist who is concerned to help formulate policy rather than to maintain his theoretical purity will, however, seek to make rough corrections for all the imperfections that he can think of in closely related markets and will then go on to make his recommendations. Thus the level of unemployment in fishing ports and subsidies or tariffs on meat and eggs are likely to draw his attention. He will, on the other hand, remain indifferent to the effects upon fishing costs of restrictive practices among suppliers of marine paint or to the effect upon the demand for fish of resale price maintenance in the confectionery trades.[9]

Even the purists, however, will agree that on resource-allocation grounds there are two things that can be recommended unreservedly. One is a shift to the eumetric curve and the other, as an alternative, is a reduction in fishing effort whenever the fishery is on a falling yield curve. Either of these will raise the catch and/or reduce costs, providing that "something for nothing" which alone appears to gladden the hearts of our erudite welfare economists.

There are complications, however, Even when cost, demand, and natural conditions remain constant through time so that steady state equilibrium can be achieved, the introduction of the regulation of mesh size or fishing effort will initially reduce the catch. This initial, tempo-

[9] Cf. E. J. Mishan [5].

rary loss clearly has to be weighed against the eventual, permanent gain. A rate of social time discount is required to make such weighing-up possible. All this complicates matters, but introduces no interesting new principles.

X. *Alternative Methods of Regulation*

There is probably no fishery conforming to our simplifying assumptions. It is not worth while removing them one by one in a general discussion such as this, however, since the particular assumptions that should replace them will vary from one fishery to another. Thus, instead of going on to complicate the analysis,[10] I shall conclude by mentioning other forms of regulation than those dealt with so far. The main alternatives that have been applied in practice are: limitation of total catch, closed seasons, closed areas, minimum size limits, and the prescription of particular types of gear. These have in common that they involve no restraint on entry to the industry so that, taken alone, none of them results in the creation of a rent of the fishery. Despite this imperfection, such measures may often be better than no regulation at all, so that if one of them is all that is politically or administratively practicable there is a suboptimum to be found.[11] In some fisheries, two of these measures, size limits and the closure of nursery grounds, are fairly close substitutes for the regulation of mesh size, so that the type of analysis presented above can be applied. With any of them, however, total revenue and total costs will be equal in equilibrium so that the social gain to be had from the fishery is limited to the sum of consumers' and producers' surpluses: $G = S + R$.[12]

An important conclusion follows from this, as it did in the case of suboptimizing mesh size. Any measure that raises weight of catch without affecting the total cost curve will raise G. This is not unambiguously an improvement in a second-best world if resources are drawn into the industry; but if demand has an elasticity of less than unity, so that the increased catch lowers total revenue, resources will leave the industry, and the matter is simpler. Unless the immediate loss outweighs the long-run gain, the only possible objection to getting more fish at a lower cost would be that the resources in the industry deserved the rents they would lose.

[10] For an example, see Beverton and Holt's discussion of fisheries based on two species caught by the same gear but having different eumetric yield curves [1, pp. 388, 421, ff.].

[11] Furthermore, existing measures have initiated cooperation and involved a start on some of the necessary research, without either of which there is no hope of further developments.

[12] We are still assuming that the fishery is only a small part of the economy.

XI. *Conclusion*

I have not attempted any general survey of the economics of fishery conservation, though I hope that this article may serve as an introduction to Professor Scott's paper on the subject and Professor Crutchfield's case study of the Pacific halibut fishery, both in the proceedings of the Ottawa conference [8]. Although my exposition is mainly in terms of trawl fisheries, it can be applied more generally to other fisheries when selectivity is controllable. My purpose has been to provide an interesting and important example of the notions of optimization and suboptimization and of the proposition that coping with external diseconomies will sometimes involve interfering with the nature as well as the scale of private productive activities [6].

REFERENCES

1. R. J. H. BEVERTON AND S. J. HOLT, *On the Dynamics of Exploited Fish Populations.* London 1957.
2. H. S. GORDON, "The Economic Theory of a Common Property Resource," *Jour. Pol. Econ.,* April 1954, *62*, 124-42.
3. J. A. GULLAND, *On the Fishing Effort in English Demersal Fisheries.* London 1956.
4. G. HEMPEL AND D. SAHRHAGE, "Recent Model Concepts on the Dynamics of Demersal Fish Stocks," *Ber. Deutsche Wiss. Komm. für Meeresforschung,* XVI:2.
5. E. J. MISHAN, "Second Thoughts on Second Best," *Oxford Econ. Papers,* Oct. 1963, *14*, 205-17.
6. RALPH TURVEY, "On Divergences between Social Cost and Private Cost," *Economica,* Aug. 1963, *30*, 309-13.
7. ———— AND J. WISEMAN, ed., *The Economics of Fisheries.* Rome 1956.
8. F. A. O. Fisheries Reports No. 5, *Economic Effects of Fishery Regulation.* Rome 1962.

[7]

PARZIVAL COPES

Factor Rents, Sole Ownership and the Optimum Level of Fisheries Exploitation*†

In the literature of fisheries economics there is a noticeable pre-occupation with the phenomenon of resource rent dissipation. The common property nature of most fishery resources—with the attendant free entry of labour and capital—gives rise to 'problems' of 'overfishing'. If at any given level of fishing effort the resource should yield a rent to the marginal operator, additional factor inputs of labour and capital will be attracted that will depress the catch per unit of effort and lower returns to all operators. This process will continue until the revenue per unit of fishing effort is reduced to the level of its marginal opportunity cost. Thus the rent attributable to the resource, that formerly accounted for the excess of revenue over marginal opportunity cost, is eliminated.

Gordon, in his seminal article that launched the theory of fisheries economics, identified optimum utilization with maximization of rent. His analysis held that at ". . . the optimum intensity of effort . . . the resource will . . . provide the maximum net economic yield . . . which can be regarded as the rent yielded by the fishery resource".[1] In accordance with this perception economists have been inclined to suggest institutional measures that would prevent or reduce 'over-fishing' and retain or recapture all or some of the rent that a fishery could yield. Commonly, the proposed remedies have involved some regime of single or central management based on sole ownership or control of the resource, that would limit labour and capital inputs.[2]

The purpose of this article is not to deny the essential correctness of the existing formal analysis regarding rent dissipation, so much as it is to question the weight of its importance. This paper will attempt, by the application of conventional criteria of welfare maximization, to balance the consideration of resource rent with that of competing social benefits which may derive from the exploitation of a fishery. In so doing the analysis will focus on the significance of consumers' surplus and producers' surplus, the latter in the form of rent (or 'quasi' rent, if one prefers) accruing to factors other than the fishery

resource itself.   Commentary will also be given on major implications
for resource management policy in respect of resource ownership
forms.

This paper will be limited to a steady state analysis following the
models used in articles by Turvey and Copes.[3]   It will abstract from
'second best' problems.   It will assume optimal techniques and fixed
combinations of factors other than the fishery resource.   Externalities will not be considered, except where they follow directly from
bio-economic interaction through variation in the level of fishing
effort.   The income redistribution effects of price changes and factor
income changes will be ignored in the process of optimizing welfare.
Total social benefit (in gross terms), or social 'revenue', will be
measured by the maximum consumers are prepared to pay for
goods—in this case fish.   Social cost will be measured in terms of
the opportunity costs of labour and capital.

The analysis that follows will be illustrated in conventional
price/output diagrams.   The long-run supply curve of a given fishing
industry (S in Figure 1) is the locus of steady state output equilibria

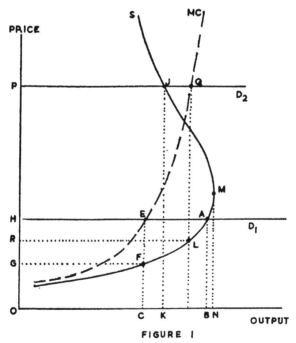

FIGURE I

at various price levels.[4]   It shows, for each level of output in the
fishery concerned, the unit cost of production, as determined by the

marginal opportunity costs of labour and capital, being the factors employed other than the fishery resource itself. For expository purposes initially it will be assumed that all fishing units (consisting of fixed combinations of labour and capital) have equal opportunity costs, and are of equal efficiency, so that their costs per unit of output will be equal. In other words, initially it will be assumed that there are no intramarginal fishing units with opportunity costs lower than those of marginal fishing units. Accordingly the supply curve will measure both the average social cost and the average private cost of fishing units. The assumption of equal opportunity costs will be removed later in the paper. Typically this supply curve will be backward bending, the maximum abscissa measuring the maximum sustainable yield of the fishery (*ON*). The various points on the supply curve correspond to various levels of fishing effort. Starting from a low level of effort at the lower left end of the curve, successively higher levels of effort are represented by successively higher points on the curve. In moving upwards along the curve through points representing successively greater effort per time period, the corresponding output measured on the horizontal axis initially will increase (though at a diminishing rate) until the maximum sustainable yield is reached (at *M*), after which further increases in effort will result in successively smaller total outputs represented by points on the backward slope of the curve. Each higher point on the supply curve represents a higher level of sustained fishing effort as well as a lower total fish stock in equilibrium.

I

To start with, the rudiments of Gordon's theory of fisheries economics will be translated into a price/output diagram as shown in Figure 1. Assuming a perfectly elastic demand (i.e., for a relatively small fishery contributing to a relatively large market) and a regime of free entry, the demand schedule D_1 will produce a long-run equilibrium at *A* with output *OB* per time period. As *AB* measures both unit cost and unit revenue, no surplus is earned: the resource yields no rent. If a curve (*MC*) is drawn marginal to the supply curve (which here is the average social cost curve), a socially optimum output level (*OC*) may be identified at *E*.[5] This output will be achieved by limiting fishing effort to the level represented by point *F* on the supply curve. At this point a maximum net social benefit (*EFGH*) may be obtained, constituting the difference between total (social) revenue (*ECOH*) and total (social) cost (*FCOG*). This net benefit is the rent yielded by the resource.

In the foregoing presentation OC measures the optimum sustainable yield of the fishery in terms of an economically defined social optimum. Accordingly any level of sustained fishing effort higher than that represented by point F on the supply curve constitutes 'overfishing' in economic terms. However, the biological potential of the fish stock permits physically a larger sustainable yield—with a maximum of ON. A sustained effort higher than that given by point M on the supply curve will result in a smaller sustained yield being taken from the fish stock and would thus constitute 'overfishing' in biological terms (as well as economic terms) . The demand given by schedule D_2 would lead, in an open entry fishery, to overfishing in both economic and biological terms. With equilibrium at J, the sustainable yield would be equal to OK, while unit revenue of JK would be entirely absorbed by unit costs. However, if by entry limitation fishing effort would be held back at L (which would bring demand and marginal cost in equilibrium at Q), a maximum rent of $PQLR$ could be realized.

II

By assuming perfectly elastic demand, Gordon's analysis avoided consideration of consumers' surplus in determining the socially optimum level of output of a fishery. In Figure 2 that assumption

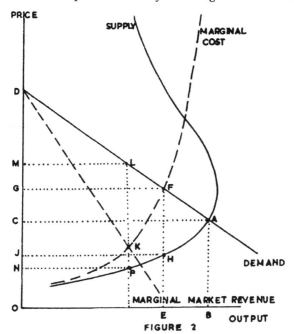

FIGURE 2

has been relaxed.   In this case the output of the fishery concerned
is not marginal to the market, so that any perceptible change in
output will affect the amount of consumers' surplus generated.   With
free entry the competitive equilibrium established in the fishery at
A will result in an output of *OB.* generating a consumers' surplus
of *ACD.*[6]   In this instance the market revenue of the fishery (*ABOC*)
is fully absorbed by social cost which equals the private opportunity
costs of fishing units.[7]   The consumers' surplus, being in excess of
market revenue, and thereby in excess of social cost, represents a
net social benefit.   It is possible, however, to interpose between
market revenue and the cost of employing fishing units a return to
the resource in the form of rent, which would equally constitute a
net social benefit.   The socially optimum output, then, must maximize the combined total of rent and consumers' surplus.   The
socially optimum output, of course, will be found at the point where
marginal social cost and marginal social revenue are equal.   In
Figure 2 the supply curve has already been identified as an average
social cost curve, with a matching marginal curve.   The demand
curve is a marginal social revenue curve, measuring as it does the
addition to total social revenue for each additional unit of output.
The socially optimum output now is found at *F* and maximum net
social benefit is given by consumers' surplus (*FGD*) plus rent (*FHJG*).[8]
In moving from output *OB* to the optimal *OE* the price that will
clear the market rises from *OC* to *OG*.   This will bring with it a
reduction in consumers' surplus, which is more than offset by the
gain in rent.   The latter could be acquired by the state in the form
of production-related taxes or license fees (and possibly redistributed
to consumers), or it could accrue to the group that is privileged to
prosecute the restricted fishery.

   The granting of sole ownership rights in respect of a fishery is
often advocated as one way of approaching optimum exploitation.[9]
Whether the owner is the state or a private agency, it is argued, the
owner would have an interest in maximizing the return from the
fishery.   The dissipation of rent by excessive effort that occurs with
free entry could be prevented.   The consideration of consumers'
surplus, however, does raise the question whether a sole owner would
be correctly motivated toward the achievement of a socially optimum
level of exploitation.   In particular, there may be a difference in the
respective motivations of public and private agencies.   A private
firm, granted the sole right to exploit a fishery, presumably would
have no interest in consumers' surplus.   To such a firm the demand

curve would not be a marginal revenue curve but an average revenue curve. A corresponding marginal ('market') revenue curve may be determined. But as the private firm would have to pay for capital and labour inputs at the rate of the opportunity costs of marginal units, the supply curve would remain the average cost curve for the firm. The intersection of the marginal market revenue and marginal cost curves at K in Figure 2 shows the maximum rent obtainable by the private sole owner as $LMNP$. But maximizing rent and consumers' surplus together is bound to yield a greater net social benefit than maximizing rent alone (except where the demand curve is perfectly elastic, when consumers' surplus will be zero). A public sole ownership agency, if instructed to observe marginal cost pricing, would operate at the socially optimum output level of OE, charging consumers a price of OG and maximizing the combined net social benefits of consumers' surplus (FGD) and rent ($FHJG$).

III

Turvey observed that the "maximum gain" that may be derived from the exploitation of a fishery contains three elements.[10]   Not only consumers' surplus and resource rent are involved, but also producers' surplus. The latter consists of the rent that intramarginal inputs of labour and capital enjoy to the extent that their opportunity costs per unit of output lie below those of the marginal inputs of these factors.[11]   For to that extent intramarginal factor inputs will receive revenue in excess of their opportunity costs; at least if there is no price discrimination and all factor inputs are remunerated according to the opportunity costs of the marginal inputs. In much of the theoretical literature of fisheries economics the consideration of producers' surplus is avoided by assuming the equality of market cost and social cost and by assuming the equality of the cost of all units of fishing output. Such were also the assumptions of the foregoing discussion in this paper. The consequences of the removal of these simplifying assumptions will now be explored.

There are generally two reasons for which intramarginal factor inputs may have opportunity costs below those of marginal factor inputs in a fishery. In the first place, differences in alternative employment and income opportunities for various fishermen and their gear may exist. But what is probably far more important is the difference in efficiency of various fishing units. To accommodate these phenomena in the model of this paper the following understanding is needed. Inputs of labour and capital will be considered combined into units of fishing effort of equal productive capacity.

Such units of fishing effort then will be equal in terms of amount of catch produced, but not necessarily in terms of amount of equipment used, numbers of crew members employed, extent of fishing time or level of operating cost. Translated into a practical example this means that if one vessel-crew combination produces twice as much fish as another, the former will be considered to represent two units of fishing effort as against one unit for the latter. This will be so even where the vessels would be identical physically and have equal numbers of crew members with individually equal opportunity costs, equal hours of fishing time and equal operating expenses. In such a case the opportunity costs in absolute terms for the two vessel-crew combinations would be the same. But for the former vessel they would be distributed over two units of fishing effort and would therefore be shown at half the level of the opportunity costs for the one unit of fishing effort represented by the latter vessel.

Close observation of catch statistics for many fishing fleets reveals considerable variation in landings for different vessels, even where these vessels represent roughly equal capital investments and where they are manned by crews of equal size. The catch variations, then, are a *prima facie* indication of considerable differences in opportunity costs per unit of fishing effort, as defined above. The differences in catch result in part from chance factors—fishermen's luck. For another part they result from variations in vested positions, such as port location, traditional or inherited rights to prime berths for the setting of gear, or similar rights obtained by other non-market means of allocation. Some of these considerations, strictly speaking, are questions of partial ownership rights. But for many fisheries the major differential in productivity, and thus in opportunity costs, simply results from differences in fishing skill and knowledge among various fishing crews. A selection process often takes place that maintains or widens these differences. Fishermen are commonly compensated on a share-of-catch basis, which means that the more efficient a crew is, the higher will be the rewards for all its members. Multiple vessel fishing companies desirous of retaining or attracting the most efficient captains and crew members, commonly allow the best captains first choice in picking the best crew members. Thus the differential in overall efficiency between vessels is maintained with resulting large differences between 'high-liners' and vessels at the bottom of the efficiency scale.

Economists looking from afar at the problems of low productivity that appear to characterize most of the fishing industries of industrially advanced nations, are often given to generalize that all fishermen earn little at their trade. In his leading article Gordon remarked: "By and large, the only fisherman who becomes rich is one who makes a lucky catch or one who participates in a fishery that is put under a form of social control that turns the open resource into property rights".[12] This notion has confirmed economists in their belief that they have found the explanation of this poverty in the peculiar common property nature of the fishery that leads to the dissipation of the resource rent. But the economists' notion is challenged by many practical men in the fishing business who know that significant numbers of fishermen and fishing company operators earn large incomes and that some gain small fortunes. They also know that the high earners, who operate competitively in fisheries with. unrestricted entry, do not depend on luck, but perform consistently at much higher levels of efficiency than do their rivals.

The difference in a fishery between marginal opportunity costs, which match revenue, and the lower opportunity costs of intramarginal fishing units is a form of rent (or 'quasi' rent). This rent is the substance of the producers' surplus. It is not attributable to the nature of the resource as such, but is primarily related to the efficiency of the intramarginal fishing units. Nevertheless, this rent must be considered a feature of the fishery as the efficiency from which it derives is an efficiency of labour and capital that is peculiar to their specific application to the fishery resource. The same units of labour and capital may or may not have any efficiency advantage in alternative employment. A good fisherman may not be any good at any other trade. The notion that fisheries allowing free entry yield no rent therefore requires qualification, for considerable rents may be earned by intramarginal factor units specifically as a result of their employment in the fishery. The literature of fisheries economics has tended to ignore the significance of this rent, perhaps because it was thought of as a matter of small differences in terms of alternative employment opportunities instead of large differences in terms of operational efficiency.

IV

The introduction of producers' surplus into the analysis requires the differentiation of social cost from market cost. This is illustrated in Figure 3. The *AMC* curve represents the supply curve described in previous figures. It measures the average market cost at each output level as determined by the opportunity costs of marginal units of fishing effort. For a competitive fishery this is the operational

Optimum Fisheries Exploitation   **153**

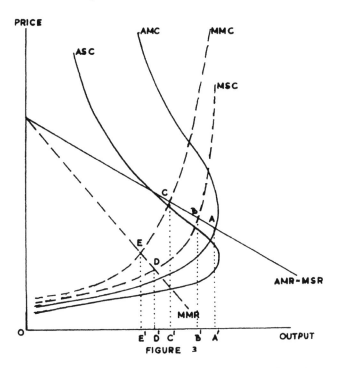

FIGURE 3

long-run supply curve that will indicate output equilibria for various price levels. The *MMC* ('marginal market cost') curve has been drawn marginal to the *AMC* curve. It measures marginal cost per unit of output to any operator possessing sole ownership rights in respect of the fishery resource (e.g., a state fishing monopoly or a private fishing company with an exclusive licence), but hiring other factors (labour and capital) at rates determined by the marginal opportunity cost of fishing effort. The *ASC* ('average social cost') curve is obtained by subtracting the average producers' surplus from the average market cost and thus represents the average opportunity cost per unit of output for labour and capital. The *MSC* ('marginal social cost') curve is drawn marginal to the *ASC* curve.[13] Also in Figure 3 a demand curve is shown. It functions both as an 'average market revenue' (*AMR*) curve and a 'marginal social revenue' (*MSR*) curve, in so far as it measures at any output the additional value enjoyed by consumers at the price the marginal buyer is prepared to pay. A 'marginal market revenue' (*MMR*) curve has been drawn marginal to the demand curve to show the marginal revenue enjoyed by sellers collectively (or by a monopoly seller) for the output of the fishery.

TABLE 1

CHARACTERISTICS OF EQUILIBRIUM LEVELS OF OUTPUT AND PRICE FOR DIFFERENT FORMS
OF FISHERIES MANAGEMENT

Market organization	Categories of benefits maximized	Incidental social benefits	Social benefits eliminated	Relation to socially optimum output	Relation of consumers' price to social optimum	Relation to socially optimum level of effort
(A) Free entry and free market	—	Consumers' surplus and producers' surplus	Resource rent	Indeterminate	Indeterminate	Overfishing
(B) State control	Combined total of consumers' surplus, resource rent and producers' surplus	—	—	Optimal	Optimal	Optimal
(C) Consumers' monopsony	Combined total of consumers' surplus and resource rent	Producers' surplus	—	Below optimum	Above optimum	Underfishing
(D) Producers' monopoly	Combined total of producers' surplus and resource rent	Consumers' surplus	—	Below optimum	Above optimum	Underfishing
(E) Resource owners' monopoly	Resource rent	Consumers' surplus and producers' surplus	—	Below optimum	Above optimum	Underfishing

Optimum Fisheries Exploitation 155

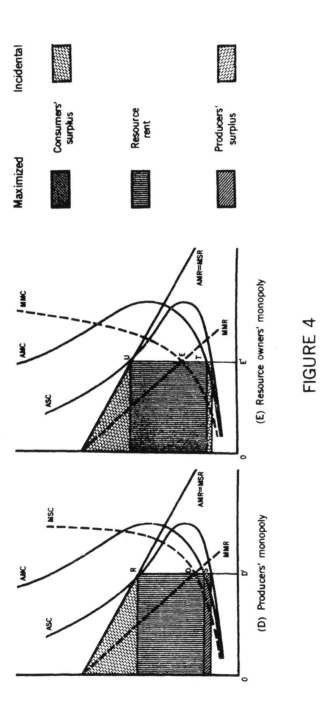

FIGURE 4

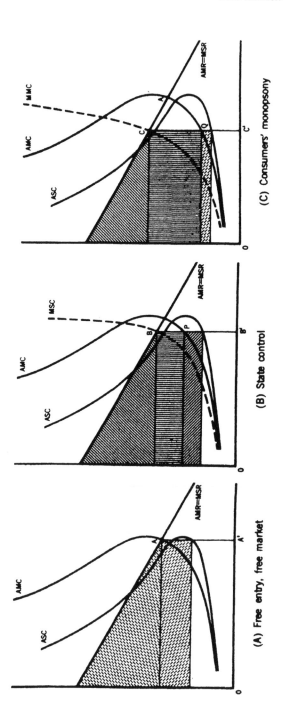

(A) Free entry, free market

(B) State control

(C) Consumers' monopsony

Figure 3 may now be used to identify the different equilibrium positions that relate to control over the fishery resource by various interest groups. Generally, three distinctly different interest groups may be recognized. Consumers would benefit from a regime that would maximize consumers' surplus. Private owners of the fishery resource would want to maximize resource rent. Finally, the owners of other factors (labour and capital) would profit from the maximization of their factor rents in the form of producers' surplus. Maximizing equilibrium positions are identified by the intersection of appropriate marginal revenue and marginal cost schedules. Having identified two marginal revenue and two marginal cost schedules, the permutations permit four distinct equilibrium positions. Three of these represent, in turn, the primacy of each one of the three interest groups, while the fourth represents the 'optimum' position of social control in the joint interest. There is also a fifth equilibrium position that is non-maximizing. It results from the free interplay of market forces. Major characteristics of these five positions have been summarized in Table 1, while Figure 4 illustrates graphically the relationship of the three categories of social benefits in each equilibrium position. Further comment on the five equilibrium positions follows:

(A) The intersection of the market demand and supply curves (A) produces an equilibrium output OA' that will apply in an open entry fishery supplying a competitive market. This is the *laissez-faire* position that is approached in most common property fisheries. It results in the dissipation of rent through overfishing, but allows for incidental amounts of consumers' and producers' surplus. Price and output equilibria may be above or below the socially optimum levels, depending in part on whether the equilibrium is located on the forward or backward slope of the supply curve.

(B) In terms of the definitions and assumptions of this paper, the socially optimum equilibrium position is given by the intersection of the marginal social revenue and marginal social cost curves at point B in Figure 3. This position allows for the largest total net social benefit by maximizing the combined total of consumers' surplus, resource rent and producers' surplus. It requires centralized control (presumably by government) to limit output to OB' by holding fishing effort to the level determined by point P on the supply curve, as shown in Figure 4(B). The price charged to consumers will be BB', while options remain open for disposition of the rent.

(C)   If consumers by collective action would acquire monopsony
control over the market for the output of a fishery, they would
be in a position to capture the rent available from the resource.
They could then maximize the combined total of consumers'
surplus and rent to their own advantage.   In this category one
could think of a consumers' co-operative as a monopsonist purchasing the entire catch from a group of competing fishermen.
The optimum output for the monopsonist co-operative would be
determined by the intersection of the MSR and MMC curves,
shown at C in Figure 4(C).   The monopsonist concerned with
maximizing consumers' benefits would be guided by the marginal
social revenue schedule that relates to values enjoyed by consumers, rather than by the marginal *market* revenue schedule
that would relate to money receipts by producers.   Conversely,
its unconcern for the welfare of producers would cause it to be
guided by marginal *market* cost that relates to the cost of its
purchases, rather than by the marginal *social* cost that relates to
the opportunity costs of producers.   To appropriate the resource
rent the co-operative would have to use its monopsonist power to
hold its buying price down to the opportunity cost of marginal
fishing units as measured by QC'.   It would then capture a rent
of QC per unit of product.   An incidental amount of producers'
surplus would still accrue to intramarginal fishing units.   In
disposing of the rent, the co-operative would have to avoid paying
dividends to its members in proportion to the amount of fish
purchased by these members.   For this would effectively lower
the price of fish to members, causing them to increase purchases.
The culmination of this process would be the establishment of
equilibrium at A in the free market position where rent would be
dissipated.   The co-operative would thus lose the advantage of
its monopsonist position and the total benefits accruing to it and
its members would diminish.   If the co-operative were confronted in the market not by a number of competing fishermen,
but by a monopolist seller, the division of the rent between
monopsonist buyer and monopolist seller would be indeterminate.
Another example of a monopsonist buyer maximizing consumers'
surplus plus resource rent could be a government marketing
monopoly selling fish in a national market and contracting with
foreign operators to deliver the fish.   The government, presumably, would have an interest in capturing a maximum of
consumers' surplus and rent for its citizens, but would have no
interest in the foreign producers' surplus.

(D)   The case of the producers' monopoly is symmetrical to that of the consumers' monopsony.   A co-operative or marketing association of all operators in a fishery (or a single operator representing all inputs in a fishery), when facing a multiplicity of buyers, could appropriate both resource rent and producers' surplus.   The optimum position for the producers' monopoly as indicated in Figure 4(D), would be determined by the intersection of the MSC and MMR curves at D.   The monopolist's concern for producers' incomes would be guided by the marginal *social* cost schedule that is related to the opportunity costs of producers.   But on the sales side the marginal *market* revenue schedule would give the appropriate guidance, demonstrating unconcern for the *social* revenue of consumers.   The monopolist seller would want to set the price at RD' to appropriate a rent of RS per unit of product.   This would maximize the resource rent and producers' surplus accruing to the producers' monopoly, while leaving an incidental amount of consumers' surplus.

(E)   The final case to be considered is that of sole ownership of a fishery resource where ownership is divorced from both consumer interests and the interests of other producing factors, i.e., capital and labour.   One could think of a 'landowner' (water owner in this case) renting the right to exploit a fish pond or oyster bed.   The resource owner, in charging the maximum rent for which he can find takers, will cause an equilibrium output to be established at the level for which marginal *market* revenue and cost coincide.   This is shown at point E in Figure 4(E).   Incidental amounts of consumers' surplus and producers' surplus will accrue.   Marginal operators will find that they can just cover their opportunity costs of TE' per unit of output by charging the market price of UE' and meeting the rent payment of UT per unit of output.

There are some common features of the three cases of private sole ownership or control of the resource—i.e., cases (C), (D) and (E)— when contrasted with case (B), which represents social control in the general interest.   The private sole ownership arrangements all result in higher prices and lower outputs.   They represent cases of 'underfishing'.   Indeed, a restriction of output is what one would expect in monopoly situations.   In each of these cases only one or two of the three categories of social benefit are maximized, so that the total net social benefit achieved is less than that of the full social control case.

Socially the least profitable is the case of the private resource owners' monopoly which maximizes only one category of social benefit, viz., resource rent. Of the four cases it provides for the lowest output and the highest price.

It is clear that the equilibrium position in the competitive market with free entry to the fishery will be non-optimal. Resource rent will be dissipated. But there is nothing in the analysis to suggest that the incidental accretions of consumers' and producers' surplus in this case could not be in excess of the total net social benefits generated under any of the private sole ownership arrangements. *A priori* there is then no reason to advocate as a matter of policy the granting of private sole ownership rights in lieu of free entry to the fishery. However, in the case where a fishery can be divided into a large number of small fishing grounds that are not interacting biologically (perhaps in an oyster fishery), separate sole ownership rights might be given in respect of individual grounds. Owner-operators would have an interest in maximizing on each ground the producers' surplus plus rent. If they were then competing with the sale of their output in a perfectly competitive market, they would not detract from consumers' surplus. Under these circumstances fishermen would resemble wheat farmers, maximizing returns from their individual holdings while selling in a perfectly competitive market. In that case sole ownership would be compatible with the social optimum and would thus be superior to free entry.

V

The analysis of this article has focused on the significance of consumers' and producers' surpluses in determining the socially optimum level of fisheries exploitation. There are two points that bear particular note as they run counter to past emphasis in the literature : (1) Rents yielded by fishing activity include rents to intramarginal inputs of labour and capital, which are not dissipated by unlimited entry to the fishery, and which could be of equal or greater importance than resource rents. (2) Management of a fishery by a regime of private sole ownership is generally non-optimal from a social standpoint and is not inherently superior to unlimited entry.

There is nothing in this paper to suggest that the social optimum does not require some restraint on entry to the fishery. But its major conclusions tend to diminish the weight of importance that has been attached to such entry limitation. They also suggest that limitation through private sole ownership arrangements, which is often advocated, in the general case would overshoot the mark and result in underfishing. The analysis then gives reason to concede some ground to those biologists and government administrators who are sceptical of economic theory and are intuitively drawn to the biological criterion of the maximum sustainable (physical) yield. It is clear that the fishing effort required for a social optimum that maximizes consumers' and producers' surpluses along with resource rent, will come closer to the fishing effort of free entry and of maximum sustainable yield, than will the more strictly limited effort that results from maximization of resource rent only.

The foregoing analysis has some practical implications for resource management policy. In the industrially advanced nations, generally, a serious degree of overfishing tends to take place. Improved technical efficiency coupled with low mobility of labour and capital out of the industry has increased pressure on the fishery resource while it has tended to depress factor incomes. This in turn has pressured governments into subsidizing the fishery. However, an increasing awareness of the nature of the problem has caused some governments to start imposing entry limitations on the fishery—at least in cases where a fishery comes under a single national jurisdiction or under an international treaty arrangement. In deciding how far to proceed with entry limitation it behoves governments not only to pursue an increase in resource rent, of which the capacity to raise license fees would be a measure. It is also necessary to watch for the retention of producers' surplus, particularly where the low incomes of producers are so central to government concern. The greater the level of fishing effort, of course, the larger will be the number of intramarginal fishing units that may earn a producers' surplus (though the total amount of producers' surplus will not necessarily increase). From the analysis it should be clear then that the optimal entrance limitation is less stringent than what the single-minded pursuit of resource rent capture would require.

The execution of a resource management policy along the lines suggested above naturally will encounter problems. The enforcement of entry limitation where freedom of entry has traditionally prevailed will likely entail political and social difficulties in phasing out surplus labour and capital. In determining the appropriate level of limitation there is also the problem of calculating the marginal social cost schedule—or the average social cost schedule from which it may be derived.   But the calculation of a reasonably realistic approximation of such a schedule should not be beyond the wit of economists and statisticians, when they do not blanch at carrying out cost-benefit analyses that require them to estimate the intangible values of re-creation or the multifarious externalities of pollution.   (These latter matters could indeed intrude into the social cost calculations of some fisheries).   The critical ingredients of social cost calculation in respect of a fishery would seem to consist of an estimation of the value of alternative employment opportunities for factors engaged in the fishery and a measurement of the varying efficiency of different fishing units.   These tasks should not be beyond the survey capa-bility of the sophisticated research units of present day government agencies.

PARZIVAL COPES.

Simon Fraser University.

NOTES

* This article is a by-product of a research project regarding the Newfoundland fishing industry supported by the Canada Council.

† Manuscript received 13/7/71.

1. H. Scott Gordon, 'The economic theory of a common property resource: the fishery,' *Journal of Political Economy* Vol. 62, April 1954, p. 130.

2. An early important contribution is Anthony Scott, 'The fishery: the objectives of sole ownership,' *Journal of Political Economy* Vol. 63, April 1955, pp. 116–24.

3. Ralph Turvey, 'Optimization and suboptimization in fishery regulation,' *American Economic Review* Vol. 54, pp. 64–76. Parzival Copes, 'The backward-bending supply curve of the fishing industry,' *Scottish Journal of Political Economy* Vol. 17, February 1970, pp. 69–77.

4. Copes, *op. cit.* p. 71.

5. The MC curve drawn matches only the forward sloping portion of the supply curve. One could devise a marginal curve matching the backward sloping portion, which would measure increases in total cost with each *reduction* in output. However, the MC curve is needed only to assist in determining the optimum level of output. As is explained below, no optimum output position will be associated with any point on the backward slope of the supply curve, so that no matching MC curve need be considered.

6. The consumers' surplus may be shown graphically as (a) the area above the level of market price and below the *marginal* social revenue curve, or (b) the rectangular area inscribed between the level of market price and the *average* social revenue curve. While the latter method would have provided for symmetry in presentation (rectangular areas inscribed between average curves are used for resource rent and producers' surplus), it would have required the drawing of an additional curve (average social revenue), adding to the complexity of the graphs. Therefore the former method of presentation has been used for consumers' surplus.

7. It is assumed throughout this article that there is no market discrimination, so that all units of output are sold at the same price. The 'market revenue' therefore will always be given by the product of output and the uniform price established at the relevant point on the demand curve.

8. *Ibid.* p. 76. In a similar situation in that article the rent generated was inaccurately identified as "maximum". It is, of course, rent plus consumers' surplus that is maximized.

9. *Cf.* Scott, *op. cit.* p. 117, who states: "... I wish to show that *long-run* considerations of efficiency suggest that sole ownership is a much superior regime to competition ...."

10. *Op. cit.* p. 71.

11. Various definitions of producer's surplus are possible and various components may be specified. *Cf.* E. J. Mishan, 'What is producer's surplus?,' *American Economic Review,* Vol. 58, December 1968, pp. 1269–83. Mishan proposes that reference to producers' surplus be dropped and that various forms of rent be defined instead. In this article producers' surplus is considered synonymous with the rent enjoyed by intramarginal units of capital and labour to the extent that their opportunity costs fall below those of marginal units of these factors.

12. *Op. cit.* p. 132.

13. It may be noted that the MSC curve will lie below the MMC curve, because increases in social cost will be less than increases in market cost to the extent that additional intramarginal rents accumulate. The MSC curve will be above the AMC curve. The latter measures at each output only the opportunity costs per unit of output for marginal fishing units. The former measures (per unit of output) these opportunity costs of marginal fishing units plus the increase in total opportunity costs related to the decrease in output per unit of fishing effort attendant on the increase in total fishing effort.

[8]

1972]

Externalities, Factor Proportions and the Level of Exploitation of Free Access Resources[1]

By J. R. GOULD

Although legal definitions of property rights are often complex and subtle, for present purposes some rough distinctions will suffice. A "private-property resource" is to be understood as a resource from which a single firm can exclude all others, and a "common-property resource" as one to which more than one firm has right of access. Within the category of common-property resources, two cases can be distinguished according as to whether access is limited to a few firms, or whether all have right of free access. The reason for this second distinction is that limited access raises the possibility of co-operation between the firms,[2] while free access inhibits co-operation for much the same reason that free entry inhibits the formation of cartels—namely that the gains from co-operation would be eroded by new entrants.

Familiar instances approximating to free-access resources occur in fishing, grazing, hunting, road transport and so on. At one time, such cases were regarded as giving rise to only minor imperfections in the market mechanism and as the province of the specialist; now, concern with environmental problems such as air and water pollution has stimulated a more general interest.

It is widely-accepted doctrine that free-access resources are over-exploited—that is, that they attract more co-operating resources than is required for allocative efficiency. Theoretical foundations for this belief can be found in the "cost controversy" of the 1920s,[3] particularly in the contributions of Pigou and Knight. Pigou [10], for the purpose of establishing the proposition that increasing-cost industries tend to expand beyond the socially optimal output, analysed an example of alternative roads between two points, one poorly surfaced but broad and subject to constant costs, the other well surfaced but narrow and subject to increasing costs from traffic congestion. Traffic would distribute itself between the roads so that in equilibrium the average cost of travelling by either road would be the same. Since Pareto-efficiency requires equality of the marginal costs, the road subject to increasing

[1] I am indebted to F. H. Hahn, H. G. Johnson, E. A. Kuska, E. J. Mishan, M. Perlman and J. D. Sargan for valuable comments on a previous draft of this paper.
[2] Although formulated in terms of consumers, Buchanan's theory of clubs [2] would appear to be relevant to this case. References in square brackets are listed on pp. 401–2, below.
[3] Some important contributions to this debate are reprinted in Part II of [13].

costs would be over-utilized; and Pigou recommended a tax on vehicles using it to achieve an efficient allocation of traffic.

Knight's critique [6] argued that Pigou's result hinged on free access to a resource and, while correct for the case analysed, should not be generalized to all increasing-cost industries. However, it is not this aspect of the debate—later further clarified by Viner's [15] distinction between technological and pecuniary externalities—which is to be the focus of attention here: the objectives of this paper are to examine and to develop further Knight's model of a free-access resource.

Several features of Knight's formal model are of great interest, and invite elaboration and extension. Whereas Pigou had worked with average and marginal costs, Knight reformulated the analysis in terms of average and marginal products. Although the latter approach when limited to a single variable factor—as in Knight's discussion—is equivalent to the former, it is potentially more fruitful because it can be extended to the analysis of factor proportions. Second, Knight's analysis is based on an industry aggregate production function and an industry equilibrium condition: there is virtually no discussion of the individual firm and thus no assurance that the analysis does not entail absurd or arbitrary behaviour by the firm. Third, most discussions of related problems begin by postulating various kinds of direct inter-action, and go on to deduce their consequences; in contrast, Knight does not postulate externalities—indeed he reaches his results without the aid of this concept.[1] It is, however, an interesting feature of Knight's approach that the existence of externalities can be predicted from the model and certain restrictions can be placed upon them.

In Section I, I discuss Knight's analysis of the single variable-factor case, and elaborate it by making explicit some implicit assumptions about production functions, by clarifying the role of externalities in the model, and by suggesting an appropriate theory of the firm. The conclusions are that, although Knight's implicit assumptions are somewhat restrictive, the elaborated model is logically consistent, easy to manipulate, and confirms Knight's result that free-access resources are over-exploited. In Section II the model is extended to the case of two variable factors. The main analytical difference from the previous case is that now the problem of efficient factor proportions arises; and the most important result—a consequence of inefficient factor proportions—is that the over-exploitation theorem no longer holds.[2] Free-access resources may be over- or under-exploited, and accordingly, increasing-cost industries may be over- or under-expanded.

---

[1] Mohring and Boyd [9] have recently contrasted the "direct interaction" approach, which stems from Pigou, with the "asset utilization" approach, which stems from Knight. They argue that the latter is the more fruitful. Other compari-sons of Knight and Pigou can be found in Mishan [7] and Cheung [4].

[2] Mishan ([8], p. 8) and Goetz and Buchanan ([5], pp. 888–9) have noted that external diseconomies may result in inefficient factor combinations and that the general equilibrium combination of outputs may lie inside society's production possibility frontier. They point out that possibly less than the Pareto-optimal

The analysis of free-access resources can pose some awkward problems in specifying the quantity of the resource used by the firms operating it. To simplify the exposition, these questions are avoided in Sections I and II by assuming that, in some sense, all of the fixed quantity of the resource is used, and that there are diminishing returns to the variable factors. Discussion of the possibility of increasing returns to the variable factors is left to Section III.

Throughout the paper, I abstract from transactions costs, the costs of administering corrective taxes, the costs of maintaining property rights, imperfections of competition, and problems of second best.

I. Model with One Variable Factor

Pigou's two-road example, while adequate for the purpose of illustrating that free-access resources are inefficiently exploited, is an unnecessarily cumbersome device if the objective is also to derive the supply curve of the industry. It will be convenient to specify an alternative model which contains all that is essential to Knight's argument.

Thus, it is assumed that a product X is produced by a variable factor L, in perfectly elastic supply at price W, combined with a free-access resource M, specific to the production of X and in fixed supply; that employment of L is subject to diminishing returns; that there is perfect competition with P, the price of X, regarded as a parameter by all firms; and that the rest of the economy is free from distortions. In these circumstances, a requirement for a Pareto optimum is that the price of the variable factor equals the value of its marginal product.

Knight's analysis

A version of Knight's argument appropriate to the simplified model can be related to Figure 1. The curves X/L and X_L are respectively the average and marginal products of L, and represent an aggregate production function. If access to the resource is free, competitive producers will be attracted so long as receipts exceed factor payments, and *vice versa*. In equilibrium, factor payments must equal receipts— that is, $WL = PX$ or $W/P = X/L$. Thus firms would employ in aggregate L_2—which is greater than the socially efficient quantity L_1, where $W/P = X_L$.

Knight's crucial point is that such inefficiency would not arise if M were privately owned. For, Knight argued, the private owner could

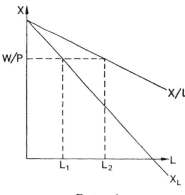

FIGURE 1

either hire L and exploit the resource himself, or make a charge to competitive firms for right of access. In the first case, profit maximization requires that $W/P = X_L$, and the private owner would hire L_1. In the second case, he could charge firms $P(X/L - X_L)$ per unit of L employed, or alternatively $P(X/L - X_L)L/X$ per unit of output, both charges being evaluated at L_1. These charges—the second of which is the same as the tax recommended by Pigou—by raising the effective price of L, or lowering the effective price of X, induce an equilibrium output of L_1. Which of these methods is employed is a matter of indifference to the private owner and to society; in each case he can be said to be maximizing the rent attributable to M, and in each case the result is Pareto-optimal.

This argument served to refute the generalization which Pigou had drawn from his analysis of road congestion. For while Knight agreed that exploitation of free-access resources is carried beyond the optimal point, he showed that it was a mistake to generalize to all cases where expansion of the industry forces up the costs of the constituent firms. He argued that where, as is usually the case, all relevant resources are privately owned and competitive firms bid up factor prices as industry output expands, the increased costs are rents rather than social costs and the effect is precisely that required for efficient resource allocation.

Although his argument is decisive as a criticism of Pigou, there are nevertheless several points at which Knight's model of a free-access resource is obscure, largely because he conducted his analysis in terms of an aggregate production function and suppressed the role of the individual firm. The rest of this section is concerned with clarification of these points.

Externalities

In contrast to Pigou, Knight made no explicit use of the concept of externalities in arriving at his conclusions. As a result, his discussion can convey the impression that the inefficient exploitation of a free-access resource results simply because a scarce and valuable resource is treated as a free good, and that externalities play no essential role in explaining the inefficiency.[1]

However, this impression is mistaken, for externalities are implicit in Knight's analysis. Knight's argument hinges on two propositions:

(a) that the exploitation of the resource is carried into the region of diminishing average product,

(1) $X/L > X_L$;

(b) that for the industry to be in equilibrium profits must equal zero, or

(2) $W/P = X/L$.

Although Knight makes no use of the equilibrium condition for the profit-maximizing firm, it is clear that each firm must equate W with the private value of the marginal product of L, or, writing the private marginal product as X_L',

(3) $W/P = X_L'$.

For an equilibrium to exist in Knight's model, it follows from (1), (2) and (3) that $X_L' > X_L$. Thus, whereas Pigou postulated external economies to arrive at his result, the discrepancy between social and private marginal products by which external diseconomies in production are defined can be deduced from Knight's formulation supplemented by the equilibrium condition for the firm. Looked at from this point of view, the many reported instances of external diseconomies in the operation of free-access resources can be regarded as empirical confirmation of the model.

Once attention is drawn to interdependence of production functions external to the price mechanism as a central feature of the exploitation of free-access resources, it is natural to ask whether external benefits, as well as costs, might be observed. There is a good *a priori* reason to believe that they will, for it is often difficult for an individual firm to appropriate all the benefits of the marginal contribution to output of some factors. For example, if a grazier spreads artificial fertilizer on common pasture, or an oil producer pumps gas into the pool to increase the flow, or a fisherman feeds the fish in a lake, part of the benefits in each case will accrue to the other exploiters of the resource.

---

[1] Cheung [4] appears to argue strenuously for the irrelevance of externalities and for the importance of effective contractual arrangements. However, it seems to me that his attack on the concept of externality is essentially terminological. His own analysis demonstrates that imperfections in contracts produce discrepancies between private and social marginal products. Current usage is to call such discrepancies "externalities", and there seems little point in abolishing or changing this label. Emphasis on this terminological point distracts attention from Cheung's more important point that economists have tended to neglect the examination of the causes of externalities—in contrast to their consequences—and that this task can usefully be approached by the analysis of contracts.

It might seem at first sight that Knight's model rules out such external benefits since, as we have seen, the logic of diminishing returns and the equilibrium conditions requires that $X_L' > X_L$. However, this is not the case. It is quite possible for a factor to generate both external costs and benefits; by way of illustration, a fishing boat not only reduces the fish population available to others but also may reduce their search costs when it sights a shoal; and cattle both consume and fertilize common pasture. The magnitude and sign of the difference between social and private products is the net effect of such external benefits and costs. If external costs outweigh external benefits in equilibrium, the condition $X_L' > X_L$ will be met.

In short, the restrictions placed on externalities are that the use of L must impose external costs, may confer external benefits, and, in equilibrium, external costs must dominate external benefits so that $X_L' > X_L$. As we shall see, the result that the variable factor must impose net external *diseconomies* in equilibrium is a peculiarity of the single variable-factor model, and does not hold for all factors when several are variable.

The role of externalities deserves emphasis because the distinction between free resources and free-access resources is not always carefully drawn.[1] Free-access resources are a special case of resources which have zero market prices but for Pareto-optimality have positive shadow prices. The essential distinction is between resources which are free but exclusive to the user, and those which are free but whose use is non-exclusive. If steel ingots were supplied to firms without charge, no interdependence of production functions need be involved: exploitation of a lake with free access generates externalities—and as a corollary which becomes clear in Section II, inefficiency will generally not be cured simply by making a charge for access.

Production functions

Knight used the same average and marginal product curves to represent, on the one hand, the aggregate output of a large number of competitive firms, and on the other hand, the output of a single private owner exploiting the resource himself. This procedure entails the implicit assumption that the relation between the aggregate output of X and the aggregate input of L is independent of the structure of the industry, that is, of the number and sizes of the individual firms exploiting the resource.

The assumption can be rationalized by supposing that there are no differences in entrepreneurial efficiency, nor any entrepreneurial economies or diseconomies of scale, nor other considerations, which would imply that one firm, or size of firm, was more efficient than another. Each firm's technology can be specified by the same relation $X = X(L, M)$, which is homogeneous of degree one when both L and M are variable and the firm exploits M as private property.

---

[1] Even by writers such as Mishan ([8], p. 3) and Winch ([16], pp. 102–3).

For a single private owner of the resource, or an industry with free access, the production function would then be $X = X(L, \bar{M})$, with diminishing returns to L wholly explained by the fixed quantity of M. Turning to one of the many firms exploiting a free-access resource, we must distinguish between production functions as technological relations and as behavioural relations. In accordance with the assumption that the aggregate production function is independent of industry structure, we assume that the industry employs the same technology as would a single private owner. However, the production function on which the

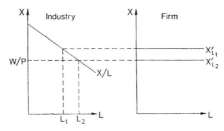

FIGURE 2

firm bases its decisions involves a behavioural hypothesis in that it describes how the firm sees the relation between its own inputs of L and outputs of X, bearing in mind that it is but one of many firms exploiting the same resource.

The behavioural hypothesis must be consistent with the technological assumptions above, and, preferably, should conform with traditional theory.[1] In standard competitive theory, the firm takes the price of the product as a parameter determined by market demand and supply on which its own influence is negligible. Analogously, it can be assumed that each of a large number of firms exploiting a free-access resource takes the average product of L as a parameter determined by the aggregate input of all firms. If there are no indivisibilities or entrepreneurial economies or diseconomies of scale, the firm will regard its private average product as independent of its own employment of L, assuming this to be negligible in relation to total employment. Thus, in Figure 2, when the industry is in total applying L_1, each firm sees its private aggregate and marginal product curves as the horizontal line X'_{L_1}.[2]

[1] There is more than one way of doing this. Compare Cheung ([4], section III) who assumes that the firm takes as given the quantity of L applied by other firms. Referring to Figure 1, suppose that initially the industry is applying L_1 units. Cheung assumes that a new entrant would see his private average product curve as the segment of the X/L curve to the right of L_1 read with reference to axes with origin at L_1. This hypothesis and his subsequent analysis are, as Cheung remarks, analogous to Cournot's classic treatment of oligopoly with freedom of entry. In equilibrium, the number of firms tends to infinity, the size of the firm tends to zero, and private marginal products tend to industry average product.

[2] The hypothesis that firms in a competitive industry see their private production functions as of constant returns to scale and subject to parametric shifts brought about by changes in industry aggregate employment is used by Chipman [3].

390   ECONOMICA   [NOVEMBER

Equilibrium of the firm

The firm will expand or contract output according as to whether X'_L is greater or less than W/P. Equilibrium requires both profit maximization and zero profits, and is consistent with aggregate employment of L_2. The industry equilibrium is stable, notwithstanding that each firm is in neutral equilibrium and of indeterminate size: an increase in industry output would reduce private average and marginal products below W/P (and *vice versa*), which tends to correct the perturbation.

Industry supply curve

The industry supply curve is readily derived from Figure 1. As P rises with constant W, W/P falls and the employment of L rises. As long

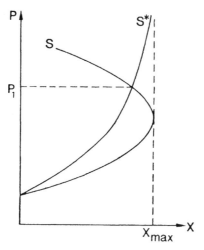

FIGURE 3

as the marginal product of L is positive, industry output rises with price. If the marginal product of L becomes negative after a point, further increases in P increase the employment of L but reduce the output of the industry. The shape of the supply curve of the industry exploiting a free-access resource is illustrated in Figure 3 by the curve S.

The curve S^*, the locus of Pareto-optimal outputs related to P, is derived from Figure 1 by equating W/P with X_L, and approaches X_{max} asymptotically as P increases. It is also to be interpreted as the supply curve if the resource is under private ownership.

At prices below P_1, the intersection of S and S^*, the free-access resource will produce more than, and at prices above P_1 less than, the amount of X which is Pareto-optimal. However, as is clear from Figure 1, at any price the resource is over-exploited in the sense that too much L

Summary

In this section I have argued that Knight's analysis implies externalities, and that it can be elaborated, by explicit assumptions about the production functions and an analysis of the individual firm, to form a logically consistent model. For purposes of contrast with the following section, the main results from the one variable-factor model can be summarized as four propositions, none of which holds when the model is extended to more than one variable factor: (i) if exploitation is in the region of diminishing returns, the private marginal product of the variable factor must exceed its social marginal product in equilibrium; (ii) the resource is over-exploited in the sense that more of the variable factor is employed than is Pareto-optimal; (iii) if the social marginal product of the variable factor is positive in equilibrium, the output of X is greater than the Pareto-optimal output; (iv) the Pareto-optimal output can be induced by taxes on either input or output.

II. MODEL WITH TWO VARIABLE FACTORS

In this section, it is assumed that two variable factors K and L, in perfectly elastic supply at prices R and W respectively, are applied to the resource M. Otherwise, the model is similar to that used in the preceding section; in particular, it is assumed that output of X depends only on the aggregate quantities of K and L employed and is independent of the number or sizes of the firms. The aggregate production function, now written $X = (K, L, M)$, is subject to constant returns when all three factors are variable, and to diminishing returns to scale variations of K and L when M is fixed.

Externalities

The equilibrium condition of zero profit is no longer equilvalent to equality of the value of the average product and the price of the factor; for if this equality held for any one factor, total payments to that factor would exhaust the value of the total output. In equilibrium, the values of the average products must exceed the respective factor prices —or $R/P < X/K$ and $W/P < X/L$—if, as will be assumed, neither the quantities nor the prices of K or L are equal to zero.

The result that $X_L' > X_L$, that net external diseconomies must be imposed on other firms, can no longer be deduced as it was in Section I. Corresponding to conditions (1), (2) and (3), we have in the case of two variable factors:

(1A) $X/L > X_L$

(2A) $W/P < X/L$

(3A) $W/P = X_L'$.

these relations, and a similar argument holds for X_K and X'_K. It would appear possible that in equilibrium individual factors could generate either net external economies or net external diseconomies or neither.

However, given that the underlying production function $X = X(K,L,M)$ is subject to constant returns to scale, it can be shown that externalities must be generated by an industry with free access to the resource, and that while net external economies are possible, net external diseconomies must necessarily be displayed by at least one factor.

From Euler's theorem

(4) $\qquad KX_K + LX_L < X;$

the zero-profit equilibrium condition can be written

(5) $\qquad X = KR/P + LW/P;$

and profit maximization requires

(6) $\qquad R/P = X'_K$ and $W/P = X'_L.$

From (4), (5) and (6),

(7) $\qquad K(X_K - X'_K) + L(X_L - X'_L) < 0.$

The propositions stated above follow from (7). First, $(X_K - X'_K)$ and $(X_L - X'_L)$ cannot both equal zero if two factors are employed: exploitation of the free-access resource must entail some externalities in equilibrium. Second, it is possible for either $(X_K - K'_K)$ or $(X_L - X'_L)$ to be greater than zero: one factor may generate net external economies. Third, $(X_K - X'_K)$ and $(X_L - X'_L)$ can both be less than zero but they cannot both be greater than zero: both factors can generate net external diseconomies but both cannot generate net external economies.

In short, net external economies may occur, but net external diseconomies must occur. It is worth noting that this argument and result are generalizations of those in the single variable-factor model, and they extend in a straightforward manner to any number of variable factors.

Production functions

In accordance with the assumption that the aggregate production function $X = X(K, L, M)$ is independent of the structure of the industry, it is assumed that all firms, and all sizes of firms, are equally efficient.

The specification of the firm's private production function is not quite as straightforward as in the case of the single variable factor. It cannot be assumed that the firm takes the average product of its factors as equal to the industry average products and independent of the quantities of its own inputs, for this would lead to the absurd consequence that the private average product of each factor was independent of private factor proportions. The appropriate analogous assumption for the case of two variable factors is that the firm believes that if it employs factors in the same proportions as the industry, its private average products will equal the average products for the industry.

This assumption conveys nothing about the private production functions when the firm uses factors in proportions different from the industry average. However, all that need be said for the purposes of the following argument and for consistency with the rest of the model is: that the private production function is subject to constant returns to scale; that, because M is fixed, the firm's output depends inversely on the aggregate inputs of K and L; and that the firm regards these aggregate inputs as parameters on which its own variations in inputs have negligible effect.

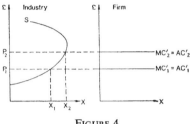

FIGURE 4

Equilibrium of the firm

The private production function and the factor prices determine an expansion path for the firm which generates a horizontal private average and marginal cost curve, the height of which depends on the industry's aggregate inputs of K and L.

For given aggregate K and L, the firm expands or contracts output according as to whether P is greater or less than its private MC. If $MC' < P$, all firms try to expand output, aggregate employment of K and L increases, and each firm finds its average and marginal cost curve rising. When private AC and MC have risen to equal P, the industry is in stable equilibrium, although the individual firms are in neutral equilibrium and of indeterminate size.

The relation between firm and industry equilibrium is illustrated in Figure 4, where S is the industry supply curve—the locus of industry outputs at which $P = AC' = MC'$. S bends backwards in the region of negative social, but positive private, returns to scale. It should be noted that in contrast to the single variable-factor case, S has not been derived directly from the aggregate production function.

Factor proportions

In general, the aggregate combination of factors employed on a free-access resource will not be that which minimizes the cost of any given output. For minimum aggregate costs, the social marginal rate of factor

394   ECONOMICA   [NOVEMBER

substitution X_K/X_L must equal the ratio of the factor prices R/W. But each firm will equate R/W with the private marginal rate of factor substitution X_K'/X_L'. Since social and private marginal products must differ for at least one factor, except in the special case where the ratio of social to private marginal products is the same for both factors, social and private rates of factor substitution will not be equal.

The point is illustrated in Figure 5. The isoquant X_1 shows aggregate combinations of K and L which produce a fixed level of output. The combination of factors which minimizes aggregate costs is C_1, where the isoquant is tangent to the isocost line AB. Suppose that $X_K'/X_L' < X_K/X_L$. Then, if firms were in aggregate employing C_1, it would seem

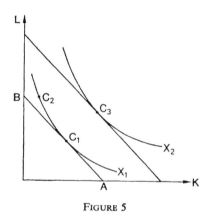

FIGURE 5

to each firm that its private costs were not minimized because $R/W = X_K/X_L > X_K'/X_L'$. Each firm would try to reduce its private costs by employing a higher proportion of L to K. Conversely, individual firms can only be in equilibrium producing an aggregate output of X_1 if $R/W = X_K'/X_L' < X_K/X_L$—that is, at a point on X_1 to the north-west of C_1 such as C_2, and at an aggregate cost greater than that incurred for the combination C_1.

The industry supply curve

Given the aggregate production function and factor prices, the minimum social costs of any output can be determined. In Figure 6 the curve AC^* shows the minimum average social cost for given outputs, and MC^*, marginal to AC^*, approaches the maximum feasible output X_{max}^* asymptotically as P increases. The curve MC^* is the locus of Pareto-optimal outputs, and it is also the supply curve S^* if the resource is owned privately by a firm which takes P as a parameter.

The supply curve of an industry with free access to the resource, along which price equals average cost, is represented by S. Except in the special case that $X_K'/X_L' = X_K/X_L$, it follows from the argument of the

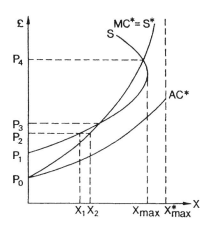

FIGURE 6

preceding section that S will lie everywhere above AC^*. X_{max} is shown smaller than X^*_{max} because inefficient factor combinations prevent the industry from attaining the maximum feasible output.

S, as it is illustrated in Figure 6, intersects S^* twice—output is greater than the Pareto-optimal output that a private owner would produce for prices in the range P_3 to P_4; otherwise, except below P_0, equilibrium output would be less than the Pareto-optimal output.

In the single variable-factor case, the possibility that an industry with free access would produce less than the Pareto-optimal output stems from the possibility of operation in the region of negative marginal social products. By contrast, with two variable factors, notwithstanding that both marginal social products are positive, output can be less than Pareto-optimal because of inefficient factor combinations.

Factor quantities

The comparison between equilibrium and socially optimal factor quantities can be analysed in terms of a "substitution" effect and an "output" effect. The method is illustrated in Figures 5 and 6. Suppose that the price of X is P_2, so that industry equilibrium output X_1 is less than the Pareto-optimal X_2, and that $X'_K/X'_L < X_K/X_L$. In Figure 5 the efficient factor combination for the Pareto-optimal output X_2 is indicated by the point C_3; the factor combination actually employed by the industry to produce X_1 is indicated by C_2; the point C_1 indicates the minimum cost combination for producing the industry output X_1.

The differences between the Pareto-optimal combination at C_3 and the equilibrium combination at C_2 can be split into two parts: the output effect, the difference between C_3 and C_1, which is the difference between the minimum cost combinations for the Pareto-optimal and the equilibrium outputs; and the substitution effect, the difference between

C_1 and C_2, that is, the difference between the minimum cost and the actual combinations for producing the equilibrium output. If the expansion paths are upward sloping, the output effect from C_3 to C_1 reduces the quantities of both K and L to below the Pareto-optimal levels. Given that $X'_K/X'_L < X_K/X_L$, there will be a substitution effect against K and in favour of L. Combining these two effects, the employment of K must fall below the Pareto-optimal level; but the employment of L may rise (as indicated in Figure 5) or fall, depending on the relative magnitudes of the output and substitution effects. By a similar argument, if the price of X were between P_3 and P_4 and output greater than Pareto-optimal, the quantity of L would be greater than, while the quantity of K could be greater or less than, the Pareto-optimal quantities.

It is sometimes said that the employment of factors which confer external benefits is too small (see for example [4], p. 53), and of factors which impose external costs is too large (see for example [5], p. 889). These assertions would appear to be based on the substitution effect and to neglect the output effect; and the contraries are quite possible.

The analysis of factor quantities shows that the proposition that free-access resources are over-exploited lacks a sound theoretical basis. When two variable factors are employed, output may be greater or smaller than is Pareto-optimal; and, compared with the Pareto-optimal quantities, more of both factors or less of both factors, or more of one and less of the other, may be employed.

Pigovian taxes

In contrast to the model with one variable factor, a tax on output will not serve to correct for externalities when there are two factors (unless $X'_K/X'_L = X_K/X_L$). The reason is, of course, that taxes must correct not only output but also factor proportions.

The corrective tax to be applied to each factor is the value of the difference between private and social marginal products at the Pareto-optimal factor combination, that is, $P(X'_K - X_K)$ per unit of K and $P(X'_L - X_L)$ per unit of L. For a factor which confers net external benefits, this expression will be negative and a subsidy must be paid—the government must make up the privately unappropriated part of the marginal social product.

An apparently paradoxical case can occur when a Pigovian tax on both factors will increase industry equilibrium output even though both social marginal products are positive. Take the case illustrated in Figures 5 and 6, where at price P_2 equilibrium output is less than the Pareto-optimal output, and suppose that $X'_K > X_K$ and $X'_L > X_L$. The positive Pigovian taxes have two effects: by improving factor proportions, they lower the net-of-taxes average cost at the optimal output X_2 to that indicated by AC^*; and they raise the gross-of-taxes average cost at X_2 to equal P_2, so that the zero-profit equilibrium condition is satisfied.

It is worth noting that a private owner wishing to exploit the resource by charging competitive firms for right of access, rather than by employing factors directly, would have to adopt a pricing policy similar to the system of Pigovian taxes. In order to maximize the rent of the resource, he would have to induce the competitive firms to employ combinations of factors which minimize *aggregate* cost of any given output, as well as induce them to produce the Pareto-optimal output. This would require him to pay bounties on some factors and make charges for others, equal to the Pigovian subsidies and taxes. Of course, other arrangements are possible—for example, the private owner of grazing land could fertilize the land himself, and make a charge for each head of cattle grazed on the land.

Multiplying inequality (7) by $-P$ gives $KP(X_K' - X_K) + LP(X_L' - X_L) > 0$. This can be interpreted as saying that, if both corrective taxes and subsidies are necessary, the total amount collected in taxes will exceed that paid out in subsidies. The net revenue equals the maximum rent that a private owner could extract from the resource.

III. INCREASING RETURNS

Thus far it has been assumed that average products decline for all quantities of the variable factors. This assumption, while quite possibly valid in some cases, seems unduly strong in others—for example, traffic must build up to some appreciable volume before road congestion occurs.

The possibility that variable factors display a range of non-decreasing average products raises problems of the degree of utilization of the resource which have been avoided until now. Some of the issues can be illustrated by contrasting Knight's treatment of the single variable-factor case with that of Worcester. In Knight's paper, the average product curve is drawn with a horizontal section before diminishing returns set in, as illustrated by ACD in Figure 7. Worcester [17] draws a curve such as BCD, which displays an initial range of increasing average product.

It is well known that, given the production function $X = X(L, M)$ is subject to constant returns to scale, an increasing average product of L implies a negative marginal product of M. As a matter of definition, we can say that the resource M is used to capacity if, at a given quantity of L, $X_M \geq 0$ when evaluated at $\bar{M}$, the fixed quantity of the resource; if otherwise, there is excess capacity.[1]

The difference between the views of Knight and Worcester can be explained in terms of different implicit assumptions about the divisibility of the resource when there is excess capacity. If M is divisible and $X_M < 0$, a private owner would reduce the amount co-operating with any

---

[1] This definition, when translated into terms of average costs, corresponds with the familiar Chamberlinian concept.

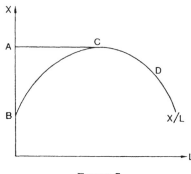

FIGURE 7

given quantity of L until $X_M=0$,[1] maintaining X/L at its maximum value—the average product curve would appear as ACD. On the other hand, if M is indivisible, the average product curve is constrained to follow the path BCD.

For an industry with free access, Worcester, implicitly assuming indivisibility, derives a U-shaped supply curve from his inverted U-shaped average product curve. Assuming that the industry faces a downward-sloping demand curve which cuts the supply curve from above,[2] a stable equilibrium is possible with falling supply price and increasing average product of L. In this range $X_L > X'_L$, use of L confers net external benefits—and a subsidy on L is called for.[3]

If access is free and M is divisible, the problem is less straightforward. It might seem at first sight that, as assumed by Knight, the average product curve for an industry with free access would coincide wth ACD, that of a private owner. However, this overlooks the point that if the whole of $\bar{M}$ need not be used, the industry has, in effect, to make a decision about the quantities of both L and M—a one variable-factor case when it is assumed that the whole of $\bar{M}$ is employed becomes a two variable-factor case when the quantity of M can be varied up to $\bar{M}$; and the possibility of inefficient factor combination arises.

Whether or not factors are combined inefficiently depends on whether

[1] In writing X_M as the partial derivative of $X = X(L, M)$, I am assuming that the degree of utilization of $\bar{M}$ is quantifiable in the same way that M is quantifiable, for example, that using half a lake is equivalent to using a lake of half the size. This assumption obviously begs some difficult questions.

[2] If the industry exploiting the resource is a price-taker, as assumed in the rest of this paper, equilibrium on the downward-sloping branch of the supply curve would be unstable and would be matched by a stable equilibrium on the upward-sloping branch.

[3] Worcester's prescription for correcting the misallocation is a subsidy on land (the resource) and no tax or subsidy on labour. Since land is fixed and (implicitly) indivisible, no tax or subsidy can affect the amount of land used. Worcester's analysis does not include a theory of the firm. Thus, from his correct statement that in equilibrium the (social) marginal product of labour is above the wage rate, he fails to draw the conclusions that it is also above the private marginal product of labour and that a subsidy on labour is necessary for efficiency.

social and private marginal rates of factor substitution differ, which in turn depends on whether externalities occur before the resource is used to capacity. Although, as we have seen, capacity utilization with decreasing average product of the variable factor is a sufficient condition for externalities when there is free access, it is not necessary. Given the contiguity of factors employed by different owners on a free-access resource, there seems no obvious *a priori* reason to rule out externalities in any region of the production function.

Some restrictions on externalities when there is excess capacity can be derived. Corresponding to the relations (4), (5) and (6) we have, from Euler's theorem

(4A) $L . X_L + M . X_M = X$;

for zero profits

(5A) $X = L . W/P$;

and for maximum profits

(6A) $X'_L = W/P$ and $X'_M = 0$.

From (4A), (5A) and (6A), $M . X_M = L(X'_L - X_L)$. Remembering that we are concerned with the case of excess capacity where $X_M \leqslant 0$, the restrictions on externalities can be put as two possibilities: either $X_M = 0$, in which case $X'_M = X_M$ and $X'_L = X_L$; or $X_M < 0$, in which case $X'_M > X_M$ and $X'_L < X_L$. In words, either there are no externalities in the use of M and L, or alternatively, use of M imposes net external diseconomies and use of L confers net external economies.

Private and social marginal rates of factor substitution necessarily differ if there are externalities with excess capacity; for since $X'_M > X_M$ and $X'_L < X_L$, it follows that $X'_M/X'_L \neq X_M/X_L$. A socially inefficient amount of M will be combined with any given amount of L, and X/L will not be maximized.[1]

The average product curves ACD and BCD in Figure 7 can now be viewed as limiting cases. In Knight's case, ACD, M is divisible and there are no externalities before $\bar{M}$ is used to capacity: the industry with free access will employ the socially efficient amount of M with any given quantity of L, and X/L will be maximized. Industry equilibrium output in the range of excess capacity will be Pareto-optimal, as illustrated in Figure 8(a) where the broken line MC^* is the locus of Pareto-optimal outputs and S is the industry supply curve.

In Worcester's case, BCD in Figure 7, the industry must employ the whole of the indivisible $\bar{M}$ with any given quantity of L. Output, as illustrated in Figure 8(b) and employment of L are less than the Pareto-optimal amounts, and a subsidy on L is required for efficiency.

The intermediate cases occur when M is divisible and there are

---

[1] The Pigovian corrective measures are a tax on land and a subsidy on labour, in each case equal to the value of the difference between social and private marginal products at the Pareto-optimal output.

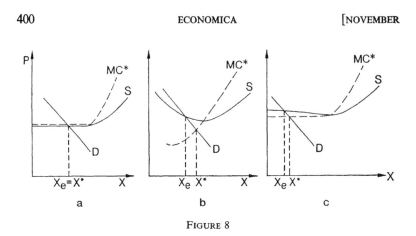

FIGURE 8

externalities. Because private and social marginal rates of factor substitution differ, the industry will combine a socially inefficient amount of M with any given quantity of L, and X/L will not be maximized. However, the quantity of M utilized, though inefficient, could well be less than $\bar{M}$; in this case, the X/L curve for the industry would lie between ACD and BCD in Figure 7. Accordingly, industry average cost will exceed minimum average social cost in the range of excess capacity.

Equilibrium and Pareto-optimal outputs are illustrated in Figure 8(c). Comparing equilibrium with optimal factor employment, there will be an output effect against both L and M, and a substitution effect against L and in favour of M. Employment of L must be less than is Pareto-optimal; but whether this is so of employment of M depends on the balance between output and substitution effects. The corrective measures are a tax on M and a subsidy on L.

IV. CONCLUSION

Pigou's contention that increasing cost industries tend to produce more than the socially optimal quantity was based on the proposition that, in equilibrium, price would equal average social cost rather than marginal social cost. Even after Viner, following the work of Knight and others, had drawn the distinction between technological and pecuniary externalities, over-exploitation of free-access resources remained as an important residual theorem.

Extension of Knight's model from one to two variable factors exposes a further gap in Pigou's argument. Pigou implicitly assumed that the industry would produce any given output at minimum social cost, neglecting the possibility that discrepancies between private and social products would result in inefficient factor combinations. If factor combinations are inefficient, the industry supply curve will lie above the locus of minimum average social cost, and could well lie above the associated marginal social cost curve. Thus, there is no theoretical presumption that free-access resources are over-exploited.

Finally, a word about the policy implications of this paper. In a general way, it confirms the view that free-access resources are exploited inefficiently as compared with a system of private property rights. However, this conclusion abstracts from the costs of maintaining property rights; and it may well be that when these costs are taken into account rights of free access are socially efficient in some cases, for example, ocean fisheries. Again, the analysis suggests that Pigovian taxes and subsidies may be difficult to assess and administer. But, on the other hand, quite crude approximations could be rewarded by significant improvements in efficiency. In short, given the level of abstraction of the model, the analysis does no more than suggest some considerations which might usefully be borne in mind by an applied economist investigating a particular case.

The London School of Economics.

REFERENCES

[1] Bottomley, A., "The Effect of Common Ownership of Land upon Resource Allocation in Tripolitania", *Journal of Land Economics*, vol. 39 (1963), pp. 91–5.

[2] Buchanan, J. M., "An Economic Theory of Clubs", *Economica*, vol. XXXII (1965), pp. 1–14.

[3] Chipman, J. S., "External Economies of Scale and Competitive Equilibrium", *Quarterly Journal of Economics*, vol. 84 (1970), pp. 347–85.

[4] Cheung, S. S., "The Structure of a Contract and the Theory of a Non-Exclusive Resource", *Journal of Law and Economics*, vol. 13 (1970), pp. 49–70.

[5] Goetz, C. J. and J. M. Buchanan, "External Diseconomies in Competitive Supply", *American Economic Review*, vol. 61 (1971), pp. 883–90.

[6] Knight, F. H., "Some Fallacies in the Interpretation of Social Cost", *Quarterly Journal of Economics*, vol. 38 (1924), pp. 582–606; reprinted in [13].

[7] Mishan, E. J., "Reflections on Recent Developments in the Concept of External Effects", *Canadian Journal of Economics and Political Science*, vol. 31 (1965), pp. 3–34.

[8] ——, "The Post-war Literature on Externalities: An Interpretative Essay", *Journal of Economic Literature*, vol. IX (1971), pp. 1–28.

[9] Mohring, H. and J. H. Boyd, "Analysing Externalities: 'Direct Interaction' vs. 'Asset Utilization' Frameworks", *Economica*, vol. XXXVIII (1971), pp. 347–61.

[10] Pigou, A. C., *The Economics of Welfare*, 1918.

[11] Schall, L. D., "Technological Externalities and Resource Allocation", *Journal of Political Economy*, vol. 79 (1971), pp. 983–1001.

[12] Scott, A., "The Fishery: the Objectives of Sole Ownership", *Journal of Political Economy*, vol. 63 (1955), pp. 116–24.

[13] Stigler, G. J. and K. E. Boulding (eds.), *Readings in Price Theory*, Homewood, Ill., 1952.

[14] Turvey, R. "Optimization and Suboptimization in Fishery Regulation" *American Economic Review*, vol. 54 (1964), pp. 64–76.

402　　　　　　　　　ECONOMICA　　　　　　　　[NOVEMBER

[15] Viner, J., "Cost Curves and Supply Curves", *Zeitschrift für Nationalökonomie*, vol. 3 (1931); reprinted in [13].

[16] Winch, D. M., *Analytical Welfare Economics*, Harmondsworth, 1972.

[17] Worcester, D. A., "Pecuniary and Technological Externality, Factor Rents and Social Costs", *American Economic Review*, vol. 59, (1969), pp. 873–85.

Extinction

[9]

Extinction of a Fishery by Commercial Exploitation: A Note

J. R. Gould

London School of Economics and Political Science

A recent synthesis and development of the literature by Smith (1969) has greatly increased the power and generality of the economic theory of commercial fishing. In contrasting his own analysis with previous work, Smith suggests that the most serious deficiency in the received doctrine is that it "is not able to handle the situation in which a species may be exploited to the point of extinction. The theory implies that short of zero unit costs the equilibrium yield to competitive exploitation is never zero with an extinct population."

As Smith observes, evaluation of this implication depends on whether one believes that commercial fishing has ever, or could ever, cause the extinction of a fishery—a matter over which there is some controversy.[1] The theoretical question is one of importance from both positive and normative points of view. On the one hand, it would appear that, if convincing evidence of commercial extinction were discovered, then the received theory of fishing would be decisively refuted. On the other hand, policy makers cannot wait indefinitely for the results of future research, especially when concerned with irreversible changes such as the extinction of species. Their attitudes toward proposals for the regulation of fishing are bound to be colored by untested conclusions of current theory. For these reasons, it would seem important to reexamine the inference which Smith draws from the received theory, particularly because Smith does not explain which features of the theory are responsible for his conclusion.

The "Traditional" Theory of Fishing

Standard theory is formulated in terms of long-run, or steady-state, equilibrium, and the following is a summary account.[2] Following the direction

---

I am grateful to Anthony D. Scott for his comments on an earlier draft.

[1] See Christy and Scott (1965, pp. 80–84) for a sketch of the theoretical arguments. There is evidence to suggest that fisheries have been seriously depleted. For example, regulation of fishing seems to have reversed the decline in the halibut and salmon populations of the Pacific coasts of Canada and the United States (see papers by Marion E. Marts and Richard Van Cleve in Crutchfield [1965]). To my knowledge, there is no overwhelming evidence that fishing has caused the extinction of a population.

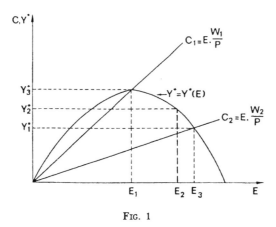

FIG. 1

set by Gordon's seminal paper (1954), models of fishing examine the interaction between the fishing industry and the fish population. The economic equilibrium of the fishing industry depends on the size of the fish population, and the biological equilibrium of the fish population depends on the size of the fishing industry. Full, or bionomic, equilibrium requires simultaneous equilibrium of both the fishing industry and the fish population.

Biological theories suggest that, given certain parameters—such as water temperature, food supply, and the level of activity of natural *and* human predators—the fish population will grow to a definite equilibrium size. At this population level, there will be a determinate steady-state yield of fish to the given level of fishing effort. This is the "sustainable yield," in the sense that, in the long run, this rate of yield to the fishing industry will be exactly offset by the natural increase in the population, and both population and yield will remain stable if none of the parameters change.

If the level of fishing effort is very low, the population will grow to a large equilibrium size, but the sustainable yield will be small. At an intermediate level of fishing effort, the sustainable yield will be larger, even though the equilibrium population is smaller. At some high level of fishing effort, the population will have been so depleted that, once again, the sustainable yield will be low. Thus, it is argued, the sustainable yield is at first an increasing, then a decreasing, function of fishing effort as depicted in figure 1, where $Y*$ stands for sustainable yield and E for fishing effort.

For simplicity of exposition, we shall assume that fishing effort is in perfectly elastic supply at cost W, and that the demand for fish is per-

---

[2] See Smith (1969) for a recent survey of the theoretical literature. Christy and Scott (1965) provide a comprehensive introduction to the whole subject.

fectly elastic at price P, assumptions which are consistent with a small fishery drawing on a large pool of fishermen and serving a large market.

A characteristic feature of fisheries is that they are "common property" resources: no fisherman possesses rights of ownership which enable him to exclude others from exploiting the fishery.[3] In these circumstances, fishermen will enter the fishery as long as the average product of effort exceeds their transfer earnings. In equilibrium, the average product of effort equals the real wage, and the rent attributable to the fishery is zero —that is, total revenue equals total cost.

Bionomic equilibrium, or simultaneous equilibrium of both the fish population and fishing effort, is illustrated in figure 1. The yield of fish in response to effort is $Y^*(E)$, and the total cost curve is expressed in terms of fish by the equation $C = EW/P$. The zero-profit equilibrium condition, $PY = WE$, is represented by the intersection of C and $Y^*(E)$. When the supply price of effort is W_1, the total cost curve is C_1, and the equilibrium effort E_1. At a lower supply price for effort W_2, the total cost curve is C_2, and the equilibrium level of effort E_3.[4]

The diagram indicates that, as the cost of effort falls below that consistent with C_1, the equilibrium yield will fall. But as long as the cost of effort is positive, the equilibrium sustainable yield remains positive— which implies a population greater than zero. This is the reasoning behind Smith's inference that the received theory rules out commercial extinction of a fish population, except at zero supply price of effort.

An Analysis of the Sustainable Yield Curve

As in the above discussion, the sustainable yield curve is usually introduced informally into the analysis of fisheries. To illustrate the nature of the implicit assumptions crucial to Smith's theorem, we now derive the sustainable yield curve from more fundamental relations. Although the interactions between fishing effort and population dynamics may be very complex, the essential point can be made with a simplified model.

We postulate, following Smith (1969), that the rate of growth of an *unexploited* fish population depends on its size, given certain *natural* parameters, according to the relation $\dot{X} = f(X)$. In figure 2, X is the size of the population measured by aggregate weight, and $\dot{X}$ its time rate

---

[3] See Gordon (1954) for a detailed discussion of this feature.

[4] The main use of the analysis has been to illustrate the overexploitation of a common-property resource (see, for example, Gordon 1954). Efficient resource allocation requires maximization of the rent of the fishery and occurs where the vertical distance between $\dot{Y}^*$ and C is a maximum (given that the Pareto conditions are met in the rest of the economy). Scott (1957) pointed out that static equilibrium analysis is not fully adequate for specification of optimal policies since it excludes from discussion, for example, the strategy of intensive exploitation followed by a "fallow" period in which effort would be diverted to other fisheries.

1034   JOURNAL OF POLITICAL ECONOMY

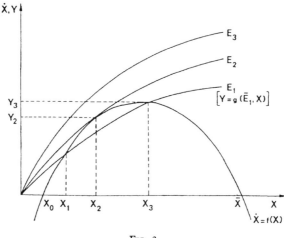

Fɪɢ. 2

of change. At levels between X_0 and $\bar{X}$ the population increases, and at levels greater than $\bar{X}$ it decreases, so that a stable equilibrium is reached at $\bar{X}$. Population X_0 is the minimum biologically viable population; at levels below X_0 population decreases because of increased vulnerability to disease and natural predators. The curve $\dot{X} = f(X)$ will be called the "natural increase curve."

The rate of yield of the fishing industry Y is assumed to depend on the amount of fishing effort E and the size of the fish population. Thus the production function can be written $Y = g(E,X)$.

We assume that the marginal product of effort g_E is positive,[5] that the marginal product of population g_X is positive,[6] and that marginal and average products of both effort and population are subject to diminishing returns.[7]

If we assume no interaction between fishing effort and the rate of natural increase,[8] the fish population is in equilibrium when the fishing industry is catching fish at the same rate as nature is replacing them—

---

[5] After a sufficiently high level of effort is reached, the marginal product of effort may be negative, and a competitive industry could be in equilibrium in this range. See the discussions of "crowding externalities" in, for example, Scott (1957) and Smith (1969). The present analysis can easily be amended to account for this possibility.

[6] Because, say, a larger population is more dense, which facilitates fishing.

[7] This form of production function is assumed in most of the literature. Turvey (1964) and Smith (1969), following the eumetric theory of Beverton and Holt (1957), introduce mesh size as an additional independent variable in the production function. Gordon (1954) argues that the usual a priori arguments to support the hypothesis of diminishing returns are difficult to apply to the case of fisheries. Of course, in the last analysis, the curvature of the fishery production function is an empirical question.

[8] Such interaction may occur because, for example, the gear selects fish of certain sizes. This may affect both the fertility and the rate of growth of the remaining stock.

that is, when $\dot{X} = Y$. The sustainable yield as a function of fishing effort is derived by solving for Y in terms of E in the equations:

$$\dot{X} = f(X), \tag{1}$$

$$Y = g(E, X), \tag{2}$$

$$\dot{X} = Y. \tag{3}$$

A diagrammatic derivation is illustrated in figure 2. Superimposed on the natural increase curve $\dot{X} = f(X)$ are a series of curves, E_1, E_2, etc., displaying the production function $Y = g(E,X)$. For fixed level of effort, say E_1, the line E_1 show- Y as a function of X.

Consider the level of effort E_1. If the fish population is above X_3, the yield exceeds the natural increase and population declines. For $X_1 < X < X_3$, $\dot{X} > Y$ and population increases. For $X < X_1$, $\dot{X} < Y$ and population declines to zero. Thus with effort at E_1, the population is in stable equilibrium at X_3 and the sustainable yield is Y_3. In general, the sustainable yield for any given level of effort is indicated where the relevant E curve cuts the natural increase curve from below.

The important point to note is that E_2 is the highest level of effort for which the sustainable yield is positive.[9] At, for example, $E_3 > E_2$, $Y > \dot{X}$ for all values of X; if E_3 is maintained the population is eventually extinguished and the sustainable yield is zero.

The sustainable yield curve consistent with the natural increase curve and the production function in figure 2 can be illustrated with figure 1. For values of E up to E_2 the sustainable yield is positive; at E_2 there is a discontinuity in the sustainable yield curve; for values of $E > E_2$ the sustainable yield is zero.

Inspection of figure 2 indicates that, if the minimum viable population X_0 is positive and the marginal product of population is always positive, then the sustainable yield curve must have a discontinuity. If either of these conditions does not hold, it is possible, although not necessary, that the sustainable yield curve varies continuously.

It is clear that Smith's theorem can only be deduced in a straightforward manner if the sustainable yield curve varies continuously, and this must be regarded as a special case. If the total cost curve (for example, C_2 in fig. 1) goes through the discontinuity, intuition may suggest (correctly) that, at the corresponding (positive) supply price for effort, the fish population would be wiped out. However, it is evident that the sustainable yield curve is a far from satisfactory analytical tool for the discussion of population extinction. The basic model can easily be reformulated in a manner more suited to this purpose.

---

[9] At (X_2,Y_2) the sustainable yield is stable from above but unstable from below.

1036   JOURNAL OF POLITICAL ECONOMY

An Alternative Formulation of the Model

The model employed in the previous section is summarized in equations (1)–(3), plus the zero-profit equilibrium condition:

$$PY = WE. \tag{4}$$

The diagrammatic analysis consisted in solving equations (1)–(3) in figure 2 to get Y in terms of E; this relation was then combined with equation (4) in figure 1 to solve for equilibrium values of Y and E. An alternative method, which better displays the possibilities of population extinction, is to eliminate E from equations (2) and (4) to get Y in terms of X, and to combine this relation with equations (1) and (3) to solve for equilibrium $\dot{X}$ and X.

Figure 3 can be used to derive equilibrium levels of fishing effort and yield for different population levels by solving equations (2) and (4). The X curves display the production function $Y = g(E,X)$, this time treating E as the independent variable and X as a parameter. Thus, with population X_2 and effort E_1, yield is Y_1.

The fishing industry is in equilibrium for a *given* population when $PY = WE$, that is, at the intersection of $Y = EW/P$ with the relevant X curve. Thus, given X_2, the equilibrium yield is Y_1, and so on. From figure 3 we derive the curve Y_1^{**} in figure 4, showing the equilibrium yields at different population levels, which we shall call the "catch locus" to avoid confusion with the sustainable yield curve.

Superimposed on the catch locus Y_1^{**} in figure 4 is the natural increase curve $\dot{X} = f(X)$. The *net* rate of population increase equals the natural rate of increase minus the catch, so that for the fish population to be in equilibrium Y_1^{**} must equal $\dot{X}$. Moreover, since by assumption the industry is in equilibrium on the catch locus, the point (X_e, Y_e^{**}) is a position

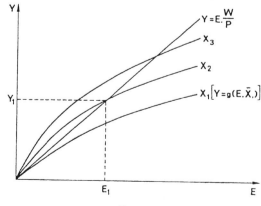

Fɪɢ. 3

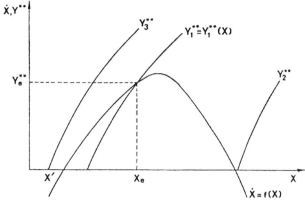

Fɪɢ. 4

of full steady-state bionomic equilibrium. It is also a stable equilibrium in the sense that at $X > X_e$, $Y_1^{**} > \dot{X}$, that is, more fish are being caught than nature is replacing, and population decreases; while at $X < X_e$, $Y_1^{**} < \dot{X}$, and X increases.

From figure 3 it is clear that the position of the catch locus in figure 4 depends on the parameters of the production function and on the values of W and P. For example, with a low P and a high W, the catch locus could be that depicted by Y_2^{**}, with the fishery not sufficiently profitable to warrant exploitation. The catch locus Y_3^{**} illustrates the case of the extinction of the fish population by the fishing industry. The fishing industry's catch exceeds the natural rate of increase at each level of population. If we start with a newly discovered fishery, the catch, the amount of fishing effort, and the population decline continuously. Eventually, at population X', the industry size reaches zero, at which point natural forces eliminate the remaining population.

It has been argued that the commercial extinction of a fish population is inconceivable since fishing would become unprofitable long before the population was wiped out. The concept of a minimum biologically viable population is, in effect, a counter argument; it dispels the image of fishermen scouring the oceans in profitless search for the last male or female fish. However, a minimum viable population greater than two is not a necessary condition for commercial extinction. This could be illustrated by amending figure 4 so that both the natural increase curve and the catch locus start from the origin, but the latter is steeper than the former in that neighborhood. As a possible example, whales are so gregarious and so highly visible, that it is quite conceivable that if a handful of whales were left in the world they would be hunted and captured. On the other hand, cod, for example, are sufficiently difficult to find as virtually to guarantee their preservation from extinction by commercial fishing with present technology.

To summarize: the traditional theory of fishing, when supplemented by an analysis of the sustainable yield curve, is consistent with the extinction of a fishery by a fishing industry. The formulation in figures 3 and 4, where economic considerations are summarized in the catch locus, and biological factors in the natural increase curve, provides a convenient— albeit simple—model for elucidating the conditions for extinction.

References

Beverton, R. J. H., and Holt, S. V. *On the Dynamics of Exploited Fish Popu-
    lations.* Fisheries Investigations, vol. 19, ser. 2. London: Ministry of Agricul-
    ture, Fisheries and Food, 1957.
Christy, F. T., and Scott, A. *The Commonwealth in Ocean Fisheries.* Baltimore:
    Johns Hopkins Press, 1965.
Crutchfield, James A., ed. *The Fisheries: Problems in Resource Management.*
    Seattle: Univ. Washington Press, 1965.
Gordon, H. Scott. "The Economic Theory of a Common-Property Resource:
    The Fishery." *J.P.E.* 62 (April 1954): 124–42.
Scott, Anthony. "Optimal Utilization and Control of Fisheries." In *The Eco-
    nomics of Fisheries,* edited by Ralph Turvey and Jack Wiseman. Rome: Food
    and Agriculture Org., 1957.
Smith, Vernon L. "On Models of Commercial Fishing." *J.P.E.* 77 (March/April
    1969): 181–98.
Turvey, Ralph. "Optimization and Suboptimization in Fishery Regulation."
    A.E.R. 54 (March 1964): 64–76.

[10]

The Economics of Overexploitation

Severe depletion of renewable resources may result from high discount rates used by private exploiters.

Colin W. Clark

Renewable resources, by definition, possess self-regeneration capacities and can provide man with an essentially endless supply of goods and services. But man, in turn, possesses capacities both for the conservation and for the destruction of the renewable resource base.

Indeed, man's increasing capacity to seriously deplete the world's natural resources appears to be reaching a critical stage (*1*); if this is not imminent for the nonrenewable resources (*2*), it certainly appears so for many of the renewable ones (*3*). The problems of environmental pollution that loom so large today, for example, often result from a process of overexploitation of the regenerative capacity of our atmospheric and water resources. Economists lately have devoted much attention to environmental questions (*4*), and most are agreed that "externalities"—that is, effects not normally accounted for in the cost-revenue analyses of producers—are the leading economic cause of pollution and the destruction of natural beauty.

Animate resources, or biological resources, are also subject to serious misuse by man. An accelerating decline has been observed in recent years in the productivity of many important fisheries (*5*), particularly the great whale fisheries and the famous Grand Banks fisheries of the western Atlantic, as well as the spectacularly productive Peruvian anchovy fishery (*6*). As technology improves and demand increases, so the pressure on renewable resources grows more severe. The long-recognized need for effective international regulation of fisheries has never been so pressing as it is today.

A prerequisite for effective regulation is a clear understanding of the basic reasons for overexploitation, and in this regard the outstanding article by Hardin (*7*) on "The tragedy of the commons" has been a positive asset, even though economists have long been aware of the common property problem in fisheries (*8*). Indeed, in concentrating their attention on the problems of competitive overexploitation of fisheries, economists appear to have largely overlooked the fact that a corporate owner of property rights in a biological resource might actually prefer extermination to conservation, on the basis of maximization of profits (*9*). In this article I argue that overexploitation, perhaps even to the point of actual extinction, is a definite possibility under private management of renewable resources.

The implications of this argument for successful international regulation would seem to be that, if it is assumed that society wishes to preserve the productivity of the oceans and to prevent the extermination of valuable commercial species, control of the physical aspects of exploitation is essential. In particular the popular idea of maximum sustainable yield should be generally adopted, at least in the sense of setting an upper limit on the allowable degree of exploitation. Only a dire emergency in local food supply should be considered as a valid reason for temporarily running down the basic stock of a biological resource.

Antarctic Blue Whale Fishery

In developing the economic theory of a biological resource, I take as an example the Antarctic blue whale population. No economic analysis of whaling as such has yet been published, to my knowledge. Certainly, the complete failure of the International Whaling Commission to carry out its mandate to protect and preserve the whale stocks has not been convincingly explained on economic grounds.

A committee appointed by the International Whaling Commission (*10*) estimated in 1964 the net reproductive capacity, in terms of net recruitment of 5-year-old blue whales, as a function of the breeding stock of this species. Their graph, which except for the lower end from 0 to 30,000 whales was little more than an educated guess, is shown in Fig. 1. It appears to indicate a maximum sustainable yield of about 6000 blue whales per annum, but more recent information suggests that this estimate may have been somewhat too high (*11*).

Figure 2 shows the annual blue whale catch (*12*), which expanded rapidly in 1926 following the construction of the first modern stern-slipway factory ships, and ended officially in 1965 when the International Whaling Commission agreed to protect the species. At that time the remaining population was believed to be less than 200 whales, but later estimates have been more optimistic, with the stock in 1972 estimated at about 6000 blue whales (*11*). I return to the case of the blue whales after a general analysis of the economics of biological resources.

Economic Rent

The most commonly encountered proposal for managing a biological resource is to maximize the sustained yield. Indeed, this was the management scheme suggested by the committee to the Whaling Commission (*10*): "The greater the reduction of the present quota, the more rapidly will whale stocks rebuild to the level of maximum sustainable productivity." Economists, however, have taken exception to such proposals (*8*): "Focusing attention on the maximization of the catch neglects entirely the inputs of other factors of production which are used up in fishing and must be accounted for as costs."

Indeed, economists have generally suggested adopting the maximization of economic rent as a management policy. The term economic rent refers to the regular income derived from an endurable resource; it refers to net income, or excess of revenue over costs. Since there is a variety of management possibilities for most resources, it is

The author is professor of mathematics at the University of British Columbia, Vancouver 8, Canada.

worthwhile to enquire which policy will produce the maximum rent.

In order to obtain a simple mathematical model, suppose that the net recruitment to a particular resource stock of size x is given by a quadratic expression:

$$y = f(x) = Ax(\bar{x} - x) \qquad (1)$$

where $A > 0$ is a constant, and $\bar{x} > 0$ represents the natural equilibrium population. The blue whale curve (Fig. 1) has roughly this form, which is related to the logistic equation of theoretical biology:

$$dx/dt = Ax(\bar{x} - x)$$

where t is time.

We also suppose that the net recruitment is the same as (or proportional to) the sustainable yield from a population of size x.

The economic components of our model consist of a constant price $p > 0$ per unit of harvested stock, and a unit harvesting cost $C(x)$ that depends on the population size x. The simplest assumption is that this unit harvesting cost is proportional to the density of the population; in the case of pelagic or demersal fish that are more or less uniformly distributed over their range, this assumption would mean simply that $C(x)$ varies inversely with x. Thus, the total cost of harvesting the sustainable yield $y = f(x)$ would be (approximately)

$$C = By/x = AB(\bar{x} - x) \qquad (2)$$

where B is the unit cost coefficient. More general forms of the cost function are considered below.

What sustainable yield, at what population x, gives rise to the maximum rent? Since rent is the difference between revenue R and cost C, the problem is to maximize the expression

$$R - C = pAx(\bar{x} - x) - AB(\bar{x} - x) \qquad (3)$$

The maximum occurs when $x = \hat{x}$, where (see Fig. 3)

$$\hat{x} = \frac{\bar{x}}{2} + \frac{B}{2p} \qquad (4)$$

provided this expression is less than the equilibrium level $\bar{x}$. (The case $\hat{x} > \bar{x}$ corresponds to the case of negative rent $R - C$ for all populations x; in this case the resource is of no economic value.)

It is clear from Eq. 4 that the rent-maximizing population $\hat{x}$ is greater than the level $\bar{x}/2$ of maximum sustained yield. It is this observation that seems to have led to the belief that a

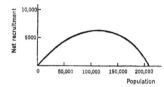

Fig. 1. Recruitment curve for blue whale population.

private resource owner would necessarily attempt to conserve his resource stock. I return to this question after discussing the common-property problem.

Since maximizing rent appears to be the same thing as maximizing profits, the question now arises, why in practice do fisheries and other resource industries never seem to attain this result? Economists have studied this question in detail; their solution was described by Gordon (*8*):

In sea fisheries the natural resource is not private property; hence the rent it may yield is not capable of being appropriated by anyone. The individual fisherman is more or less free to fish wherever he pleases. The result is a pattern of competition among fishermen which culminates in the dissipation of the rent . . .

To summarize the argument for dissipation of rent, suppose first (see Fig. 3) that the fishery is actually operating at the rent-maximizing level $\hat{x}$. Then, observing that the working fishermen are making a profit, new fishermen will be attracted to the industry. Fishing intensity will increase and the fish population will decrease, as will the total rent. As long as any rent remains, the process continues. The fishery will expand until in the end the population reaches the level x_0 of zero economic rent. Thus, in a competitive

situation, the rent will be entirely dissipated and economic efficiency will vanish.

In practice, fishermen will no longer be attracted to a fishery when they can earn a greater income in some alternative employment. This alternative income determines what economists call the opportunity costs of labor in fishing, and these costs are normally included in the total cost function. In cases of high unemployment, opportunity costs for fishermen may be nearly zero, so that the rent dissipation argument would be particularly forceful in explaining the overexploitation of fisheries.

So runs the standard economic argument for the overexploitation of resources, neatly laying the blame on open competition, particularly among the impoverished and the powerless. Yet the most spectacular and threatening developments of today, such as the reduction of the whale stocks and of the demersal fisheries on the Grand Banks, can by no means be attributed to impoverished local fishermen. On the contrary, it is the large, high-powered ships and the factory fleets of the wealthiest nations that are now the real danger. Poor and wealthy nations alike, however, may suffer unless successful control is soon achieved.

Economists themselves have begun to question the adequacy of the rent-dissipation argument to explain current developments (*13*). The fact that (as in the above model) extinction is theoretically impossible has been called "one of the more serious deficiencies of the received doctrine" (*14*). But the principal shortcoming of the existing theories is their disregard of the time variable, both biologically and economically.

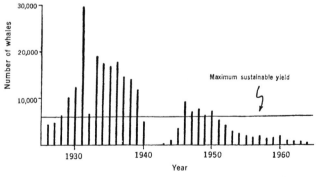

Fig. 2. Annual blue whale catch, 1925 to 1965. [Data compiled from (*12*)]

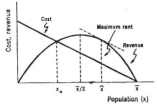

Fig. 3 (left). Economic rent.

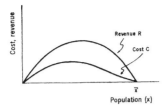

Fig. 4 (right). Cost-revenue curves (extinction feasible).

On the one hand, biological populations take time to respond to harvesting pressures, and only approach a new equilibrium after several seasons. On the other hand, but equally important, the value of monetary payments also possesses a time component due to the discounting of future payments. It denies the fundamental principles of economics itself to overlook the latter effect, and that is just what the rule of maximizing economic rent does.

The fact that maximization of rent and maximization of present value are not equivalent has long been recognized in agriculture (15) and forestry (16), and some analyses of fishing have also recognized the difference (17). The latter, however, have invariably utilized advanced mathematical techniques from the calculus of variations and optimal control theory, and are consequently somewhat beyond the level of intuitive understanding.

In the remainder of this article, I first show how the possibility of extinction can easily be included in the analysis, and then discuss the question of maximization of present value in resource management. The principal outcome will be that if extinction is economically feasible, then it will tend to result not only from common-property exploitation, but also from the maximization of present value, whenever a sufficiently high rate of discount is used (18). Generally, high rates of discount have the effect of causing biological overexploitation whenever it is commercially feasible.

The question of the cause of high discount rates is a complex one; it is sufficient to remark that at any time the discount rate adopted by exploiters will be related to the marginal opportunity cost of capital in alternative investments. In a technologically expanding economy, this rate could be quite large.

When applied to the Antarctic blue whale, the analysis indicates that an annual discount rate between 10 and 20 percent would be sufficient for extinction to result from maximization of the present value of harvests, assuming that extinction is commercially feasible. Such rates are by no means exceptional in resource development industries.

The question of the feasibility of extermination of the whale stocks is an interesting one. Gulland (19) has pointed out to me that fishing for the Antarctic blue whale probably would have become uneconomical several years earlier had it not been for the simultaneous occurrence of finback whales in the same area. It appears likely that the whalers agreed to a moratorium on blue whales in 1965 because they did not anticipate any significant further profits from the species.

These considerations raise serious doubts, in my opinion, about the wisdom of assuming that corporate resource exploiters will automatically behave in a socially desirable manner (20). There is no reason to suppose that the fishing corporations themselves desire regulations designed to conserve the world's fisheries. The governments of the world will fail in their responsibility to their citizens unless they succeed in formulating effective international conservation treaties in spite of pressures from these corporations.

Possibility of Extinction

The fact that populations can be driven to extinction by commercial hunting hardly needs to be emphasized. Only a minor change in the model described above is required in order to include the possibility of extinction in a reasonable way.

In Eq. 2 we made the assumption that harvesting costs vary inversely with population x. It thus appears that

costs become infinite as x approaches zero. The variable x, however, is in reality restricted to integral values ($x = 1, 2, 3, \ldots$), and the cost of extinction is actually the cost of a unit harvest when $x = 1$. The simplest way to adjust the model to admit the possibility of extinction is to replace Eq. 2 by

$$C = \frac{By}{x+1} = \frac{ABx(\bar{x}-x)}{x+1} \quad (2')$$

In this formula, the coefficient B represents the cost of extinction, that is, the cost of a unit harvest which reduces the breeding population from one to zero. If B is less than the price p, then the cost curve C will lie below the revenue curve $R = pAx(\bar{x}-x)$ for all values of x, as in Fig. 4. In this case the zero rent population x_0 equals zero, and rent dissipation will lead to extinction.

In practice, extinction may not require the actual extermination of the last member of the population. Biologists speak of a minimum viable population such that survival is impossible, or highly improbable, once the population falls below this level (21). Such a possibility is easily included in our model by replacing Eq. 1 for the net recruitment by

$$y = A(x - \underline{x}) \, (\bar{x} - x) \quad (1')$$

where $\underline{x}$ represents the minimal viable population. Note that there is no sustainable yield when $x < \underline{x}$. In this case extinction is again economically feasible provided the cost coefficient B of Eq. 2 or 2' is small (22).

Henceforth for the sake of definiteness I adopt the model described by Eqs. 1 and 2', so that extinction is feasible if the extinction cost B is less than the price p. (The more general case of Eq. 1' can be treated by a similar analysis, or can be reduced to the previous case by shifting the origin of the population axis to the point $\underline{x}$ below which extinction becomes automatic.) Hence the rent function $R - C$ is given by

$$F(x) = pAx(\bar{x}-x) - \frac{ABx(\bar{x}-x)}{x+1} =$$
$$f(x)\left[p - \frac{B}{x+1}\right] \quad (5)$$

Maximization of Present Value

The concept of economic rent as discussed so far is time-independent. A more general understanding of the concept, as it applies to agricultural land economics, has been given by Gaffney (*15*), who identifies several categories of economic rent. Some of these do not apply to the case of fisheries or other wild animal resources, but his categories of conservable flow and expendable surplus are relevant in general.

In Gaffney's words, the expendable surplus is "that portion of virgin fertility whose emplaced value is less than its liquidation value." In other words, the immediate profit obtained from expending this surplus exceeds the present value of revenues that could be obtained in perpetuity by conserving it. Conversely the conservable flow refers to that portion of fertility whose emplaced value is greater than its liquidation value.

The expendable surplus thus provides a temporary contribution to rent, and disappears once it is expended, leaving the conservable flow as the enduring rent. Obviously, the expendable surplus and the conservable flow are complementary quantities; how much of virgin fertility is assigned to each category depends critically, as we shall see, on the rate of discount utilized in computing present values.

In our own case, let $\hat{x}$ now denote the economically conservable breeding population. The problem is to determine the value of $\hat{x}$. The conservable flow equals the rent $F(\hat{x})$ from Eq. 5, and the emplaced value of this rent is just the present value of a (continuous) annuity $F(\hat{x})$, namely

$$P_1(\hat{x}) = \int_0^\infty F(\hat{x})\, e^{-\delta t} dt = \frac{1}{\delta}\, F(\hat{x}) \qquad (6)$$

where $\delta > 0$ is the adopted discount rate. A high discount rate corresponds to a low emplaced value, and vice versa.

To derive the value of the expendable surplus, suppose that the population is originally at its natural equilibrium level $\bar{x}$. The surplus is therefore $\bar{x} - \hat{x}$, and this can produce an immediate gross revenue of $p(\bar{x} - \hat{x})$, at a harvesting cost given by

$$\int_{\hat{x}}^{\bar{x}} C(x)\,dx$$

where, as in Eq. 2', $C(x) = B/(x + 1)$ is the unit harvest cost at the population level x. Thus, the value of the surplus is equal to

$$P_2(\hat{x}) = p(\bar{x} - \hat{x}) - B \log \frac{\bar{x} + 1}{\hat{x} + 1} \qquad (7)$$

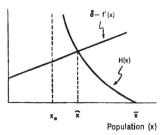

Fig. 5. Maximization of present value (extinction not feasible).

The value of $\hat{x}$ is now determined by maximizing the total present value $P_1(\hat{x}) + P_2(\hat{x})$ (*23*). From Eqs. 6 and 7 we obtain, except in the end-point cases ($\hat{x} = 0$ or $\bar{x}$), the necessary condition

$$\frac{1}{\delta}\, F'(\hat{x}) = p - \frac{B}{\hat{x} + 1} = p - C(\hat{x}) \qquad (8)$$

Equation 8 is a marginal condition of the type familiar in economic analysis. The right-hand expression, $p - C(\hat{x})$, represents the additional, or marginal, net revenue obtained by harvesting one unit from the population $\hat{x}$. The left-hand expression, $\delta^{-1}F'(\hat{x})$, is the marginal increase in the present value of the annuity $F(\hat{x})$ that results from leaving this additional unit of population to contribute to net recruitment. Neglecting exceptional cases, we must have equality of these marginal values at the optimal population $\hat{x}$.

Since by Eq. 5 we have $F(x) = f(x)[p - C(x)]$, a simple calculation reduces Eq. 8 to

$$\delta - f'(\hat{x}) = \frac{-C'(\hat{x})f(\hat{x})}{p - C(\hat{x})} \qquad (9)$$

[Equation 9 can be derived generally for an arbitrary recruitment function $f(x)$ and unit cost function $C(x)$.]

In analyzing Eq. 9 there are two cases to consider, depending on whether extinction is feasible or not. If $p < B = C(0)$, then extinction is not feasible. Let

$$p = C(x_0) \qquad (10)$$

so that x_0 represents the population at which price equals unit harvesting cost. Thus $F(x_0) = 0$, that is, x_0 is the "zero rent" level, which we found would be the level resulting from common-property dissipation of rent. Since $F(x) < 0$ for $x < x_0$, it is clear that the desired equilibrium population $\hat{x}$ must be $\geqslant x_0$.

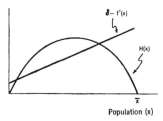

Fig. 6. Maximization of present value (extinction feasible).

Let $H(x)$ denote the expression on the right side of Eq. 9. Then (Fig. 5) $H(x) > 0$ for $x > x_0$ and the graph of $H(x)$ is asymptotic to the line $x = x_0$. The left side of Eq. 9 is a linear function with a positive slope. Consequently Eq. 9 has a solution $\hat{x}$ lying between x_0 and $\bar{x}$. The value $\hat{x}$ is the conservable population.

The two special cases $\delta = 0$ and $\delta = +\infty$ deserve comment. When $\delta = 0$, Eq. 8 implies that $F'(\hat{x}) = 0$. Thus, a zero discount rate corresponds to the maximization of economic rent and results in the largest possible conservable flow. When δ is infinite, on the other hand, we see from Fig. 5 that $\hat{x} = x_0$. In this case $F(\hat{x}) = 0$ and there is no conservable flow; the entire profitable portion of the virgin population is expendable surplus: it is completely dissipated. Rent maximization and rent dissipation thus occur mathematically as two extreme cases of maximization of present value.

Let us turn to the case $p > B$, in which extinction is feasible. In this case $H(x)$ is a positive bounded function (Fig. 6). Depending on the value of δ, there may be one, several, or no solutions to Eq. 9. As before, the case $\delta = 0$ corresponds to the maximization of rent. Now, however, the rent will be dissipated (and the population exterminated) not only for an infinite discount rate, but also for any sufficiently high rate. The following theorem is proved in the Appendix.

THEOREM. *Assume extinction is feasible* ($p > B$). *Then extinction will indeed occur as a result of the maximization of present value, whenever* $\delta > 2f'(0)$.

Note that $f'(0)$ represents the maximum reproductive potential of the population.

Let us return at last to the blue whales. Figure 1 indicates a maximum reproductive potential of about 10 percent per annum [and more recent reports indicate an even smaller rate, perhaps 4 to 5 percent (*11*)]. If in their calculations of profit and loss, the owners of the whaling fleets were to utilize an annual rate of discount of 20 percent or greater, they would therefore opt for complete extermination of the whales—at least as long as whaling remained profitable. This would occur whether they were competing, or cooperating, in the slaughter (*24*).

Summary

The general economic analysis of a biological resource presented in this article suggests that overexploitation in the physical sense of reduced productivity may result from not one, but two social conditions: common-property competitive exploitation on the one hand, and private-property maximization of profits on the other. For populations that are economically valuable but possess low reproductive capacities, either condition may lead even to the extinction of the population. In view of the likelihood of private firms adopting high rates of discount, the conservation of renewable resources would appear to require continual public surveillance and control of the physical yield and the condition of the stocks.

Appendix

To prove the theorem stated above, we will show that Eq. 9 has no solution in case $\delta > 2f'(0)$ and $p > B$; this implies that $\hat{x} = 0$ maximizes the total present value, since $\hat{x} = \bar{x}$ would give both zero rent and zero present value.

Since $H(x)$, the right-side expression in Eq. 9, is a decreasing function of p, it suffices to consider the case $p = B$. Then by the generalized mean value theorem of elementary calculus,

$$H(x) = -C'(x)\,\frac{f(x)}{C(0) - C(x)}$$

$$= \frac{C'(x)f'(\xi)}{C'(\xi)} \qquad (0 < \xi < x)$$

$$< f'(\xi) < f'(0)$$

Thus $\delta - f'(x) > 2f'(0) - f'(x) > f'(0) > H(x)$, so Eq. 9 has no solution as claimed.

References and Notes

1. J. Forrester, *World Dynamics* (Wright-Allen, Cambridge, Mass., 1971); D. H. Meadows, D. L. Meadows, J. Landers, W. W. Behrens III, *The Limits to Growth* (Universe, New York, 1972).
2. R. Gillette, *Science* **175**, 1088 (1972).
3. E. F. Murphy, *Governing Nature* (Quadrangle, Chicago, 1967), pp. 8–11.
4. For a recent survey see R. M. Solow, *Science* **173**, 498 (1971).
5. F. T. Christy and A. D. Scott, *The Common Wealth in Ocean Fisheries* (Johns Hopkins Press, Baltimore, 1965), pp. 80–86.
6. T. Loftas, *New Sci.* **55** (No. 813), 583 (1972).
7. G. Hardin, *Science* **162**, 1243 (1968).
8. H. S. Gordon, *J. Polit. Econ.* **62**, 124 (1954).
9. See J. Crutchfield and A. Zellner, *Economic Aspects of the Pacific Halibut Fishery* (Government Printing Office, Washington, D.C., 1963), pp. 19–20.
10. *International Whaling Commission, 14th Annual Report* (International Whaling Commission, London, 1964), appendix 5, pp. 32–83.
11. J. Gulland, *New Sci.* **54** (No. 793), 198 (1972).
12. Committee for Whaling Statistics, *International Whaling Statistics, Nos. 1–49* (Norwegian Whaling Council, Oslo, 1930–1963).
13. See, for example, S. N. S. Cheung, in *Economics of Fisheries Management: A Symposium*, A. Scott, Ed. (Institute of Animal Resource Ecology, University of British Columbia, Vancouver, 1970), pp. 97–108.
14. V. L. Smith, *J. Polit. Econ.* **77**, 181 (1969).
15. M. M. Gaffney, *Natur. Resour. J.* **4**, 537 (1964).
16. A. D. Scott, *Natural Resources: The Economics of Conservation* (Univ. of Toronto Press, Toronto, 1955).
17. C. G. Plourde, *Amer. Econ. Rev.* **60**, 518 (1970); *West. Econ. J.* **9**, 256 (1971); A. Zellner, in *Economic Aspects of Fishery Regulation* (Food and Agriculture Organization of the United Nations, Rome, 1962), pp. 497–510; also see (*13*).
18. See also C. W. Clark, *J. Polit. Econ.*, in press.
19. J. Gulland, personal communication.
20. That oligopolistic development of resources may be the most destructive of all possibilities has been recognized in the conservation literature. See S. V. Ciriacy-Wantrup, *Resource Conservation: Economics and Policies* (Univ. of California Press, Berkeley, 1952), pp. 190–198.
21. For further details see K. E. F. Watt, *Ecology and Resource Management* (McGraw-Hill, New York, 1968), pp. 54–73.
22. J. R. Gould, *J. Polit. Econ.* **80**, 1031 (1972).
23. The Pontrjagin maximum principle [L. S. Pontrjagin, V. S. Boltjanskii, R. V. Gamkrelidze, E. F. Mishchenko, *The Mathematical Theory of Optimal Processes* (Pergamon, Oxford, 1964), pp. 18–19] can be invoked to prove that maximization of the present-value integral

$$\int \exp(-\delta t)\, h(t)\,\{p - C\,[x\,(t)]\,\}\, dt$$

for $h(t) \geq 0$ is equivalent to maximization of the simple expression $P_1(x) + P_2(x)$ described here.
24. This possibility was also suggested by D. Fife [*Environment* **13** (No. 3), 20 (1971)].
25. Among the many friends and colleagues whose ideas have contributed to this article, I would especially like to thank P. Bradley, P. Pearse, A. Scott, and all other economists who have patiently suffered my errors.

Disaggregated Models

[11]

On Models of Commercial Fishing

Vernon L. Smith

University of Massachusetts

I. An Industry Model with Externalities

Commercial fishing is characterized by three key economic and techno-logical features that are relevant to the formulation of an economic theory of fish production.

1. A fishery resource, although conceivably exhaustible, is replenish-able; that is, it is subject to laws of natural growth which define an en-vironmental biotechnological constraint on the activities of the fishing industry.

2. The resource and the activity of production from it form a stock-flow relationship. The new growth in the population fish mass depends upon the harvest rate relative to natural recruitment to the stock. If the harvest rate exceeds the recruitment rate, the stock declines, and vice versa.

3. The recovery or harvesting process is subject to various possible external effects all of which represent external diseconomies to the firm: (*a*) Resource *stock externalities* result if the cost of a fishing vessel's catch decreases as the population of fish increases. (*b*) *Mesh externalities* result if the mesh size (or other kinds of gear selectivity variables) affects not only the private costs and revenues of the fisherman but also the growth behavior of the fish population. (*c*) *Crowding externalities* occur if the fish population is sufficiently concentrated to cause vessel congestion over the fishing grounds and, thus, increased vessel operating costs for any given catch. All of these various types of externalities arise fundamentally because

The preparation of this paper was supported by National Science Foundation grant GS-1835 to Brown University. It is one of a series of papers dealing with the economics of production from extractive natural resources (Smith, 1968). I am in-debted to Anthony Scott for many helpful comments on an earlier draft.

of the "common property," unappropriated (Gordon, 1954; Scott, 1955) character of most fishery resources, especially ocean and large lake fisheries.

The literature of fishing economics has drawn attention, in some form, to each of these characteristics, and has initiated the development (Crutchfield and Zellner, 1962) of a formal dynamic industry model of the productive process and the interaction of the exploiting industry with the exploited population. But there seems to be a need for generalization, explication, and integration of this previous work. Toward this end we will treat the case of a homogeneous industry composed of K identical fishing vessels or firms, each producing at a catch rate x pounds per unit time. The total harvest rate is then Kx.

The purpose of the model is to provide one example of a descriptive theory that transforms any specific pattern of assumptions about cost conditions, demand externalities, and biomass growth technology into a pattern (conceivably observable) of exploitation. The model or variations on it would appear to have much wider possible applications, such as (1) a theory of bionomic equilibrium in primitive hunter cultures and (2) possibly the rudiments of an economic theory of species extinction, both historical and modern.

We consider a single fish species with population mass X in pounds. In the absence of predation by man, following Lotka (1956) (see also Christy and Scott, 1965, pp. 7–8, 81), we assume a recruitment rate or growth function $dX/dt \equiv \dot{X} = f(X)$. We posit that $f(X)$ has the properties $f(\underline{X}) = f(\bar{X}) = 0, f'(X^\circ) = 0, f''(X) < 0, X \geq 0, 0 \leq \underline{X} < X^\circ < \bar{X}$. The equilibrium population in the natural state is $\bar{X}$, and populations below $\underline{X}$ are assumed not to be viable because of vulnerability to disease, parasites, or predators, or to inadequate fecundity. By setting $\underline{X} = 0$ we get the special case when such considerations are not relevant. The solution to the differential equation $\dot{X} = f(X)$ provides the law of growth for the species. When $f(X)$ is quadratic the result is the popular logistics law of growth (Lotka, 1956, pp. 64–66). For individual fish the empirical law of growth is an asymmetrical sigmoid curve with the inflection point at a weight below one-half the asymptotic weight (Beverton and Holt, 1957, pp. 31–35, 96–135). However, there is experimental evidence to suggest that some life forms follow the logistics law in a constant trophic environment. By postulating $f(X)$, without any attempt to deal analytically with the components of mass growth—the birth process, individual member growth, death and capture processes—we are electing to take an aggregative approach to the biotechnology. Beverton and Holt (1957, pp. 329–30) have discussed the function and limitations of such a "sigmoid curve" theory of population growth.

When the population is exploited by a fishing industry employing mesh size m so that harvesting is confined to those members whose size is not

COMMERCIAL FISHING MODELS 183

below m, a natural generalization of the growth hypothesis is[1]

$$\dot{X} = f(X, m, Kx). \tag{1}$$

In (1) I assume that $f(X, m, Kx)$, as a function of X for any given m and Kx, exhibits the inverted "U" properties specified above. Figure 1 provides an illustrative mapping of the recruitment function. Also in (1), it will be assumed that $f_3 < 0$, that is, any increase in harvest will lower net recruitment. In general, we assume an interaction between the harvest and the productivity of the stock. If there is no interaction, then $\dot{X} = f(X, m) - Kx$; that is, an additional ton of catch reduces instantaneous growth by a ton.

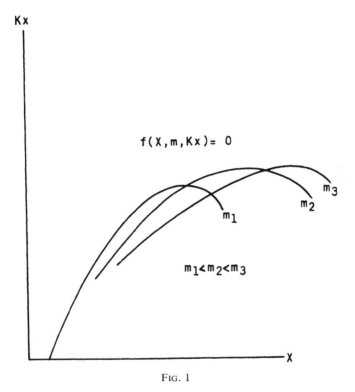

Kx

f(X,m,Kx)= 0

m_3

m_2

m_1

$m_1 < m_2 < m_3$

X

FIG. 1

[1] Various special forms of such a recruitment–rate function are implicit in the fishing literature. All treatments known to me assume steady-state conditions ($\dot{X} = 0$). Thus, in Gordon's (1954) pathbreaking analysis of "bionomic" equilibrium, the harvest is assumed to depend upon population size and "fishing effort," E, or $Kx = F(X, E)$ in my notation. His model further assumes that population declines with harvest, $X = X(Kx)$. Turvey (1954), dealing with the economics of mesh control, assumes $Kx = F(E, m)$, in my notation. Turvey follows Beverton and Holt (1957) in assuming that population size does not interact with fishing effort in determining the harvest rate (or total fishing mortality) under steady-state conditions. I dispense with any use of the concept of "fishing effort," since it is adequately and more familiarly measured by operating cost. All the above authors assume $C = C(E)$—cost is an increasing function of effort.

The most general hypothesis governing the long-run operating costs of a fishing vessel must account for both stock and crowding externalities. If we let $\hat{\pi}$ be the minimum rate of profit required to hold a vessel in the industry, then total cost per unit time is assumed to be given by $C = \phi(x, X, m, K) + \hat{\pi}$ for an individual fisherman and fishing vessel. Given the size of the fish population, mesh size, and number of vessels, cost increases with the vessel catch rate, x. Similarly, for any given catch rate, mesh size, and industry size, cost decreases with X. A *ceteris paribus* increase in mesh size is assumed to require fishing "effort" (in particular, the number of a vessel's netting motions, due to escapement), and therefore costs, to increase if the same weight of catch is to be maintained. Thus, m enters as a private cost factor in ϕ. As an externality it enters indirectly via equation (1). Finally, an increase in the size of the industry, with x, X, and m fixed, will increase each vessel's operating costs due to crowding. Either of the externality variables K or X may be absent from the cost function in particular fisheries. Crowding may only rarely be a factor, and for some species cost may not be affected significantly by population size. Where population size is very large relative to the industry, resource stock externalities are likely to be negligible. Hence, we assume that $\partial C/\partial x \equiv C_1 > 0$, $\partial C/\partial X \equiv C_2 \leq 0$, $\partial C/\partial m \equiv C_3 > 0$, $\partial C/\partial K \equiv C_4 \geq 0$.[2]

Industry revenue $R(Kx, m)$ is assumed to depend upon both the harvest Kx and the mesh size m used by all K vessels in the industry.[3] Increases in m, for a given harvest, raise the average size of fish caught. The result will be an increase in revenue for a species whose larger members are in demand, while revenue will decrease if only the smaller members are desired. Hence, profit for the individual fisherman can be written $\pi = p(m)x - C(x, X, m, K)$, where $p(m) = [R(Kx, m)]/(Kx)$. We assume that the individual fisherman desires to maximize this profit, but that he perceives only x and m as decision variables, with x not affecting price. His price may be affected by m because species size is priced as a quality in the market very much as "long grain" and "short grain" rice bring different prices to the competitive rice farmer. Thus, for given m, the individual fisherman perceives a fixed price $p(m) = [R(Kx, m)]/(Kx)$ at which he can sell unlimited quantities of fish, x, giving him a revenue $p(m)x$. His profit-maximizing decision rules are therefore

$$p(m) \equiv \frac{R(Kx, m)}{Kx} = C_1(x, X, m, K), \tag{2}$$

$$p'(m)x \equiv \frac{R_2(Kx, m)}{K} \leq C_3(x, X, m, K), \text{ if } <m = \underline{m}. \tag{3}$$

---

[2] Anthony Scott has called my attention to the possibility that $C_4 < 0$. If more vessels make it easier to find fish, then fish discovery appears as an external economy.

[3] For simplicity it is assumed that all vessels use the same size mesh, whereas in practice, for some species, different vessels might specialize in the capture of different sizes of fish. A treatment of this case would require a relaxation of the assumption that firms are identical.

Condition (2) requires the perceived price to equal marginal catch cost, and (3) requires the marginal revenue from varying the composition of the catch (mesh) to equal its marginal cost. If in the latter case marginal cost exceeds marginal revenue at the maximum, then $m = \underline{m}$; that is, we make mesh as small as possible.

In Figures 2 and 3 we illustrate possible partial equilibrium solutions for $m = m°$. In Figure 2 $p'(m)$ is assumed to be positive, and above $C_3(x°, X°, m, K°)$, for some values of $m > \underline{m}$, with (x, X, K) given. We have then a unique optimal mesh size $m° > \underline{m}$ for each fisherman. Figure 3 illustrates the case in which each fisherman has no private incentive to harvest only the older and larger members of the species and proceeds to use the smallest practicable mesh size, which is, by definition, $\underline{m}$.

Finally, we assume free entry (and exit) in proportion to profit (and loss), with the exit-speed coefficient not necessarily equal to the entry-speed coefficient. That is, letting $dK/dt \equiv \dot{K}$, we have

$$\dot{K} = \begin{cases} \delta_1 \pi, & \text{if } \pi \geq 0, \\ \delta_2 \pi, & \text{if } \pi < 0, \end{cases} \tag{4}$$

with $\delta_1 \geq \delta_2$, so that vessels might enter the particular fishery in response to profit at a more rapid rate than they would leave in response to loss. This would be the case if vessels were relatively specialized and durable. On the other hand, if fishing vessels can easily be used for the capture of other species, or even for non-fishing activities, then we might have $\delta_1 = \delta_2$.

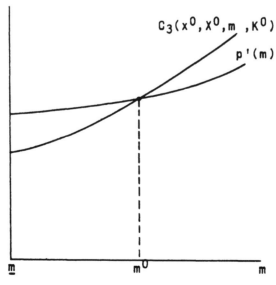

FIG. 2

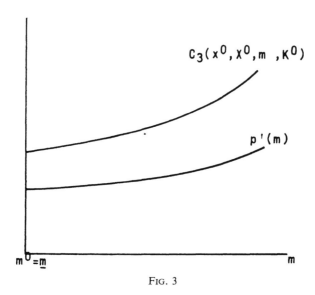

$C_3(x^0, x^0, m, K^0)$

$p'(m)$

$m^0 = \underline{m}$ m

Fig. 3

If we assume further that equations (2) and (3) together provide unique values of x and m for every (X, K) pair, the system (1)–(4) reduces to two non-linear first-order differential equations:

$$\dot{X} = F(X, K), \tag{5}$$

$$\dot{K} = I(X, K). \tag{6}$$

When $\dot{X} = 0$, (5) defines those combinations of (X, K) that produce ecological equilibrium between the fish mass and the exploiting industry. When $\dot{K} = 0$, (6) defines those combinations of (X, K) that produce equilibrium between the exploiting industry and alternative uses of capital in the economy as a whole.

The system (1)–(4) assumes a single large fishery exploited by a competitive industry. A special case of this system occurs when there are numerous fisheries in a competitive world and price can be regarded as given to the individual fishery. Then, the profit function is again $\pi = p(m)x - C(x, X, m, K)$, but now Kx is total output from an individual fishery and $p(m)$ is the world price independent of the output of any one fishery. In this case, corresponding to (1)–(4):

$$\dot{X} = f(X, m, Kx), \tag{1'}$$

$$p(m) = C_1(x, X, m, K), \tag{2'}$$

$$p'(m)x \leq C_3(x, X, m, K), \quad \text{if} <, \quad \text{then } m = \underline{m}, \tag{3'}$$

$$\dot{K} = \begin{cases} \delta_1 \pi, & \text{if } \pi \geq 0 \\ \delta_2 \pi, & \text{if } \pi < 0 \end{cases}. \tag{4'}$$

II. A Quadratic Illustration

The above model is not mathematically simple. It is very rich in possible solutions, given only the stated qualitative restrictions on the cost, revenue, and recruitment rate functions. This can be demonstrated clearly in a very simple illustration which assumes no crowding externalities and which abstracts from mesh size considerations.

Assume a quadratic total revenue function, $R(Kx) = (\alpha - \beta Kx)Kx$, with $\alpha, \beta > 0$, and a quadratic recruitment rate function, $f(X) = (a - bX)X$, with $a, b > 0$. Notice that we have $\underline{X} = 0$, $\overline{X} = a/b$, $X^\circ = a/(2b)$. Finally, let total cost be $C = [(\gamma x^2/X) + \hat{\pi}]$, with $\gamma > 0$, $\hat{\pi} > 0$. The equation system corresponding to (1)–(4) becomes

$$\dot{X} = (a - bX)X - Kx, \tag{7}$$

$$\alpha - \beta Kx = \frac{2\gamma x}{X}, \quad \text{or} \quad x = \frac{\alpha X}{2\gamma + \beta KX}, \tag{8}$$

$$\dot{K} = \begin{cases} \delta_1\pi, & \pi \geq 0 \\ \delta_2\pi, & \pi < 0 \end{cases} \quad \pi = (\alpha - \beta Kx)x - \frac{\gamma x^2}{X} - \hat{\pi}. \tag{9}$$

With $\dot{X} = 0$, equations (7) and (8) define the resource stock equilibrium curve $F(X, K) = 0$ shown in Figure 4. With $\dot{K} = 0$, (8) and (9) define the industry investment equilibrium curve $I(X, K) = 0$ in the same figure. The

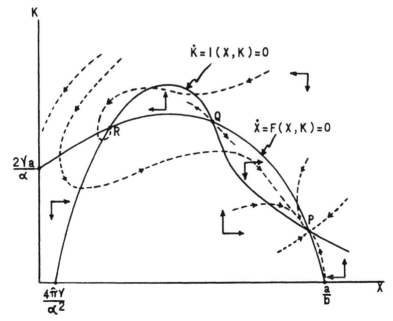

FIG. 4

directions of motion of a point in phase space (X, K) are indicated by the perpendicular arrows in the six partitions of the non-negative quadrant that are formed by the two curves. Points above the fishery resource equilibrium curve represent states in which the total harvest exceeds the recruitment of new stock, c..using a net decline in the stock, while points below correspond to states in which recruitment exceeds the harvest, causing the fish stock to rise. Points above the industry investment equilibrium curve are states in which profits are negative, causing an outflow of capital (if $\delta_2 > 0$) from the industry, while points below represent states of positive profit, causing an inflow of capital into the industry.

The dashed curves in Figure 4 illustrate various possible dynamic paths in phase space on the assumption that $\delta_1 = \delta_2$. Beginning at any initial point $[K(0), X(0)]$ on such a path, the system moves along the path in the direction indicated. Thus, if $K(0) = 0$, $X(0) = a/b$, corresponding initially to an unexploited fishery and non-existent fishing industry, investment and the fish population mass move along the indicated path to the stationary stable equilibrium, point P. Another stable equilibrium point is illustrated by R, toward which convergence is along a cyclical path while Q is an unstable equilibrium.

Figure 5 illustrates a case with a single equilibrium point in the positive quadrant. The path OST is intended to illustrate a possible dynamic path with $\delta_1 = \delta_2 > 0$, while OST' is a possible outcome if $\delta_1' = \delta_1 > \delta_2' > 0$.

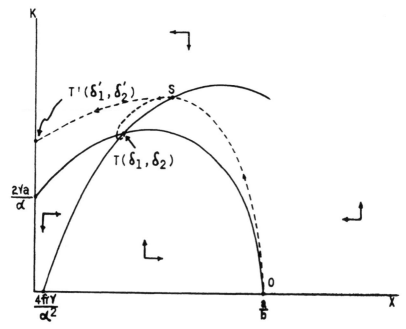

FIG. 5

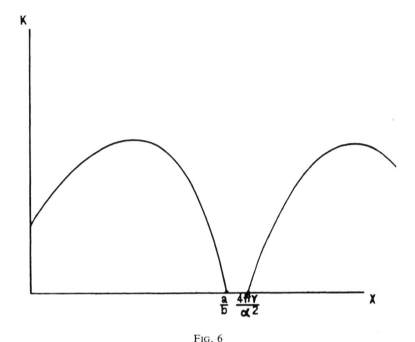

FIG. 6

Figure 6 illustrates a case in which $a/b < (4\hat{\pi}\gamma)/\alpha^2$, and exploitation of the fishery is not economically viable. This results from a combination of relatively (i) low market value (α), (ii) high required rate of return on vessels ($\hat{\pi}$) (as, for example, might occur if risks are unusually high), (iii) high operating costs (γ), or (iv) low natural-state population mass (a/b). This set of conditions is, of course, the most common case, since the overwhelming majority of ocean and freshwater animal species are not commercially recoverable.

III. Sole Ownership or Right of Access to a Fishery

For purposes of contrast and comparison with decentralized competitive exploitation, consider next the case of a fishery exploited by a sole owner or a firm to whom an exclusive right of access has been granted. Although such a state is rare in modern capitalistic societies, it was and is very common, where feasible, in societies judged to be more "primitive."

Indeed, Gordon (1954, p. 134) reports that common tenure is very rare even in hunting societies, and then only in those cases where the hunted resource is migratory over such a large area that it becomes unfeasible for the society to regard the resource as husbandable. Apparently, the "invisible hand" has an ancient history of endowing societies with an economic wisdom in traditions of fisheries resource utilization that have not always been continued in modern societies.[4]

Whatever may be the proper interpretation of anthropological data on the granting of property rights, the economic function of such rights is clear: The appropriation of a resource by a sole owner, or its legal equivalent, internalizes or "privatizes" the social costs associated with the three types of externalities discussed above.

Consider a sole owner exploiting a single fishery that is small in relation to the world supply, so that price $p(m)$ is given to the fishery. His profit is $\pi = p(m)Kx - KC(x, X, m, K)$, which he desires to maximize with respect to the choice of (x, m, X, K) subject to the constraint $f(X, m, Kx) = 0$. The Lagrange function is then $\phi = p(m)Kx - KC(x, X, m, K) + \lambda f(X, m, Kx)$, and the first-order conditions for a constrained maximum can be put in the form

$$p(m) = C_1 - \lambda f_3, \tag{10}$$

$$xp'(m) + \frac{\lambda f_2}{K} \le C_3, \quad \text{if } <, \text{ then } m^\circ = \underline{m}, \tag{11}$$

$$\lambda = \frac{KC_2}{f_1}, \tag{12}$$

$$\frac{\pi}{K} = p(m)x - C = KC_4 - \lambda f_3 x, \tag{13}$$

$$f(X, m, Kx) = 0. \tag{14}$$

The Lagrange multiplier, λ, is the marginal profitability of recruitment. The marginal profitability of the fleet harvest is $-\lambda f_3$, which, in profit equilibrium, must equal the marginal "social" cost of the harvest, $(KC_2 f_3)/f_1$, due to resource stock externalities, from (12). The term "social" cost refers to costs external to the individual vessels and therefore external to decentralized competitive firms, but of course they are private "user" costs to the sole owner. Condition (10), in the form $p(m) = C_1 - (KC_2 f_3)/f_1$, requires the vessel catch rate to be adjusted until price equals direct plus user catch cost. Condition (13) requires the net marginal direct return on investment in a vessel, $p(m)x - C$, to equal the marginal social cost, $KC_4 - (KC_2 x f_3)/f_1$, of adding the vessel to the fleet. The term KC_4

---

[4] However, no such economic wisdom is apparent during the Pleistocene period in the husbanding of those native North American mammals whose adult body weight exceeded one hundred pounds. Some 70 per cent of all such species (mammoth, mastodon, horses, camels, ground sloths, oxen, antelope, saber-toothed tiger, giant beaver, and so on, totaling over one hundred species) became extinct during a period of only one thousand years following the arrival of the Paleo-Indians about twelve thousand years ago. For a forceful marshaling of evidence in favor of the hypothesis that man, through the use of fire, the stone-tipped spear, and the communal hunting party, succeeded in a gigantic "overkill" of these megafauna, see Martin (1967).

reflects external costs due to crowding, while $-(KC_2xf_3)/f_1$ reflects catch externalities by the additional vessel. Finally, (11) requires the marginal private plus social revenue from mesh adjustment (varying catch composition) to not exceed the (private) cost of the adjustment. The term $\lambda f_2/K = (C_2f_2)/f_1$ measures social gains (savings in external catch costs) produced by an increase in mesh size, which in turn increases population recruitment.

IV. Competitive versus Centralized Recovery

By comparing the system (1′)–(4′), $\dot{X} = \dot{K} = 0$, with (10)–(14), it is seen that the conditions governing exploitation under the competitive and centralized organizations are the same except for the above social costs. Several propositions follow from such a comparison.

Note first that the sole owner will never deplete a fishery to a population mass at which $f_1 > 0$; that is, in Figure 1, for an optimal mesh, he will always operate to the right of the maximum sustainable yield point. To prove this, assume the contrary and let the system (10)–(14) be satisfied by a point (x^*, X^*, m^*, K^*) with $f_1(X^*, m^*, K^*x^*) > 0$. Then from the properties of the function f, there must exist an $X^{**} > X^*$ such that $f(X^*, m^*, K^*x^*) = f(X^{**}, m^*, K^*x^*) = 0$, and $f_1(X^{**}, m^*, K^*x^*) < 0$. But since $C_2 < 0$, it follows that the point (x^*, X^{**}, m^*, K^*) produces the same total catch, satisfies the constraint (14), and provides a larger profit. Profit cannot be a maximum unless $f_1 < 0$. Hence, from (12), $\lambda > 0$.

Second, under sole ownership, a pure profit or rent is provided by the fishery: Since $C_4 > 0$ and $\lambda f_3 < 0$, from (13) we must have $\pi > 0$ in equilibrium. This positive rent would eventually be absorbed in higher costs by free entry under a competitive organization of production.

The literature of fishery economics contains several discussions of the effect of competition versus sole ownership on capital requirements and output (or sustainable yield) in the stationary state. Thus, Gordon (1954, p. 141) states, "The uncontrolled (competitive) equilibrium means a higher expenditure of effort, higher fish landings, and a lower continuing fish population than the optimum [sole ownership]." Christy and Scott (1965, p. 9) say, "Eventually [under competition], the fishery may arrive at an equilibrium . . . which is likely to be marked by a relatively large amount of effort, a low population, and a low sustainable yield," as contrasted with sole ownership. Also (*idem*), "There will tend to be an excessive amount of capital and labor applied to the fishery."

It does not follow from the models of the present paper that capital requirements are greater or that output (sustainable yield) is unambiguously either larger or smaller under competition than under sole ownership. This is best demonstrated by a counterexample. Consider a simple case of the above models in which $f_2 \equiv C_3 \equiv C_4 \equiv 0$ and $\dot{X} = f(X) - Kx$

                       JOURNAL OF POLITICAL ECONOMY

(no mesh or crowding externalities). Also assume that the output capacity of each vessel is fixed (for example, marginal operating cost might be constant up to some capacity limit), say $x = \bar{x}$, and that $\delta_1 = \delta_2 = \delta$. Then the cost function for a vessel is $C(\bar{x}, X)$ and exploitation is defined by $\dot{X} = f(X) - K\bar{x}$ and $\dot{K} = \delta[p\bar{x} - C(\bar{x}, X)]$. At a stationary equilibrium,

$$p\bar{x} = C(\bar{x}, X^*), \tag{15}$$

$$K\bar{x} = f(X^*), \tag{16}$$

where X^* is the unique fish population at which the industry is just normally profitable. The heavy-lined curves of Figures 7 and 8 illustrate the functions (15) and (16) in phase space and an equilibrium of both population and capital investment at (X^*, K^*).

The sole owner's profit function is $\pi = pK\bar{x} - KC(\bar{x}, X)$, and his static equilibrium is defined by

$$K\bar{x} = \frac{(p\bar{x} - C)f'}{C_2}, \tag{17}$$

$$K\bar{x} = f(X), \tag{18}$$

where (17) is implied by the first-order conditions for a constrained maximum. In this model, comparing the two systems of production

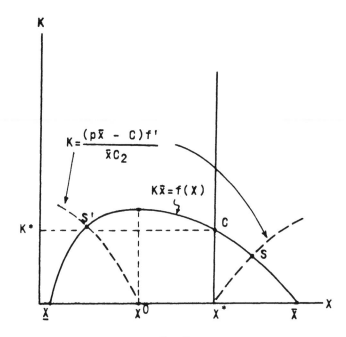

FIG. 7

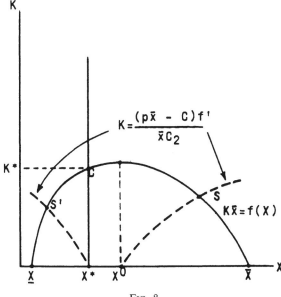

FIG. 8

organization becomes a matter of comparing equations (15) and (17). In Figures 7 and 8 the dashed lines through S and S' represent, in general, the function (17) on the assumptions $C_2 < 0$, $C_{22} > 0$ and the previously stated properties of $f(X)$.

Thus, in Figure 7 is depicted a competitive solution with $X^* > X^\circ$ and a sole owner's maximum at S. To show why, we compute

$$\bar{x}\frac{dK}{dX} = \frac{(p\bar{x} - C)f''}{C_2} - f' - \frac{(p\bar{x} - C)f'C_{22}}{(C_2)^2} \qquad (19)$$

and observe from (15) that $p\bar{x} - C \lesseqgtr 0$ according as $X \lesseqgtr X^*$. Now, if $\underline{X} \le X < X^\circ$, then $p\bar{x} - C < 0, f' > 0$, and it follows from (17) and (19) that $K\bar{x} > 0$ and $\bar{x}(dK/dX) \lesseqgtr 0$, as shown by the dashed curve through S' in Figure 7. At $X = X^\circ, p\bar{x} - C < 0, f' = 0$, and $K\bar{x} = 0$, $\bar{x}(dK/dX) < 0$. For $X^\circ < X < X^*$, $K < 0$, so this segment of (17) is omitted in Figure 7. At $X = X^*$, $p\bar{x} - C = 0$ and $f' < 0$, and it follows that $K = 0$ and $\bar{x}(dK/dX) = -f' > 0$. Finally, on $X^* < X < \bar{X}, p\bar{x} - C > 0, f' < 0$ and $K > 0$, while $\bar{x}(dK/dX) > 0$, as shown by the dashed curve through S.

Similarly, in Figure 8, for $\underline{X} \le X \le X^*$, $K \ge 0$, and $\bar{x}(dK/dx) \lesseqgtr 0$; for $X^* < X < X^\circ$, $K < 0$; and for $X^\circ \le X < \bar{X}$, $K \ge 0$, $\bar{x}(dK/dX) > 0$, as indicated by the dashed lines. In both figures, point C represents the competitive equilibrium; S and S' represent points satisfying (17) and (18) for the sole owner, but only S can be a global maximum and therefore a profit equilibrium point. Comparing C and S in Figure 8, it is clear that

Fisheries Economics I

the competitive equilibrium may require a larger or smaller amount of capital than sole ownership; also, the harvest, which is the same as the population yield ($K\bar{x}$), may be larger or smaller at C than at S. That point C may imply a smaller harvest (and capital requirements) than S is made obvious by the fact that we could choose $p\bar{x}$ so that X^* equals (or is near to) $\underline{X}$, where the competitive harvest is zero, yet $\bar{x}(dK/dX) > 0$, $K > 0$, for $X° \leq X < \bar{X}$ so that at S, $K\bar{x} > 0$.

These results, insofar as they stand in contrast to the earlier literature, are due to the explicit hypothesis that population reduction increases operating cost, while at first increasing, then decreasing, sustainable yield.

V. Regulation of Competitive Recovery

An alternative to centralized management as a means of achieving efficient fishery production is to regulate appropriately the competitive process. This has been the attempt in practice. In theory the objective of regulation is to induce the decentralized competitive industry to behave like a sole owner. One way to achieve this is for the social costs appearing in the system of behavior equations (10)–(13) to be imposed by the regulating authorities upon the decision-making units of the industry. Equations (10)–(13) exhibit three kinds of social costs which must be reflected in each competitive fisherman's profit criterion: (i) A unit catch cost, $\lambda = KC_2/f_1$, reflecting the effect on fishing cost of a reduced population caused by an additional unit of catch. This social cost can be imposed on each fisherman by levying an extraction fee $U = -\lambda f_3$ on each pound of catch docked by a vessel. (ii) An annual vessel operating cost, KC_4, which measures the external crowding cost caused by an additional vessel in the industry. This charge is most easily levied by an annual license fee $L = KC_4$ on each operating vessel. (iii) Finally, as a control on mesh size, we impose a penalty cost, k, on the mesh employed by a vessel, where

$$k = \begin{cases} P, & \text{if } m \lessgtr m° \\ 0, & \text{if } m = m° \end{cases}.$$

The P is a fine large enough to dominate net revenue, so that it never pays the fisherman to choose $m \lessgtr m°$, where $m°$ is the optimal mesh satisfying the system (10)–(14).[5]

---

[5] It might be supposed that only a minimum mesh size restriction need be imposed, with a fine levied on the harvesting of smaller members of the population. In most cases this would probably be sufficient. However, for those species whose smaller members are prized, or whose medium-sized members are preferred by consumers, maximum or intermediate mesh restrictions would be in order. The strong-fine condition, k in the text, covers all these possibilities.

Hence, the individual fisherman's profit function for a vessel is $\pi = p(m)x - C(x, X, m, K) - Ux - L - k$, and the conditions for long-run industry equilibrium are

$$p(m) = C_1 + U, \tag{20}$$

$$m = m^o, \quad k = 0, \tag{21}$$

$$\pi = p(m)x - C - Ux - L = 0, \tag{22}$$

$$f(X, m) = Kx. \tag{23}$$

In (20) each fisherman is assumed to adjust his catch rate until price equals marginal cost inclusive of extraction fees. Due to the high penalty for $m \lessgtr m^o$, m^o is each fisherman's optimal mesh size. Equation (22) defines industry investment equilibrium when firms are subject to extraction and license fees. Hence, the equations for equilibrium of regulated competitive exploitation are the same as for the sole owner, (10)–(14), provided that $U = -\lambda f_3$ and $L = KC_4$ are fixed at optimizing values satisfying (10)–(14) and k has the form specified above (see Turvey, 1964).

VI. Comparison with Current Theory

Briefly stated, current fishing theory (Gordon, 1954, p. 136; Scott, 1955; Schaefer, 1957; FAO Fisheries Reports No. 5, 1962; Christy and Scott, 1965, chap. ii; Crutchfield, 1965) is as follows: Total cost for the industry is a function of fishing effort, $C = C(E)$. The sustainable yield or harvest by the industry, Y, is a function of effort, $Y = g(E)$. Total revenue is a function of yield, $R = R(Y)$. Hence, net return is

$$N = R(Y) - C(E) = R[g(E)] - C(E), \tag{24}$$

with effort E the only adjustment variable. Where mesh size is introduced as a decision variable (Turvey, 1964, pp. 66–67), it is assumed that $Y = g(E, m)$, that g has a maximum with respect to m for each E, and that revenue and cost are independent of m. Mesh size determination is then a separable suboptimization problem.

Often it is assumed that effort per fisherman is constant, so that total effort can be measured by the number of fishermen (K in my model). Net return is expressed as a function either of effort, as in (24), or of K. Figure 9 reproduces the diagram most often used as an illustration. That is, most authors assume constant long-run cost per fisherman so that C is proportional to E or K. Also, fish price is usually assumed to be constant, so that total revenue simply follows the inverted U-shaped sustainable yield curve. In Figure 9, the sole owner does not expand exploitation beyond OS, where net revenue is a maximum. Under decentralized, unregulated exploitation, the equilibrium effort or number of fishermen is at OD and all the rent of the fishery is absorbed in cost.

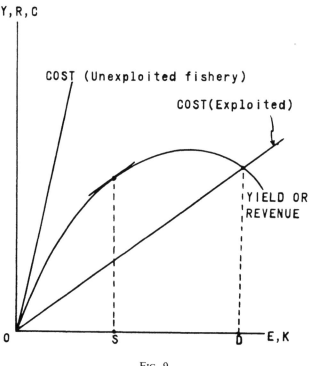

Fɪɢ. 9

   This theory is able to account for the situation in which it is not commercially feasible to exploit a fishery; that is, cost may be everywhere
above revenue for a particular species (see Fig. 9). However, it is not
able to handle the situation in which a species may be depleted to the point
of extinction. The theory implies that short of zero unit cost, the equilibrium yield to competitive exploitation is never zero with an extinct
population. (Such a solution is possible in [1']–[4'], for example, as in
Figure 8 if $X^* < \underline{X}$.)
   This is perhaps one of the more serious deficiencies in the received
doctrine.[6] In addition, the standard analysis does not provide a dynamic
theory;[7] it is not explicit about the various types of externalities that may

---

   [6] If one believes in the boundless potential and inexhaustibility of the seas, then
this is not a serious deficiency. But the myth of boundless potential, in my view, has
been effectively destroyed by the arguments of Christy and Scott (1965)—in fact the
present paper is in large measure an attempt to bring the economic theory of fishing
into line with the persuasive arguments of their valuable and stimulating work.
   [7] An important exception to this is to be found in the Mathematical Appendix to
the book by Crutchfield and Zellner (1962, pp. 112–17). Their Appendix models are
in the same dynamic spirit as those of the present paper, but use the quadratic form
of the biological differential equation constraint such as I have used in the illustration

arise; nor does it explicitly distinguish the effect of such variables as vessel catch rate, fish population mass, investment, and mesh size.

The models developed in this paper can be adapted for purposes of comparison with Figure 9. Returning to the illustration used for the counterexample in Section IV, we have the vessel catch rate fixed, and total industry cost is $KC(\bar{x}, X)$. Total revenue is $R = pK\bar{x}$, and net revenue to the industry is

$$N = pK\bar{x} - KC(\bar{x}, X). \tag{25}$$

The conditions for maximum N to the sole owner are:

$$p\bar{x} = C + \frac{KC_2\bar{x}}{f'}, \tag{26}$$

$$K\bar{x} = f(X). \tag{27}$$

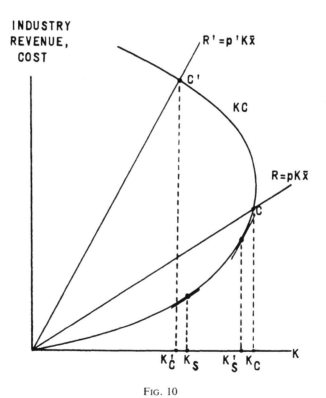

FIG. 10

of Section II. The present paper generalizes their biological constraint, but more importantly, couples this constraint with a dynamic model of the exploiting industry. They also treat the problem of optimal fishery management over time, which I expect to consider in a separate paper.

Competitive equilibrium results when (25) and (27) are satisfied, with $N = 0$.

Figure 10 illustrates these solutions and provides the counterpart of Figure 9. Total revenue is proportional to fishermen (or the number of vessels). Total cost as a function of K only is determined simultaneously by $KC(\bar{x}, X)$ and condition (27). The slope of industry cost, in these terms, is $[d(KC)]/dK = [(KC_2\bar{x})/f'] + C$. Notice that the industry total-cost function does not have an inverse, because $f(X)$ does not. At price p the number of vessels operated by the sole owner is K_s, while the competitive industry operates $K_c > K_s$ vessels. If price is $p' > p$, as shown, the sole owner operates K'_s vessels, while $K'_c < K'_s$ are operated under competition. Under the assumptions of Figure 9, the theory of this paper implies *linear revenue* and *backward bending cost*, rather than linear cost and the humped revenue function of Figure 9.

References

Beverton, R. J. H., and Holt, S. V. *On the Dynamics of Exploited Fish Populations.* ("Fisheries Investigations," Vol. XIX, Ser. 2.) London: Ministry of Agriculture, Fisheries, and Food, 1957.

Christy, F. T., and Scott, A. *The Common Wealth in Ocean Fisheries.* Baltimore: Johns Hopkins Press, 1965.

Crutchfield, James A. (ed.). *The Fisheries: Problems in Resource Management.* Seattle: University of Washington Press, 1965.

Crutchfield, James A., and Zellner, A. *Economic Aspects of the Pacific Halibut Fishery.* ("Fishery Industrial Research," Vol. I, No. 1.) Washington: U.S. Department of the Interior, 1962.

Food and Agricultural Organization Fisheries Reports No. 5, *Economic Effects of Fishery Regulation*, Rome, 1962.

Gordon, H. Scott. "The Economic Theory of a Common-Property Resource: The Fishery," *J.P.E.*, LXII, No. 2 (April, 1954), 124–42.

Lotka, A. J. *Elements of Mathematical Biology.* New York: Dover Publications, 1956.

Martin, Paul S. "Pleistocene Overkill," *Nat. Hist.*, LXXVI, No. 10 (December, 1967), 32–38.

Schaefer, M. B. "Some Considerations of Population Dynamics and Economics in Relation to the Management of the Commercial Marine Fisheries," *J. Fisheries Res. Board Canada*, XIV, No. 5 (September, 1957), 669–81.

Scott, A. "The Fishery: The Objectives of Sole Ownership," *J.P.E.*, LXIII, No. 2 (April, 1955), 116–24.

Smith, Vernon L., "Economics of Production from Natural Resources," *A.E.R.*, LVIII, No. 3 (June, 1968), 409–31.

Turvey, Ralph. "Optimization in Fishery Regulation," *A.E.R.*, LIV, No. 2, Part I (March, 1964), 64–76.

[12]

The Relationship Between Firm and Fishery in Common Property Fisheries[†]

Lee G. Anderson[*]

INTRODUCTION

The purpose of this paper is to show the relationship between firm and industry in a common property fishery and to compare and contrast it to the standard analysis. Most of the previous work in this area [Gordon 1954; Scott 1955; Crutchfield and Zellner 1961; Copes 1970; Gould 1972a; Anderson 1973, 1975a, 1975b; and Clark 1973] has been in aggregate terms only. In the model developed by Smith [1968, 1969] and in the interchanges between him and Fullenbaum, Carlson, and Bell that these works generated [Smith 1971, 1972; and Fullenbaum, Carlson, and Bell 1971, 1972], however, the operation of the firm is directly taken into account. Because neither Smith's model nor the traditional literature deal adequately with changes in the level of firm output in the context of obtaining an industry equilibrium, Fullenbaum, Carlson, and Bell conclude, "The integration of the classical theory of the firm and the traditional theory of commercial fishing remains an unfinished task" [1972, p. 768].

I propose to complete this integration in a model that focuses attention on the level of fishing effort as the output of the individual vessels rather than on the catch rate as do Smith and Fullenbaum, Carlson, and Bell. This change in the frame of reference will make the analysis of vessels in a fishery strictly analogous to the standard firm-industry model and more important, more logical and empirically more useful. It is more logical because, as the discussion to follow will show, while vessels can directly control fishing effort, they can only indirectly control vessel catch rate. This change then allows for a more explicit model. It is empirically more useful because it allows for the use of biological and physical information in a form already provided by fishery scientists and fishery management organizations. These strong statements notwithstanding, the generalized model that results from the analysis to follow, although significantly different, is a modification of the basic model as presented by Smith. The benefits from completing this integration are a description of vessel operation while in the process of reaching a long-run industry equilibrium and a more complete analysis of regulatory problems.

The paper will begin with a brief review of the traditional fisheries model

---

[†]I am grateful for useful comments on an earlier draft from my colleagues Richard J. Agnello, Lawrence P. Donnelley and from Ivar Strand and two referees, but the usual disclaimer applies. This study was begun while I was at the University of Miami, and was completed at the University of Delaware. It was sponsored by their Sea Grant Institutional programs which are administered by the National Oceanic and Atmospheric Administration of the United States Department of Commerce.

[*]Department of Economics and College of Marine Studies, University of Delaware.

which will be used as the basis for a simple diagrammatical analysis of the vessel and fishery. Then a formal mathematical model will be presented and compared with earlier models. The analysis will be static in the sense that no consideration will be given to maximization over time, although there will be some brief discussion of the process of obtaining a static open-access equilibrium.

THE TRADITIONAL AGGREGATE MODEL

Following Schaefer [1954, 1957, 1959] and Beverton and Holt [1957, pp. 30, 31], but making some small adjustments to make the transition to the analysis to follow easier, the traditional theory of commercial fishing can be expressed in a simplified form as follows.[1] The rate of growth of the fish stock is a function of its size:

$$\dot{X} = \dot{X}(X) \qquad [1]$$

where X is the size or biomass of the fish stock and $\dot{X}$ is the natural growth rate. From the biologist's point of view, fishing mortality or yield per period of time can be expressed as:

$$F = fX \qquad [2]$$

where f is the fishing mortality coefficient. As long as the distribution of the stock remains uniform, f is proportional to effort expressed in standardized units of fishing time or, in some cases, fishing activity [Beverton and Holt 1957, p. 30]. The process of standardization takes into account differences in tonnage and gear types of different vessels. For example, a day's fishing by a 25-ton side trawl vessel is not equivalent to one of a 100-ton side trawler. The same holds true for a 100-ton side trawler and a 100-ton stern trawler. For an example of

how this standardization process is carried out see Brown et al. [1975].

Given this proportionality between effort and fishing mortality, annual yield can be expressed as:

$$F = a\bar{E}X \qquad [2']$$

where a is the constant of proportionality, and $\bar{E}$ is total effort. Total effort is the product of K, the number of vessels, and E, the average amount produced by each one. This expression, derived from biological theory, is subject to the normal economic interpretation of a production function. It makes explicit a very important point—the same amount of standardized effort will obtain a different yield according to the stock size. Or put somewhat differently, effort, which can be measured in standardized days fishing, is defined in terms of the manmade mortality on the stock, and its correspondence to catch is directly related to stock size.

There will be a biological equilibrium in the fishery, i.e., the net rate of growth in the stock will equal zero, when its natural growth rate just equals the industry catch rate.

$$\dot{X} - a\bar{E}X = 0 \qquad [3]$$

From this biological equilibrium condition it is possible to derive an equation for the equilibrium stock size as a function of total effort.[2]

$$X^* = X^*(\bar{E}) \qquad [4]$$

---

[1] To keep the analysis straightforward, mesh size or other selective size controls and crowding externalities will be ignored in this section, although they will be included in the mathematical formulation of the integrated model.

[2] One of the areas of contention between Smith and Fullenbaum, Carlson, and Bell is whether or not this is a single-valued function. Subsequent work by Gould [1972a] and Southey [1972] has demonstrated the conditions under which it will be double-valued and the repercussions that follow. For our purposes, we will assume that it is always single-valued.

That is, a given level of effort will cause the stock to reach an equilibrium at that size where the natural growth rate equals the harvest obtained by that combination of effort and stock size. Substituting [4] into [2'] obtains sustainable yield as a function of effort only.

$$F^* = a\bar{E}X^*(\bar{E}) \qquad [5]$$

A sustainable yield occurs when the total yield obtained by a particular level of effort equals the growth rate of the stock when it is in equilibrium for that level of effort. Because of the effect of fishing effort on the equilibrium stock size, increases in effort will eventually cause sustained yield to decrease, i.e.:

$$\frac{\partial F^*}{\partial E} > 0, \; 0 < \bar{E} < \bar{E}_m$$

$$\frac{\partial F^*}{\partial E} < 0, \; \bar{E}_m < \bar{E}$$

Maximum sustainable yield occurs at $\bar{E}_m$.

The sustained yield curve is ordinarily used as the long-run production function of the industry. Therefore, the traditional analysis is necessarily long run in that it assumes complete stock adjustment to changes in the level of effort.

Assuming that the price of fish (P_F) and the cost of producing a unit of effort (c) are constant,[3] the profit function for the fishery is:

$$\bar{\pi} = P_F F^* - c\bar{E} \qquad [6]$$

As Gordon [1955] has pointed out in his now classic article, the unregulated fishery will, under normal circumstances, reach an equilibrium where industry profit is equal to zero. As long as a positive industry profit exists, vessels will be encouraged to enter the fishery and utilize the stock, and with common property there is nothing to prevent

them from doing so. In this fixed-price static model, the fishery's contribution to the economy is maximized, however, at that level of effort where the marginal revenue of effort is equal to its marginal cost, i.e., where industry profit is maximized. At this point, called maximum economic yield (MEY), the positive rent being earned is that due to the uniqueness of the fish stock.

Figure 1 contains the typical graphical representation of the traditional analysis. The total revenue curve is a "monetized" sustainable yield curve which reaches a maximum at $\bar{E}_m$, and total cost is a linear function of effort. The open-access equilibrium will occur at $\bar{E}_2$ and MEY occurs at $\bar{E}_1$. Note that because of the position of the intersection of the cost and revenue curves, not only will the open-access fishery dissipate the rent of the fish stock, but in this case it will be operating in a situation where a reduction in fishing effort will actually increase yield.[4]

THE FIRM AND THE INDUSTRY

Description of the Model

As a first step in the integration of this analysis with the classical theory of

---

[3] For a discussion of the problem with a variable price of output, see Copes [1970] and Anderson [1973]. The assumption of a fixed price is analytically more simple, and it rules out monopoly problems, yet still captures the essence of the analysis.

[4] To be formally precise, of course, the goal of regulation should be to maximize net revenues over time. Recent work has shown that a dynamic MEY will occur somewhere between the open-access equilibrium and static MEY; as the discount rate approaches zero, dynamic MEY approaches static MEY. Also a dynamic MEY may occur beyond $\bar{E}_m$ and in some cases may involve destruction of the stock if the discount rate is relatively high and the price of output is higher than the cost of the effort necessary to harvest the last unit of the stock. For a detailed discussion of these points see Brown [1974], Clark [1974], and Neher [1974].

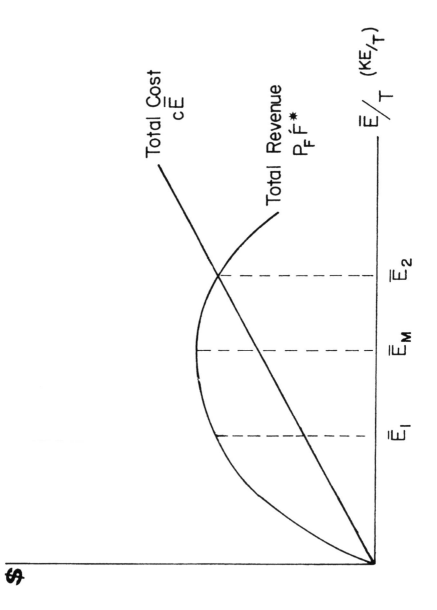

FIGURE 1

the firm, consider the fact that production, and hence revenue, from the fish stock are essentially industry-wide phenomena. Total sustainable yield is a function of total effort and stock size. But since the latter is a function of effort, sustainable yield is determined solely by total effort given the ecological parameters. Therefore, as long as there are many vessels, each of which can provide only a small percentage of total effort, the vessels must take average catch per unit of effort as a parameter determined by the fishery as a whole. This is analogous to competitive firms taking price as a parameter determined by the industry as a whole. On the other hand, cost is a firm phenomenon. Neglecting crowding externalities, the cost of providing effort is determined by the scale (size of vessel) and level of operation (days of fishing, etc.) of the firm. The reason the traditional analysis uses a constant cost of providing effort is that it is essentially a long-run analysis which assumes that any change in the level of effort is the result of a change in the number of efficiently operating vessels [Crutchfield and Zellner 1961, p. 14].

Bearing these two points in mind, consider Figure 2, which is the standard firm-industry diagram modified to fit the peculiarities of a common property fishery. Part A contains the average and marginal cost curves for effort of the representative vessel. Cost is expressed in terms of effort, E, because it is the variable over which the vessel has control. The individual boat has no direct control over its catch in the same way a normal firm has control over its output. The vessel can, however, directly control its production of effort which, given the average catch per unit of effort as set by the interaction of the total fishery-wide

level of effort and the size of the fish stock, will determine its total catch. All of this, of course, assumes that each boat cannot significantly affect the amount of total effort.

The production function for the vessel can be represented as $E = E(a_1, a_2 \ldots a_n)$ where the a's represent the n inputs used (i.e., boat, nets, fuel, ice, men, etc.). As pointed out above, effort can be measured in standardized days fished, but in some cases, other measures such as days out of port, number of times nets are set and retrieved, number of traps that are tended, etc., each properly standardized, would be more appropriate.

The cost curves derived from this production function have the normal shape. It should be stressed that these curves are in terms of standardized effort so as to be additive and hence comparable to the right side of the diagram.[5] With a

---

[5] To make this transition it is only necessary to know the equivalence ratios for different types of boats and gears. For example if the average cost functions for three types of vessels of differing tonnage and/or gear types in terms of calendar days (D_i) fished were:

$$C_A = \alpha_1 - \beta_1 D_A + \psi_1 D_A{}^2$$
$$C_B = \alpha_2 - \beta_2 D_B + \psi_2 D_B{}^2$$
$$C_C = \alpha_3 - \beta_3 D_C + \psi_3 D_C{}^2$$

and if the relative effects of fishing mortality are such that:

$$D_A = 3D_B = 5D_C$$

then using D_A as a standard of reference (i.e., $D_A = E$), these cost curves can be expressed as:

$$C_A = \alpha_1 - \beta_1 E + \psi_1 E^2$$
$$C_B = \alpha_2 - \tfrac{1}{3}\beta_2 E + \tfrac{1}{9}\psi_2 E^2$$
$$C_C = \alpha_3 - \tfrac{1}{5}\beta_3 E + \tfrac{1}{25}\psi_3 E^2$$

The above transformation allows the cost curves for the different vessels to be expressed in units that are comparable with respect to their effect on the stock.

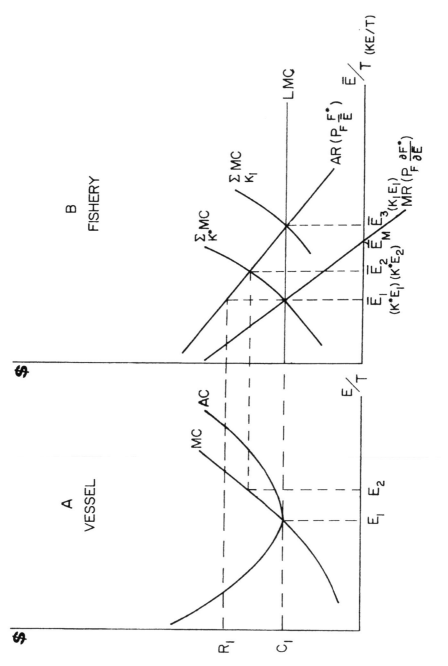

FIGURE 2

given sized boat and other fixed gear, the average annual cost of a unit of effort at first decreases as the amount of effort increases mainly due to the spreading of fixed costs. After a certain point, however, increasing the output of effort per period will result· in increased average cost because of diminishing returns to the variable· factors and other items such as an increased proportion of the maintenance that must be performed at sea, increased handling costs of catch, etc. By ruling out the possibility of crowding externalities, the cost curves of the representative firm will not change with the number of firms in the industry. As usual the marginal cost curve serves as the supply curve (for effort) for the vessel.

The average and marginal revenue curves for effort for the fishery as a whole, $\bar{E}$ which equals KE, K being the number of vessels, are depicted in part B of Figure 2. They can be derived directly from the total revenue curves in Figure 1. Fishery-wide average revenue can be thought of as the derived demand curve for effort. Notice that as the total amount of effort in the fishery increases, the long-run average return to effort (i.e., the market's willingness to pay for it) decreases. This results not from a change in the price of fish (although if price were variable this would aggravate the problem as long as increases in effort lead to increases in yield) but rather to the bio-technological phenomenon of decreasing long-run returns to effort. The fishery-wide marginal revenue curve can be thought of as the social marginal value of effort. Because of the fact that $(\partial F^*)/(\partial \bar{E})$ is negative for industry effort greater than $\bar{E}_m$, the social marginal value of effort is also negative in this range even though the derived demand is positive.

In the normal fashion, the horizontal sum of the supply curves of the firms is the supply curve of effort for the fishery. For example, $\sum_{K_1} MC$ is the fishery-wide supply curve for effort when there are K_1 firms.

The Analysis

Using this format, it is now possible to study the interaction of the fishery as a whole and the individual vessel. Very briefly, the intersection of the fishery-wide supply curve for effort and the average revenue curve determines the average return per unit of effort (assuming a biological equilibrium is reached). This average revenue is the product of the average yield per unit of effort, which is an industry-wide phenomenon, and the price·of fish. Each vessel takes this as its perceived price or marginal revenue of effort (because this is what they will earn by selling the catch obtained by each unit of effort) and operates where the marginal cost of producing effort equals this amount. Of course, all of this occurs simultaneously. Each vessel operates individually according to a perceived price which is set by all vessels operating in concert.

Open-access equilibrium of the fishery will occur at $\bar{E}_3$, with K_1 vessels each producing E_1 units of effort. At this point the average return to effort is just equal to the minimum average cost of producing effort by the representative vessel. If there were fewer vessels, the fishery-wide supply curve would shift to the left increasing the average return to effort. Each vessel would respond to this by increasing its level of effort until marginal cost equaled perceived price. The existence of greater than normal profits would, however, encourage entry of new vessels until average industry return is pushed down to the minimum average cost.

Maximum economic yield is where the value produced by the marginal unit of effort is just equal to the long-run marginal cost of producing it. Assuming no crowding externalities, the long-run marginal cost curve of the fishery will be constant and equal to the minimum of the average cost curve of the representative firm, i.e., additional effort can be assumed to be provided by additional boats each operating at the minimum of their average cost curves.[6] MEY, therefore, occurs at $\bar{E}_1$, where the long-run industry marginal cost curve intersects the fishery-wide marginal revenue of effort curve. Utilizing the normal interpretation of short-run and long-run cost curves, this means that in order to achieve MEY, the number of vessels must be reduced to K^* *and* that each of them must operate at E_1. At this point each vessel will be earning a rent of E_1 $(R_1 - C_1)$ which is its share of the rent of the fish stock. It is not a monopoly rent. To see this, notice that at MEY the marginal cost of effort, $(\partial Cost)/(\partial \bar{E})$ is equal to the marginal return to effort, $(P_F)(\partial F^*)/(\partial \bar{E})$. This can easily be transferred to $P_F = (d\ Cost)/(dF^*)$; i.e., the price of fish is equal to the marginal cost of increasing sustained yield.

As a final point, note that this model allows one to describe the long-run cost curve of the fishery. Open-access equilibrium effort will be produced as efficiently as possible because ultimately, any expansion will take place in the form of new units producing in an optimal fashion. (The common property aspect of the problem will not affect the long-run cost of producing effort, but it will ensure that an improper amount will be used.)[7] And while it generates the

same long-term results as the traditional analysis, at the same time it provides an analysis of how the vessel operators will, in the short run, alter their production levels in response to differences in their perceived price of effort and marginal cost, thus making the cost of effort increase at non-equilibrium points. It is this extra insight that gives rise to the restriction on vessel output listed above as necessary to achieve MEY. Note from the figure that if the number of vessels is restricted to K^* but no controls are placed on vessel output, an industry equilibrium will occur at $\bar{E}_2$ with each vessel producing E_2 units of effort. This is the point where the perceived marginal return to effort is equal to marginal cost of effort given the limited number of vessels. At this point each remaining vessel will be earning more profit than at E_1. This increase, however, comes at the expense of efficiency in the economy as a whole because at $\bar{E}_2$ (K^*E_2) the long-run marginal cost of producing effort is greater than its social marginal value. This means, of course, that optimal regulation will require constraints on effort per boat as well as on the number of boats.

---

[6] To be formally precise, the long-run industry marginal cost curve would be jagged (unless there is a homogeneous production function) and would only approach a horizontal line as the minimum cost output from each vessel approached zero.

[7] Gould [1972*b*] argues that in situations where free-access resources are exploited with more than one variable input, the factor proportions utilized will not be the one that minimizes cost. This model, which I believe to be a useful description of a fishery for management purposes, is not in disagreement with these conclusions. It assumes that there is only one variable input, effort. Granted it is a composite input, manufactured with variable input proportions, but there is no reason to believe the firms that produce it will operate above their average cost curves. And as the analysis in the text shows, competition among them will force each to operate at the minimum of that curve.

Since the traditional analysis ignores variation in this effort-level of individual vessels, this policy recommendation does not follow. (As first pointed out by Turvey [1964] and as will be demonstrated in the next section, a general analysis of a fishery also requires regulation of mesh size.)

An interesting example of this problem is the limited entry program in the British Columbia salmon fishery. The number of vessels was frozen with plans for a government buy-back program to reduce effort. (For a complete description of the program, see Morehouse [1972] and Pearse [1972]). Initially, new boats could be constructed only if another was retired. In the first years of the program, however, the average tonnage of the new vessels was three times that of the ones they replaced. Subsequently, the program managers have instituted a ton-for-ton replacement rule but are still finding that changes in the amount or composition of gear on each vessel is changing so as to increase the actual amount of fishing effort produced. In terms of this model, a proper replacement rule should be in terms of the ability to produce standardized units of effort.

A MORE FORMAL MODEL

Basic Description

While this simple diagrammatical model with its fairly restrictive assumptions (instantaneous biological equilibrium, no crowding or mesh size consideration) does present the essence of the fishery and vessel analysis, a more rigorous presentation is possible by formulating it in mathematical terms as

follows. All the variables not specified in the equation titles are defined as above and in addition m is the mesh size, $\hat{\pi}$ is the minimum profit necessary to keep the vessel in the fishery during the period of analysis, and $P_F(m)$ is a relationship between the price of fish and their average size at capture which is determined exogenously to this particular fishery.

Net Rate of Growth of Stock:

$$\frac{dX}{dt} = f(X, m, KE) \qquad [7]$$

Yield:

Fishery $\qquad F = g(X, m, KE) \qquad [8]$

Vessel $\qquad F = (\dfrac{\bar{F}}{KE}) \cdot E \qquad [8']$

Cost:

Vessel $\qquad C = \phi\,(E, m, K) + \hat{\pi} \qquad [9]$

Fishery $\qquad \bar{C} = KC \qquad [9']$

Profit Equations:

Fishery $\qquad \bar{\pi} = P_F(m)\,\bar{F} - \bar{C} \qquad [10]$

Vessel $\qquad \pi = P_F(m)\,(\dfrac{\bar{F}}{KE})E - C \qquad [10']$

Economic Equilibrium Conditions:

Fishery $\qquad P_F(m)\bar{F} - \bar{C} = 0 \qquad [11]$

Vessel

effort $\qquad P_F(m)(\dfrac{\bar{F}}{KE}) - \dfrac{\partial C}{\partial E} = 0 \qquad [11']$

mesh size $\qquad (\dfrac{\bar{F}}{KE}) \dfrac{dP_F}{dm} - \dfrac{\partial C}{\partial m} = 0 \qquad [11'']$

Dynamic Adjustment of Vessels:

$$\frac{dK}{dt} = \delta_1 \bar{\pi} \text{ if } \bar{\pi} \geqslant 0$$

$$\frac{dK}{dt} = \delta_2 \bar{\pi} \text{ if } \bar{\pi} < 0$$

[12]

The net rate of growth of the stock equation is a generalization of equation [3]. With the introduction of mesh considerations, a change in the size at capture may affect the age composition and size of the stock, and as a direct result, total yield, and so the net growth rate may not be the simple difference between natural growth and catch.

The yield of the fishery as a whole is a function of stock size, mesh size, and the total amount of effort. Yield of a single vessel is the product of the average catch per unit of effort (which is determined by the fishery as a whole and which the vessel considers a parameter at any point in time) times the level of effort the boat produces. Vessel cost is a function of the amount of effort produced, and, to be completely general, of the mesh size and the number of boats in the fishery. Mesh size may affect cost in that smaller sizes may be more prone to tangles and may increase the energy necessary to complete a tow. The number of vessels may affect cost if they interfere with each other. Total costs for the fishery as a whole is the product of the number of vessels and individual vessel cost. Note the fishery-wide costs are derived from the individual vessels and that vessel yield is derived from fishery-wide yield.

The profit equations follow directly from the respective definitions of price, yield, and cost. The open-access fishery as a whole will be in equilibrium when total profit is zero. (Division of both sides of [10] by $\bar{E}$ will express the equilibrium condition in terms of fishery-wide revenue and cost per unit of effort as is pictured in part B of Figure 2.) Each vessel will be in equilibrium when the perceived price of effort is equal to its marginal cost and when additions to revenue from changes in mesh size equal the resultant change in cost, assuming no constraints on the range of m.

The dynamic adjustment equation is an acknowledgment that, like stock size, K does not achieve an instantaneous equilibrium either and, specifically, that the speed of entry and exit is directly related to the size of industry profits.

For a dynamic analysis of the fishery, the model can be collapsed into equations [7], [11'], [11''] and [12]. As Smith has demonstrated, by assuming that the vessels achieve instantaneous equilibrium with regard to individual levels of effort and mesh size, it is possible to study the possibility of, and the process of obtaining a bionomic equilibrium—a point where the rate of change of both X and K are equal to zero simultaneously. It is interesting to note that by dropping the assumption of instantaneous vessel adjustment and specifying how the level of effort and the mesh size used by the representative vessel varies with the marginal net profitability of each, bionomic equilibrium then requires a simultaneous zero rate of change in X, K, E and m.

Comparison of Earlier Works

This model builds upon the Smith model and the Fullenbaum, Carlson, and Bell adaptions to it in the following way. Smith's production function was implicit in his cost function which was dependent, among other things, upon the catch rate. (See below.) Fullenbaum,

Carlson, and Bell introduced a fishery-wide production function $[Kx = Kg(X, m, K)$ where x is the vessel catch rate] but it is built up from the catch functions of the individual vessel $[x = g - (X, m, K)]$ which have no specification for changes in effort at the vessel level. Accordingly, vessel cost is a fixed opportunity cost $(c = \hat{\pi})$ which means that industry cost is Kc. The present model utilizes the production or yield curve from the traditional analysis which is an industry-wide phenomenon, i.e., equation [8]. Vessel production is derived from it according to the industry-wide average catch per unit of effort *and* the amount of effort the vessel produces, i.e., equation [8′]. Accordingly, vessel cost is a function of the amount of effort it produces as well as mesh size and the number of vessels, if there are crowding externalities, i.e., equation [9].

Note that focusing attention on the level of effort and introducing the yield function makes it possible to construct an explicit revenue function for both the fishery as a whole and the individual vessel. This makes it possible to specify exactly how stock size externalities are related to the firm. As previously mentioned, Smith's production function is implicit in his cost function, $C = \phi (x, X, m, K) + \hat{\pi}$, where again x is equal to the vessel catch rate. (In the present model, the vessel catch rate can be expressed as F^*/K under the assumption that all vessels operate at the same level of effort.) Consequently, Smith relied on an implicit revenue function, $R = R(Kx, m)$ [Smith 1969, p. 184]. The population size, X, is a legitimate argument in the cost function *when* it is expressed in terms of the catch rate. The larger the stock size, the greater catch per unit of effort and so the lower the cost of obtaining any given catch rate. But since

each vessel can only indirectly control catch rate in a way that is directly related to the sustainable yield function, it makes sense to explicitly take it into account. By concentrating on the level of effort, the current model demonstrates that stock size affects the firm through its perceived price of effort; it does not really affect the cost of operating the boat. In addition, the introduction of these equations makes the model empirically more meaningful because there is considerable information on or about the yield curves of many commercially exploited fish stocks. In addition, there is also time-series data of levels of effort for these fisheries.

Analysis of Maximum Economic Yield

It is a simple matter to derive the first-order conditions necessary to achieve a static maximum economic yield from the fishery using this model. To do so, however, it is necessary to again rely on the concept of sustainable yield. Since, in the general model, equilibrium stock size is a function of total effort and mesh size, the profit equation for the fishery as a whole may be written:

$$\bar{\pi} = P_F (m) g [X(KE, m), m, KE]$$

$$- [K \phi (E, m, K) + K \hat{\pi}]$$

The variables to be controlled are K, E, and m; X is only indirectly controlled. The first-order conditions for a profit maximum are:

$$\frac{\partial \bar{\pi}}{\partial K} = P_F(m) \left[\frac{\partial g}{\partial X} \left(\frac{\partial g}{\partial KE} \right) E + \frac{\partial g}{\partial KE} E \right]$$

$$- \left[\phi + \bar{\pi} + K \frac{\partial \phi}{\partial K} \right] = 0 \qquad [13]$$

$$\frac{\partial \overline{\pi}}{\partial E} = P_F(m) \left[\frac{\partial g}{\partial X} \left(\frac{\partial X}{\partial KE} \right) K + \frac{\partial g}{\partial KE} \right] K$$

$$- K \frac{\partial \phi}{\partial E} = 0 \qquad\qquad [14]$$

$$\frac{\partial \overline{\pi}}{\partial m} = \frac{dp_F}{dm} g + P_F(m) \left[\frac{\partial g}{\partial X} \left(\frac{\partial X}{\partial m} \right) + \frac{\partial g}{\partial m} \right]$$

$$- K \frac{\partial \phi}{\partial m} = 0 \qquad\qquad [15]$$

Each of these conditions has the normal interpretation that the marginal revenue of the variable involved must equal its marginal cost. As [13] indicates, the marginal revė ̇ ̇e of K depends upon the effect an extra vessel will have on equilibrium stock size and hence indirectly on yield as well as its direct effect on yield. The marginal cost is that of adding the extra vessel plus any increased cost that must be borne by existing vessels. Similarly, from [14] the marginal revenue of E in each boat depends upon an indirect stock size effect on yield and upon a direct effect on yield. The interpretation of its marginal cost is straightforward. From [15] it can be seen that the marginal revenue of changes in mesh size depends upon the change due to changes in the price of fish plus the direct and indirect effect mesh size will have on yield. Again, the interpretation of marginal cost in this case is straightforward.

By comparing equations [13], [14] and [15] with [11], [11'] and [11''], it is possible to see the difference between the MEY and the open-access operation of a fishery. The basic difference is that in open-access average returns rather than marginal returns are used and no consideration is given to stock or crowding externalities.

SUMMARY

The model presented here allows for the analysis of the relationship between the firm and the industry in a common property fishery. The basis for the analysis is to consider fishing effort as the output of the vessels. Formulated in this way, the model provides for a more explicit and hence empirically meaningful model. Use of the model can provide insight that is necessary for the proper regulation of a fishery composed of independent vessels.

The same concept applies to other common property resources as well. For example, in cases of utilization of oil or water from subterranean pools where productivity depends upon total pumping effort and the spacing of the wells, the proper variable for analysis is the amount of pumping effort of each user in context of its location. The amount of product each user obtains depends upon its operation level and the productivity of the common pool which is a function of total pumping effort. Another example that is only slightly different is the utilization of a common property air or watershed as a receptacle for waste products. The pollution that results is a function of the temporal and geographical dispersion of the total amount of wastes produced. From the regulator's point of view, the proper variable is the amount of waste produced by each firm in context of their dispersion characteristics. The pollution caused by each operator is a function of his waste products and the condition of the air or watershed which is a function of the dispersion of the wastes of all operators.

References

Anderson, L. G. 1973. "Optimum Economic Yield of a Fishery Given a Variable Price of Output." *Journal of the Fisheries Research Board of Canada* 30(4): 509–518.

———. 1975a. "Optimum Economic Yield of an Internationally Utilized Common Property Resource." *Fishery Bulletin* 73(1): 51–66.

———. 1975b. "Analysis of Open-Access Commercial Exploitation and Maximum Economic Yield in Biologically and Technologically Interdependent Fisheries." *Journal of the Fisheries Research Board of Canada* 32(Oct.): 1825–1842.

Beverton, R. J. H., and Holt, S. J. 1957. *On the Dynamics of Exploited Fish Populations.* London: Her Majesty's Stationery Office.

Brown, B. E., et al. 1975. "The Effect of Fishing on the Marine Finfish Biomass in the Northwest Atlantic from the Eastern Edge of the Gulf of Maine to Cape Hatteras." International Commission for the Northwest Atlantic Fisheries Res. Doc. 75/18, serial no. 3470, Dartmouth, Canada.

Brown, G., Jr. 1974. "An Optimal Program for Managing Common Property Resources with Congestion." *Journal of Political Economy* 82(1): 163–174.

Clark, C. W. 1973. "Profit Maximization and the Extinction of Animal Species." *Journal of Political Economy* 81 (July/Aug.): 950–961.

Copes, P. 1970. "The Backward-Bending Supply Curve of the Fishing Industry." *Scottish Journal of Political Economy* 17: 69–77.

Crutchfield, J. A., and Zellner, A. 1961. "Economic Aspects of the Pacific Halibut Fishery." *Fishery Industrial Research*, vol. 1. Washington, D.C.: U.S. Department of the Interior.

Fullenbaum, R. F.; Carlson, E. W.; and Bell, F. W. 1971. "Economics of Production from Natural Resources: Comment." *American Economic Review* 61 (June): 483–487.

———; ———; ———. 1972. "On Models of Commercial Fishing: A Defense of the Traditional Literature." *Journal of Political Economy* 80 (July/Aug.): 761–768.

Gordon, H. 1954. "The Economic Theory of a Common Property Resource." *Journal of Political Economy* 62(2): 124–42.

Gould, J. R. 1972a. "Extinction of a Fishery by Commercial Exploitation: A Note." *Journal of Political Economy* 80 (Nov./Dec.): 1031–1038.

———. 1972b. "Externalities, Factor Proportions, and the Level of Exploitation of Free Access Resources." *Economica* 39(4): 383–402.

Morehouse, T. A. 1972. "Limited Entry in the British Columbia Salmon Fishery." In *Alaska Fisheries Policy,* ed. Arlon R. Tussing, et al. Institute of Social, Economic, and Government Research, University of Alaska.

Neher, P. A. 1974. "Notes on the Volterra-Quadratic Fishery." *Journal of Economic Theory* 8(1): 39–49.

Pearse, P. H. 1972. "Rationalization of Canada's West Coast Salmon Fishery." In *Economic Aspects of Fish Production.* Paris: Organization for Economic Co-operation and Development.

Schaefer, M. B. 1954. "Some Aspects of the Dynamics of Populations Important to the Management of Commercial Marine Fisheries." *Inter-American Tropical Tuna Commission Bulletin* 1: 25–56.

——— 1957. "Some Considerations of Population Dynamics and Economics in Relation to the Management of the Commercial Marine Fisheries." *Journal of the Fisheries Research Board of Canada* 14(5): 669–681.

Scott, A. 1955. "The Fishery: The Objectives of Sole Ownership." *The Journal of Political Economy* 63 (April): 116–124.

Smith, V. L. 1968. "Economics of Production from Natural Resources." *American Economic Review* 58 (June): 409–431.

———. 1969. "On Models of Commercial Fishing." *Journal of Political Economy* 77(2): 181–198.

Part III
Regulation

[13]

Development of Economic Theory on Fisheries Regulation[1]

ANTHONY SCOTT

Department of Economics, University of British Columbia, Vancouver, B.C. V6T 1W5

SCOTT, A. 1979. Development of economic theory on fisheries regulation. J. Fish. Res. Board Can. 36: 725–741.

A survey of the economic literature of fisheries regulation shows that little of analytical value for the comparison of alternative regulatory techniques has emerged. The suggestion that the general literature on regulation, and on public choice, has something to contribute to the understanding of alternative regimes produces eight criteria. These are applied to the choice between two systems of restricting entry: a tax, and quotas. The transactions costs of the two systems are also investigated. The hypothesis is formed that the eight criteria, plus expected transactions costs, give the edge to a quota system; but this is only illustrative of the approach.

Key words: regulation, management, costs, quotas, taxes, revenue, licensing

SCOTT, A. 1979. Development of economic theory on fisheries regulation. J. Fish. Res. Board Can. 36: 725–741.

D'un survol de la documentation économique sur le contrôle des pêches émergent peu de choses de valeur analytique permettant de comparer diverses options de contrôle. A supposer que la documentation générale sur le contrôle et sur le choix public puisse contribuer à la compréhension de différents régimes, on peut en déduire huit critères. Ces critères sont appliqués au choix entre deux systèmes de limitation de l'entrée: taxe et contingents. Nous examinons les coûts d'opération des deux systèmes. Nous énonçons l'hypothèse que les huit critères, plus les coûts anticipés des opérations, donnent l'avantage au système des contingents; cela ne sert toutefois qu'à illustrer la méthode d'approche.

Received October 16, 1978
Accepted March 26, 1979

Reçu le 16 octobre 1978
Accepté le 26 mars 1979

I. Introduction

SERVING as a background paper, this essay has an unavoidable bibliographic flavor. Its subject is not the emergence of fisheries regulation, a subject yet to be tackled by academics, but the development of the economic theory related to that subject.

My general impression is that this theory has not progressed very far since it was taken up with enthusiasm in the 1960s. However, now that extended fishing zones, out to 200 miles or beyond, are to become the rule, there is a new imperative to combine the original theme of fishery economics with the economic theory of regulation in industrial organization literature. This paper's reminders of what has already been done may enable students to take up the challenge. There is now a widespread demand, among all L.O.S. participants, for advice on methods of regulation and sharing. In the absence of appreciation of the progress and mistakes already made, the danger is that the industry will relapse into the labyrinth of ad hoc, contradictory, mercantilistic, and ineffective regulation of the not-so-distant past. Inertia may then make it impossible

to move to a clearer set of institutions and rules.

The paper is organized as follows. In Section II I present a brief historical survey of fishery economics to identify the four main themes with which the broad literature (since World War II) has been concerned. In Section III I examine more closely the regulatory literature. In Section IV I turn to a small part of regulation, that having to do with restriction of entry. And in Section V I narrow down to the economic literature on property rights in fisheries. All this is reinforced by a long bibliography, pieced together from both direct consultation and secondary sources. It is weak on official reports on proposed or actual regulatory schemes; these are hard to locate at the best of times and I have had to work from an Australian base where the "tyranny of distance" has compelled me to make do with what diligent and helpful librarians have been able to collect.

II. Themes in Fishery Economics

It is now over 50 yr since Warming, followed 25 yr later by Gordon and others, and assisted by fisheries biologists and administrators, lifted the study of fisheries economics out of its previous obscurity. Till then it had been the occasional subject of academic and policy papers (see especially MacGregor 1949), many of them applying the concepts of such economic fields as industrial relations and labor markets (Jamieson and Gladstone 1950; White 1954); agricultural economics

[1]This paper forms part of the Proceedings of the Symposium on Policies for Economic Rationalization of Commercial Fisheries held at Powell River, B.C., August 1978.

both econometrically and in studies of supply organization (in the studies and routine reports of official fisheries' departments); industrial organization, especially marketing (Gregory and Barnes 1939); and, of course, economic history (Innis 1940).

Much of this previous work was brought to bear on the common-property insights of Gordon, Popper, and the rest in September 1956, in the exciting I.E.A./F.A.O. Fisheries Economics Round Table assembled in Rome. Its proceedings were ambitiously entitled, justifiably as it then seemed, *The Economics of Fisheries*, and were brilliantly edited by R. Turvey and J. Wiseman. In addition to those whose names are listed therein as contributors and discussants, the Round Table had the good luck to attract, at least for a few days, E. A. G. Robinson, G. Haberler, E. Lindahl, and, although I am less sure about this, some additional heavyweights such as D. H. Robertson and B. F. Haley, who had been the leaders of the I.E.A.'s first congress (held a week before in Rome). These great men knew little or nothing about the new field for applied economics. But they did know about Knight's theoretical work on external economies and property (Knight 1924) and they usefully prevented a tendency among new enthusiasts to blame every unusual feature of the fishing industry, anywhere, on the ubiquitous common-property characteristic.

It is not necessary, or interesting, to set out the history of fisheries' economics since that time. One good reason is that specialized accounts have recently been published by Pearse (1978), Butlin (1975), and Peterson and Fisher (1977). V. Smith's new anthology (1977) includes several articles with good bibliographies up to their date of publication. Most F.A.O. conference proceedings include accounts of what has gone before; for examples see the earlier background papers by Scott and Dickie in the Hamlisch volume (1962). And always, there are adequate histories of the subject attached to available official reports and background working papers.

The exciting new point of departure, mentioned above, was common property. F. Knight and Pigou had already debated this subject, but it had not been applied in any interesting analytical way by economists to such other resources as common ranges or pastures, oil wells, water tables, the oceans, or the atmosphere — although of course many applied economists and legal scholars recognized the problems for centuries.

S. Gordon touched off a wave of theorizing. Carefully applying Knight's two-road example to two fishing grounds, he demonstrated that on any fishing ground the unrestricted entry of units of effort would lead to an unusual *bionomic* equilibrium (Gordon 1954; Cheung 1971), in which the resource would yield less fish than its capacity, and labor and capital would be overapplied so that their marginal product in fishing would be below opportunity cost. Gordon and others offered these results to help explain poverty among fishing peoples; and the prevalence of the share or lay

system to distribute fishermens' incomes of vertical integration between fishing vessels and buyers or packers (Cassady and others in Turvey and Wiseman 1957). Another group, viewing the "dissipation of rent" as a loss of natural capital, stressed instead that common property was inefficient in that it prevented controlled allocation of stock or catches over time, either within the season or between generations (Scott 1955, 1972, and in Turvey and Wiseman 1957; see also Ciriacy-Wantrup 1952).

The evils of common property having been identified, what should economists recommend? This was the second theme that got its start in the 1950s. Those who followed Gordon and Knight naturally tended to favor the invention and application of new property rights, similar to those that had brought to an end common land-use. Those who had been brought up on Pigovian welfare economics favored taxes or subsidies to bridge the gap between private and social net marginal products.

These methodological differences, however, soon gave way to the more practical issues of the choice among taxes, charges, and some feasible form of property rights. The first lesson being taught was that existing ad hoc fisheries restrictions (of gear, open season, etc.) were more inefficient than any of property rights, subsidies, or license fees. None of the major policy instruments made much legislative headway; instead, whatever progress was made was in convincing policymakers that many restrictions were ineffective or wasteful, whereas other, novel, restrictions (such as "eumetric" net-mesh controls) might increase every fisherman's potential catch (Hamlisch 1962; Turvey 1964); Crutchfield and Zellner 1962; Crutchfield 1959, 1961, 1965; Beverton and Holt 1957; Pontecorvo 1966).

What did become disconcertingly clear was that among biologically and legally trained administrators, diplomats, and decision-makers, the "economics" of the fishery was regarded as applicable to distributive, "for-whom?" questions. The fishery economist expected to justify his client's fishery goals, such as effectively excluding potential fishermen from one social or economic grouping and conserving or protecting the stock for exploitation by another.

This distributive bias was also revealed in the 1950s debates into the international law of ocean fisheries, and the doctrines ("freedom of the seas"; the "abstention principle"; coastal states' rights; and so on) invoked in the great issues debated during the first U.N. Law of the Sea conferences (1958). Although expounded in Benthamite law-reform rhetoric, the doctrines usually collapsed to catch-phrases favorable to fishing rights for particular nations. (Christy and Scott 1965; Crutchfield 1966; McDougal and Burke 1962). Even the great campaigns of the 1970s to keep the oceans from being enclosed into extended national territory or exclusive resource zones were advanced, and received, more as distributional principles ("the common heritage of mankind") than on efficiency grounds (Christy 1973).

As the second Law of the Sea conference approached, the efficiency principles (strongly advocated by biologists) of keeping known biological stocks under unified management faded away: ethical and distributive arguments (usually about "rights") replaced them almost completely. Even today, when more economists have reinforced the original group who worked on alternative regional and international fishing policies and institutions, it apparently comes as a surprise to lawyers, diplomats, and politicians that fisheries' economics is not exclusively devoted to assembling the economic aspects of nationalistic and protective briefs.

If fisheries economics is to be saved from such a fate, its salvation will have come from bioeconomics, to give it the title of C. W. Clark's monograph on the dynamic management of renewable resources (1976). In my bibliography, the selection of items listed will suggest how venturesome this new breed of fishery economist has become (Clark 1973; Clark et al. 1977; Spence 1973).

As S. Gordon indicated in his 1954 article, the mathematical contributions of biologists pointed the way to economic analysis. Not only was the form of modeling familiar to those economists who had some familiarity with demography and capital theory, but the methodology, with its strong deductive approach (perhaps arising out of a paucity of data) to broad generalizations, promised a welcome collaboration between the two groups (Schaefer 1954). This probably reached its height during the period when economists were studying the works of Beverton and Holt (1957; see Dickie in Hamlisch, 1962, and Turvey, 1964). Since then, one feels that the mathematical bioeconomists have pressed ahead, beyond their data, more than is familiar or acceptable in biology. This has been particularly true as policy-oriented economists have veered from prediction to optimization, employing successively calculus, the calculus of variations, programming and control-theory formulations (Crutchfield and Zellner 1961; Plourde 1971; Gould 1972; Clark and Munro 1976; Quirk and Smith 1970; Scott 1970; Smith 1968, 1969; Neher 1974; Bell 1972; Lewis 1977). What *has* held the groups together is the flourishing use of the computer and the joint application of its program to the newer ecological and management assignments of fishery scientists, and to the numerical solution of policy problems involving both biological and economic submodels (Hannesson 1974a; Royce et al. 1963).

III. The Economics of Fishery Regulation

In this section I will briefly review the economic theoretic literature on forms of regulation of fishing. I can be brief because this literature does not, it now seems to me, constitute a coherent body of theory at all. The majority of models introduce controls in a purely formal, or negligently offhand manner, changing a net price or adding a constraint. Thus, most of the theoretical "regulatory literature" simply arises in the process of examining models' comparative static or dynamic properties. It is not part of a literature on detailed, multiobjective, real-world, regulatory options.

On the other hand, there is practical literature, much of it by well-known economists, on instruments of regulating fisheries, with some reference to some of the probable consequences of each instrument. Reflecting their authors' wisdom and experience, this literature does explore some expected consequences of various proposals. But it is very partial; rarely does it start with a set of criteria and compare the alternatives systematically. Furthermore it is not placed in a general-equilibrium framework. In these respects, it has been dramatically overtaken by the new economic literature on pollution abatement and environmental protection. (The best-known systematic comparisons [Scott in Hamlisch 1962; Crutchfield 1961; Brown 1974]; do escape some of these flaws, but they were written nearly 20 yr ago).

Why are fisheries regulated? Caution comes with age, and I am not so quick as I was in the early 1960s to explain motives. Furthermore, because any public intervention into a market economy, even into a faulty-market sector like the marine fishery, will probably induce further interventions to prevent evasion, or forestall unwelcome adjustments, observed action or apparent effect of a regulation may not tell us much about the motives that originally inspired it.

However, we are not deaf and blind, and we do know what has been claimed for some new regulations when they have been proposed. Furthermore, we do know something of the situations that existed, or were feared, before regulations were instituted.

First, regulations have been instituted to encourage what is now called "X-efficiency." Declining catches, stocks, and profits have indicated that the industry had fallen from its national production possibilities path, and an "overfishing" problem has appeared. To condense a well-known theory, the pressure of fishing has reduced the stock below the level that would offer larger catches.

The practical measures for controlling existing fishing mortality depend for their effectiveness on reducing the "effort" to which stock is exposed. Since effort is the arithmetical product of the number of units (usually, vessels), times the length of the period they are in use, times their "catching power" (proportion of the stock they are capable of catching per unit time), effective mortality regulation depends on reducing one or more of these effort components while preventing the others from making a compensating increase.

Among the best-known techniques for reducing these components are reducing the number of vessels (a relatively new control); reduction in their size or motive power (thus reducing the period of time the vessels can remain in contact with the stock); closing for certain seasons in particular areas; and prohibiting the use of certain effective or "destructive" types of gear (thus reducing vessels' catching power).

Obviously, the effectiveness and cost of these controls on the components of effort differ. We shall discuss controlling the number of vessels later in this survey; it has been advocated for a long time and has economic aspects that are more interesting than the others. Control of the size, speed, or fishing power of vessels is a very common device, being relatively easy to enforce, and having popular distributive aspects. Closed areas and seasons are of doubtful effectiveness, especially if the fish do not remain to be caught in the open seasons, or do not migrate from the protected grounds to be caught in an open area.

Under favorable circumstances, all these measures can increase the yield by increasing recruitment or reducing mortality. They compel the industry to make an investment in the stock, cutting down immediate catches to obtain larger catches later on. Whether such investments are justified by their eventual benefits has recently been extensively studied. The application of bioeconomics has shown how control theory can be applied to planning the extent of the original sacrifice and the timing and form of the eventual gain (see Crutchfield and Zellner 1962; Quirk and Smith 1970; Southey and Scott 1970; Plourde 1971; Clark and Munro 1976, and many others). Numerical methods have, by the use of the computer, also been used in these studies (Hannesson 1974a; Loose 1978; and a very large number of unpublished official studies).

Compulsion is justified because the fishery is common property, and fishermen will not be willing to sacrifice a present catch for a later benefit, if they believe the benefit will not be captured by themselves. Compulsion is also sometimes justified because the stock is a public good, in that locating (and sometimes catching) fish is easier for all the fleet when the stock is conspicuous and dense, than when it is small.

However, the net benefit of regulation can easily tend to zero, if the benefits are reduced by increasing fishermen's outlays on doing whatever is permitted to take the larger catch. These permitted adjustment costs may include expenditures not only on vessel and gear, moving closer to ports, storing fish on board longer, but also on reequipping their vessels to make them usable in other seasons in other fisheries.

The same increased costs result from attempting to ignore the components of effort, and instead setting overall catch quotas. This technique, pioneered for halibut on the West Coast, allows the fleet freedom to combine the components of effort, fishing wherever and however the owner chooses. However, the larger catches due to quotas have encouraged fishermen to increase the number of vessels, improve their speed, gear, and so forth. Consequently, the authorities, to protect their quota, have been forced to shorten the catching period.

In the end, therefore, overall quotas have led to a swelling of costs just as have control of effort components.

Cost is the chief theme of economists' criticisms of all controls to deal with overfishing. Although they do increase total catch (and consumers' surplus) the associated rent invites a cost increase of the same amount. Crutchfield has pointed to other defects, including distorting the market by shortening the landing season, preventing "progessivity" by prohibiting the new techniques or vessel types and, as already mentioned, forcing owners to build adaptable rather than specialized vessels, and to base themselves on nearby, rather than home, ports.

It follows that the need for the authorities to multiply their regulations increases exponentially. "Overfishing" regulations that reduce one component of fishing effort induce further controls to suppress increases in other components. Also, because it is becoming obvious that the setting of an overall quota encourages private investment to outwit or anticipate competitors, rather than to increase the quality, value, or amount of the catch, administrators are tempted to regulate or forbid these investments as well.

A second type of regulation, already referred to, is to increase the quality, weight, or size of the individuals in the annual catch. Most of those in use are connected with the timing of the catch, and have the ultimate effect of postponing harvesting of an individual until a later season or year. For example, a closure may prevent fishermen, competing with one another for the available catch, taking individual fish when they are too small, or in poor condition, or so remote from the port that they must be held in value-reducing storage or board. This kind of regulation, applied to seasonal timing, is common for crustaceans and molluscs, and for migratory fish. Applied to interseasonal deferment, its application has been gradually broadened following studies of the possibilities of "eumetric" fishing, allowing small fish to grow in weight, quality, and value. Other types of regulation may work to reduce the costs of fishing rather than to increase the quality. For example, the costs of catching a fish inshore may be less than when it is intercepted on the high seas.

These controls take the form of a social investment, and must be evaluated as such. They have an immediate cost in sacrificing landings, to reap an ultimate gain in net landed value. Because of the common property characteristic of fisheries, individual fishermen will not make such investments (sacrifices) unless forced to do so.

Third, fishery regulation is often redistributive, or, to describe the same motivation, it is protective of chosen groups. This is well-recognized in the literature. The Scott and Crutchfield surveys mention the protection of fishermen by government regulations that specify maximum size of vessel, permitted gear, motive power, closed season, area of catch, or permitted species. In particular local circumstances, such specifications tend to exclude competition from particular outside fishermen. Alternatively, those who may, or may not, fish are sometimes mentioned explicitly. We have heard of protected groups described in terms of their age, race, language, and of course nationality.

This protective goal obviously conflicts with any efficiency goal and, I will argue, is probably the more powerful. Its attractiveness will be tempered only by recognition of the costs it imposes on others. In some c ses these may be consumers, in which case we might have a situation analogous to farmers or utilities using natural or contrived monopoly to increase wholesale and retail prices (Weitzman 1974). Political means may be used to regulate such behavior. But, in most cases, the victims will be the excluded fishermen. Since they will be potential, not actual, entrants, without "historic rights" or visible connection with the fishery, they find it difficult to muster political support for relaxing the protective rules.

Water, the fishery habitat, is itself of course a multiple-use resource, and "protection of fishing" includes also protection of incumbent fishermen against foreigners, and against those who would seek other fish (mentioned at the end of this section). Furthermore, much policy is invoked and refined to allow the sharing of estuaries, rivers, and lakes between fish and commercial fishermen and the energy industries; waste-disposal interest; and sport fishermen (Marts and Sewell 1959; Crutchfield and Zellner 1962; Stephenson 1977).

Fourth, fisheries may be regulated to prevent the waste of labor and capital and promote the efficient allocation of resources. Governments have long been aware that in the fishing industry the number of fishing units tends to increase beyond the amount warranted by the expected catch. Three results were said to follow. Catch per unit effort being low, boat owners were unable to obtain the expected return from their investment. These low returns flowed (through the share system) to fishing hands, reducing their incomes to the "poverty" level (something less than their opportunity incomes elsewhere). Finally, with new entrants diluting the catch, existing vessels tended to reequip with faster motors and smaller-mesh nets and to sail from ports closer to the fishing grounds.

These results are often visible, but they depend on special conditions. In economist's jargon, free access to the common-property resource tends only to increase the number of multiinput units of effort until, at the "bionomic equilibrium" each unit adds less to the value of the harvest than in alternative sectors. Thus resources are allocated inefficiently among sectors.

But this does not mean that take-home rewards must be lower than elsewhere. Special factor supply conditions are necessary to explain low fishery incomes. Among those that have been suggested are asymetrical mobility (easy come, never go) and willingness to accept risk or variability in earnings. Such necessary conditions

are often the basis of beliefs that identify those fishermen have special individual and social preferences. Alternatively, it is sometimes argued that many fishermen have such a strong comparative advantage (in endurance and skills) in fishing that their alternative incomes would be even lower (see Copes 1972 and earlier contributions on the Newfoundland fisherman; this subject was repeatedly discussed in the comments in Turvey and Wiseman 1957 and Hamlisch 1962).

Thus low incomes do not prove that labor and capital are wasted in the industry; fishermen may just be low-income people. The usual evidence of waste advanced to justify regulation is more pragmatic: the catch could be found and brought in, with the same personal cost and equipment wear and tear, with fewer units of effort. Crutchfield and others have frequently made this observation with respect to salmon — Alaska's salmon catch could be taken with about 25% of the 1960s levels of effort; British Columbia's with 50%; Puget Sound with less than 50%. The latest numbers come from a new Ph.D. thesis: in the Skeena River, a sole owner who could not vary the number of boats in use over the season could dispense with about 15% of the present gillnet fleet (Crutchfield and Pontecorvo 1969; Loose 1978).

Economists' surveys of fishery regulation make much of the power of government action to remedy the wastes of common property and similar market failure. But I do not believe that an increase in efficiency in resource allocation very often enters into politicians' motivation for intervening in the fishery, or in any other sector. Indeed, as long as some of the other goals mentioned here are satisfied, politicians, if they understood that it existed, would probably approve of the redundant absorption of labor and capital into fishing. In any case, they are understandably unenthusiastic about measures that may displace people from jobs, close down plants, or empty ports.

This disenchantment is reinforced by growing confidence in some of the approach of the general economic theory of regulation, especially as advanced by G. Stigler and his associates, and its cousin the theory of public choice, as represented by J. Buchanan, G. Tullock, and W. Niskanen.

While most public utility regulation is concerned with the costs and price setting of existing firms, some of it is concerned with entry of new competitors. Regulation is not treated as a free good, but as a commodity that is supplied by politicians (ultimately) and administrators and tribunals (more directly). There are two large parties whose demand is supplied: the members of the regulated industry, and the public. A common conclusion is that when price or costing choices must be made, industry is favored over the public: in the long run the regulatory institution finds itself protecting the controlled industry.

More recent development of this theory, emanating from the public-choice approach, start with the prefer-

730   J. FISH. RES. BOARD CAN., VOL. 36, 1979

ences and constraints of the decision-maker. Why does he "protect" the producer, or allow himself to be "captured" by the industry he is supposed to be policing? (See Peltzman et al. 1976.)

This new development, searching for both maximand and constraint, seems to me to point the way for profitable research into fishery regulation, and for collaboration in the fields of industrial organization and the economics of law. Clearly, rate-setting for Bell Telephone is not a very similar process to determining the stumpage rate in public forests, or the price of butter, and is even more unlike the determination of closed seasons and areas in fisheries. Yet in all these processes, we have similar procedures: information gathering; hearings, negotiation and bargaining; and monitoring and enforcement. And we have bureaucrats who gain by a proliferation of the demand for these procedures, institutionalized as expanding departments or commissions. The decision-maker himself could easily be shifted from one job to another. What uniformities would we see in his decisions? What differences in constraints would alter his behavior if he became a fishery regulator?

I am not aware that much progress has been made on this question, except perhaps with respect to studying the behavior of fisheries' spokesmen at L.O.S. preparatory meetings and caucuses. (Although these could not really be called "regulatory" procedures, they do provide clues. Christy and I were once criticized for identifying the little group of fishery diplomats (1965), but we did not proceed to analyze their group behavior. And I was once included in a Chicago University conference on industrial regulation, to the indignation of the invited audience, who could not see enough similarity between fisheries' and telephone regulation problems to justify combining them on the same panel.)

Probably the difficulties of applying regulatory theory of fisheries are these: (1) in many regulatory actions, the public interest is not represented, or even known to exist; (2) the industry is not always split between two industrial classes, but between fishermen living in different areas, or using different gears; (3) technical information is not provided by the adversaries, but by the regulatory commission; (4) some decision-makers in fisheries, like those in forestry, are able to resist capture by the industry because their own utility functions, deriving from their training and profession, inclines them to attempt to impose their own ideas of biological "conservation" and perhaps "economic efficiency" on an industry where no party is advocating these aims. Existing studies of the U.S. forest service, in their *regulatory* capacity, might support this generalization.

IV. Restriction of Entry

In the preceding section, it was shown that open-access fisheries, subject to depletion of stock, deterioration of quality, and falling incomes, have been proposed

for and subjected to several kinds of direct regulation. Four lessons emerged.

First, regulation is most acceptable when it protects fishermen's private investments. The best instances of these private benefits are the exclusion of competition from vessels other than those who, having crews with the same social or economic characteristics as the incumbent fishermen, or using the same gear as those already exploiting the stock. To a lesser degree, it is also welcome if it helps to reduce the costs of fishing, or raises the quality and value of the fish.

Second, the control of quantity and quality, through the control of the application of effort, loses its value as a social investment as fishermen adjust to the chosen regulation, by costly investment of their own. This is partly because the number of vessels is not controlled.

Third, in general, neither the government nor the participants have any concern about the inefficiency of the exploitation of common-property industries. The two points mentioned above suggest that political support would be given to restrictions on the overapplication of inputs if, in the form of restrictions in entry, they would work to the advantage of incumbent fishermen. They might do this by cutting down the necessity to intercept or anticipate their rivals on the fishing grounds, cutting their costs, or need to invest in costly equipment, increasing their catches, or increasing the value of the fish. But there is no political interest in the general-equilibrium point that common property leads to a distorted allocation of labour and capital between fishing and other industries.

Fourth, even less, because it is rarely discussed, is there any interest in the idea that the maximization of the present value of a fishery might take the form not only of sacrifice of present catches for increased future yields, but also take a "pulse" or cyclical form. This would take advantage of noncovexities in fishing effort costs, in revenues, or in the timing of the most rapid rate of increase of stocks after they have been heavily fished (Gordon 1957, 1958).

These points suggest that, although economists may favor entry restriction, neither the practising politician nor the individual fisherman is particularly interested in such a policy. Groups, as opposed to individuals, among processors, vessel owners, and crew members may have leaders whose own positions would strengthen their own strategic bargaining position over fish prices, taxes, subsidies, or other aspects of their position vis-a-vis other interest groups or the government, if *particular* schemes of entry restriction are adopted. For example, a union or co-op might approach closed-shop status; and companies might achieve a buyer's ring or cartel. But *individual* fishermen or vessel owners, like individual voters, would not stand to gain much from entry restriction. Thus we must look to statesmanlike economists and biologist administrators to convince politicians that efficient entry restriction is economically desirable.

In this section, I pull together some lessons, concerning restriction of entry, arising from experience with, attitudes to, and theory concerning, methods of entry restirction. Before doing so, however, we should note that there are many other problems of uncontrolled fishing than those that have emerged in discussing the motives for regulation. Most existing types of regulation provide no answers to these problems; they remain as serious as before the era of regulation. Because they remain as challenges to any system of entry restriction, I will attempt to take them into account in what follows.

First among these is the misallocation of fishermen and efforts between grounds, the original problem posed by Knight and Scott Gordon. This may not be a problem if the fleets have a short range, and so need not choose between grounds. (One may ask, however, why a short-range fisherman would stay with a marginal ground when he can move to another port and use the richer fishery.) But, if vessels have a long range, spatial allocation (or dispatching) seems to be necessary.

The second is the allocation of the catch over time. The measures to be mentioned in this section can be used to close seasons, and could be used to close whole years to fishing. But goals for "pulse" (or cyclical) fishing have not yet entered the discussions of fisheries policies. This dynamic problem will be linked below with the dynamic problem of finding an entry-restricting system that will guide the fishery to a chosen equilibrium.

Third is the choice of methods to keep a balance between stocks of fish that are caught together (e.g. by seining or trawling). Levels of effort that may allow one fish to flourish may either overexploit or neglect other stocks. In the limit, one may be extinguished. This is getting attention today in the tuna–dolphin problem, but it has been of concern to biologists for some years, particularly where large factory-ship-base expeditions sweep up nearly every type of fish from the ocean floor. New types of policy, not yet proposed, are needed to bring selectivity to low-cost fishing techniques.

Fourth, there is the problem of the choice of size of stock, catch, timing, or species in international or trans-frontier fishing grounds. The aims of domestic policies thus far have been simply to exclude foreign fishermen. But it is now possible that, even with exclusive extended fishing zones, a preferable policy might be to compromise between the implicit aims of all those nations exploiting the same stocks or region. Especially if the foreigner must pay for the right to fish, he may offer more if the country in whose waters he fishes allows him to seek, and take, the species he wants at the right stage in their growth at his own pace. Policies to deal with such questions are on the agenda of today's diplomats, but they have not yet emerged as distinct goals.

To avoid having to deal with the devising of detailed systems that will handle the need not only for economic efficiency and administrative workability, but also the four extra problems mentioned above, I shall narrow the main discussion in this section to the choice between a fee (or tax) on the one hand, and a transferable-rights

system on the other. In the early pages I shall also avoid complex (multiground, multispecies, multination, or multigear) fisheries, but the practical reader should observe that I am not unaware of them and am struggling to introduce them systematically.

By confining myself to the tax-versus-right's question, I am putting to one side more arbitrary or discriminatory entry restriction systems, especially the "grandfather" systems. These discriminatory systems are usually proposed as a feasible scheme for confirming the fishing rights of existing vessels, crews, or owners, and reducing them by attrition. Scott (1962) and Crutchfield (1961) have discussed the characteristics and the advantages and disadvantages of this means of gradually phasing out excessive effort from an "overfished" stock (Pearse 1978). It should be seen as a type of transitional process. From a political point of view, it saves the government from forcibly ejecting "marginal" participants from the industry. From the point of view of economists (familiar with the adjustment problems of tariff reduction or industry restructuring), it reduces the obvious administrative and ethical problems of whether and how to pay compensation to those whose livelihood is removed by a change of policy. Compensation of a kind is indeed received: those who stick it out the longest find that their catches and earnings increase; while those who leave find equivalent reward in the blessings of retirement or of the Hereafter. However, these seem advantages of expediency only. From an ethical point of view, a grandfather system has the distributional morals of a tontine (Scott 1962).

From an efficiency point of view, as well, the grandfather scheme has little to recommend it. The fleet composition, aging, and changing as vessels and crews drop out, will (except by accident) differ from that which would under a permanent regime, seem optimal. Thus there may be a prolonged transitional period *after* the last grandfathers have departed, as the equilibrium fleet's distribution of vessel lengths, storage capacities, speed, weather-kindness, and gear types and combinations are phased in. Given the durability of vessels and the impracticality and danger of using them without strict maintenance, which also prolongs their service life, the "bang-bang" program prescribed by Clark and others may encompass more years than is usually realized.

There are many conceivable alternative discriminatory systems: entry can be rationed by race, color, creed, etc.; by bribery of officials; by queuing; and by lottery. The arbitrary expulsion of part-time and "sport" fishermen with low catches (in the opening stages of more elaborate schemes) should take a prize for high-handed, inefficient discrimination (Pearse 1971). Some of these do amount to an informal price or market system; but apart from emphasizing that all of them permanently inhibit participation by some specialists whose comparative advantage lies in fishing, the literature of fishing has had little to say about them.

In the earlier fisheries' literature, the influence of Pigou made it seem obvious that a tax on the catch or a fee on inputs was needed to equate marginal social and private costs. By the 1960s, however, consideration of the potentialities of the existing system of licenses (mostly for administrative information and revenue), and of the political drawbacks of introducing fisheries' taxes, had opened the literature to comparisons of the two systems. Existing administrative fisheries' licenses, like pollution permits, indeed made it possible to graft on either system: license fees could be increased until they rationed entry; or the number of licenses could be restricted (Sinclair 1960; Hamlisch 1962; Campbell et al. 1974).

Fee systems, however, seemed to have little hope of general political or constitutional acceptance. In some countries, the power or right to administer a charge high enough to ration entry to the right level would be considered as a right to levy an indirect tax, or perhaps an export tax. This power could not easily be delegated to a fisheries' administrative agency; it might instead be the prerogative of a different level of government; of a representative legislative assembly, not an appointed agency; and be subject furthermore to a crippling "uniformity of taxation" rule. On the other hand, the accepted power of government agencies to set or negotiate the financial aspects of oil leases and timber tenures did suggest that ways could also be found around these legal barriers to a price-rationing system, even though the resulting institution would appear to the lawyer closer to a system of property rights than of prices. The lawyer would be right, of course: it is difficult to distinguish the two systems analytically.

There was also scathing criticism (particularly from Crutchfield) of the political naiveté of arguing for a higher fee to be imposed on low-income, under-employed participants in a depleted fishery. Even if in the long run no restricted-entry system could offer higher incomes than opportunity earnings, it at least ought to avoid beginning by squeezing out excessive workers by reducing them below the opportunity level. This argument, however, must meet the counterargument that in many countries fishing incomes are still high, and overfishing is not yet a problem: in these places fees or taxes would not come as a crushing burden.

Marketable fishing rights are no longer the academic panaceas that they seemed in the early 1960s. Authors then were forced to take their examples from the rights to pasture animals on public land, membership in partnerships or stock exchanges, and transferable quotas to deal in certain farm products, foreign exchange, or imported goods. These examples were not all helpful. In any case, they did not meet the fundamental objection that fishing was regarded by many as an essential civil liberty, not simply as free access to the public domain. In the course of persuasion, it became clearer to both audiences and exponents that "fishing rights"

were not a unique instrument. They were a bundle that might have all or some of several characteristics: they might be *limited*; they might be *transferable*; they might convey *several* privileges and responsibilities; they might have a fixed *duration*; and they might be priced and so *marketable*. As time progressed, actual examples emerged, usually grafted on to existing license systems. Reviewed today, the early literature seems very unorganized and vague. The authors were struggling for a priori generalizations about license limitations among what gradually emerged as a near-infinity of combinations of the characteristics *underlined* above. As is so often the case with a novel proposal put forward under academic auspices, the authors seemed to have no sense of which side-effects of the scheme should be regarded as merely interesting, which as real or costly difficulties, which as insuperable political or constitutional constraints, and which as economically inconsistent or contradictory (see Scott, Crutchfield, Sinclair, Popper, and other contributors and commentators in Hamlisch 1962; Crutchfield 1961; Sinclair 1960; and Christy and Scott 1965).

As with earlier types of regulation, hesitation to adopt marketable rights stemmed from unknown and unresearched distributive questions. It was soon perceived that it would be just as difficult with rights as with taxes to devise a permanent payment scheme in which fishermen got some share of the rent. Scott had pointed to this difference between fishing and farming in 1955 and 1956, implying that only sole ownership would place fishermen, like farmers, in a position to share in the rent, or at least in changes in the rent, of resources. His 1962 paper on regulation returned to the same theme: if marketable rights were given away or auctioned, the rent would be captured either by the first holder (in a lump sum as the net present value of expected future harvests) or by the government. A fisherman's cooperative, once set up as sole owner for the entire stock, could guarantee its members a dividend that included the rent. But either the price of a membership in such a cooperative would rise so that only the original members obtained rent, or the membership would rise indefinitely, pushing the individual dividend toward zero, or at most the rate of normal profits. (The same would be true of a corporation.) Thus a property-rights system (like a tax system) offered no attractive distributive advantages over open access.

In recent years, this no-gain for fishermen argument has been set aside, if not denied, by writers who emphasized that the average income in a fishery included producers' rent (Copes 1970 and 1972; Turvey 1964). By this they have meant the differential returns to exceptional skill or endurance, relative to the marginal vessel or crew member. It is difficult to follow the arguments of some who have invoked this element in earnings; in some forms, it seems to me to be wrong. If I understand what is being argued, it is that producers' surpluses tend to be dissipated when there is open entry or crude regulation of the types reviewed in Parts II and III of this paper. But either a fee or a property-rights system (by restricting entry to those who, because they can earn the most producers' rent, can pay the highest fees or offer most for fishing rights) will allow such persons to get a greater surplus over their opportunity income. How are we to understand this argument? First, I may have misinterpreted it. Second, it may be that special skills are less useful under the restricted-entry (fee or rights) conditions; it may be, for example, that differential returns in the former situation are entirely explained, and entirely absorbed, by investment in capital intensity. But I do not believe these are the facts. Fisheries' experts repeatedly speak of durable groupings of skippers, vessels, and crews according to the size of their catch or earnings, year in and year out. The needed abilities are to find the fish, bring it in quickly, keep it in good condition, keep down crew size by cooperation and hard work, and land it and return to the grounds swiftly. These are certainly needed under open entry (with crude regulation); there appears no reason why the same abilities should not be valuable with or without regulation.

Third, it may be that under open access, the differential return does exist, but the average catch is so small even for the skilled fisherman that he gets a very small producers' rent. Under limited entry, however, the skilled fishermen would be free to expand their shares of the catch, and so their producers' rent. This amounts to arguing that the supply curve (Turvey and Copes) or the total-cost-of-harvesting-effort-curve (e.g. in Christy and Scott 1965) has a different intramarginal shape under open access than it has under closed entry. If this is so, it is possible that the differential going to a skilled fisherman is so much higher under the restriction system that, *after* taxes or the purchase of a property right, *average* incomes to *all* surviving fishermen are higher than under open access.

This third argument seems to be plausible, but only possible. The work of Pearse on the buyback scheme is relevant to the underlying problem of producers' rent and opportunity cost, as is the continuing statistical–economic–biological effort to agree on nonquestion-begging systems of measuring necessary effort (see both Pearse and Rothschild contributions to OECD 1972; Scott 1960).

A second source of hesitation, and of trial-and-error, has been the question whether to limit access to the fishery by restricting inputs or outputs. This question applies equally to fee systems and rights systems, and the actual problems were well forseen in the 1961–62 literature. In the environmental literature, the same question crops up as the problem of whether to tax (or subsidize) inputs into the water-pollution-creating process, or the output of pollutant. Consequently, little need be said. Pearse's 1971 paper, Munro's report of 1977, and doubtless many official reports have kept the British Columbia buyback experiment before the professional eye. (For a recent survey see McKellar 1977.)

The theory is simple enough. If profit is to be made by exploiting some production function in which input characteristics are substitutable, then restricted access or high price of one input or characteristic will cause the entrepreneur to use more of others. Thus if a tax or restriction is placed on the number of vessels, it will pay to design vessels larger, faster, or more labor- or gear-intensive. This substitution, weakening the attainment of the original purpose of the restriction, will then induce the authorities to increase the tax or to tax or restrict another feature, such as number of tons, length, or some weighted index of vessel effort characteristics. All such formulae must induce costly adjustments, both in terms of the extra investment to make the most of the narrow design restrictions, and in terms of the extra amount of other inputs employed to make up for its deficiencies. B. Campbell, Pearse, and Munro suggest that official studies have shown that in spite of the retirement of many vessels, the amount of capital (in deflated dollar units) is now greater than before license limitation and buyback began. This need not be the case, of course; it should not be assumed that input restrictions must invariably produce unacceptably high costs.

Such observations have led to a new interest in restricting the catch. One method, proposed by Jamieson (1958) and many others in the 1960s, was to restrict the output of the harvesting process, instead of the inputs. In countries where the fish were priced by prior agreement between buyers and sellers (represented by associations and unions), or where the crew was paid in a share of the gross proceeds of the trip, the sales slip at the dock became a standardized, accepted document. At this point it was possible to measure the size and composition of each vessel's catch (other details were often included, such as time and place of catch) (see Crutchfield 1955; Gerhardsen, Zoetweij, and others in Turvey and Wiseman 1957, and Rothschild and other participants in OECD 1972).

In these countries at least catch was measured and described, and so was available as a unit for restriction. Either a tax on landings or a fixed catch-quota per input unit would do. Elements of both systems are mentioned in the literature in the late 1950s and by the time of Crutchfield and Zellner 1961, Royce et al. 1963, Crutchfield 1961 and Scott 1962 were becoming commonplace.

Furthermore, it was seen at once, both for a stationary harvesting model and for dynamic models, that at a sufficiently high level of abstraction, the systems were identical. In the stationary model, one could indifferently price the fish (fee) and let the market decide on the catch, or set the total and individual quota, and let the market determine the value and the fees. The same was true for a dynamic model, *except* that both fees and the number of fish in an individual quota (or the number of quotas) would have to change as the model moved toward some stationary or level or time-path.

It was seen that this system could well be tied in with

734   J. FISH. RES. BOARD CAN., VOL. 36, 1979

the *national* quotas being set for an expanding number of international fishing conservation conventions. If each nation were to be given (or sold) the right to catch a given number and quality of fish, would it not be easiest to break this down into regional, and then into vessel quotas? Indeed, could not an international trade in quotas emerge, not only between the national authorities but also between the individual vessels? (See Christy and Scott 1965; see also the subsequent 1966–68 debates at the Rhode Island Law of the Sea meetings; Christy 1973a, b.)

Thus began a general inclination away from restricting vessels or fishermen, in favor of restricting total catch by taxation or quotas. But any illusion about the seeming simplicity of this approach is soon shattered by realizing that output is not a unique product, but usually consists of a number of products. As foreshadowed earlier, the problems (and decisions that must be made) about this jointness can be classified as follows: (1) the catch from one stock usually consists of fish of several ages, sizes, and qualities; (2) the catch from one place may contain fish of more than one stock; and (3) more than one species; (4) the catch landed at one place is caught in more than one place; and (5) all can be caught in more than one season or year.

These five complexities mean that, strictly speaking, any scheme of preventing excessive entry by rationing the right to take fish can achieve efficiency in the allocation of inputs among outputs (that is, bring about equalities of marginal rates of substitution of inputs and outputs with their respective prices) only if it assigns prices (fees) to each fish caught according to its age, stock, species, place, and time; or, alternatively, if it distributes separate marketable rights for each of these. Furthermore, because fishing has the public-goods "externality" of reducing the density of its own stock and thus reducing finding and catching costs, but increasing congestion costs, these prices or rights should not be fixed, but should vary with each stock's density. Finally, because there are ecological interdependencies between the rates of growth and the spatial migrations of the various stocks and species, the prices or rights should vary with the damages the catching of fish from each stock imposes on the future availability of other stocks.

While this list of conditions for a perfect economic system of controlling the catch is not always as daunting as it seems, its complexities do reduce the apparent superiority or regulating catch over restricting inputs. The technology does not exist, as yet, for allowing fishermen to adjust to fishing fees or rights by substitution among stocks and species in one place, so that joint capture must remain the rule. But technology and capacity does exist for fishermen to choose the size, place, and time of catching fish. It follows that either (1) prices (fees) must be set for each of these characteristics of output, (2) quotas assigned to particular fishermen with specific stipulations about each characteristic, or (3)

general regulations must be made about the permitted catch of all fishermen to be divided among sizes, places, and times. All these modifications are possible, although the information requirements of the specific pricing or quota alternatives seem *very* high.

Such considerations reinforce what I discern to be general tendencies both to move towards the control of outputs (catch) rather than inputs (vessels and gear), and to use property rights or taxes. In the following pages, we shall attempt a more formal comparison of the two methods of restricting output. The reader should bear in mind that, as has already been recognized, at the level of formal analysis, price restriction is the dual of quantity restriction. If there were no "difficulties" about information and transactions costs, and no other aim but efficiency, there would be no reason for preferring either. To suggests where theory should go now, therefore, we must continue in this paper to remain suspended between the pure theory of the single fish stock and the ragbag of technical administrative and political difficulties that confront actual fisheries' management agencies.

V. Property Rights in the Catch (Quotas)

The main approach of this section will be to explore under eight main headings the theory of fishing rights (quotas) by comparing them with taxes. In the previous section it was recognized that, at the level of abstract stationary or dynamic theory, the two systems are equivalent. Three important omissions must be mentioned first.

The first omission is the use of subsidies. A fisherman or vessel could be bribed *not* to catch a fish, instead of being charged for catching it. Furthermore, he could be presumed to have a property right in catching fish, and be bribed to sell it. As Coase and others have shown, it often does not matter which of these two techniques is chosen in dealing with a spillover or other transaction between two parties. However, in the case of the fishery, there are also public-good effects (stock effects) to worry about; and there are problems about deciding whether potential fishermen, who demand a bribe, are serious potential entrants. To avoid prolonged analysis of these problems (which may be soluable) bribes (negative fees) will not be discussed. In the real world, fishermen are bribed to *enter* the fishery, not to leave it, by vessel subsidies, tax concessions, and unemployment insurance. These measures offset the entry-restriction measure to be discussed here; they are presumably to be justified on distributional grounds. We will not discuss them further.

The second omission is distributional, having to do with public revenue. Fisheries are part of the public domain ("the common heritage of mankind," in the L.O.S. formulation). Thus it seems to many people that fishermen ought not to make money without sharing the rent with the public. This way of looking at the matter can lead to the amount of revenue becoming the criterion for the choice between entry-restrictions systems: tax revenue versus price revenue. Economists are often disconcerted at this reappearance of the distributional aspects of policy choice into what they had pared down to an allocational issue. (Other examples are the surprising prevalence of politicians to regard the protective tariff, and the pollution tax, either as promising sources of revenue — which indeed they may be — or as a public claim that ought to be acknowledged even if it is unremunerative. See Campbell 1976; Fox and Swainson 1975; Australia, I.A.C., 1977, p. 34.)

We shall say little about revenue here, acknowledging that if it is an important goal, some systems of licensing must be rejected. In almost every country today, unregulated fisheries are a drain on the economy's production possibilities; furthermore, they are a drain on the public purse. That this situation continues suggests that revenue production does not yet rank high among public goals for fishery regulations.

The third omission concerns alternative, dynamic, objectives. In the literature it is usually assumed that restrictions are to be used to change the size of the fleet and *hold it in a steady state*. Thus in the absence of random perturbations, the annual catch is to be guided along a chosen path until the stock and the catch reach the desired level. It follows that the transitional tax or level of rights may differ from the final rate or level. Furthermore, the new interest in pulse or cyclic fishing, and the possibilities of a demand for a gradually rising or falling trend, responding to external changes in factor supply, substitutes, preferences (and so on) all suggest that gradually changing tax rates, or rights, may be in order. Finally, "flexibility" of rates and rights may be sought for special reasons, varying from changing attitudes to revenue to the expected need to make room for other uses of the ocean or for foreign fishermen. In this section, I cannot deal with the proliferation of types of tax and right that would have the right degree of flexibility to be acceptable to fishermen on the one hand and cyclic or spiral on the other. Instead, I will discuss institutions designed to produce the correct steady amount of effort, recognizing the likelihood that information is often wrong, technology changes, ocean currents and climate shift, and demands rise and fall. How a fishery approaches this fixed (stochastic) level is too complicated a question for this survey.

In this comparison of fishing quotas with fees or taxes, we have in mind the following systems.

The tax system would consist of a charge, levied at the landing place, on each fish. The complications of age, stock, species, place and time, and of interdependencies and of stock externalities already mentioned, might be handled by having a different rate of tax for fish of each characteristic, variable with respect to the rate of depletion of the stock, and with respect to the impact of the catch on other stocks. Alternatively, supplementary regulations and controls could govern these problems. Assuming that the aim is to hold the

stock and catch at a certain (stochastic) level, the basic tax would, in equilibrium, be at a fixed, proportional rate per unit of catch. So far as I know, no regulatory tax or royalty on the catch is anywhere in effect today.

The quota system would look a little like some present restricted rights or license systems, except that a quota would not convey the privilege of going fishing, nor of using a vessel; it would permit the holder to land a certain number of fish. Quotas would be transferable, marketable, and of fixed duration. As with the tax system, the complications of age, stock, species, place and time, interdependency and stock externalities might in part be dealt with by special individual quotas for fish of each characteristic and in part by general laws about the conditions under which all rights were valid.

How are these rights to be distributed and redistributed? We have already discussed a "grandfather system" to approach the desired number of participants. Another, perhaps better, system is to give every active fiserman a quota certificate; however, the size of the catch going with each quota may be very small; and fishermen will be given an incentive to buy and sell quotas to assemble rights sufficient to catch a profitable amount (Australia 1977, p. 35).

Both these systems make the quotas a free gift to the original holders. Other important alternatives are that they be initially free but taxable (not as a royalty, but as a capital asset); that they be repurchased, repossessed, or periodically reauctioned; and that they be redistributed by lottery (Scott 1962; Crutchfield 1961). (Furthermore, they could be used for positive redistributive purposes, such as guaranteeing a livelihood to isolated communities, or to protected groups.)

None of these distributive aspects affects the working of the quota system, so long as quotas are transferable, marketable, and of fixed duration. Any holder's behavior during the life of his quota should be independent of how he acquired it.

There is a literature on quotas, much of it unpublished papers that deal with managing extended fishing zones. (Anderson 1977 is a good example of published work.) However, seen from a distance, little of an analytic nature has been written since the 1960s (Hamlisch 1962), except concerning international waters (Christy and Scott 1965 inspired a number of L.O.S. contributions, many of them in the legal literature. See, however, Crutchfield 1966, Crutchfield and Pontecorvo 1969; Pearse 1971, Christy 1973, Scott and Southey 1970, and Tussing 1971. The extensive literature proposing or analyzing pollution property rights is essentially untapped by fisheries economists, although it is by now mostly common currency among environmental economists. Readers would gain by reexamining Ackerman 1977, Scott and Branson 1972, Dales 1968, and Tietenburg 1974 and subsequent papers.) I now commence the examination of the quota system, under eight headings.

a) *Directness:* For an overfished stock, taxes would

work through their discentive effect. At a certain tax rate, stock effects would increase fishing costs until the taxed fish was no longer valuable enough to justify continuing fishing. Taking into account the fixed costs of vessels and gear, the tax would make fishing completely unprofitable for some fishermen, and potential fishermen.

Whether the tax rate struck was correct, however, would be a matter for trial and error. As with charges on pollution, it is difficult to predict how much fishing will persist, in the short run and in the long run, at a given rate, and whether this will bring the catch and stock to their target levels. A new rate may have to be tried. The resulting uncertainty about the permanence of any rate would be a disadvantage of the system in itself, quite aside from the waste of time in over- or under-fishing at the wrong tax rates.

I believe that a quota or property-right system would have less of this uncertainty about fisherman response. When it has been decided how large the catch (for the relevant period) is to be, pro-rata shares will be assigned to each fisherman quota. He *knows*, for that period, how many fish he can catch. If it were not for stock effects, furthermore, he could choose his own pace of bringing in his quota. Even with stock effects, he need not race for the catch to the same extent as a taxed fisherman must. It follows that the catch will be more spread out over the season, to suit the preferences of the fisherman and of the market.

b) *Stock and catch uncertainty:* Of course, even when the amount that may be taken by each fisherman is determined and announced, it may turn out that the official calculations were wrong. There may be, say, fewer fish available than when the tax rate of individual quota were set. In these circumstances, the authorities must change the tax rate per fish under the first system, and the allowable catch per individual quota under the second.

The tax system, depending on a predictable relationship between the after-tax value of fish and the amount of effort expended by the fleet, is probably capable of bringing about the desired direction in change in effort. But, with many participants, it may be affected by lags, and by too much variety in individual responses, to achieve precision in the change; short-run over- or under-reactions are therefore to be expected. Furthermore, unexpected changes in tax rates may create fisherman expectations of further changes, in either direction, thus aggravating tendencies to over- or under-react.

The quota system is more precise in this respect. A change in the number of any kind of fish that may be taken with a unit quota can be quickly signalled to all quota holders. With experience, the authorities should be able to bring about precisely the desired change in catch. Expectations are a problem here too, however. If fishermen expect that the catch-value of their quota is likely to be reduced, they will tend to accelerate their catching effort. (It does not appear that an expected in-

crease in the value of the quota would have a symmetrical decelerating effect. Good discussions of uncertainty with water licenses and rights are to be found in Roberts 1975, Roberts and Spence 1974, and Fox and Swainson 1975).

c) *Progressivity:* Economists who write about fisheries are very tense about the rate of innovation. Laymen and politicians often believe that fishermen are unduly conservative, reluctant to adopt new techniques. This observation is reinforced when subsidies on new equipment, or the working of the crew share system, is biassed toward traditional methods. Furthermore, unprogressive fishermen may have obtained protective regulations that prevent the introduction of new techniques.

Such evidence is not convincing. Many fishermen, like farmers, obtain pleasure from planning and acquiring ever more elaborate vessels, gear and communications. Furthermore, Scott and others have argued that the race for fish (in the many-fishermen case) encourages too-rapid innovation in techniques that allow a fisherman to intercept or anticipate his rivals (1962). This tendency may not be serious. In general, as long as vessels race for the fish, neither a tax system nor a quota system will prevent lop-sided over-investment in speed and capacity. Only sole ownership can do this.

d) *Interception: Allocation of fishing within the season:* As Christy and Scott (1965) bring out, a system of national catch quotas fails to remove one costly aspect of unregulated fishing. I refer to the tendency, as over-fishing becomes more serious, for fishermen to capture a good share of the available catch by intercepting and capturing the fish before their rivals. This leads (spatially) to pushing effort out to sea into the path of migratory fish, and (seasonally) to advancing effort into winter months or earlier weeks than usual. The result is more remote, dangerous and costly fishing, more concentrated marketing, and lower quality of the product.

Both tax and quota systems can deal with this problem. The tax rate can be progressive with respect to the place, time, or aggregative rate of fish landings, or some combinations of these. Quotas can be located and dated to discourage excessive fishing under high-cost, high-concentration conditions.

Either of these modifications would have high administrative costs. Costs could be avoided by a system of dispatching individual vessels, or by simply closing the fishery in certain seasons and places. Dispatching usually, but need not, involve sole ownership or queuing as taxicab dispatching shows (see Loose 1978).

e) *Divisibility:* We turn to two closely associated characteristics of the two systems: divisibility and complexity of the units under control.

Little need be said about tax divisibility: taxes can be levied on individual fish, tons, or boat loads. There is a danger that catch quotas may, for administrative purposes, be in larger units. Such indivisibility can be costly. Pearse (1971) writing about input restrictions has shown this convincingly. When (vessel) fishing rights are indivisible, potential fishermen with relatively low catching ability but also relatively low opportunity costs, are unable to enter sufficiently high bids for licenses for whole vessels even though they could bid more *per fish* than the successful highliners. Thus they are shouldered out because the unit for which they bid is bigger than they need. The answer is obviously to permit them to bid for smaller units. If this can be done with catch quotas, taxes and quotas would be equally successful in allocating those with a comparative advantage in fishing to the industry.

In my opinion, the same comment applies to well-know allied work on pollution by Buchanan and Tullock (1975). They prefer a pollution tax to a pollution license when the overall policy is to reduce industry output. They argue that an appropriate tax allows all firms remaining to adjust to optimum scale on their long-run cost curves, whereas pollution licenses leave too many firms operating at a scale less than optimum. This criticism of licenses, however, is valid only if they are indivisible. If they can be subdivided and marketed, their effect on industry behavior should be the same as that of a tax (Buchanan and Tullock 1975).

f) *Complexity:* I have said, several times, that because the typical fleet is composed of a variety of vessel sizes using a variety of gear, and because it is searching for fish from a variety of stocks and places, there is great difficulty in deciding on a single unit of entry restriction.

In the existing B.C. *vessel*-restriction system (Pearse 1972) as with that described by Crutchfield and Pontecorvo (1969, p. 178–80), it is recommended that aggregate fleet fishing power be kept constant by making licenses applicable among various sizes and types of vessel in proportion to their respective officially accepted fishing power. This recommendation is analogous to a recommendation for transferable pollution effluent permits: each should be useful for alternative set *equi-damage* rates of waste disposal of the respective waste substances, thus allowing the holder of a permit to flexibly change location or industrial process and so the damage from his effluent (Dales 1968; Scott 1972). In fishing, similar fixed-ratio systems can be used either for tax systems, with factors to lower or raise tax rates according to species, place, and time of landings, or for quota systems, with factors to convert fish harvested from various species, places, or periods to set fractions of the standard quota.

But such factors or ratios are extremely difficult to set, and change rapidly over time as knowledge accumulates and conditions change. In spite of the extra paperwork involved, it would probably be better to assign each vessel a packet of rights, one for each species, stock, place, and time. These would be exchangeable, so fishermen could specialize. They would also tend to discourage inadvertent joint catching of stocks, by forcing a fisherman to acquire from other vessels (for money, or in exchange), rights for fish not

sought or wanted. In a fleet with many participants, each type of right would acquire its own market value. The authorities, by "open market operations," (buying in or selling extra particular rights) could guide the fleet away from overfished stocks or places and into underutilised fisheries with rather more enforcement costs. A tax system could work the same way.

The information and enforcement costs of either of these detailed systems, in a complex fishery, would be very high. In both, the basic source of information would be fishermen's own declarations about the source and nature of their catch. Checking up on these statements would be difficult and costly. Unlike more primitive systems (such as closed areas or closed seasons), "self-enforcement," by vessels keeping an eye on each other, might not work well. This is because each vessel, under a pure tax or a pure quota system could choose freely when and where to fish, so long as each paid the appropriate fee or used the appropriate part of its total quota for fish taken from an area with a small aggregate, total-fleet, permitted harvest. Thus the tendency of others to jealously prevent a vessel poaching or contravening other regulations, is weakened when tax and quota systems allow any given vessel to do anything, if it is willing to pay the price.

For less complex fisheries, the number of types of license, or fee, might be quite small. Meany (1977) suggests that in many fisheries each species, if not each stock within that species, would need its own specialized gear. He recommends a different fee for each gear, thus in effect also recommending an easily enforceable fee for each stock, species, or place. This suggestion also applies to quotas/rights.

However, for very large (and international) fleets, the administrative costs of multiregion, multipurpose, multispecies catch or vessel licensing would be severe. It is probable that a system that set out to bring about efficiency in the allocation of labor and capital to fishing many species, and places, with one or more fleets, would attempt to reduce the costliness of administering and enforcing one-right-per-vessel-for-each-stock-and-place systems (or their equivalent tax systems) by removing some economic choices from fishermen and replacing them with blanket, enforceable, prohibitions or controls.

To conclude the tax–quota choice, a complex fishery is bound to induce high transactions (information and enforcement) costs under either system of regulation. Considering that specific quotas would have to be produced at the dock to permit the sale of each type of fish, and that the market in specific quotas would keep their specificity in the public eye, I am inclined to argue that the enforcement and compliance costs of quotas would be somewhat less than for a system of specific taxes for each stock, species, and place (Eckert 1973). Under the tax system, there would always be a temptation for the vessel to claim that all its catch came from one or a few specific categories. This would be impos-

738   J. FISH. RES. BOARD CAN., VOL. 36, 1979

sible for a vessel that, during a year, relied on all its subquotas.

g) *Joint action, collusion, and interest groups:* Support or opposition to rights or warrant systems in fisheries frequently arise from beliefs that instead of the flexibility and versatility of individually held quotas, the sector would wind up with the inflexibility of corporate, oligopolistic, or union-held blocks of quotas. This concentration or combination is certainly a possibility, perhaps more probable with a system of rights than with taxes.

The general purpose would be to reduce the number of participants, vessels, owners, or fishermen. This would allow those who had purchased or combined rights to exact some control over the fishery additional to those administered officially.

In negotiating the price of fish, such combination in the hands of the companies (for the unions) would give them extra bargaining strength, although this strength would not necessarily be greater than that now held by fishing associations or unions. These possibilities have affected attitudes to limited fishing quotas by focusing attention on their potentiality for upsetting a known set of institutions, rather than on their specific merits.

A second application of the power of concerted action would be for groups to acquire blocks of quotas for their own use (as with the officially protected groups mentioned in Part II — here the protected holders might be native Indians, or old-age pensioners). Such acquisitions could be at public expense, for redistributive purposes. Alternatively, self-appointed groups could acquire licenses to promote other activities by *preventing* fishing. For example, they might, holding quotas, wish to augment or reinforce official efforts at fish or mammalian conservation, reduce permanently a stock they disliked or feared, obtain for themselves more stock and elbowroom for water sport or sportfishing (compare Dales 1968), use the waters for waste-dumping or land fill, thus killing the fish for which they held quotas; or reduce the presence of commercial fishermen in certain places (oil fields, hazardous navigational narrows, or holiday areas).

Because of the public-goods aspects of fisheries, it is unprofitable for a single person to acquire rights for any of these purposes that affect the whole fleet: joint action is needed by associations or firms. The work of M. Olsen gives clues to the benefits and costs of forming joint or collusive clubs or associations (1965). The more specific each quota or right, however, regarding to species, place, or time, the less is the need for joint action. On the other hand, if an outsider merely wishes to catch a certain number of fish, he has merely to buy a quota; joint action is not required unless there are economies of scale in sportfishing (congestion, provision of marinas, selection of certain subspecies for preservation).

h) *Flexibility:* It is frequently said that one of the worst flaws in most resource-management systems is their inflexibility. P. H. Pearse in his Royal Commission Report on British Columbia's forest policy has strongly criticized this characteristic, and the same theme is to be found in Campbell et al. on water policy in Ontario (1974) and other studies to which he has contributed. A system of rules, or system of tenure, that prevents smooth adjustment to exogenous changes is likely increasingly to waste natural resources, labor, and capital, and to distribute rents and earnings away from the Crown and others who originally shared in the bounty of nature.

But flexibility is a dangerous characteristic in a policy or social institution. Like individual voters and organized pressure groups in political life, members of the fishing industry will not stand passively by while government administers a flexible scheme for restricting entry. Instead, it can form its own political pressure groups and coalitions, attempting to obtain a policy that meets their demands. The more "flexible" a policy, the more it will attract political action (and indeed social movements) to change it. Furthermore, flexibility also invites less savoury attempts to overcome an unfavorable policy: excluded or injured industry members can try bribery and corruption to buy their way into the industry, as individuals or as groups. Clearly, every person has three investment choices: he can buy the rights he needs (or pay the equivalent tax); he can signal politically for a change of general policy; or he can pay individual legislators or bureaucrats to change a policy, or at least to change its application to him. No matter how much he may believe in the rule of law, the second type of investment will seem just as moral as the first; and the third, in extreme conditions, may promise an attractively high yield relative to the first pair. If such should be the fate of the system, it would obviously both destroy the legitimacy of the regulatory system and create dangerous cost-increasing uncertainty among vessel owners.

With these possibilities, for the typical fishery, I favor a system of property rights rather than a fiscal device such as tax or subsidy. I have two reasons for this.

In the first place, to sell or give or give a marketable quota to a fisherman is to associate it in politicians' and citizens' minds with other rights, privileges, obligations, and instruments issued by the government: annuities, bonds, and titles, and currency for example. All enter their owners' portfolios of explicit, formal, negotiable assets. Thus the temptation to attack the system for personal gain, through political or corrupt activities, should be moderated by general legal and juridicial attitudes in our society that tend to preserve the integrity of the government's individual undertakings and to resist proposals that they be regarded as subject to the same changeability as, say, the rate·of property tax, income tax, royalty rate, or public insurance premium. Such solidity, integrity, and certainty is highly valuable in a system in which a large number of fisheries will be regulated by public servants under delegated powers (Buchanan and Tullock 1975; Chisholm 1978).

In the second place, a measure of allocational flexibility is, nevertheless, possible by government "open-market operations" including a buyback policy of quotas and similar market operations to change the relative value of the special privileges or prohibitions that apply to particular species, times, places, etc. If regulators were in general to use such market operations more, and uncertainty-creating, arbitrary changes in controls and prohibitions less, the willingness to comply with the revealed policies would increase. (As Scott and Southey 1970 and Crutchfield 1961 argued, it would be very costly to run a fishery on individual property rights alone. Coase disagreed with this. But some supplementary controls do seem less costly than either full internalization [sole ownership] or a finely tuned complex of individual quotas, with every characteristic marketable.)

Admittedly, this "open-market operations" policy does prevent a flexible change in the share of the public in the rent of fisheries. But my own preferences are not to strive for a system that captures rent directly for the Crown, but for a system that allows rent (net present value) to exist at all. In any case, as various commentators have reminded me, it is possible to tax a fishing right as an asset, the rate of this tax being a matter for political decision. I have some fears that frequent changes in this rate would lead to tax-avoidance behavior that, in some fisheries, could be deleterious, for example, a subquota tax can be changed. Thus the quota alternative does not require that public revenues be abandoned, or unchanging. (I do fear that if every subquota, described above, were subject to a changeable wealth or property tax, having income effects, evasive or socially wasteful behavior might result. But careful model-building and empirical research is needed to elucidate this question.) As with pollution rights, so with fishing quotas, there is no need to regard tax or fee method of entry restriction as uniquely suited to distributing much of the rent to the public.

Examination of my discussion under all eight of the headings above will suggest my own final preference for a rights system. It is perfectly true that, as abstract theory would insist, a tax on the catch, and a property-rights system can work in similar fashion. Whatever one can do, for efficiency, equity, and conservation, the other can be made to do. What then should be the final basis for choice?

The final choice should be based on the minimization of transaction and administration costs. The survey has shown that both systems can achieve the same ends; therefore the only difference between them is the "difficulty" of obtaining information under one system or the other; fuller examination of this problem suggests that we are talking about the inputs needed for information about the fishery and about the fleet, for decision-making in the interests of efficiency and equity, and for enforcement, monitoring, and feedback. Economics tells us that all these activities have costs, and that we

may in fact compare the two systems by comparing the costs of using them (not, to repeat, the costs of fishing, which can be the same under either system).

The editors would not want me to commence a comparison of the levels of these costs, under the eight headings, for the two systems. In any case, we need facts and not opinions. So I will conclude with a hypothesis, which I have formed by reexamining the eight main criteria in this Section: the total costs of administration, transactions, marketing, information, and compliance (and of evasive behavior as well) would be lower under a quota or rights system than under a tax system that achieves the same present value (net of such costs). Both systems would be very costly techniques for dealing with some regulatory important details of fishery management. In these respects, they ought to be supplemented with blunt, crude, controls of timing, spacing, place, and so forth. When this has been done (when each system is "optimized") I would predict that one based on quotas would have lower overall transaction and administration costs than a fee (or royalty) system.

Acknowledgments

Bibliographic help has been received from Ms R. Forward, Centre for Resource and Environmental Studies, A.N.U.; Mr F. Meany; and the collection of the Fishery Division, Australian Department of Primary Industry. My colleagues P. Pearse and H. Campbell have commented on parts of this version. B. Pope and D. Garriott have helped in revising and typing the two versions.

ACKERMAN, S.-R. 1977. Market models for water pollution control. Public Policy 25: 383–406.

ANDERSON, L. G. [ed.]. 1977. Economic impacts of extended fisheries jurisdiction. Ann Arbor Science, Ann Arbor, Mich.

AUSTRALIA, INDUSTRIES ASSISTANCE COMMISSION. 1977. Fish and fish processing: Part A: fish. (Draft Report), I.A.C., Canberra, Australia.

BELL, F. W. 1972. Technological externalities and common property resources: an empirical study of the U.S. northern lobster fishery. J. Polit. Econ. 80: 148–158.

BEVERTON, R. J. H., AND S. J. HOLT. 1957. On the dynamics of exploited fish populations. HMSO, London.

BROWN, G. M. 1974. An optimal program for managing common property resources with congestion externalities. J. Polit. Econ. 87: 163–173.

BUCHANAN, J. M., AND G. TULLOCK. 1975. Polluters profits and political response: direct controls vs. taxes. Am. Econ. Rev. 65: 139–147.

BUTLIN, J. 1975. Optimal depletion of a replenishable resource: an evaluation of recent contributions to fisheries economics. Chapter 6. In D. W. Pearce [ed.] Economics of natural resource depletion. Macmillan, London.

CAMPBELL, H. F. 1976. Rent vs. revenue maximisation as an objective of environmental managements, p. 185–201. In A. Scott [ed.] Economics of fisheries management: a symposium. H. R. MacMillan Lectures in Fisheries. University of British Columbia, Vancouver, B.C.

740   J. FISH. RES. BOARD CAN., VOL. 36, 1979

CAMPBELL, R. S., P. H. PEARSE, A. SCOTT, AND M. Uzelac. 1974. Water management in Ontario — an economic evaluation of public policy. Osgoode Hall Law J. 12: 475–526.

CHEUNG, S. N. S. 1971. The structure of a contract and the theory of a nonexclusive resource. J. Law Econ. 13(1).

CHISHOLM, A. H. 1978. Choice of instruments for environmental protection (unpublished conference paper, A.N.C.).

CHRISTY, F. T. JR. 1973a. Alternative arrangements for marine fisheries: an overview, programme of international studies of fisheries arrangements, Paper No. 1, R.F.F. Washington, D.C.

1973b. Fisherman catch quotas, Ocean Dev. Int. Law J. 1: 121–140.

CHRISTY, F. T. JR., AND A. SCOTT. 1965. The common wealth in ocean fisheries. Johns Hopkins, for R.F.F. Baltimore, MD.

CINACY-WANTRUP, S. VON. 1952. Resource conservation — economics and policies. University of California Press, Berkeley, Calif.

CLARK, C. W. 1973. Profit maximisation and the extinction of animal species. J. Polit. Econ. 81: 950–961.

1976. Mathematical bioeconomics: the optimal management of renewable resources. Wiley, New York, N.Y.

CLARK, C. W., F. H. CLARKE, AND G. R. MUNRO. 1977. The optimal exploitation of renewable resource stocks: problems of irreversible investment. PNRE Paper 8, University of British Columbia, Vancouver, B.C.

CLARK, C. W., AND G. R. MUNRO. 1976. The economics of fishing and modern capital theory: a simplified approach. J. Environ. Econ. Manage. 2: 92–106.

COPES, P. 1970. The backward-bending supply curve of the fishing industry. Scott. J. Polit. Econ. 17: 69–77.

1972. Factor rents, sole ownership and the optimum level of fisheries exploitation. Manchester School 40(2).

CRUTCHFIELD, J. A. 1955. Fishermen's union and the anti-trust statutes: the economic issues. Ind. Labour Relations Rev. 8: 531–536.

[ed.]. 1959. Biological and economic aspects of fisheries management. University of Washington Press, Seattle, Wash.

1961. An economic evaluation of alternative methods of fishery regulation. J. Law Econ. 4.

[ed.]. 1965. The fisheries: problems in resource management. University of Washington Press, Seattle, Wash.

1966. International fisheries. Am. Econ. Rev. (May)

CRUTCHFIELD, J. A., AND G. PONTECORVO. 1969. The Pacific salmon fisheries: a study of irrational conservation. Johns Hopkins for R.F.F. Baltimore, Md.

CRUTCHFIELD, J. A., AND A. ZELLNER. 1962. Economic aspects of the Pacific halibut fishery. (Fishing Industrial Research Vol. 1). U.S. Dept. of the Interior, Washington, D.C.

DALES, J. 1968. Pollution, property and prices. Toronto, Ont.

ECKERT, R. D. 1973. On the incentives of regulations: the case of taxicabs. Public Choice 14(83).

FOX, I., AND N. SWAINSON. [ed.]. 1975. Water quality managements: the design of institutional arrangements. University of British Columbia Press, Vancouver, B.C.

GERHARDSEN, G. M. 1952. Production economics in fisheries. Rev. Econ. Lisbon.

GORDON, H. S. 1954. The economic theory of a common property resource: the fishery. J. Polit. Econ. 62: 124–142.

1957. Obstacles to agreement on control in the fishing industry, p. 65–72. In R. Turvey and J. Wiseman [ed.] The economics of fisheries. F.A.O., Rome.

1958. Economics and the conservation question. J. Law Econ. 1: 110–121.

GOULD, J. R. 1972. Extinction of a fishery by commercial exploitation: a note. J. Polit. Econ. 80: 1031–1039.

GREGORY, H. E., AND K. BARNES. 1939. The North Pacific fisheries, (studies of the Pacific, No. 3) Am. Inst. Pac. Rel. San Francisco, Calif.

HAMLISCH, R. [ed.]. 1962. Economic effects of fishery regulation (FAO fisheries reports no. 5) F.A.O., Rome.

HANNESSON, R. 1974a. Fishery dynamics: a North Atlantic cod fishery. Can. J. Econ. 8: 151–173.

1974b. Economics of fisheries: some problems of efficiency. (Lund Ph.D. thesis) Lund, Student Litteratur.

INNIS, H. A. 1940. The cold fisheries: the history of an international economy. Ryerson Press, Toronto, Ont.

JAMIESON, S. M. 1958. The fishery. In J. Deutsch et al. The natural resources of British Columbia. (Unpublished ms)

JAMIESON, S. M., AND P. GLADSTONE. 1950. Unionism in the fishing industry of British Columbia. Can. J. Econ. Polit. Sci., p. 4.

KNIGHT, F. H. 1924. Some fallacies in the interpretation of social cost. Q. J. Econ. 38: 582–606.

LEWIS, T. R., AND R. SCHMALENSEE. 1977. Nonconvexity and optimal exhaustion of renewable resources. Int. Econ. Rev. 18: 535–552.

LOOSE, V. 1978. Optimum exploitation of a salmon fishery: a simulation approach. Ph.D. thesis, University of British Columbia, Vancouver, B.C.

MARTS, M. E., AND W. R. D. SEWELL. 1959. The application of benefit-cost analysis to fish preservation expenditures: a neglected aspect of river basin development. Land Econ.

MEANY, T. F. 1977. License limitation in a multi-purpose fishery. Aust. Fish. (November) p. 8–11 and 19.

McDOUGAL, M. S., AND W. T. BURKE. 1962. The public order of the oceans: a contemporary law of the sea. Yale University Press, New Haven, Conn.

McGREGOR, D. C. 1949. The economist looks at the oceans. Proc. R. Soc. Can. 43: 173–181.

McKELLAR, N. B., AND ASSOCIATES. 1977. Restrictive licensing as a fisheries management tool, F.E.R.U. Occas. Pap. Ser. No. 6, White Fish Authority, Edinburgh, Scotland.

NEHER, P. 1974. Notes on the Volterra-Quadratic fishery. J. Econ. Theory 8: 39–49.

OECD. 1972a. Economic aspects of fish production. Proc. Symp. Fish. Econ. 1971 OECD, Paris, France.

1972b. Problems in transfrontier pollution. Record of a seminar on economic and legal aspects. OECD, Paris, France.

OLSEN, M. JR. 1965. The logic of collection action. Harvard University, Cambridge, Mass.

PEARSE, P. H. 1971. Rationalization of Canada's west coast salmon fishery: an economic evaluation. p. 172–202. In OECD. Proc. Symp. Fish. Econ. 1971.

1972. Rationalization of Canada's West Coast salmon fishery: an economic evaluation. In Economic Aspects of Fish Production. Paris, OECD. 1972.

1978. Approaches to economic regulation of fisheries. CIDA/FAO/CECAF workshop on fishery development and planning. (February 6017) Lome, Togo.

PELTZMAN, S., ET AL. 1976. Toward a more general theory of regulation. J. Law Econ. 19: 211–248.

PETERSON, F. M., AND A. C. FISHER. 1977. The exploitation of extractive resources: a survey. Econ. J. 8: 681–721.

PLOURDE, C. G. 1971. Exploitation of common-property replenishable natural resources. West. Econ. J. (September 9): 256–266.

PONTECORVO, G. 1966. Optimization and taxation: the case of an open access resource, the fishery. In M. Gaffney [ed.] Conservation, taxation and the public interest in extractive resources. University of Wisconsin Press, Madison, Wisc.

QUIRK, J., AND V. SMITH. 1970. Dynamic economic models of fishing. *In* A. Scott [ed.] Economics of fisheries management: a symposium. H. R. MacMillan lectures in fisheries, University of British Columbia, Vancouver, B.C.

ROBERTS, M. J. 1975. Environmental protection: the complexities of real policy choice. *In* I. Fox and N. Swainson [ed.] Water quality management: the design of institutional arrangements. University of British Columbia Press, Vancouver, B.C.

ROBERTS, M. J., AND M. SPENCE. 1974. Effluent charges and licenses under uncertainty. Rev. Econ. Stud. 41: 477–491.

ROTHSCHILD, B. J. 1972. Definition of fishing effort, p. 257–271. *In* OECD Proc. Symp. Fish. Econ. 1971. OECD, Paris, France.

ROYCE, W., D. BEVAN, J. CRUTCHFIELD, G. PAULIK, AND R. FLETCHER. 1963. Salmon gear limitation in northern Washington waters. University of Washington publications in fisheries II, I. Seattle, Wash.

SCOTT, A. 1955. The fishery: the objectives of sole ownership. J. Polit. Econ. 63: 116–124.

1960. The economists' needs in fishery statistics. *In* FAO/ICES Proceedings of fishery statistics conference, Edinburgh, Scotland.

1962. The economics of regulating fisheries, p. 25–65. *In* R. Hamlisch Economic effects of fishery regulation. (FAO Fisheries Reports No. 5) FAO, Rome.

1965a. Fisheries development and national economic development, 18th annual season, Proceedings of the Gulf and Caribbean Fisheries Institute. (November)

1965b. The valuation of game resources: some theoretical aspects. Can. Fish. Rep. 4: 24–47.

[ed.]. 1970. Economics of fisheries management: a symposium. H. R. MacMillan lectures in fisheries, University of British Columbia, Vancouver, B.C.

1972. Natural resources, the economics of conservation. 2nd edition, McClelland & Stewart, Toronto, Ont.

SCOTT, A., AND C. BOBRAMSEN. 1972. Pollution certificates: a specific system of long-term resource-quality co-operation, appendix II of draft guideline principles concerning transfrontier pollution, p. 311–312. *In* OECD Problems in

transfrontier pollution. Record of a seminar on economic and legal aspects. OECD, Paris, France.

SCHAEFER, M. B. 1954. Some aspects of the dynamics of populations important to the management of the commercial marine fisheries. Inter-Am. Trop. Tuna Comm. Bull. 1(2).

SINCLAIR, S. 1960. License-limitation — British Columbia: a method of economic fisheries' management. Fisheries Department, Ottawa, Ont.

SMITH, C. L. 1977. The failure of success in fisheries management. Environ. Manage. 1: 239–247.

SMITH, V. L. 1968. Economics of production from natural resources. Am. Econ. Rev. 58: 409–431.

1969. On models of commercial fishing. J. Polit. Econ. 77: 181–198.

[ed.]. 1977. Economics of natural and environmental resources.

SOUTHEY, C., AND A. SCOTT. 1970. The problems of achieving efficient regulation of a fishery, p. 47–60. *In* Economics of fisheries management: a symposium. H. R. MacMillan lectures in fisheries, University of British Columbia, Vancouver, B.C.

SPENCE, M. 1973. Blue whales and applied control theory. Stanford Inst. Math. Stud. Rep. 108.

STEPHENSON, J. [ed.]. 1977. Economic incentives and water control management. University of British Columbia Press, Vancouver, B.C.

TIETENBERG, T. 1974. The design of property rights for air pollution. Public Policy 22: 275–292.

TURVEY, R. 1964. Optimization and sub-optimization in fishery regulation. Am. Econ. Rev. 54: 64–76.

TURVEY, R., AND J. WISEMAN. [ed.]. 1957. The economics of fisheries. F.A.O., Rome.

TUSSING, A. [ed.]. 1971. An economic study of Alaska fisheries, Studies for the Institute of Social, Economic and Government Research College, Alaska.

WARMING, J. 1911. Om grundrente af fiskegrunde. National økonomisk Tadsskrift 49: 499–505.

WEITZMAN, M. L. 1974. Free access vs. private ownership as alternative means of managing common property. J. Econ. Theory 8: 225–239.

WHITE, D. J. 1954. The New England fishing industry. Harvard University Press, Cambridge, Mass.

[14]

Economic and Social Implications of the Main Policy Alternatives for Controlling Fishing Effort[1]

J. A. CRUTCHFIELD

Department of Economics, University of Washington, Seattle, WA 98195, USA

CRUTCHFIELD, J. A. 1979. Economic and social implications of the main policy alternatives for controlling fishing effort. J. Fish. Res. Board Can. 36: 742–752.

Discussion of the need for economic rationalization has led to increasing interest in alternative strategies to control fishing effort in efficient ways. Three basic alternatives are considered: taxation, control of fishing inputs ("limited entry"), and direct limitation of output through individual fisherman quotas. Each is analyzed in terms of specified elements of economic efficiency, distribution effects, freedom to choose fishing methods and strategies, and administrative feasibility. It is noted that in practice all would be used in combination with other direct measures to assure flexible control over fishing mortality. A number of common objections to economic rationalization of open access fisheries (e.g. creation of monopoly power, unemployment, failure to reduce consumer prices, and nonmonetary values associated with fishing) are analyzed and rejected as largely invalid. While the inherent short-term instability of commercial fishing and data problems preclude any "maximizing" model of a rationalized fishery, the prospects for workable programs leading to improved economic performance without adverse societal impacts are excellent.

Key words: fishery economics, taxation, limited entry, quotas, open access fishery, sociological aspects

CRUTCHFIELD, J. A. 1979. Economic and social implications of the main policy alternatives for controlling fishing effort. J. Fish. Res. Board Can. 36: 742–752.

L'analyse du besoin d'une rationalisation économique eut pour effet de susciter un intérêt accru à l'endroit de stratégies visant à contrôler l'effort de pêche par des moyens efficaces. Trois options fondamentales sont considérées: taxation, contrôle des entrées et limitation directe des sorties par le biais de contingentements de pêcheurs individuels. Nous analysons chacune de ces options en fonction d'éléments spécifiques d'efficacité économique, effets de distribution, liberté de choisir méthodes et stratégies de pêche et possibilité administrative. On fait remarquer que, dans la pratique, toutes ces options seraient combinées avec d'autres mesures directes visant à assurer un contrôle souple de la mortalité due à la pêche. Nous analysons et rejetons comme invalides, pour la plupart, un certain nombre d'objections communément soulevées à l'endroit de la rationalisation économique de pêcheries dont l'accès est libre (e.g. création de monopoles, chômage, impuissance à réduire les prix au consommateur et valeurs non monétaires associées à la pêche). L'instabilité à court terme inhérente à la pêche commerciale et les problèmes relatifs aux données ne permettent pas de construire de modèle d'une pêcherie rationalisée à rendement maximal. Mais quand même, on a bon espoir de pouvoir élaborer des programmes pratiques qui amélioreront le rendement économique sans effets nuisibles sur la société.

Received October 16, 1978
Accepted March 13, 1979

Reçu le 16 octobre 1978
Accepté le 13 mars 1979

AT this stage of the game it should not be necessary to justify the position that an open access fishery will produce unsatisfactory time rates of use of the resource. "Unsatisfactory" covers a multitude of ills ranging from modestly excessive use of labor and capital inputs to total extinction of fish populations. Both theory and a long series of empirical studies lead to the following unpalatable but firm conclusions. Excessive quantities

---

[1]This paper forms part of the Proceedings of the Symposium on Policies for Economic Rationalization of Commercial Fisheries held at Powell River, B.C., August 1978.

Printed in Canada (J5399)
Imprimé au Canada (J5399)

of capital and labor will enter the fishery, and overcapacity will worsen as real prices rise. Biological depletion will usually occur unless fishing mortality is curtailed by measures that raise costs of fishing. The costs of management, research, and enforcement will fall on the general taxpayer.

Actual performance is often considerably worse, since these outcomes rest on a static formulation with a zero-cost adjustment mechanism. In real fisheries the physical functions linking yield and effort are highly unstable, for reasons rooted in the complex environment of the sea which are poorly understood and rarely quantifiable at acceptable cost. Coupled with this yo-yo model of physical availability is an exaggerated degree of asymmetry between entry and exit in an unrestricted fishery. While it may take considerable time to become a skilled fisherman, neither human nor physical capital requirements for a viable unit are high. Hence every "big year" — whether in terms of abundance of fish, high prices, or both — produces a burst of new construction and entry. But declines in incomes in poor years do not produce an equivalent reduction in capacity. Factor mobility in any industry is rarely that responsive, and the sociocultural setting of fishing communities makes them even more sluggish in reacting to low incomes. Also, the nearly universal share system makes the supplier of labor a coventurer in the fishing enterprise and levels out the effects of price or catch fluctuations on returns to capital. The inherent instability of the biological determinants of short-run input–output functions and the typical producer response, biased toward an excess of entry over exit, produce returns to both labor and capital that are more often below than at or above opportunity levels.

Clearly society could reasonably expect better performance from the industries utilizing living resources of the sea. In the most general terms rationalization means the "reallocation of resources . . . under open access to a controlled system designed to maximize the net value of production from the economy as a whole" (Anderson 1977). From the standpoint of economy-wide efficiency the following specific elements emerge:

1. The right level of catch — at which the marginal social value of the harvest is equated to the incremental social cost required to take it (including management costs).

2. The right size (age) composition of catch. No net economic gains can be obtained by allowing smaller fish to grow (i.e. where marginal increments to revenue from growth are just offset by marginal losses to natural mortality and costs of program implementation).

3. The right number and configuration of vessel–gear–fishermen units to minimize the aggregate real cost of taking any given catch (i.e. optimal factor combinations).

4. Optimal fleet deployment; no increase in yield and/or reduction in cost can be achieved by altering the area or time fished.

Management measures that increase net economic returns in any or all directions are regarded as steps toward rationalization.

But economic efficiency alone is not enough; the literature on natural resource management is full to overflowing with warnings about the multiple objective problem. Economic rationalization of fisheries will usually be constrained by conflicts with other social objectives. These constraints warrant special consideration since, though legitimate, they have so often been perverted into cleverly concealed arguments for the maintenance of the status quo and the peaceful life of an unharried government decision-maker (Crutchfield 1973).

Two basic assumptions of fact underly the discussion that follows. First, it is assumed that real prices of fish will increase over time (particularly in the fisheries regarded as prime candidates for gear reduction programs). The potential gains from rationalization will therefore continue to grow even where physical production cannot be expanded further. Second, it is assumed that management designed to achieve more rational use of inputs will be dealing with fisheries already overcapitalized and subject to all the ills of open access. There simply aren't enough "new" fisheries on the horizon to bother with the easier question of developing a rational harvesting and management program from the outset. (Antarctic krill and blue whiting are notable exceptions, but it is difficult to identify many others.) Moreover, public acceptance of the need for fishery management of any kind rarely precedes serious overfishing, and this is likely to be even more of an obstacle when nontraditional measures are proposed.

It is abundantly clear, from theory and experience, that no management program that does not include control over inputs to fishing (or induce fishermen to do so) can offer much lasting improvement in economic performance over open access (see Crutchfield in 1961, Scott 1962, and Anderson 1977 for examples.) Thus any discussion of the economic aspects of rationalization must focus on means of reducing total inputs for any given catch level and on the extent to which least-cost combinations of inputs can be encouraged. This rapidly reduces to three measures: taxes or fees; direct limitation of numbers of fishing units, achieved by licensing boats, fishermen or both; and, at the hypothetical level, the establishment of rights to specific quantities of fish for individual operators. As indicated in the following discussion, it is misleading to categorize them so neatly, since it is argued that combinations of two or more make better sense than any one alone.

It is taken for granted that these measures to rationalize fishing would be backed up by necessary authority to take fast, direct action to deal with emergency situations in the environment. Fluctuations in natural abundance, seasonal availability, and concentration are inherent characteristics of marine fishery resources, and forecasting capabilities are inherently limited. Rationalization measures are therefore regarded as the backbone of management programs designed to improve economic benefits, but short-term flexibility must come from the power to cut off fishing by time, area, or species when developing evidence of trouble requires quick action.

Before turning to the alternatives that promise improved economic performance, a word of caution about data is in order. In evaluating alternative methods for rationalizing marine fishing, the technical difficulty and cost of accumulating necessary information on a timely basis must be weighed very carefully. The more elegant formulations of fishery management in a dynamic framework of optimal control theory (e.g. Brown 1974; Clark 1977a, b) assume a level of information on a forecast basis which is not only unavailable at present but almost certainly will be unavailable at any conceivable cost in the future.

Apart from the technical problems of interpretation of CPUE data, and of problems associated with other estimating techniques such as cohort analysis, the fishery manager deals with lag problems that would make Professor Friedman cringe. For all practical purposes, he uses last year's data this year to prepare management plans for next year's fishery. A 2-year lag in data is simply unacceptable for refined management techniques, and in the case of pelagic and anadromous species that are peculiarly susceptible to serious depletion with even a single year's overfishing.

The Role of Taxes[2]

Taxes play a dual role in the formulation of an integrated program to rationalize overcapitalized fisheries. The two are not inherently incompatible, but in the pursuit of one it is quite possible to lose some of the benefits to be achieved from the other — hence the necessity of distinguishing them clearly. The first is the use of taxes as a means of offsetting any adverse effects on the distribution of wealth, income, or employment, as a result of an otherwise efficient management regime. Tax incidence is a tricky area of economic analysis, and the tendency to regard taxes and transfers as a simple way of redressing undesirable income inequality is seriously misleading. Nevertheless, it remains true that a highly successful limited entry program unaccompanied by any tax measure to capture a portion of the economic rent created may produce transfers of income and wealth in a direction that would be unacceptable to most people. With due acknowledgment of the philosophical nature of judgments about the goodness or badness of income distributions, management measures that confer substantial gains on groups of private enterprisers and their employees at the expense of the general taxpayer are not considered good form. A major advantage of a tax on remaining participants in an efficient fishery would be the ease with which it could be used to convert the social costs of management to an explicit charge on the productive activity that gives rise to them.

The second use of taxes in a rationalization program is to influence both the level and composition of effort directed at a particular stock. As suggested initially by Scott, and developed in considerable detail by Smith (1969) and others, taxes can be devised that will convert the several types of externality that plague an open

access fishery into contractual costs to the individual decision-maker. In theory, then, the fishery could be left entirely to the market without fear of biological depletion, excessive inputs in general, or incorrect combinations of inputs.

Unfortunately the phrase "in theory" is terribly confining. Most of these analyses, explicitly or implicitly, are carried out in long-term equilibrium terms, or assume that a dynamic analysis involves modifications only of degree rather than kind. The latter statement may be true in theory, but it is likely to be of small comfort to the fishery manager on the firing line.

Consider the following plausible scenerio. The scientists of the Fishery Management Council of Region X, gathered in solemn conclave, have decided that a reduction of fishing effort on a given herring stock is essential. They are asking for 50%, hoping for 35%, but being practical men of science will settle for 25% after the political process is completed. The fishery manager, having received this technical advice, then proceeds to make two simultaneous announcements. One is that herring stocks have been reduced (for whatever natural or man-induced reasons) to a point that requires a 25% reduction in catch. This will bring sadness to the fleet, since the income loss from a 25% reduction in catch of herring is unlikely to be offset by switching to any other fishery (most of which are equally hard-pressed), and alternative employment for either vessels or men is very restricted. Perhaps they will be comforted, however, by the manager's second announcement, which calls for a sharp increase in landings taxes — enough to bring about the desired decrease in effort either by redeployment or idleness on the part of the vessels affected.

How sharp is "sharp enough"? No general answer can be offered, of course, but in our not-so-hypothetical example factor mobility is not great in the fisheries involved, and it would take a very deep cut indeed to drive out enough effort to achieve the management objective. The cut would have to be even deeper than in conventional industries, since the share agreement, by design, buffers the impact of any reduction in vessel earnings on those who supply capital and entrepreneurship — perhaps the most mobile of the inputs in fishing. Acceptance of the medicine might be even more difficult if the target catch (and tax burden) must fluctuate widely from year to year.

None of this should be interpreted to mean that whatever other measures may be devised to reduce excess inputs would not benefit from an intelligently formulated tax program that would shape incentives in the desired direction. As has been thoroughly developed in the literature, the main requirement is that taxes be as factor-neutral as possible. A tax on landings is probably the best proxy that can be devised for a tax on effort, which is really the target. An ex-vessel landing tax is simple, straightforward in application and collection, relates the fisherman's use of the resource

---

[2]It has been suggested that the term "royalty" describes more accurately the charges to be levied and may raise fewer hackles. I agree but prefer to use terminology already used in the literature.

to his tax bill in an understandable fashion, and provides the widest possible latitude for individual adjustment to the cluster of management measures to which the vessel owner is subject. In addition, there may be merit in recent suggestions (e.g. McConnell and Norton 1978) that differential landings taxes in a mixed species fishery could improve economic output significantly by making use of self-interest of the fisherman and his limited ability to alter the species mix in his catch.

What I am most anxious to drive home is that taxes cannot be viewed as a short-term adjustment mechanism. Rather, a tax structure that pushes fishermen in the right direction with respect to total effort, the composition of effort, and deployment of gear in multiple species fisheries should be regarded as an essentially stable framework element in the overall management program. A number of other measures, fast-response in nature, must be at hand to deal with the inevitable perversity of nature and the unpleasant surprises that must be expected as long as management decisions must be based on data of uncertain quality, usually 1–2 yr out of date.

Limited Entry: Direct Control of Inputs

The second alternative (complementary rather than competitive) is to reduce total inputs directly, by restricting fishing to the holder of a legal right of access — a license, permit, or other legal evidence that a particular vessel and crew may use the resource. Again, the need for brevity suggests that we eliminate repetition of the controversy over licensing of vessels vs. licensing of fishermen. The purpose of the control is to reduce the number of fishing units that can participate. If fishermen are the constrained input some control inevitably must be exercised over vessel capacity if serious factor distortion is to be avoided. If vessel numbers alone are controlled, the same problem of restricting the application of additional labor and ancillary gear will arise, as indicated below, but at a lower level. Since the vessel is the core of the input system that makes up a fishing enterprise, logic suggests that it is the proper starting point for any restricted licensing system.

Even less purpose is served by debate over transferability. Efficiency in use, continuity of operation, and ease of administration are eloquent arguments for relatively free and costless transferability of limited fishing rights. The price they will come to carry is both an economic barometer and an allocative device that no amount of government administrative effort could match.

What, if anything, would be accomplished by a simple closure of entry, with all vessels that demonstrated a "reasonable level of activity" in the base period assured of a permit? Very little, in and of itself — but a great deal indeed if followed by a vigorous program to contain up-grading of inefficient boats, reduction in the total number of units, and other measures to induce the use of efficient vessel sizes and gear. At this point I simply wish to emphasize the critical need to freeze at least some key dimensions of an overcapitalized fleet to a past base period. Otherwise, the time-consuming process of developing a good multifaceted rationalization program (and guiding it through an indifferent or hostile legislature, usually in the face of well-organized opposition) leaves the speculative gate wide open.

Whether or not to reduce some excess capacity in the initial establishment of a base period and cut-off date is an operational question. In theory it would seem desirable to weed out, as far as possible, "dead" licenses and those taken out purely for speculation (since advance talk of a moratorium is bound to develop well in advance of action). Alaska attempted to combine a base period freeze with an elaborate point system to identify a smaller number of "worthy recipients." Whether or not the gain was worth the uproar and administrative effort is unclear, though Adasiak (1979) suggests that the intensity of the clamor was out of all proportion to the actual number of true hardship cases.

A less sophisticated but much easier way of accomplishing an initial reduction would be to establish minimum landings for qualification as "actively engaged" in the base period. If one were tough enough, a great deal of potential capacity could be eliminated at one stroke. Unfortunately, the cutting edge of such a measure is not as sharp as it appears at first glance. It would doubtless amputate a good deal of deadwood that could become effective capacity in skilled hands at a later date. But it would also eliminate an uncomfortably large number of efficient part-timers, and there might also be difficulties with highly productive fishermen operating on a well-reasoned schedule of limited participation in a number of fisheries in sequence. This could be resolved, however, by allowing credits for participation in what are recognized as complementary fisheries. On balance, there is much to recommend a fairly tough minimum landings requirement as an integral part of an initial freeze on new entry, particularly if backed up by a hearing board mechanism to deal with legitimate special cases.

Let us assume, then, that the moratorium is followed by a period of continued increases in demand and rising real prices. Whether or not any reduction in numbers of vessels has been undertaken, returns to vessel owners and share fishermen remaining in the fishery will eventually rise above opportunity levels. How far and how fast the rent component of income increases depends on the rate of increase in fish prices relative to fishing costs and the effectiveness of any efforts to reduce the number of licenses (via buy-back, minimum landing requirements, or techniques such as the British Columbia limited term nonreplaceable Class B license).

For the purpose of this enquiry assume that the rent component of gross fleet proceeds has assumed really juicy proportions. With an active market for licenses, and reasonably informed judgment among potential buyers and sellers as to income prospects, the capitalized value of the license should yield a price at which the purchaser could earn a competitive rate of return on investment only by fishing as efficiently as possible with optimal vessel and gear combinations. The only assumption required to make this completely general would be that each individual license holder, actual or potential, views his long-run marginal cost function as upward sloping.

But this would hold true only if the share of each restricted licensee in the fishery resource were somehow specified. If not, each vessel owner would probably perceive a range within which his share of the total catch could be increased by increasing the catching capacity of his unit, with the value of his expected increase in catch exceeding the increment in costs. If everyone plays the same game, of course, such expectations must turn out to be incorrect. Individual efforts to increase catches are mutually offsetting, and the only result is a general increase in average unit costs. Since the license is regarded as a sunk cost, there is no reason for the situation to be self-correcting. The lower rent will be accepted, since no single vessel owner could improve his position by reverting to a more efficient vessel/gear combination without losing catch volume that would more than offset the saving.

This has been worked over in detail by Pearse (1972), Fraser (1979), and others, and the theoretical aspects need not be elaborated further. It is important, however, to look at empirical evidence to determine how serious such induced imbalance in fishing inputs might be. Fraser, for example, suggests that the end result in the British Columbia salmon case may be as bad or worse than under open access.

The argument is persuasive, but the very pessimistic tone of its final conclusion is negated to some extent by the continued existence of very high prices for licenses in the British Columbia salmon fishery. (The same phenomenon of unexpectedly high license prices also shows up in the Alaska salmon fishery and even in Washington, which has done no more than institute a moratorium.) This can be explained in two ways: either there continues to be a net saving in real factor inputs compared to that which would prevail in open access; or fishermen have continued to entertain, for 10 yr, erroneous expectations. The latter seems a little hard to swallow. While it is true that the high returns earned by British Columbian salmon vessels and the correspondingly high license prices are a result of the extraordinary rise in salmon prices rather than the modest reduction in the number of licensed vessels, the price stimulus would certainly have induced further investment in new boats except for entry limitation. Indeed, the ultimate absurdity of the open access situa-

tion is the proliferation of additional fishing units solely as a result of increased prices in the face of an operative biological constraint on further catches.

Evidence that total investment in the British Columbian salmon fishery has increased, despite a decline in numbers of boats, is persuasive, though, as pointed out by Pearse and Wilen (1979), measurement of the change in total factor costs as a result of the limited entry program is extremely difficult. Unquestionably, windfall gains have accrued to the lucky holders of licenses blessed with legitimacy under the new program. But a substantial number of the licensees will have paid a lump sum which will permit a normal competitive rate of return on investment (and a normal rent for superior entrepreneural ability). It seems likely that fishermen are sophisticated enough, in addition, to recognize that a license initially acquired free of charge has an opportunity cost that must be included in the investment base on which the anticipated rate of return is weighed. The combination of these two factors will surely funnel licenses into the hands of better qualified, more professional entrepreneurs whose major occupation is fishing. Whether this is good, bad, or indifferent involves social and philosophical questions that are not at issue here. From an efficiency standpoint, it does imply that some of the increases in the prices of licenses reflect the shifting of licenses to better qualified personnel — surely a desirable outcome.

While it seems clear that there remains an incentive for operators to expand capital investment against a tonnage or length constraint, would this proceed to a degree that would dissipate all or most of the potential economic gain from the entry limitation program? Obviously this is an empirical question, not readily answered with available data. Perhaps the most important reason for my skepticism rises from the simple mechanics of the fishing operation. The vessel is, after all, only a platform that carries harvesting equipment. There are obvious limitations on the extent to which additional capital investment, whether in propulsion, navigation and depth finding gear, or any other variable can increase catching power if key proxies for increased fishing power such as tonnage and length are constrained. Hence any expansion in catching power from a given hull size will certainly be achieved at rising marginal cost. How rapidly costs rise is an empirical question that will vary, fishery to fishery, but they certainly will increase faster than if additional capacity could be provided by more vessels. A tax on landings, desirable on other grounds as well, would reinforce the effect of rising marginal costs from "capital-stuffing" a fixed-size vessel. Both sides are clearly making assertions; what is needed is careful quantitative measurement of the cost of inefficient factor combinations induced by the scheme.

Though it is impossible to quantify the proposition, there seems to be some merit in Newton's argument (1979) that investment in safer, more comfortable, and better equipped vessels is, at least in part, a movement in the direction of more desirable factor combinations, made possible because rising real prices for salmon and limited entry increased fishermen's incomes and, in the process, provided improved access to the capital market. (Adasiak and others indicate that bank financing of boats and gear has also become easier in Alaska and Washington.) To a limited extent, therefore, the observed trends in British Columbia may simply represent a reversion to a fleet composition that might have existed in the absence of such arbitrary restrictions, and, equally acceptable, rational choices by fishermen now enjoying better earnings and borrowing power to improve their own comfort and safety.

It should also be emphasized that any tendency toward excessive capital investment against a numbers–tonnage–length limitation is not necessarily related to the failure to siphon off all or most of the economic rent created by the program. A license fee of appropriate size would do that job, but would have no effect on the marginal cost-increased catch comparison that drives the individual vessel owner to overinvest. A tax on landings would help — not because it reduces retained rent, but because it adds to the incremental cost of additional catches. Efficient regulation does not require full adherence to Henry George!

Distribution Effects

A great deal of concern has been expressed over the possibility of undesirable distribution of benefits and costs of limited entry (and, by inference, any other intervention to rationalize a fishery). A professional version of this argument is presented by Bromley and Bishop (1977). In large part their argument is a polished restatement of the familiar proposition that optimal factor allocation will be different for each distribution of income, on the reasonable assumption that neither preference patterns nor the marginal utility of present versus future consumption will be identical at opposite ends of the income distribution. It also rests on the equally acceptable proposition that many fisheries are not suitable candidates for limited entry programs because the cost of data acquisition and the management framework necessary to utilize it efficiently would exceed any benefits that might accrue.

These arguments can be found in textbooks running far back indeed. Leaving aside for the moment the crucial analytical issue — how one determines that one income distribution is better than, or equal to, or worse than another — two practical problems dilute the impact of their criticism of the limited entry concept. First, once income distribution is introduced as one element of a complex objective function, other methods of achieving a desired distribution effect must be evaluated. Second, it seems unlikely that the differences in income distribution achieved by partial changes in such a small fraction of the total labor force would be

of real significance, barring the "once for all" windfall gain that will accrue to the initial owners of licenses when a limited entry program is initiated. If the windfall to the original holders causes too much of a lump in the throat, it can be readily reduced by a tax on landings (and a consequent reduction in present value of the license).

Most programs aimed at reducing redundant inputs in the fishery affect income distribution in rather different ways than those with which economists have usually been concerned. The most important effect is to change incomes and employment opportunities by user group (i.e. gear types, fishermen in different areas, or commercial versus recreational users) rather than by income levels within the fishery as a whole. Most rationalization programs, carried on in a sensible way and at a sensible pace, need not create unacceptable income differentials within user groups or among fishermen on the one hand and other competing users of capital and labor within the regional economy. Although a landings tax would be preferable, almost any type of license, fee, or tax on favored participants would suffice to minimize this type of inequality. The more serious problems are not matters of egalitarian concern but rather the political pressures associated with changes in the position of competing fishermen groups, each with a vested interest in maintaining some aspects of the status quo.

Good examples abound. Reduction in the Pacific coast salmon troll catches would result in significant increases in both tonnage and gross value of total landings, since the trollers take large numbers of immature fish which would provide growth in excess of natural mortality if harvested later. But a reduction in the number of trollers would produce very small increases in catches to those remaining; the bulk of the benefits would accrue to inside net fisheries and to anglers. Similarly, reductions in catches of small cod in the North Atlantic would produce larger aggregate returns but at a cost of net losses to some nations and very significant gains to others.

A more important issue is raised by the choice of a base for comparison of distribution impacts. It was pointed out in the opening section of this paper that many open access fisheries characterized by market prices that are high relative to harvesting costs are likely to overrun long-term equilibrium positions where all inputs earn opportunity incomes. It is not particularly useful to assess the effect of any fishery rationalization scheme on the basis of relative income distribution effects within the fishery alone. The more relevant scale would be the regional economy, and in this setting reduction of excess inputs would seem virtually certain to reduce one pocket of persistently low incomes.

I cannot generate much concern over the allegation that limited entry schemes discriminate against potential new entrants. Their only "loss" would appear to be the opportunity to share the capital gains accruing to those grandfathered into the system. Thereafter, any current or potential operator of a fishing unit (assuming that the former accounts properly for the opportunity cost of his license and capital) faces precisely the same cost and revenue functions, and new entry is possible on the same basis as it would be in most other natural resource industries: purchase or lease of a right to harvest. It is surprising to find the fisheries singled out for special castigation because of the burdens imposed on potential new entrants. Why not equal concern over the plight of the young man who wishes to enter retailing, only to find that the cost of purchasing or leasing good retail sites (with a very high rental component) has risen astronomically in recent years? A quick glance at a semilog plot of farmland prices yields the same concern with respect to potential new farmers.

If the rationalization program includes, as every economist would hope, provisions for transfer of licenses with minimum transaction costs, entry is no more prohibited in the controlled fishery than in any other small business area. While this assumes reasonably free access to the capital market, the prices of licenses and the necessary physical equipment, although higher than fishermen have been used to in the past, are still very low relative to the requirements for entry into virtually any other field of selfemployed activity; and they reflect income prospects that have made conventional financing of boats and gear much easier. To the extent that the very young, untrained, and handicapped find entry more difficult because of the higher capital investment required, one can only reply that it is unwise to be poor, ignorant, or unwell — the same restrictions on entry that apply everywhere. A properly designed limited entry program does no more than prevent new entrants from inflicting additional costs on other members of society — fellow fishermen and the general taxpayer.

The Part-Timer Problem

One of the more perplexing aspects of both efficiency and distribution effects of a rationalization program is the treatment of part-time participants. Obviously, there are many fisheries in which part-time participation is dictated by the availability of fish, weather conditions on the grounds, or concentrations of fish sufficiently dense to permit them to be harvested economically. Most analyses of entry limitation, my own included, have tended to assume, uncritically, that economic efficiency would be improved if the fishery were to shift more and more to professional, full-time fishermen. While the basic argument still seems sound, in isolated communities a working schedule that combines the seasonal harvesting of certain species of fish with complementary seasonal activities in agriculture or forestry would be more efficient. The situation is muddied by the likelihood that fishing methods employed by a professional fishermen group, harvesting

one species seasonally and then moving to others, would differ from those employed by a true part-timer shifting from a seasonal fishery back into agriculture, teaching, or some other low paid occupation. We can only fall back on the old standby — each case would require specific analysis of relative economic efficiencies under different configurations of the labor force.

Another dimension of the part-timer problem involves in-season mobility among different fisheries. Efficiency calculations and analysis of distribution effects based on the assumption of optimal factor combinations for units completely specific to one fishery neglect the real possibility that combination vessels, with or without additional fishing equipment, might earn larger net returns by shifting among different fisheries, not only between seasons but within seasons. Such flexible multifishery operation would influence not only the number of units that would be "optimal" for any one fishery, but almost certainly would alter the size, design, and equipment carried by the individual fishing vessel.

While I know of no empirical evidence to draw on, it is quite conceivable that license prices under limited entry would impede the development of efficient multiple fishery operations simply because the vessel license is indivisible. On the other hand, if license prices were reduced by an ex-vessel landing tax, the cost of part-time participation would be less of a deterrent.

Individual Fish Quotas

The final alternative considered — a rather old idea that has recently taken on new life (e.g. Christy and Scott 1965; Christy 1973) — would create rights to specific quantities of fish (individual quotas) rather than rights to participate in the fishery evidenced by a vessel or personal license. The notion is actually more flexible than this, since it could readily be modified to include percentage shares in a variable overall quota, or multiple species quotas with variable "exchange ratios" for fish taken by nonselective gear. Indeed the number of possible variants is one of its most attractive features.

The advantages of this method of management make it worthy of careful consideration. Since it works back from a predetermined total catch to the amount available to the individual operator, it offers much tighter control over each season's catch than either tax or licensing controls on inputs (both of which would have to be backed by area and/or time closures to achieve effective short-term control over the total harvest). In-season and annual adjustments would be somewhat easier to institute. Individual quotas would allow a much greater degree of entrepreneural freedom to the individual fisherman. If the system is tightly enforced, there is no reason why his choice of vessel, gear, area deployment, or fishing time should be closely regulated. This is particularly important if multifishery operation is attractive, since it enables the fisherman to exercise a wide range of options as to time, area, and intensity of participation in each. The investment in rights would be limited to that quantity actually required for each fishery. The system should encourage technological development and innovation; better gear and techniques would simply raise the price of existing quotas without subjecting the resource to a surge of new fishing mortality.

There is no apparent incentive under an individual quota system to overinvest in the vessel and gear — or, to be more precise, to select anything but the least cost combination and deployment of inputs. Income effects would be much the same as under a license limitation scheme.

There are, of course, some thorns on the rose, and though they are more administrative than economic in the narrow sense, they are no less real in implementing a rationalization program. The first is enforcement. In theory a fish purchase ticket of the type now widely used for both business and statistical convenience should provide an accurate check on compliance with an individual quota. But unless an inspector stands at the shoulder of each person preparing the ticket, odd things can happen (e.g. unless most fish are under the quota systems, simply reporting a controlled species as an uncontrolled item opens the door to widespread evasion). Moreover, for very valuable species — salmon, halibut, etc. — new distribution channels can open up with surprising speed, and they would be very hard indeed to pinpoint. A similar, though less serious factor to be considered is that the quotas will always be taken — if I fall short I can readily buy from someone else who has filled up but is still available to fish (transfering at sea if necessary).

The problems of enforcement are not unique to the fisherman quota option, of course. They would be at least as serious in a program that relied on taxes alone, but would probably be less troublesome under limited licensing.

The individual quota, freely transferable, will act as a rationalizing device on a brute force basis. In effect, the price mechanism is substituted for taxes as a way of forcing out inefficient users of inputs. Competition for quotas will force prices to levels that may impede the achievement of other social objectives, e.g. protection of small-scale operators and of isolated fishing communities whose efficiency is lower. Even if prices are allowed to ration out the least efficient users, it would probably be necessary to undertake a quota repurchase system, comparable to the buy-back efforts under license limitations; and, as in those cases, increasing taxation of remaining operators would be necessary if the buy-back scheme is to remain solvent.

Finally, introduction of a quota system on a substantial scale would require an educational program that would be both costly and time-consuming. Fishermen are not prone to jump at new ideas, particularly

if the impact on the individual is both complex and uncertain, and quotas would represent a drastic departure in regulatory techniques. Where the fishery is simple in structure (Pacific halibut or king crab, for example), it might be possible to move fairly quickly, but the reaction of fishermen and fishery managers in the complex salmon and multispecies bottom fisheries would probably be negative.

On balance, there may be less difference in implementation of license limitation and quota systems than meets the eye. Both are potentially capable of generating net economic benefits by shedding redundant capacity at a controlled rate that could demonstrate success without imposing more than minimal hardship on existing participants. Both would favor the more skillful and energetic fishermen. Both would have to be tailored to meet the special needs of the isolated, economically immobile fishing community. Both must be backed up by a tax system that pushes in the right direction with respect to factor combination and overall input usage, shifts the cost of management from the general taxpayer to the producers and users of fish, and prevents unacceptably large windfall gains. Limited licensing is more prone to distortion of factor combinations, but could be introduced more rapidly and with less confusion and hostility than individual quotas. Both create the danger of redeployment of labor and capital to other fisheries unless special measures are taken to prevent it. Both would benefit from a moratorium that has some braking effect while a detailed program is being developed, sold to politicians and the industry, and put into operation (none of which can be accomplished overnight).

Rationalization and Market Structure

The political obstacles to rationalization of marine fisheries are formidable indeed. One of the more formidable — and least defensible — is the argument that any kind of entry limitation constitutes monopolization — or at least the creation of a "privileged class."

The danger of monopolistic practices seems greatly exaggerated. Canadian and American fish are sold largely in national and international markets. Cross elasticities of demand for any given fishery are therefore high, and for most products the availability of imports establishes a firm ceiling on prices at the wholesale and retail level. Under any scheme in which access to the fishery requires purchase of a transferable license or fish quota, there is no more restriction on entry to the fishery than there would be to agriculture in an area where arable land is physically limited. It is true that a limited entry or individual quota scheme would create a body of fishermen with strong common interests in improving prices received for their product, and formation of a single bargaining unit would be correspondingly easier. The possibility of a limited

750   J. FISH. RES. BOARD CAN., VOL. 36, 1979

cartel exists. But, as with any other cartel, the ability to restrict output below levels that would otherwise prevail requires that the cartel be able to allocate shares among the band of brothers in an acceptable manner and to enforce those allocations. Otherwise, the familiar tendency in all cartels for one or more members to expand shares of the restricted output — to benefit from the "umbrella" without assuming its responsibilities — becomes overwhelming.

In the case of a fishery, the difficulties of establishing and policing allocations of a cartel-determined output lower than would otherwise have been harvested would appear to be nearly impossible. The number of fishermen involved, the variability of resource availability in the short-run, the complexities imposed by participation of individual units in more than one fishery, and a host of other circumstances attendant upon commercial fishing would make the maintenance of effective monopolistic restriction a cartel manager's nightmare. Reconciliation of the interests of competing gear types, part-time versus full-time participants, and other group objectives within the cartelized fishery would make the balancing act even more difficult.

With respect to market structure, a stronger case could be made that the creation of a group of fishermen holding private property rights would tend to redress the market power now heavily titlted in the direction of the waterfront buyer. Waterfront markets for marine fish tend to be oligoposonistic in structure, since the processing firms, though not large in absolute terms, are usually large in terms of the restricted geographic areas over which fish can economically be transported to first receivers. This imbalance of economic power has been partially offset by unionization, by the formation of fishermen's cooperatives, or by government action to eliminate control over potential new entrants at the buyer level. The addition of a limited entry program would create a new negotiating environment with respect to first receivers, but it is difficult to envisage any real threat of monopoly by fishermen or bilateral monopoly leading to restrictive joint maximization policies by fishermen and processor groups. To do so would require a degree of policing and overt enforcement that could be readily detected and attacked under existing antimonopoly legislation.

Some Invalid Objections to Rationalization

I now turn briefly to some criticisms of limited entry systems of the licensing type that are also applicable to any rationalization scheme.

One of the astonishing aspects of recent discussions of controlled access is the widespread agreement that limiting inputs is not significant as a "conservation" measure, that is, as a means of holding physical catch to some predetermined level. While other methods of restricting fishing mortality can achieve any given impact on stocks, it is equally true that reduction in effort, if pursued far enough, can achieve the same result. Moreover, in some major fisheries (Pacific salmon and halibut, for example) the effect of open access has clearly been to reduce total yield in weight over time as a result of severely unbalanced harvesting of several substocks within the total populations.

Perhaps the best way to deal with the whole matter is to dismiss the relevance of "competition" between limited entry and more direct methods of limiting fishing mortality as means of achieving a desired level, composition, and area distribution of fishing mortality. They are complementary rather than competitive techniques in real world fishery management even if, hopefully, more reliance will be placed on rationalization in the future.

Another of the myths that seems to pervade the recent literature on limited entry programs is the alleged failure to reduce prices to the consumer. To the extent that limited entry, coupled with other techniques, permits rebuilding a previously depleted stock, it is possible that increased output could exert some downward pressure on prices, but only if it more than offsets the secular increases in demand which have been evident for many decades on a worldwide basis. Rationalization of fishing is not intended as a means of increasing output (though that may be a desirable by-product) but rather of assuring that any given output will be taken at lower cost — certainly with fewer fishing units, and less certainly, with more desirable factor combinations within harvesting units. To reiterate a point that has been made repeatedly in previous discussions by economists, the principal gain to the general public from a more rational fish harvesting sector would come from increased output and lower prices of other goods.

Another of the more dubious arguments against rationalization of fishing is based on the impetus to exploitation of latent or underutilized resources provided by excessive capitalization in the more valuable fisheries. It is doubtless true that exploration and development of new tuna grounds in the western Pacific was carried out more rapidly than would have been the case had the eastern Pacific not undergone a tremendous influx of vessels following the successful introduction of the high seas purse seine. Similarly, it may be that substitution of American for foreign exploitation of pollock, hake, and other low valued groundfish of the north Pacific will be accelerated by growing excess capacity in the king crab fishery. But surely this would be an optimal use of inputs only by sheer chance. The economic rent from efficiently exploited yellow-fin tuna and king crab stocks could have provided revenue for subsidization of directed exploration and development at far lower total cost. In a narrow technical sense, it might also be pointed out that the vessels most likely to be squeezed out of an overcapitalized fishery are the older, less efficient units — hardly the ones that would be picked for risky and innovative exploratory fishing.

A considerable amount of nonsense has been written (e.g. Cicin-Sain 1979) about the impact of limited entry programs on unemployment — in the fishery, and in associated industries in coastal ports. It is hard to view this as a really serious problem in general terms. For example, the total number of fishermen engaged in the salmon fisheries of Washington is substantially lower than the quarter-to-quarter fluctuations in employment by the Boeing Company. None of the existing limited entry programs and none of those proposed, to the best of my knowledge, have ever implied that large numbers of fishermen would be forced to seek other occupations involuntarily. Most have started with a moratorium applying to existing license holders, coupled in some instances with procedures for weeding out "paper" licenses acquired speculatively in anticipation of limited entry. Any subsequent reduction in gear has been based on a buy-back program or some other method of attrition resting on voluntary relinquishment of licenses. The problem, if there is one, is to redirect the flow of potential new entrants from the fishery to other occupations. Since this involves redirection of younger men with a wide range of education and employment opportunities, it is difficult to see how sensible implementation of a gear reduction program would exert any significant effect on regional unemployment.

An important exception would be in isolated fishing communities where other employment opportunities are severely limited. It is quite possible that spreading employment in the fishery would still represent the least-cost method of achieving some desired minimum economic and social standards. But this is a matter to be determined by study, not simply assumed. Otherwise, there is a very real danger that the basic problems of inadequate regional employment opportunities and immobility of younger entrants to the labor force will be neglected, since the fishery constitutes an easy dumping ground for excess labor inputs.

The argument that secondary activities associated with the fishing industry will be adversely affected by a limited entry program is even less convincing. It is true that boat building would be adversely affected, particularly if prospects for continued increases in fish prices are realized, but is precisely what a limited entry program is intended to achieve — a reduction in the totally useless accumulation of excess capital in a fully utilized fishery.

Rationalization has also been opposed on the ground that it constitutes an obstacle to the realization of some optimal combination of income and nonmonetary satisfactions derived from the occupation itself (Orbach 1979). But surely the same combination of benefits from employment accrues to most occupations — teaching, farming, professional and managerial work, and professional sports come to mind as obvious examples. It is also worth emphasizing that open access, particularly in areas where fishermen have enjoyed rapid increases in product prices, has frequently generated a flood of entry which has reduced opportunities for the dedicated fishermen about whom the emotion centers. For example, large increases in salmon trolling and gillnetting by people with only a minor interest in or commitment to commercial fishing has made it extremely difficult for more professional fishermen, presumably those to whom the "way of life" argument applies most cogently, to enjoy that way of life and still eat regularly. It should be emphasized that the situation in isolated communities may involve a very different assessment of the way of life argument, particularly where there are strong cultural ties to the industry (as in the case of many of the Indian and Eskimo fisheries of the Pacific northwest).

Conclusion

Perhaps our legitimate concern with the problems of each major policy option for rationalization has blinded us to the very real gains that can be realized if our sights are not set too high. Any system to reduce excess capacity in a marine fishery will be suboptimal in a formal economic sense. Other nonefficiency objectives must be accommodated; fisheries will always be subject to wide, unpredictable fluctuations in natural abundance; and any real progress calls for institutional changes that will find heavy going in the traditionally conservative fishery community (industrial and governmental). But the weaker welfare test for efficiency is clearly met by any reasonably framed rationalization program that stabilizes total effort in the face of rising prices and/or reduces it with rising or constant prices. Those leaving the fishery could conceivably be fully compensated for any loss of real income (and, if it could be defined, a cash payment to compensate for noneconomic satisfactions derived from participation) while leaving sufficient gross income to provide opportunity returns or better to the owners of all inputs remaining in the fishery. To the extent that a rationalization program carried far enough to warrant such an efficiency test might also increase the physical quantity of fish that can be taken over time, reduce the impact of errors in short-term management decisions, and eliminate the need for efficiency-killing regulations, the case is even stronger. If the rationalization program is phased in time to permit reduction of unnecessary inputs on a voluntary basis, the program would come very close to meeting the stronger test that there be no losers, though certainly with unequal distribution of net benefits.

Much of the discussion of limited entry programs seems to take it for granted that rationalization will add substantially to the cost of management. This is unquestionably true during the early stages of the program, since a change of objectives and techniques of public intervention can hardly be achieved without building new data bases, creating new institutions, and, if the program's advocates have any common sense

752 J. FISH. RES. BOARD CAN., VOL. 36, 1979

at all, a major educational effort. But what of the future? Hopefully, programs can be designed that will reduce unnecessary inputs steadily over time until a reasonable level of economic efficiency is realized. Shifting a substantial part of management to the market mechanism and eliminating some of the more ludicrous (and therefore difficult to enforce) gear and vessel restrictions might be expected to lighten the administrative burden. The costs of doing nothing or of continuing to stifle the efficiency of a steadily rising amount of fishing capacity to save the resource are increasing very rapidly. The essential question is not whether rationalization programs increase management costs but whether they produce net benefits. On balance it would appear that they can do so; whether or not they succeed will depend on the determination with which industry, government, and concerned academics tackle the jobs of design and implementation.

ADASIAK, A. 1979. Experience with limited entry: Alaska. *In* Proceedings of workshops on limited entry. University of Washington, Seattle, Wash. (In press)

ANDERSON, L. G. 1977. The economics of fisheries management. The Johns Hopkins University Press, Baltimore, Md.

BROMLEY, D. C., AND R. C. BISHOP. 1977. From economic theory to fisheries policy: conceptual problems and management prescriptions, p. 281–302. *In* L. G. Anderson [ed.] Economic impacts of extended fisheries jurisdiction. Ann Arbor Science Publishers, Ann Arbor, Mich.

BROWN, G. L. JR. 1974. An optimal program for managing common property resources with congestion externalities. J. Politic. Econ. 82: 163–174.

CHRISTY, F. T. 1973. Fisherman catch quotas. Ocean Div. Int. Law J. 1: 121–140.

CHRISTY, F. T. JR., AND A. SCOTT. 1965. The common wealth in ocean fisheries. The Johns Hopkins Press for Resources for the Future, Inc., Baltimore, Md.

CICIN-SAIN, B. 1979. Evaluative criteria in making limited entry decisions: an overview. *In* Proceedings of workshop on limited entry. University of Washington, Seattle, Wash. (In press)

CLARK, C. W. 1977a. Mathematical bioeconomics: the optimal control of renewable resources. John Wiley – Interscience, New York, N.Y.
1977b. Control theory in fishery economics: frill or fundamental? *In* L. G. Anderson [ed.] Economic Impacts of extended fisheries jurisdiction. Ann Arbor Science Publishers, Ann Arbor, Mich.

CRUTCHFIELD, J. A. 1961. An economic evaluation of alternative methods of fishery regulation. J. Law Econ. 4: 131–143.
1973. Economic and political objectives in fishery management. Trans. Am. Fish. Soc. 102: 481–491.

FRASER, G. A. 1979. Limited entry: the experience of the British Columbia salmon fishery. *In* Proceedings of workshop on limited entry. University of Washington, Seattle, Wash. (In press)

McCONNELL, K. E., AND V. NORTON. 1979. An evaluation of limited entry and alternative approaches to fishery management. *In* Proceedings of workshop on limited entry. University of Washington, Seattle, Wash. (In press)

NEWTON, C. 1979. The British Columbia limited entry programme. *In* Proceedings of workshop on limited entry. University of Washington, Seattle, Wash. (In press)

ORBACH, M. K. 1979. Social and cultural aspects of limited entry. *In* Proceedings of workshop on limited entry. University of Washington, Seattle, Wash. (In press)

PEARSE, P. H. 1972. Rationalization of Canada's west coast salmon fishery: an economic evaluation. *In* OECD, Economic aspects of fish production, Paris.

PEARSE, P. H., AND J. E. WILEN. 1979. Impact of Canada's Pacific salmon fleet control program. J. Fish. Res. Board Can. 36: 764–769.

SCOTT, A. 1962. The economics of regulating fisheries. *In* R. Hamlish [ed.] Economic effects of fishery regulation. FAO Fish. Rep. No. 5.

SMITH, V. L. 1969. On models of commercial fishing. J. Politic Econ. 77: 181–198.

[15]

Towards a Predictive Model for the Economic Regulation of Commercial Fisheries

COLIN W. CLARK

Department of Mathematics, University of British Columbia, Vancouver, B.C. V6T 1W5

CLARK, C. W. 1980. Towards a predictive model for the economic regulation of commercial fisheries. Can. J. Fish. Aquat. Sci. 37: 1111–1129.

A model of the commercial fishery, incorporating the microeconomic decisions of individual vessel operation, is developed and employed to predict the consequences of various methods of regulation, including: (i) total catch quotas; (ii) vessel licenses; (iii) taxes on catch (or effort); (iv) allocated catch (or effort) quotas. Among the principal predictions of the analysis are: (a) total catch quotas do not improve the economic performance of an open-access fishery; (b) limited entry results in distortion of inputs unless every input is controlled; (c) taxes and allocated transferable catch quotas are theoretically equivalent to one another in terms of economic efficiency, and both are capable in principle of optimizing exploitation of the common-property fishery.

Key words: economics, fishery regulation, management, quotas, licenses, taxes, fishermen's quotas, common-property resource

CLARK, C. W. 1980. Towards a predictive model for the economic regulation of commercial fisheries. Can. J. Fish. Aquat. Sci. 37: 1111–1129.

L'auteur élabore et utilise un modèle de pêcherie commerciale, incorporant les décisions microéconomiques d'exploitation de bateaux individuels, dans le but de prédire les conséquences de diverses réglementations, y compris : (i) contingents de prises totales; (ii) immatriculation des bateaux; (iii) taxes sur les prises (ou effort); (iv) contingents de prises (ou d'effort) assignés. Les principales prédictions découlant de l'analyse sont, entre autres : (a) les contingents de prises totales n'améliorent pas la performance économique d'une pêcherie dont l'accès est libre; (b) l'accès limité cause une distorsion des entrées, à moins que chaque entrée ne soit contrôlée; (c) taxes et contingents transférables assignés s'équivalent en termes d'efficacité économique, et les deux peuvent en principe conduire à une exploitation optimale d'une pêcherie possédée en commun.

Received October 15, 1979
Accepted March 25, 1980

Reçu le 15 octobre 1979
Accepté le 25 mars 1980

THE literature on the economics of commercial fisheries has been particularly concerned with the problem of "dissipation of economic rent" in the common-property, open-access fishery (Gordon 1954; Smith 1969). This situation has been compared with the rent-maximizing fishery, using a theory based either on a static model (Gordon 1954; Turvey 1964) or on a dynamic model (Crutchfield and Zellner 1962; Quirk and Smith 1970; Clark and Munro 1975). Exhaustive reviews of the economics of fisheries are given in the books of Christy and Scott (1965), Clark (1976), and Anderson (1977b). On the basis of such models, it has been noted that taxes on catch (or on fishing effort), representing the "shadow price" of the fish stock, would in theory serve to correct the misallocation of resources resulting from the common-property externality.

For quite transparent reasons, however, taxes (or other charges against users) of an appropriate magnitude have seldom actually been imposed in fisheries,

although they may now become more common for foreign vessels fishing within the economic zones of coastal states. For the case of domestic fisheries, it seems likely that political considerations will continue to require that a major share of any economic benefits from the resource accrue to the fishing industry, rather than directly to the public purse. Inasmuch as the theoretically elegant tax instrument does not allow this, the investigation of the economic consequences of alternative forms of regulatory policy becomes important. Such is the purpose of this article.

Until recently, relatively few attempts have been made to forecast the economic consequences of the various practical alternatives for fishery regulation (see, however, Scott and Southey 1970; Christy 1977; Pearse 1979). Two reasons for this deficiency may be first that the theory necessarily becomes rather complex, especially in view of the various alternative hypotheses that may reasonably be adopted concerning economic, technological, biological, and behavioral aspects of the fishery, and second that the lack of exclusive jurisdiction over major fish stocks prior to the declaration of

Printed in Canada (J5808)
Imprimé au Canada (J5808)

1112 CAN. J. FISH. AQUAT. SCI., VOL. 37, 1980

200-mile zones tended to limit regulation to purely biological aspects, thereby rendering detailed economic analysis somewhat academic. (An interesting exception to this rule is the international whale fishery, which has employed an allocated quota system since shortly after World War II. It might be argued that the truly spectacular profitability of whaling was sufficient incentive to cause the relatively few participants to adopt an effective barrier to overexpansion of capacity. Although capacity did exceed the productivity of whale stocks, the excess would doubtlessly have been greater in the absence of quotas; see Clark et al. (1979).)

At any rate, inasmuch as most coastal states have by now established 200-mile zones of exclusive jurisdiction over fishery resources, our subject has acquired new significance. Several recent conferences, for example, have been concerned primarily with the new challenges presented by exclusive fisheries jurisdiction (L. G. Anderson [ed.] 1977a, P. H. Pearse [ed.] 1979; United Nations FAO 1979). Moreover, economically oriented management programs have already been instituted in various fisheries, including vessel licensing programs, fishermen's quotas, and (in at least one instance), virtual sole ownership. Licensing programs, for example, have been established in the British Columbia and Alaskan salmon (*Oncorhynchus* spp.) fisheries, in most Australian fisheries, and elsewhere (Pearse 1979). Fishermen's quotas were established in 1976 in Canada's Bay of Fundy herring (*Clupea harengus*) fishery, and are being extended to other Canadian fisheries. Finally, the prawn (*Penaeus* spp.) fishery of Shark Bay in Western Australia, which has been operated since 1963 under a single licensed purchasing company, behaves like a sole-ownership fishery (Meany 1979). These management programs together provide a useful backlog of test cases for any general theory.

Techniques of fishery management can be classified broadly into two types, according to whether they are mainly directed towards the control of the fish population so as to maintain a high level of productivity, or whether they also attempt to maintain economic efficiency of the fishing industry. The methods that have traditionally been used, such as total catch quotas, closed seasons and areas, gear restrictions, and so on, belong largely to the former category, whereas methods such as licenses, taxes or royalties, and allocated quota systems are of the second type. In practice, it may be appropriate to employ a combination of methods of both kinds (Crutchfield 1979); this possibility adds to the complexity of a general theoretical analysis, as the present paper demonstrates.

In the next section we describe our general model of the open-access fishery. This is, in essence, the traditional economic model of the fishery (one-dimensional, continuous-time, deterministic), but we develop it with considerably greater attention to certain details than is customary. Besides being necessary for our later work, this attention to detail pays off in two respects: we

obtain certain new predictions from the traditional model, and at the same time we recognize certain limitations that are seldom brought out explicitly.

Subsequent sections modify the basic open-access model in appropriate ways, to address the question of "optimal" fishing, and then to analyze the economic efficacy of the alternative forms of regulation. A principal result, which will come as no surprise to those familiar with the economics of control, is that under suitable hypotheses, taxes and allocated transferable quota systems are equivalent in terms of economic efficiency — but certainly not in terms of their distributional implications.

It must be emphasized that the present study is far from complete, as no attempt has been made to investigate formally the consequences of numerous realistic alternative hypotheses and complexities. Nevertheless I believe that the results obtained here will prove to be qualitatively quite robust, at least under the assumption of determinism. The really interesting and difficult problems arise, I expect, when one tries to handle problems of random fluctuations and uncertainty.

Open-Access Fishery

Although we develop our basic model of the unregulated, open-access fishery in somewhat greater detail than usual, the model is in essence the standard model used in the literature. In particular, it follows fairly closely the (diagrammatic) model of L. Anderson (1977b). The underlying biological model is the Schaefer (1954) model which has been employed in the management of yellowfin tuna (*Thunnus albacares*), Peruvian anchoveta (*Engraulis ringens*), and various other fisheries. This very simple biological model is used in this study to allow us to concentrate on the economic complexities of fishery regulation. Later, however, we shall introduce a somewhat more detailed seasonal fishery model, and a second variation (nonconstant catchability) will be discussed under Additional Considerations.

We suppose that there are N vessels (N is not assumed to remain constant), the catch rate of each vessel being given by

(1) $\qquad h_t = qE_t x$

where E_i denotes standardized fishing effort exerted by vessel i, x denotes fish population biomass, and $q = $ constant is the catchability coefficient. Notice that there are no direct externalities (interference or cooperation) between vessels, i.e. h_i is not affected by E_j for $j \neq i$. Over time there will of course be stock externalities, as E_j will affect x and hence also the future catch rates h_i for every i. (Direct externalities are discussed briefly under Additional Considerations.)

The i'th vessel has a net revenue flow given by

(2) $\qquad \pi_t = \pi_t(x, E_t) = pqxE_t - c_t(E_t)$

where p is the price of landed fish and $c_i(E_i)$ repre-

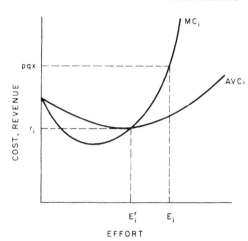

FIG. 1.   Marginal cost $MC_i = dC_i/dE_i$, and average cost $AVC_i = C_i/E_i$, for the typical vessel. When the fish stock biomass is x, marginal revenue equals pqx and vessel i employs effort level E_i shown, provided that $pqx > r_i$. Otherwise $E_i = 0$.

sents variable cost of effort. Each vessel is a price taker and has the traditional U-shaped marginal cost curve for the "supply" of fishing effort (Fig. 1).

Fish population dynamics are modeled by means of the "general production" model

(3) $\dfrac{dx}{dt} = F(x) - \sum\limits_{i=1}^{N} h_i$

where $F(x)$ represents net rate of biomass growth. In the Schaefer model per se we have

(4) $F(x) = rx(1 - x/K)$

where r, K are positive constants. More general growth functions, such as those proposed by Pella and Tomlinson (1969), are easily handled; we shall assume in any case that

$F(0) = F(K) = 0; F(x) > 0, F''(x) < 0$ for $0 \le x \le K$.

It is Eq. (3) that explicitly links the activities of different fishing vessels. To specify our model completely we will also require (a) a rule determining the level of effort exerted by each vessel, and (b) a rule specifying the conditions of entry and exit from the fishery. For the first rule, we simply assume that (unless forced by regulation to do differently) each vessel maximizes its current net revenue flow, taking price p and the fish population level x as given. Thus we have from Eq. (2)

(5) $\begin{cases} c_i'(E_i) = pqx & \text{if } pqx > r_i \\ \quad E_i = 0 & \text{if } pqx < r_i \end{cases}$

where r_i denotes the minimum average cost (Fig. 1) for the i'th vessel. The second part of Eq. (5) also gives us our exit condition:

(6)   Vessel i leaves the fishery if $pqx < r_i$.

Using the inverse ϕ_i of the marginal cost curve, we can also express Eq. (5) in the form

(7) $E_i = \phi_i(pqx)$

where in particular $\phi_i(pqx) = 0$ for $pqx < r_i$. Let E_i^r denote the effort level corresponding to minimum average cost: $C_i'(E_i^r) = r_i$. Vessel i thus never employs a positive level E_i below E_i^r.

Next, under what conditions will a vessel enter the fishery? For existing vessels, it is reasonable to assume that reentry will occur if the inequality in Eq. (6) becomes reversed, e.g. as a result of an increase in biomass or in price. If $c_i(E_i)$ includes the opportunity cost (e.g. of any alternative fisheries), this will be true virtually by definition.

Modeling the entry of additional, newly constructed vessels is more problematical, since the decision to invest in a vessel necessarily involves the vessel owner's expectations of future returns — expectations that normally involve a considerable degree of uncertainty. Although these expectations could be modeled in a variety of more or less sophisticated ways (Wilen 1979), we will adopt the simple assumption that entry of new vessels is determined entirely by current conditions in the fishery. Explicitly, a potential new vessel with revenue function $\pi_j(x, E_j)$ will enter the fishery if

(8) $\max\limits_{E_j} \pi_j(x, E_j) \ge \gamma_j$

where γ_j is a constant representing (but not necessarily equal to) the total opportunity cost of capital for the vessel in question. Assuming $\gamma_j > 0$, it is easily seen that condition (8) can be expressed as:

(9)   vessel j enters the fishery if $pqx > r_j'$

where $r_j' > r_j$.

For each individual vessel, then, there exists a "gap" between the exit and (initial) entry criteria, resulting of course from the tacitly assumed nonmalleability of vessel capital. The existence of such a gap seems very much in evidence in many actual fisheries. The existence of nonmalleability gaps also introduces serious theoretical difficulties both for predicting the development of the open-access fishery, and for determining the social optimum. (The implications of nonmalleability of vessel capital for optimal fishery management have been discussed by Clark et al. (1979). Nonmalleability of fishermen's skills may well be of equal importance to nonmalleability of vessel capital, of course.)

Returning to the analysis, let us henceforth list vessels in terms of decreasing marginal efficiency, in the sense that

(10) $r_1 \le r_2 \le \ldots \le r_N$.

1114   CAN. J. FISH. AQUAT. SCI., VOL. 37, 1980

There is no reason to suppose that the parameters r_j' occur in the same order — some vessels may have high marginal operating cost but low capital cost, and vice versa. Assume that exactly N potential vessels make the decision to enter the fishery originally; thus

$$pqx(0) > r_i' \quad \text{for} \quad i = 1, 2, \ldots, N.$$

Fishing will then reduce the biomass $x(t)$ over time. If no further entry occurs, the industry supply curve for total effort is given by

$$(11) \quad E_{\text{total}} = \Phi_N(pqx(t)) = \sum_{i=1}^{N} \phi_i(pqx(t)),$$

which is clearly an increasing function of x. The fishery will thus converge to an equilibrium $\bar{x}$ satisfying

$$(12) \quad F(\bar{x}) = \Phi_N(pq\bar{x}) \cdot q\bar{x}.$$

(We are here assuming infinite demand elasticity, $p =$ constant, for simplicity. The treatment of finite demand elasticity is fairly straightforward, although finite demand elasticity can in theory introduce instabilities and multiple equilibria into the basic fishery model (Clark 1976, chap. 5.)

What can be said about the equilibrium at $\bar{x}$? Not much unless further assumptions are made. For example, it could happen that $pq\bar{x} > r_i'$ for every i; this would occur if the existing number of vessels is small, and fishing remains profitable for all vessels, but no additional vessels enter the fishery. However, this should be an unstable condition in the open-access fishery, and it is therefore reasonable to assume at least that

$$pq\bar{x} \leq \max r_i'.$$

Alternatively, it could happen that at $\bar{x}$ we have

$$pq\bar{x} < r_i \quad \text{for some } i$$

in which case such inframarginal vessels exit from the fishery as the fish stock is reduced. In this case the marginal vessel will be operating just at the minimum of its average cost curve, $E_i = E_i^r$, but other remaining vessels will be earning positive returns. The sum total of these returns constitute producers' surplus relative to variable cost; relative to total cost, producers' surplus at $\bar{x}$ could well be negative, however. Finally, all the initial vessels could remain active in the fishery, with

$$\max r_i < pq\bar{x} < \max r_i'.$$

The existence of nonmalleability gaps thus leads to considerable ambiguity in predicting the bionomic equilibrium of the open-access fishery; the equilibrium achieved depends upon the initial biomass level $x(0)$ and the initial entry $N(0)$ (or more generally, on the history of entry), as well as on expectations. What can be predicted is that, for $\bar{N} = $ number of remaining vessels we have

$$(13) \quad \max_{1 \leq i \leq \bar{N}} r_i \leq pqx \leq \max_{1 \leq i \leq N} r_i'.$$

Notice that the ambiguity remains even if all vessels are assumed to be equivalent in terms of both variable and fixed cost.

In the special case that nonmalleability gaps do not exist, Eq. (13) implies the equilibrium condition

$$(14) \quad pq\bar{x} = r_{\bar{N}}$$

where $\bar{x}$ is determined by

$$(15) \quad F(\bar{x}) = q\bar{x} \sum_1^{\bar{N}} \bar{E}_i$$

with $\bar{E}_i = \phi_i(pq\bar{x})$. These $(N + 2)$ conditions are easily seen to determine a unique solution $\bar{x}, \bar{N}, \bar{E}_i$, the classical bionomic equilibrium of the open-access fishery (Copes 1972).

Returning to the nonmalleable case, it might be supposed that, over the long run, depreciation would take its toll, fleet capacity would decline, and $\bar{x}$ would adjust upwards as a result of reduced total catches. In practice, however, it seems likely that other changes in the fishery, such as real price shifts, technological innovations, and so on, would occur on at least as rapid a time scale as vessel depreciation. Thus Eq. (13) seems to be the most that one can say for medium-term equilibria.

Technological innovation is always a significant factor in the development of a fishery. Even under the extreme assumptions of classical "bionomic equilibrium," with all current vessels operating at zero rent $(pq\bar{x} = r_i$ for all $j)$, the incentive for cost-reducing innovation exists, since the innovator will then be able to enter the fishery and catch fish profitably. Moreover, other vessels will then be forced to adopt the innovation to fish without incurring losses.

A capital-intensive innovation, such as a vessel of new type, will of course only be introduced if the corresponding entry criterion $\max_E \pi(\bar{x}, E) > r'$ is satisfied. Thus the nonmalleability gap is relevant to any such innovation. In practice, this means that major innovations will tend to be introduced in a discontinuous fashion, each innovation leading to a new bionomic equilibrium, at a lower stock level $\bar{x}$. (What happens if the fishery is managed so as to maintain a fixed stock ("escapement") level $\bar{x}$? Our present continuous-time model is not appropriate for the analysis of this case, but it is clear that the result of innovation under open access will be a progressive shortening of the fishing season, equilibrium being determined by the corresponding increase in the cost of fishing; see A Seasonal Model.)

Because of the complexity and ambiguity involved in modeling nonmalleability, this topic will not be emphasized in the remainder of the paper. It would be unfortunate, however, if the practical implications of nonmalleability were to be ignored, and we will continue to take note of the phenomenon.

Optimization Analysis

To assess the need for regulation, and the relative

efficacy of alternative methods thereof, it is useful to have an ideal for comparison, and this is the main purpose of optimization analysis. We present here both a static and a dynamic optimization analysis of our basic fishery model. The static analysis is quite trivial, but helps to understand the dynamic analysis, which becomes particularly difficult when nonmalleability of capital is taken into consideration. Actually, however, we make no attempt here to work out the details of the nonmalleable case, but merely refer to previous work on this topic.

We begin, then, by considering "maximum economic yield" (MEY) in the static sense; this is determined by the conditions:

(16) $\quad \underset{E_i, N, x}{\text{maximize}} \sum_{i=1}^{N} \pi_i(x, E_i)$

subject to

(17) $\quad F(x) = qx \sum_{i=1}^{N} E_i.$

Use of Lagrange multipliers leads immediately to the necessary conditions

(18) $\quad (p - \lambda)qx = c_i'(E_i) \quad (i = 1, 2, \ldots, N)$

(19) $\quad \qquad = \dfrac{c_N(E_N)}{E_N}$

where λ is the multiplier. We hence obtain

(20) $\quad c_N'(E_N) = r_N$

i.e. the marginal vessel operates at its minimum average cost, and

(21) $\quad c_i'(E_i) = r_N \quad (i = 1, 2, \ldots, N).$

Furthermore, after some easy algebra, we get

(22) $\quad \dfrac{pqxF'(x)}{F'(x) - \bar{F}(x)} = r_N$

where

$\bar{F}(x) = F(x)/x.$

Equation (21) determines the effort levels $E_1, \ldots, E_N$ as functions of N (as yet unknown). Then Eq. (17) and (22) can be solved to obtain the optimal values of x and N, so that the optimal solution is completely specified. We also obtain

(23) $\quad \lambda = p - \dfrac{r_N}{qx}.$

Since $\bar{F}(x) > F'(x)$ for all $x > 0$, Eq. (22) implies that $F'(x) < 0$ at the optimum, i.e. x is greater than the "maximum sustained yield" biomass level given by

$F'(x_{\text{MSY}}) = 0.$

Equation (22) also implies that $\lambda > 0$, as expected

from its usual interpretation as a shadow price of the resource stock. These results are classical.

Assume now that nonmalleability gaps are non-existent. Then the equilibrium conditions of the open-access fishery are given by Eq. (14) and (15):

$p q \bar{x} = r_{\bar{N}} \quad \text{and} \quad F(\bar{x}) = q\bar{x}E(\bar{N})$

where $E(\bar{N}) = \sum_1^{\bar{N}} \bar{E}_i$ is total fishing effort. The (static) optimum, on the other hand, is determined by Eq. (17)–(22), which imply that

$(p - \lambda)qx = r_N \quad \text{and} \quad F(x) = qxE(N).$

Since $\lambda > 0$ it can be seen that:

(24) $\quad \bar{x} < x, \quad \bar{N} > N, \quad E(\bar{N}) > E(N), \text{ and}$
$\qquad\qquad\qquad\qquad \bar{E}_i > E_i \text{ for } i = 1, \ldots, N.$

In other words, the open-access fishery involves too many vessels, each using too much effort, resulting in overexploitation of the fish stock. (To prove Eq. (24) note that $E = \sum E_i = \sum \phi_i(pqx)$ is an increasing function of both x and p, whereas $F(x)/qx$ is a decreasing function of x. Hence the equation $E = F(x)/qx$ determines unique values of x and E, with x a decreasing and E an increasing function of p. Therefore, since $\lambda > 0$ we obtain $x > \bar{x}$ and $E < \bar{E}$. This in turn implies that $E_i < \bar{E}_i$ for every i (for which $\bar{E}_i > 0$, and also that $N < \bar{N}$.)

The foregoing analysis would be relevant to the management of a common-property fishery in situations where existing capacity is excessive (and in other cases where nonmalleability of vessels is irrelevant). Methods whereby the necessary reduction in fleet size, and in effort per vessel, might be achieved will be discussed in the next section. We emphasize that in calculating the optimal reduction of an overexpanded fishing fleet, only the actual opportunity costs of existing vessels enter the calculations, and neither depreciation nor interest on nonmalleable vessel capital are included as costs.

Next we consider the dynamic optimization model, assuming initially that vessel capital is malleable, i.e. that vessels exit and enter the fishery costlessly. The dynamic optimization problem is the following:

(25) $\quad \text{maximize} \int_0^\infty e^{-\delta t} \sum_{i=1}^{N} \pi_i(x, E_i)\, dt$

subject to

(26) $\quad \dfrac{dx}{dt} = F(x) - qx \sum_1^N E_i; \quad x(0) = x_0$

and

(27) $\quad x(t) \geq 0, \quad E_i(t) \geq 0, \quad 0 \leq N(t) \leq N_0$

where $\delta > 0$ is the rate of discount. The variables $N(t)$ and $E_i(t)$, $i = 1, \ldots, N$ are control variables, and $x(t)$ is the state variable. Applying the formalism of the Pontrjagin maximum principle to this problem we obtain the following necessary conditions for an interior maximum:

$$(28) \quad c_i'(E_t) = (p - \mu)qx, \quad i = 1, 2, \ldots, N$$

$$(29) \quad c_N(E_N)/E_N = r_N$$

$$(30) \quad \frac{d\mu}{dt} = (\delta - F'(x))\mu - (p - \mu)q \sum_1^N E_i.$$

Here $\mu = \mu(t)$ is the costate variable (shadow price of the stock $x(t)$). Equations (26)–(30) possess a unique equilibrium solution $x(t) = x^*$, $\mu(t) = \mu^*$, $N(t) = N^*$, $E_i(t) = E_i^*$, where:

$$(31) \quad c_t'(E_t^*) = r_{N^*}, \quad i = 1, 2, \ldots, N^*$$

$$(32) \qquad\quad = (p - \mu^*)qx^*$$

$$(33) \quad \frac{pqx^*(F'(x^*) - \delta)}{F'(x^*) - \delta - \bar{F}(x^*)} = r_{N^*}$$

$$(34) \quad F(x^*) = qx^* \sum_1^{N^*} E_i^*$$

$$(35) \quad \mu^* = p - \frac{r_{N^*}}{qx^*}.$$

Note that for $\delta = 0$ these equations are identical with Eq. (17)–(22) that determine the static optimum. Eq. (33) can also be written in the more familiar form

$$(36) \quad F'(x^*) + \frac{r_{N^*}F(x^*)}{qx^{*2}(p - r_{N^*}/qx^*)} = \delta$$

where the left side represents the "own rate of interest" of the resource stock (Clark and Munro 1975).

The following inequalities are easily deduced from the above equations:

$$(37) \quad \lambda > \mu^* > 0$$

$$(38) \quad x > x^* > \bar{x}$$

$$(39) \quad N < N^* < \bar{N}$$

$$(40) \quad E_t < E_t^* < \bar{E}_t, \quad i = 1, 2, \ldots \bar{N},$$

where the unadorned letters (on the left) refer to the static optimum and the overbars refer to the equilibrium of the open-access fishery, also under the assumption that vessel capital is malleable.

We saw earlier that nonmalleability introduces a "gap" separating the entry and exit conditions of the open-access fishery. A similar gap is also introduced into the optimal fishery: there exist two stock levels $x_1^* < x_2^*$, the lower level being optimal relative to exit, and the higher level to entry, of vessels. The optimal initial number of vessels N_0^* then depends on the initial biomass level $x(0)$, with larger $x(0)$ implying larger N_0^*. As in the case of the open-access fishery, it can be shown that under optimal investment and exploitation, the fish biomass $x(t)$ will be driven towards an equilibrium level x lying within the nonmalleability "gap": $x_1^* \leq x \leq x_2^*$. For further discussion see Clark et al. (1979), where this result is proved rigorously (but only under more stringent hypotheses than those adopted here).

What can be said about the optimum N_0^* in comparison with the number of vessels entering under open-access conditions? Our model does not provide a unique answer to this question, because we have made no specific postulate concerning the entry parameters r_j' — see Eq. (8) and (9). In most circumstances, however, there is every reason to expect excess capitalization under open access, since at N_0^* the average rate of return to capital will exceed the marginal rate δ. On the other hand, if individual vessel owners have high opportunity costs for (borrowed) capital, or if they are abnormally risk adverse, capitalization in the open-access fishery might be inhibited to a suboptimal level.

Finally, something should perhaps be said concerning optimization of vessel design: for example, why would vessels of different efficiency levels exist in an optimal fishery? Indeed, if one were planning an optimal fishery on an unexploited homogeneous stock, presumably all vessels would be of the same optimal design. (Of course, our model covers this as a special case.) In practice, however, one usually has very little information about an unexploited stock, and vessel design will initially involve large amounts of trial and error. Optimization — even very approximate optimization — would only become possible for a fairly well-developed fishery. Existing vessels may not be optimal, but replacing them by more efficient new vessels may not be optimal either. Thus at any given time, a variety of vessels may exist in the fishery. Of course, differences in efficiency may arise from other causes, particularly variations in the skill of skippers and crews.

Regulation

We next use our basic theoretical framework to analyze the consequences of various alternative methods of fishery regulation. The methods to be discussed are (a) total catch quotas, (b) vessel licenses, (c) taxes (royalties) on catch, (d) allocated vessel quotas, and (e) regulation of fishing effort. The first two methods are discussed later, under A Seasonal Model.

Although the various methods are discussed separately here, combinations of methods will be discussed later. Similarly, only questions that are directly related to our mathematical model are discussed here; other aspects of the subject are discussed in the concluding section.

Taxes (Royalties) on Catch

Assume now that the authorities impose a tax on the catch of fish, at a rate τ per unit catch. Otherwise the fishery is assumed to operate as an open-access fishery, without further control. We first consider the case of malleable vessel capital. The corresponding equilibrium conditions then become

(41) $c_i'(E_t) = (p - \tau)qx, \quad i = 1, 2, \ldots, N$

(42) $c_N(E_N)/E_N = r_N$

(43) $F(x) = qx \sum_1^N E_t.$

These equations are identical in form with Eq. (31), (32), and (34) for the optimal fishery, and we therefore conclude that the socially optimal equilibrium can be achieved by means of a tax

(44) $\tau = \mu^*$

where μ^* denotes the adjoint variable corresponding to the optimal equilibrium biomass x^*. More precisely, setting $\tau = \mu^*$ (a constant) drives the open-access fishery to the optimal equilibrium; however, the transitional phase resulting from a constant tax τ will not generally be optimal since the costate variable $\mu(t)$ depends on time. This will be discussed more fully below.

It is interesting to observe that a single regulatory instrument, the catch royalty, serves to achieve simultaneous optimality of vessel numbers, effort per vessel, and fish population biomass. This arises from the structure of our model, which involves only a single source of externality, the fish biomass x. In practice, exploited fish stocks are generally structured with respect to age, species, and location of fish; in such cases an optimizing tax would in theory also have to be similarly structured. (See also Additional Considerations, where externalities due to crowding of vessels are considered.)

ALLOCATED VESSEL QUOTAS

Next we suppose that N vessels are each allocated catch quotas Q_i, which allow fish to be captured, at a maximum rate equal to Q_i:

(45) $h_t \leq Q_t.$

The total quota is (Moloney and Pearse 1979):

(46) $Q = \sum_1^N Q_t.$

Quotas are assumed to be transferable in any portion among existing vessels, or to new vessels. We then assume that a perfect market for quotas becomes established; let m denote the unit price on this market.

Vessel i will clearly wish to buy (sell) additional quota units from its given allocation Q_i if

$\dfrac{\partial \pi_i}{\partial h_t}(x, Q_t) > m \quad [<m].$

Consequently the equation

(47) $\dfrac{\partial \pi_i}{\partial h_t}(x, D_t) = m$

determines the i'th vessel's demand for quotas

$D_i = D_i(x, m)$. It is easy to see that

(48) $\dfrac{\partial D_t}{\partial m} \leq 0, \quad \dfrac{\partial D_t}{\partial x} \geq 0.$

The total demand for quotas is

(49) $D(x, m) = \sum_1^N D_i(x, m)$

where the sum includes any potential entrants holding zero quotas initially. Since the supply of quotas is fixed at $\bar{Q}$, the market-clearing condition $D(x, m) = \bar{Q}$ determines a quota price $m = m(x)$. We have $dm/dx \geq 0$.

The equilibrium conditions (for the case of malleable vessel capital) can now be written down. Clearly each vessel will harvest exactly at the level of quotas held: $h_i = D_i = qxE_i$. Equation (47) thus becomes

(50) $c_i'(E_t) = (p - m)qx, \quad i = 1, \ldots, N.$

The entry/exit equilibrium condition is, as always,

(51) $c_N'(E_N) = r_N$

and we also have biological equilibrium:

(52) $F(x) = qx \sum_1^N E_t$

with, by assumption

(53) $Q = \sum_1^N D_t = qx \sum_1^N E_t.$

Since the total quota Q is assumed to be given, Eq. (50)–(53) determine unique values of the unknowns N, x, m, E_1, $\ldots$, E_N. (More precisely, the equation $F(x) = Q$ generally determines two values of x, say $x_1^* < x_2^*$. Suppose, to be explicit, that x_2^* is the desired optimal biomass. Then a quota $Q = F(x_2^*)$ will ensure that $x(t) \to x_2^*$ provided that $x(0) > x_1^*$. Otherwise a reduced initial quota will be needed for the purpose of stock rehabilitation.) If in fact $Q = Q^* = F(x^*)$, where x^* denotes the optimal biomass level, the resulting values of these unknowns will correspond to the optimal values N^*, x^*, $m = \mu^*$, E_i^*, as can be seen by comparison with the equations for optimal equilibrium, Eq. (31)–(35). We conclude, therefore, that the method of allocated, transferable vessel quotas is equivalent, in terms of economic efficiency, with a tax on catch. Vessel quotas are thus capable of achieving an optimal fishing equilibrium; it is only necessary that the total quota Q itself correspond to the optimal sustained catch rate $F(x^*)$. Under the (important) assumption of transferability, the initial distribution of quotas is irrelevant, as far as achieving an economically efficient equilibrium is concerned, since the quota market will lead to an efficient redistribution of quotas.

REGULATION OF FISHING EFFORT

In terms of our model, it is easy to see that taxes or

1118   CAN. J. FISH. AQUAT. SCI., VOL. 37, 1980

quotas on catch, respectively, are equivalent to taxes or quotas on fishing effort. This follows trivially from the assumed direct relationship between catch and effort, $h_i = qxE_i$, so that a tax τ on catch is equivalent to a tax $qx\tau$ on effort, and so on. However, as has often been noted, the administrative problems associated with regulating effort will generally be of a quite different nature than for the case of catch regulation. This question will be discussed further in the final section.

NONEQUILIBRIUM SOLUTIONS

If we assume that vessel capital is completely malleable, so that vessels can move freely into or out of the fishery, the above analysis can be readily extended to the nonequilibrium situation. Recall that (Eq. 28–30) the optimal solution is characterized by the equations

$$(54) \begin{cases} c_i'(E_t) = (p - \mu)qx, \quad i = 1, 2, \ldots, N \\ c_N'(E_N) = r_N \\ \dfrac{d\mu}{dt} = (\delta - F'(x))\mu - (p - \mu)q\sum_1^N E_t \\ \dfrac{dx}{dt} = F(x) - qx\sum_1^N E_t \end{cases}$$

A catch tax $\tau = \tau(t)$ results in the following system:

$$(55) \begin{cases} c_i'(E_t) = (p - \tau)qx, \quad i = 1, 2, \ldots, N \\ c_N'(E_N) = r_N \\ \dfrac{dx}{dt} = F(x) - qx\sum_1^N E_t \end{cases}$$

(this supposes that, under the nonmalleability assumption, vessels enter and exit so that the marginal condition $c_N'(E_N) = r_N$ holds for all time). Similarly, the allocated quota system yields

$$(56) \begin{cases} c_i'(E_t) = (p - m)qx, \quad i = 1, 2, \ldots, N, \\ c_N'(E_N) = r_N \\ \dfrac{dx}{dt} = F(x) - qx\sum_1^N E_t \\ Q(t) = qx\sum_1^N E_t \end{cases}$$

If $\mu(t)$, $x(t)$ denote the optimal values as determined by Eq. (54), we see that the dynamic optimum can be achieved by imposing either a varying catch tax $\tau(t) = \mu(t)$, or a varying allocated catch quota $Q(t) = h(t)$, the optimal harvest rate, or for that matter, a corresponding varying effort tax or quota.

As noted in the previous section, the case in which vessel capital is nonmalleable is more difficult both to model and to analyze. However, in the case that the initial situation is one of open-access equilibrium (and depreciation of vessel capital is ignored), the above analysis applies directly: the tax $\tau = \mu$, or quota

FIG. 2. A seasonal fishery model. Recruitment R_k equals harvest H_k plus escapement S_k, which provides the subsequent recruitment R_{k+1}, etc.

$Q = h(t)$, optimizes the fishery. Of course, the cost expressions $c_i(E_i)$ used in this analysis exclude both interest and depreciation of vessel capital, as noted previously.

The situation is more complex if $N(0)$ is below the open-access equilibrium. Under the nonmalleability assumption, the number of vessels $N(t)$ becomes a second state variable (Clark et al. 1979), and has a corresponding shadow price $\mu_N(t)$. Optimization thus requires two control instruments, for example a tax on catch plus vessel license fee. Because of the ambiguity surrounding entry of nonmalleable capital, it does not seem possible to model this situation in a rigorous fashion, however.

Seasonal Model

To analyze the effects of certain methods of fishery regulation such as total catch quotas, seasonal closures, annual vessel licenses, and so on, it is necessary to employ a periodic, or seasonal model of the fishery. Since in fact most fisheries are exploited on a seasonal pattern, such models are also somewhat more realistic than continuous-time models of the type used in the previous sections. The use of an alternative form of model is also convenient in allowing us to assess the robustness of the results obtained above.

The seasonal model used here is a Ricker (1954), or "metred" stock-recruitment model of the type discussed in Clark (1976, chap. 7), and illustrated in Fig. 2. Escapement in year k (or other period), denoted by S_k, and is assumed to determine the subsequent year's recruitment to the fishery, R_{k+1}:

$$(57) \quad R_{k+1} = F(S_k)$$

where F is a given stock-recruitment function. For simplicity we suppose that F exhibits neither depensation nor overcompensation; i.e.

$$F(0) = 0, \quad F'(S) > 0, \quad F''(S) < 0.$$

Fishing is assumed to take place over a relatively short season, during which natural population processes can be ignored. (It is routine to modify the model so as to include natural mortality during the fishing season, but the equations become quite unwieldy without adding any insights to our analysis of fishery regulation.) Letting $x(t)$ denote the fish population biomass during the fishing season, we then postulate

$$(58) \quad \frac{dx}{dt} = -qx \sum_1^N E_t(t), \quad 0 \le t \le T$$

$$(59) \quad x(0) = R, \quad x(T) = S$$

where T denotes the maximum season length, and R and S denote recruitment and escapement, respectively, for the given season. (The seasonal subscripts k are suppressed here, for simplicity of notation.)

OPTIMIZATION ANALYSIS

We first consider the problems of intraseasonal optimization, i.e. the optimal allocation of vessels and effort per vessel over a single fishing season. The problem is

$$(60) \quad \underset{N(t), E_t(t)}{\text{maximize}} \int_0^T \sum_{t=1}^N \pi_t(x, E_t) dt$$

subject to Eq. (58), (59), and to the constraints

$$(61) \quad E_t \ge 0, \quad N \le N_0$$

where N_0 denotes the total number of available vessels, and where

$$(62) \quad \pi_i(x, E_t) = pqxE_t - c_i(E_t)$$

with $c_i(E_t)$ denoting variable effort cost. In this intraseasonal optimization problem, R and S are assumed given; later we optimize interseasonally with respect to R and S, and also N_0.

The above optimization problem is easily solved with the aid of the maximum principle. The Hamiltonian is

$$(63) \quad \mathcal{H} = \sum_1^N \pi_t(x, E_t) - \lambda qx \sum_1^N E_t$$

and we have the following necessary conditions for optimality:

$$(64) \quad \frac{\partial \mathcal{H}}{\partial E_t} = 0 \text{ (unless } E_t = 0), \quad i = 1, 2, \dots, N$$

$$(65) \quad \frac{\partial \mathcal{H}}{\partial N} = 0 \text{ (unless } N = N_0)$$

$$(66) \quad \frac{d\lambda}{dt} = -\frac{\partial \mathcal{H}}{\partial x}.$$

Carrying out the indicated calculations, we obtain

$$(67) \quad c_i'(E_t) = (p - \lambda)qx, \quad i = 1, 2, \dots, N$$

$$(68) \quad = c_N(E_N)/E_N$$

$$(69) \quad \frac{d\lambda}{dt} = -(p - \lambda)q \sum_1^N E_t$$

(except for the cases where $E_i = 0$ or $N = N_0$). From Eq. (58) and (69) we see immediately that

$$\frac{d}{dt}\{(p - \lambda)x\} \equiv 0$$

so that

$$(70) \quad \{p - \lambda(t)\}x(t) \equiv c = \text{constant}.$$

It thus follows from Eq. (67) that the optimal effort level for each vessel remains constant throughout the fishing season, and all vessels equate marginal cost. Of course the catch rate qxE_i for each vessel declines as the biomass is reduced. (This rather sharp result is a consequence of our assumption that natural population processes such as growth and mortality of fish can be ignored during the fishing season. If, for example, natural mortality is a constant $M > 0$, so that Eq. (58) becomes $\dot{x} = -Mx - qx \sum E_i$, we obtain $d/dt[(p - \lambda)x] = -Mpx$, and hence optimal effort levels will be greater at the beginning of the season.)

To complete the solution, treat N as a parameter. For $N < N_0$, Eq. (68) fixes E_N at the minimum average cost level

$$c_N(E_N)/E_N = r_N$$

and Eq. (67) then determines all the E_i. Since by assumption r_N is an increasing function of N, each E_i also increases with N, and hence optimal total effort

$$(71) \quad E_t(N) = \sum_1^N E_i$$

increases with N, too. Thus the more vessels, the more rapidly the biomass x is reduced. The optimal N is then determined by the terminal condition

$$(72) \quad x(T) = R \exp(-qTE_t(N)) = S$$

provided such N does not exceed N_0. The optimal fleet size (for any given values of R, S) is therefore the smallest that can capture the harvest $H = R - S$ in the available fishing season $0 \le t \le T$.

If the solution N to Eq. (72) exceeds N_0, then clearly all N_0 vessels are employed, each using a constant effort E_i, satisfying

$$c_i'(E_t) = (p - \lambda)qx = c.$$

The value of c (and hence of all E_i) is determined by the requirement that $x(T) = S$. Note that in this case, all vessels operate at effort levels above the level of minimum average cost.

To summarize this solution: the optimal fleet size is the minimum needed to harvest the given "quota" $H = R - S$ during the available season. Each vessel employs a constant level of effort E_i, equating marginal cost among all vessels. If there are unused vessels available, then the marginal vessel minimizes its average cost.

Since total effort $E_t(N)$ is constant, our equation can be solved explicitly. From Eq. (72)

$$(73) \quad E_t(N) = \frac{1}{qT} \ln \frac{R}{S}.$$

Hence total rent for the season is (as expected)

(74) $\hat{\pi}(R, S) = \int_0^T \{pqxE_t(N) - c_t(N)\} \, dt$

$= p(R - S) - Tc_t(N)$

where

(75) $c_t(N) = \sum_1^N c_t(E_t)$

is the total cost rate for the N vessels. Note that N depends on R and S (provided $N < N_0$) as specified by Eq. (73).

We next wish to determine the optimal escapement level S (and hence optimal recruitment $R = F(S)$ and harvest $H = R - S$), so as to maximize the present value of net returns from the fishery. As usual, we first avoid the difficult question of nonmalleability of vessel capital, and thus assume that all costs are encompassed by the terms $c_i(E_i)$. The constraint $N \le N_0$ is then dropped. Our optimization problem is then:

(76) $\underset{\{S_k\}}{\text{maximize}} \sum_{k=1}^\infty \alpha^k \hat{\pi}(R_k, S_k)$

where α $(0 < \alpha < 1)$ is the discount factor, subject to

(77) $R_{k+1} = F(S_k), \quad k = 1, 2, 3, \dots$

with initial recruitment R_1 given.

The optimal equilibrium escapement $S = S^*$ is determined by the condition (Clark 1976, eq. (7.81))

(78) $F'(S) \cdot \dfrac{\partial\hat{\pi}/\partial S + \partial\hat{\pi}/\partial R}{\partial\hat{\pi}/\partial S} = \dfrac{1}{\alpha}.$

The optimal transitional sequence $\{S_k\}$ can be determined by dynamic programming, but we shall not go into the details.

Only variable costs occur in the above model — $c_i(E_i)$ represents the cost rate corresponding to the effort level E_i. The seasonal model allows for greater flexibility in modeling fixed costs than is possible in the continuous-time model, since both annual fixed costs ("setup" or "mobilization" costs) and permanent fixed costs (nonmalleable capital costs) can be included. Suppose, for example, that vessels are malleable on an annual basis, but incur setup costs β_i if they do enter the fishery in any particular year. The value β_i would represent such costs as preparation of the vessel for the fishing season, transportation from home port to the fishing ground, and also the opportunity cost associated with alternative fisheries that must be completely bypassed if the given fishery is to be exploited.

The intraseasonal optimization problem with setup costs is

(79) $\text{maximize}\left\{\int_0^T \sum_1^N \pi_t(x, E_t) \, dt - \sum_1^N \beta_t\right\}$

subject to the same conditions as before. The conditions characterizing the optimal solution are the same as in

Eq. (64)–(70), except that Eq. (68) becomes

(80) $\dfrac{c_N(E_N) + \beta_N/T}{E_N} = (p - \lambda)qx.$

The prescription is therefore simply that the marginal $(N < N_0)$ vessel is to operate at minimum average total cost (we assume now that vessels are listed in order of increasing average total cost).

Finally we consider briefly the nonmalleable case. Suppose for simplicity that the fishery is initially underexploited. The optimization problem is then to calculate the optimal fleet size N_0, so as to maximize net present value:

(81) $NPV = \sum_{k=1}^\infty \alpha^k \hat{\pi}(R_k, S_k) - \sum_{t=1}^{N_0} \xi_t$

where the second sum represents the total capital cost of N_0 vessels. The underlying intraseasonal and interseasonal problems are subject to the constraints $N \le N_0$. This problem (Clark et al. 1979) results in a nonmalleability gap $[S_1^*, S_2^*]$, containing the ultimate equilibrium optimal escapement level S, the exact location depending on the initial conditions. If $S = S_1^*$ the optimal fleet size N_0 will be greater than the optimal number of vessels N employed at equilibrium, but otherwise $N = N_0$.

OPEN ACCESS

For the case of the unregulated open-access fishery we assume as before that each vessel continually maximizes its net revenue flow $\pi_i(x, E_i)$. Hence

(82) $c_t'(E_t) = pqx$, or else $E_t = 0$.

Vessel i ceases fishing for the season if the biomass is reduced below the level of zero revenue flow, i.e. if

(83) $pqx \le r_t.$

With N vessels entering the fishery at the beginning of the season, we have

(84) $\dfrac{dx}{dt} = -qx \sum_1^N E_t(t); \quad x(0) = R.$

The fish biomass S at the end of the season is determined by these equations: if there are sufficiently many vessels (and the season is sufficiently long), S will simply equal the zero-revenue level r_1/pq for the most efficient vessel; otherwise $S = x(T)$ from Eq. (84).

Net seasonal revenue for the i'th vessel is given by

(85) $\hat{\pi}_t(R, N) = \int_0^T \pi_t(x(t), E_t(t)) \, dt.$

If β_i represents setup cost (including opportunity cost of alternative fisheries), the i'th vessel will enter the fishery if

(86) $\hat{\pi}_t(R, N) > \beta_t.$

This inequality indicates that the seasonal entry decision depends on the abundance of fish and on the number of other vessels entering the fishery (or, more realistically, on the vessel owner's estimate of these parameters!). Of course, prices and costs also enter the decision; these parameters also obviously influence the initial decision to acquire a vessel, or to transfer a vessel license.

We see that, under open access, the fishery experiences excessive entry of vessels, excessive use of effort by each vessel, shortening of the fishing season, as well as excessive depletion of the fishery resource. Furthermore, because of the competition for fish, effort is concentrated at the beginning of the season, rather than remaining constant throughout the season as it should. What effect do the various management techniques have on these outcomes?

ESCAPEMENT REGULATION, OR TOTAL CATCH QUOTAS

We first consider the traditional approach to fishery management whereby annual catches are regulated on purely biological (or piscatorial) grounds. The objective usually adopted is the maintenance of the fish stock in its most productive state — the MSY (maximum sustainable yield) objective — although recent practice has tended to recommend somewhat more conservative objectives (e.g. Larkin 1977).

We shall assume that the fishery is managed by the method of total catch quotas: the fishing season is opened (at $t = 0$), and fishing is allowed to take place until the total catch has reached some specified level H, the annual quota. In our deterministic model, specifying the quota H is of course equivalent to specifying the escapement level $S = R - H$ (assuming that the recruitment level R is known). Also, since we are supposing that the behavior of fishing vessels is predictable, specifying H is also equivalent to specifying the length of the fishing season. In practice the three methods will not be equivalent, because of uncertainties in the values of R, S, and H. Nevertheless, the methods have the same purpose — the maintenance of an appropriate level of the stock. (Some reference should be made to the use of mesh-size regulations, or other controls on fishing gear, which have the effect of preventing the capture of small fish. Such methods cannot be modeled without adopting a much more complex age-structured biological model (Clark 1976, chap. 8), and hence will not be discussed in this article.)

Let us suppose, therefore, that the fishery is closed by the regulating authority when $x(t) = S$, a certain selected escapement level. No other controls are employed (but will be considered later). For the individual vessel, we have, as before

$$(87) \quad \begin{cases} c_i'(E_t) = pqx & \text{if } pqx > r_t \\ E_t = 0 & \text{if } pqx \le r_t \end{cases}$$

Let $\pi_i(x) = pqxE_i(x) - c_i(E_i(x))$ denote the net revenue flow for $E_i = E_i(x)$ determined by these condi-

tions. Clearly $\pi_i(x)$ is a nondecreasing function of x:

$$\frac{d\pi_i}{dx} = pqE_i + pqxE_i' - c_i'(E_t)E_t'$$

$$= pqE_i(x) \ge 0.$$

The fish population biomass $x(t)$ follows the equation

$$(88) \quad \frac{dx}{dt} = -qx \sum_{i=1}^{N} E_i(x), \quad x(0) = R.$$

Let $x_\infty(t)$ denote the solution of the same equation (with N fixed), but with $x(0) = R_\infty$, where R_∞ denotes the level of recruitment in the unregulated open-access fishery. Given that regulation increases the level of recruitment, $R > R_\infty$, it follows from the uniqueness theorem for ordinary differential equations that

$$x(t) > x_\infty(t) \text{ for all } t.$$

Hence also $\pi_i(x(t)) > \pi_i(x_\infty(t))$, so that unless the length of the fishing season (with N fixed) actually decreases because of the escapement regulation (unlikely, but possible), we see that escapement regulation increases net seasonal revenue, in equilibrium, for each vessel in the fishery. Of course, this increase may not occur immediately, since the depleted fish stock will take time to recover to the regulated level.

But now we come to the main prediction of this model: if entry to the fishery is not limited, management by escapement control will lead to an increase in the number of vessels beyond the "bionomic equilibrium" of the unregulated fishery. In many cases, depending on circumstances, there will also result an actual further shortening of the fishing season, even though annual total catch will increase.

I believe this simple analysis indicates, more incisively than in previous studies, the economic limitations associated with the traditional biological approach to fishery management. The overall economic benefits of such a management program may well turn out to be negative if increases in fleet capacity are extensive. Indeed, many secondary effects may make matters even worse, for example, the shortened season may lead to increased processing and storage costs, and perhaps a decline in the quality of fish; the increase in fleet capacity may lead to congestion and interference among vessels at sea, which may increase the costs of management, and so on (see Additional Considerations).

Of course, there are undeniably beneficial results of the traditional type of regulation — an increased supply of fish to consumers or foreign markets, increased employment in the fishing and processing sectors, perhaps decreased variation in annual catches, and so on. All these results will be highly visible, and the fact that the fishery has become even more seriously overcapitalized than before the management program was introduced may be overlooked — at least until the inevitable poor season leads to economic disaster for the expanded industry.

LICENSES

Next we consider regulation involving a limitation on the number of participants, i.e. a vessel licensing program. In modeling this situation we encounter the problem of specifying the extent to which the rights of participants are restricted, in terms of vessel characteristics and use. Also, it must be specified whether vessel owners are subject to fees or royalties related to their fishing rights, or whether these rights are essentially free to their owners.

To begin with, let us imagine that N specific vessels are licensed, but that no fees or royalties are imposed. Each vessel has a given production function, and hence a specific cost function $c_i(E_i)$ for effort. Modification of vessels so as to alter $c_i(E_i)$ is assumed to be outlawed (how this might be accomplished is another question).

Let us also suppose that escapement is regulated at S. The optimal use of effort E_i^* over the fishing season is then determined by the conditions

$$c_t'(E_i^*) = (p - \lambda)qx = \text{constant}, \quad i = 1, 2, \ldots, N$$

with the total effort level $\sum E_i$ fixed so that $x(T) = S$. Under competitive fishing, however, effort $E_i(t)$ will satisfy

$$c_t'(E_t(t)) = pqx, \quad i = 1, 2, \ldots, N.$$

Vessels will thus employ excessive levels of effort throughout the season, particularly at the beginning. This in turn will result in a shortening of the fishing season. Nevertheless, net seasonal economic return per vessel, viz

$$(89) \quad \bar{\pi}_i(R, N) = \int_0^T (pqxE_t - c_t(E_t)) \, dt - \beta_t$$

will be positive, and if the number of vessels is sufficiently restricted, will exceed capital costs, possibly by a wide margin. (This difference, if it is positive, will come to be reflected in the price at which vessel licenses are sold — assuming transfer of licenses to be allowed.) The extent of suboptimality, $\sum_1^N (\pi_i^* - \bar{\pi}_i)$ where

$$(90) \quad \pi_i^* = \int_0^T (pqxE_t^* - c_t(E_t^*)) \, dt - \beta_t$$

can only be assessed if the cost functions $c_i(E_i)$ are known explicitly.

Crutchfield (1979) has argued that appropriate regulations controlling vessel length, tonnage, horsepower, and gear will result in sharply increasing marginal cost curves $c_i'(E_i)$ for $E_i > E_i^*$, with the result that $\bar{\pi}_i \approx \pi_i^*$. This argument is clearly correct as far as it goes. It is also clear, however, that such regulations would have the effect of restricting the flexibility of vessel operations. From our optimization analysis, we see that the optimal effort levels E_i^* depend on the recruitment level R. If R is subject to annual fluctuation, inflexibility in the application of effort may be undesirable. But these are problems that, important as

they may be, take us beyond the scope of the present article.

If vessel licenses are used as the economic component of regulation, an estimate of the optimal number of vessels is needed. Suppose for simplicity that escapement is also regulated, at some predetermined level S (in theory, the optimal values of S and N should be determined conjointly). Then clearly the optimum number of vessels will simply be the minimum number required to capture the sustained yield $F(S) - S$. If vessels exert effort greater than the optimum, this "second best" optimum will consist of fewer vessels than the optimum optimorum.

Suppose that N vessels are licensed for the fishery. The value V_i of a license is defined as the present value of net seasonal returns:

$$(91) \quad V_t = V_t(R, N) = \bar{\pi}_t(R, N)/\delta$$

where $\bar{\pi}_i$ is given by Eq. (89). (This assumes that the fishery remains in equilibrium, with constant annual recruitment R; otherwise $V_i = \sum_{k=1}^{\infty} \alpha^k \pi_i(R_k, N)$.) Clearly $V_i(R, N)$ is a decreasing function of N.

The number of vessels entering the open-access fishery when capital is nonmalleable is determined by the function $V_i(R, N)$, where N now simply denotes the number of vessels already participating in the fishery. An additional vessel (j) will then enter the fishery provided that

$$V_j(R, N + 1) > \xi_j$$

where ξ_j represents the capital cost of the vessel. An existing vessel (i) will exit from the fishery if

$$V_i(R, N) < 0$$

The entry–exit theory is thus the same as for our previous model and needs no further discussion. In general (unless vessel owners use above normal discount rates), the open-access fishery will attract excessively many vessels.

The number of vessels may be controlled either by restricting fishing rights to a specified set of licensed vessels (with or without actual license fees), or alternatively by selling (or auctioning) licenses on an open market. If λ_i denotes the license fee for vessel i (license fees may depend on vessel characteristics), then a new vessel will tend to enter the fishery only if, for this vessel

$$(92) \quad V_t(R, N) > \xi_i + \lambda_t$$

(where N = total number of licenses). On the other hand, an existing vessel will purchase a license and keep fishing, if

$$(93) \quad V_t(R, N) > \lambda_t.$$

Here we are assuming that the fee λ_i is paid once and for all; annual license fees affect "setup cost" rather than capital cost.

Note that there is no such thing as an "optimal"

schedule of license fees here, since we are assuming that the number of licenses (for each type of vessel) is fixed at its optimum. If licenses were sold at a competitive auction, however, the most efficient vessels would obtain the licenses, and total economic rent from the fishery, as capitalized in the license values, would accrue to the licensing authority. If licenses are transferable, the long-run tendency will be for license ownership to pass to the most efficient fishermen, regardless of the initial method of awarding licenses. The nonmalleability gap indicated by Eq. (92) and Eq. (93) will in general slow down this process, however.

To summarize, vessel license fees are primarily of distributional importance, and do not themselves serve the purpose of economic optimization (although auction or transfer of licenses can have this effect). This distributional aspect can be applied to the notion of "buy-back" programs, which have been employed in some overexpanded fisheries, including the British Columbia salmon fishery. License fees collected from inframarginal vessels can be used to compensate the excluded submarginal fishermen. But this again is a matter of distribution, not efficiency per se.

TAXES ON CATCH

Suppose next that catch is taxed at a fixed rate τ; assume first that no other methods of regulation are employed. Behavior of the fishery is then characterized by the open-access model (see Eq. (82)–(86)), with price p replaced by $p - \tau$. The effect of the tax is thus to reduce each vessel's input of effort over the season, and to increase escapement accordingly. Also, since marginal vessels leave the fishery, N is reduced. Although all these changes are in the right direction, it is clear from the optimization analysis that a constant tax rate fails to achieve the optimal distribution of effort over the season:

(94) tax rate τ: $\quad c_t'(E_t) = (p - \tau)qx$

(95) optimality criterion: $c_t'(E_t) = (p - \lambda)qx = $ constant.

To achieve optimality, the tax τ must thus equal $\lambda(t)$, which depends on the time t at which the fish are caught; $\lambda(t)$ decreases with t. Such a variable tax rate could in principle optimize effort, escapement (by choice of the constant in Eq. (95)), and vessel numbers. It may be doubted, however, whether continuously varying tax rates could actually be employed in practice — and there is also the problem of knowing when a particular fish was caught!

How serious is the distortion in effort caused by using a constant tax rate? For simplicity, suppose now that escapement S is separately regulated (this clearly does not change the above conclusion that constant tax rates are suboptimal). Let $\tau = \lambda(T)$, for example. It is then clear from Eq. (94) and (95) that the degree of distortion in effort $E_t(t)$ will depend on the variation in the fish stock $x(t)$ over the fishing season, i.e.

on the annual "exploitation rate"

$$\rho = \frac{R - S}{2(R + S)}.$$

For some species, including demersal species such as Atlantic cod (*Gadus morhua*) and halibut (*Hippoglossus* spp.), the optimal annual exploitation rate is quite small (in the range of 10–20%), but other species may be exploited at rates up to 50% or higher. While taxes on catch may not be appropriate for species of the latter type, they should be acceptable in other cases.

Under the assumption that the annual exploitation rate is small, so that $x(t) \cong \bar{x} = \frac{1}{2}(R + S)$, the (near) optimal tax rate can be determined as follows. From Eq. (67) to (70) we have for the optimum:

(96) $c_i'(E_t) = r_N, \quad i = 1, 2, \ldots, N$

where the number N of vessels is just the number required to catch the annual quota $H = R - S$ in time T. From Eq. (94), with $x(t) \cong \bar{x}$, the optimal tax rate is therefore given by

(97) $\tau^* = p - \dfrac{r_N}{q\bar{x}}.$

The resulting effort levels given by Eq. (87) then satisfy

(98) $c_t'(E_t(t)) = r_N \dfrac{x(t)}{\bar{x}} \cong r_N.$

ALLOCATED VESSEL QUOTAS

Suppose now that vessel i has a seasonal quota H_i. The vessel owner assumes that the stock will be fished down at some specific rate, and thus concludes that the population will be of a given size $x(t)$ at time t. (This problem might be more rigorously formulated as the competitive solution of an N-person differential game, but I shall not attempt this level of sophistication here; see Clark 1979.) His own catches constitute a negligible portion of the total. The vessel owner then formulates the following optimal control problem for his cumulative catch $y_i(t)$:

(99) maximize $\displaystyle\int_0^{t_f} (pqxE_t - c_t(E_t))\, dt$

subject to

(100) $\dfrac{dy_t}{dt} = qx(t)E_t(t), \quad 0 \le t \le t_f$

(101) $y_i(0) = 0, \quad y_t(t_f) = H_t$

(102) $t_f \le T.$

(In this formulation we are assuming that the quota restriction is binding, and hence the vessel really catches its full quota.) The problem is easily solved; the Hamiltonian is

$$\mathscr{H} = pqxE_t - c_t(E_t) + \lambda_t qxE_t.$$

The maximum principle immediately yields

(103) $c_i'(E_i) = (p - \lambda_t)qx$

(104) $\dfrac{d\lambda_t}{dt} = -\dfrac{\partial \mathcal{H}}{\partial y_t} = 0$,   or $\lambda_t = $ constant.

If $t_f < T$, the transversality condition $\mathcal{H} = 0$ at $t = t_f$ implies

(105) $E_t(t_f) = E_t^r$.

Equations (100)–(105) then determine unique solutions for λ_i, t_f, and $E_i(t)$, $0 \le t \le t_f$. If $t_f < T$, this is the optimal solution. Otherwise $t_f = T$, and Eq. (105) is deleted and λ_i and $E_i(t)$, $0 \le t \le T$, are determined by Eq. (100)–(104). Which case arises depends on the vessel's quota H_i: if H_i is small we get $t_f < T$, otherwise $t_f = T$.

The outcome of the analysis is that, given a seasonal quota H_i, the vessel owner will employ a decreasing level of effort $E_i(t)$, as given by Eq. (103), exhausting his quota in time $t_f^i \le T$. Given that the quota is binding (i.e. that the vessel owner could increase net revenue by increasing his catch beyond H_i — provided other vessels did not also try to do so!), then $\lambda_i > 0$ and $E_i(t)$ will consequently be less than it would be under, say, a total quota management system, where

(106) $c_i'(E_i) = pqx$.

These findings correspond to one's intuition: since the vessel owner is guaranteed his seasonal quota, and cannot exceed it, he does not need to enter into a competitive "scramble" at the beginning of the season. On the other hand, he can reduce his costs to some extent by concentrating his catch at the early part of the season when the stock level is high. The latter situation is suboptimal for the fishery as a whole, since vessel i's catches have an external influence on all vessel's subsequent catch rates.

We observe next that, provided each vessel quota H_i is of appropriate magnitude, the allocated seasonal quota system is equivalent (in terms of its effect on effort levels $E_i(t)$) to a constant tax on catch. It is only necessary that each quota H_i be set so that the value of λ_i determined as above coincides with the tax rate τ, for then Eq. (94) and (103) coincide. In particular, if the annual exploitation rate ρ is small, then quotas H_i can be allocated so as to achieve an approximately optimal distribution of effort over time.

It should be noted that this equivalence result does not depend upon transferability of quotas, as it did in our continuous-time model. It does, however, require that the quota allocations themselves be optimal, which was not necessary in the previous model. What will happen if seasonal quota allocations are initially suboptimal, but are transferable in whole or in part? If a competitive market for quotas is established, one would hope that the result would be an optimal reallocation of quotas. We proceed to prove that this is correct.

We continue to assume that the annual exploitation

rate is small, so that $x(t) \cong \bar{x}$, a constant. Then the equations (100)–(105) characterizing the i'th vessel's optimal effort can be solved explicitly. The optimal effort level E_i is a constant, given by

(107) $E_t = \begin{cases} E_t^r & \text{if } H_t < q\bar{x}E_t^r T \\ H_t/q\bar{x}T & \text{if } q\bar{x}E_t^r T \le H_t \le q\bar{x}\bar{E}_t T \\ \bar{E}_t & \text{if } H_t > q\bar{x}E_t T \end{cases}$

where $\bar{E}_i$ is the solution of

$c_i'(\bar{E}_i) = pq\bar{x}$.

In other words, if the quota H_i is small, the vessel operates at minimum average cost until the quota is exhausted; in this case we have

(108) $E_t = 0$ for $t > t_f = H_t/q\bar{x}E_t^r$.

At intermediate quota levels, the vessel fishes at the constant rate required to fulfil its quota. Finally, if the quota is sufficiently large, the vessel simply maximizes its net revenue flow $pqxE_i - c_i(E_i)$, and captures less than its quota, i.e. the quota is not binding in this case.

The piecewise-linear relation of Eq. (107) is shown in Fig. 3a; from this relation we can also calculate the net seasonal revenue for vessel i:

(109) $\hat{\pi}_i = \begin{cases} \left(p - \dfrac{r_t}{q\bar{x}}\right)H_t & \text{if } H_t < q\bar{x}E_t^r T \\ pH_t - c_t(H_t/q\bar{x}T)T & \text{if } q\bar{x}E_t^r T \le H_t \le q\bar{x}\bar{E}_t T \\ (pq\bar{x}\bar{E}_t - c_t(\bar{E}_t))T & \text{if } H_t > q\bar{x}\bar{E}_t T \end{cases}$

As in the continuous-time model, we now suppose that a perfectly competitive market is established for vessel quotas, and let m denote the price on this market. Vessel i's demand $D_i(m)$ for quotas is then given by:

(110) $\dfrac{d\hat{\pi}_t}{dH_t} = m$,

see Fig. 3b (note that $\hat{\pi}_i$ also depends on the stock level $\bar{x}$, so that Eq. (110) is directly analogous to

FIG. 3. Details of the allocated quota system for a seasonal fishery: (a) relationship between vessel's harvest quota H_i and effort level E_t, as given by Eq. (107) of text; (b) ith vessel's demand curve for quota allocation H_i in terms of quota price m, as given by Eq. (110) of text.

Eq. (47)). The total quota is fixed at $\bar{Q} = F(S) - S$:

(111) $\quad \sum_{i=1}^{N} D_i(m) = \bar{Q}.$

Now from Eq. (110) we see that the marginal (N'th) vessel will have

(112) $\quad p - \dfrac{r_N}{q\bar{x}} = m.$

Differentiating in (109), we also have, for $i = 1, 2, \ldots, N - 1$

$$p - c_i'(E_i) \cdot \frac{1}{q\bar{x}} = m$$

or

(113) $\quad c_i'(E_i) = (p - m)q\bar{x} = r_N.$

After the quota market has cleared, we will also have $H_i = D_i(m)$, and thus $\sum H_i = \bar{Q}$. But these are precisely the optimality conditions — see Eq. (67), (68) and the discussion pertaining to them. The equilibrium price m for quotas is also equal to the optimal catch tax as given by Eq. (97).

To summarize these results, first we have shown that, in a seasonal fishery model, a constant tax on catch is equivalent in its effect on effort to an allocated vessel quota system. Second, if intraseasonal variation in the stock level is minor (i.e. if the annual exploitation rate is small), either method can be used to achieve an approximately optimal fishery, and transferable quotas will tend to be reallocated in an optimal manner in a competitive quota market. (The reader will note that the assumption of a low exploitation rate really means that the fishery remains approximately in equilibrium at $x = S$, so that the continuous-time and discrete-time models should approximately coincide, as in fact they do.)

An interesting sidelight of the analysis is that the quotas of "part-time" fishermen will be bought up by the full-time fleet. However, since the above analysis fails to account for either setup costs or nonmalleability, this prediction should be accepted with qualifications.

In cases where the exploitation rate is high, we have seen that an optimizing catch tax τ should theoretically vary over the fishing season. Similarly, in such cases the optimizing quota "rates" should also vary. For example, vessels could possess monthly (or weekly) quotas, rather than seasonal quotas. But there are probably not many actual fisheries where such refined management systems would be necessary, although the possibility of evening out the flow of raw fish to processing plants might suggest that such techniques be considered. Indeed, it is not unusual, at least partly for this reason, for the price offered by processors to vary as the fishing season progresses.

Additional Considerations

Further study will be required to test the robustness of the results of this analysis to alternative, more realistic assumptions pertaining to the fishery. Much of this work unfortunately promises to be rather difficult. In this section I discuss briefly two rather minor, but important modifications to our basic model. Other questions will be raised (but not resolved) in the following section.

VARIABLE CATCHABILITY

The standard fishery production function

(114) $\quad h = qEx \quad (q = \text{constant})$

is generally agreed to be overly simplistic for many fisheries, particularly those based on pelagic schooling species such as anchovies and herrings (e.g. Gulland 1977). These populations often tend to maintain school size, and to contract geographically when the size of the population declines. Thus catchability (i.e. the fraction of the population captured per unit of effort) tends to increase as x is reduced; this can be modeled by:

(115) $\quad h = q(x)Ex$

where $q(x)$ is a positive decreasing function of x.

The limiting case would seem to be $q(x) \propto x^{-1}$ (otherwise catch rates would increase with decreases in the stock level!), giving

(116) $\quad h = aE \quad (a = \text{constant}).$

In this extreme case it is clear that, if fishing is profitable at all, the open-access fishery will lead to the extinction of the population (unless the number of vessels capable of fishing profitably is so small that depletion is impossible).

For this case, the solution of the dynamic optimization model (see Eq. (25)–(30)) gives the conditions:

(117) $\quad c_i'(E_i) = (p - \mu)a, \quad i = 1, 2, \ldots, N$

(118) $\quad c_N(E_N)/E_N = r_N$

(119) $\quad \dfrac{d\mu}{dt} = (\delta - F'(x))\mu.$

Hence the optimal equilibrium biomass x^* is given by

(120) $\quad F'(x^*) = \delta;$

in this case x^* is independent of the cost and price variables. Next, consider a catch tax τ; then net revenue of vessel i is:

(121) $\quad \tilde{\pi}_i = (p - \tau)aE_i - c_i(E_i)$

so that the effort level employed by the i'th vessel is determined by

(122) $\quad c_i'(E_i) = (p - \tau)a \quad (\text{unless } E_i = 0).$

Total effort $\sum_1^N E_i$ is a decreasing function of the tax rate τ, and the optimum can be achieved by means of a uniquely determined tax $\tau = \tau^* = \mu$. However, since E_i is independent of stock abundance x, the corresponding control over x is likely to be weak, unless entry to the fishery is also controlled. Allocated vessel catch quotas, on the other hand, while once again theoretically equivalent to taxes, would provide a much more direct control over exploitation.

The most serious difficulty associated with variable catchability, however, is the bias that is introduced into the catch-effort statistic as an index of stock abundance. In the extreme case of Eq. (116), catch per unit effort is not correlated at all with stock abundance x, so that the data produced by the fishery gives no indication of declining stocks. Added to this difficulty is the fact that the dynamics of the fishery are "catastrophic" relative to total fishing effort under these circumstances (Clark 1976, sec. 1.2).

CROWDING EXTERNALITIES

As noted earlier, the formulation of our basic model involves only one source of externality, the effect that one vessel's catch has on the stock subsequently available to other vessels. (This is usually referred to as the stock externality.) Another type of externality that may be important in some fisheries results from crowding of vessels and the resulting interference of each vessel with other vessels' fishing operations. The most direct way to model such externalities is to replace Eq. (2) for the i'th vessel's net rate of return by:

$$(123) \quad \pi_i = \pi_i(x, \vec{E}) = pqxE_i - c_i(\vec{E})$$

where $\vec{E} = (E_1, E_2, \ldots, E_N)$ is the vector of effort levels for all vessels in the fishery. In this convenient formulation, effort E_i is always measured in terms of its effect on the fish stock ($h_i = qxE_i$), but the cost of E_i now depends on the activities of other vessels. Assuming that the externality results in diseconomies, we then have

$$(124) \quad \frac{\partial c_i}{\partial E_j} > 0 \text{ for all } i, j.$$

(If vessels cooperate in locating fish, or in capturing large schools, externalities might actually lead to economies of production: $\partial c_i / \partial E_j < 0$ for $j \neq i$.)

From the optimization analysis (static or dynamic) one obtains the necessary conditions

$$(125) \quad \frac{\partial c_i}{\partial E_i} = (p - \mu)qx - \sum_{j \neq i} \frac{\partial c_j}{\partial E_i}$$

$$(126) \quad = c_N(\vec{E})'E_N$$

where μ is the usual (Lagrange or adjoint) multiplier. The second term in Eq. (125) reflects the presence of externalities; cf. Eq. (18) and (28).

Suppose now that the i'th vessel is taxed on its catch at the rate τ_i; we then obtain, from the assumption that $\tau_i(x, \vec{E})$ is maximized with respect to E_i:

$$(127) \quad \frac{\partial c_i}{\partial E_i} = (p - \tau_i)qx$$

$$(128) \quad = c_N(\vec{E})/E_N$$

Thus the presence of crowding externalities destroys the neat correspondence between the optimal tax rate and the multiplier. In fact we see that the optimizing tax now varies from vessel to vessel:

$$(129) \quad \tau_i^* = \mu + \frac{1}{qx^*} \sum_{j \neq i} \frac{\partial c_j}{\partial E_i}.$$

The second term on the right of this equation is a correction reflecting the marginal externality of vessel i as it affects the rest of the fleet. Under the assumption that externalities cause diseconomies, this correction is positive. (In his paper on crowding (congestion) externalities, Brown (1974) asserts that a two-tax system is needed to achieve the optimum, but later derives an equation showing that a single tax suffices. The confusion seems to arise from Brown's use of an aggregated production function, and disappears once a disaggregated model is used.)

To summarize, taxes on catch can be used in principle to optimize the fishery in the presence of crowding externalities, but different vessels must be taxed at different rates to achieve optimality.

What about vessel quotas? The market clearing conditions, Eq. (50) and (51), now become

$$(130) \quad \frac{\partial c_i}{\partial E_i} = (p - m)qx$$

$$(131) \quad = c_N(\vec{E})/E_N$$

where m denotes the quota price. If the total quota Q is taken as the optimal equilibrium catch, $Q = F(x^*)$, we see that the quota system is equivalent to a constant catch tax $\tau = m$, and hence fails to optimize effort throughout the fishery. The quota market reflects the stock externality but not the crowding externality. To achieve the optimum by means of allocated vessel quotas, it would be necessary to use in addition a correcting tax $\Delta \tau_i = \sum_{j=i} \partial c_j / \partial E_i$ on the catch of each vessel.

How important are these corrections likely to be in practice? In other words, how serious is the problem of crowding externalities? There is no question that severe crowding does occur in some fisheries, but in an open-access fishery, much of this crowding is probably largely due to the excess entry of vessels. If the fishery were to be managed in the general vicinity of optimality, using either fixed catch taxes, or allocated vessel quotas, it seems likely that crowding externalities would largely be eliminated — although clearly no strict conclusion of this nature is possible.

Summary and Perspective

Our study is based on a general model of the open-access fishery, in which the production function of each vessel is modeled individually. The only source of externality in the basic model is the "stock externality," whereby each vessel's catch affects the fish biomass and thereby the future catches of all vessels. Congestion externalities are discussed briefly later. In biological terms our basic model employs a simple "general production" model of population dynamics, and uses the traditional catch-effort relationship $h = qEx$ (q = constant); later the possibility of variable catchability, $q = q(x)$, is discussed briefly.

In attempting to identify the "bionomic equilibrium" traditionally associated with the unregulated open-access fishery, we encounter complications that center around considerations of a short-term vs. long-term nature. Unless vessel capital is perfectly malleable, the "medium-term" equilibrium (which does not allow for innovations) may depend on initial conditions. The analysis agrees, nevertheless, with the received theory in predicting both overexploitation and overcapacity of the unregulated fishery, compared with the social optimum.

Using a seasonal (discrete-time) version of the basic model, we show that the customary methods of circumventing depletion of fish stocks — total catch quotas, closed seasons, etc. — generally lead to further expansion of fleet capacity and a corresponding shortening of the fishing season. (The same can be proved for mesh-size regulations that attempt to achieve "eumetric" fishing (Clark 1976, p. 275).)

The following methods of influencing the efficiency of fishing were analyzed in the paper: (i) vessel licenses, (ii) catch taxes or royalties, (iii) taxes on effort, (iv) allocated vessel quotas. Each of these methods was analyzed separately, although in practice various combinations would be possible. The main theoretical result of the analysis was that catch taxes and allocated, transferable vessel quotas are mathematically equivalent in their effect on effort use, and hence on economic efficiency. The explanation of this equivalence is simply that quotas, being limited in total quantity, acquire a "scarcity value," which by the assumption of transferability becomes reflected in the quota market. Even a fisherman who elects not to sell his quota faces an opportunity cost, which plays exactly the same role as a tax on his catch.

Both taxes and allocated vessel quotas are capable of achieving the economic optimum in a constant equilibrium situation, but both are in theory suboptimal in a seasonally fluctuating fishery (unless tax rates, or quota "rates" can be adjusted as the season progresses). However, unless the intraseasonal variation of the fish biomass is severe, the departure from optimality would not be significant.

The theoretical equivalence of taxes and allocated quotas, while perhaps of considerable economic interest, should not be taken too literally. For, in practice, quotas obviously provide a direct control over catches, and hence over the state of the fish stock, whereas taxes act only indirectly. The precise relationship between tax rate and total catch rate would depend on the price of fish, the cost of fishing effort, and in general on the behavioral responses of fishermen, all of which are subject to serious imprecision and uncertainty.

The analysis given in this paper is readily applied to the case where both catch taxes and allocated quotas are employed. The economic optimum can be achieved via any combination; if τ denotes the tax rate and m the quota price, the basic equation (e.g. Eq. 50) becomes simply $c_i'(E_i) = (p - \tau - m)qx$. The price of quotas on the quota market is thus simply reduced by the tax τ. From the distributional point of view, this allows the government to divide the economic rents of the fishery in any desired proportion between fishermen and public revenue.

In our model, effort taxes are merely transformations of catch taxes, as a result of the assumed explicit relationship ($h_i = qxE_i$) between catch and effort for each vessel. In practice, however, catches are usually much more easily gauged than is effort. Consequently, catch taxes would normally be expected to provide more accurate control over the fishery than effort taxes. Since reasonably accurate catch data is basic to fishery management in any case, catch taxes should usually be quite feasible as part of the management program (political considerations aside).

Regarding vessel license programs, our analysis indicates that the authorized vessels will tend to exert excessive levels of effort; if total catches are also regulated, the fishing season will become unduly shortened. Moreover, vessel owners will have an incentive to increase the capacity, horsepower, and other characteristics of their vessels beyond optimal levels. An economically effective management program would have to control all these aspects, while at the same time encouraging worthwhile innovations. A license program (without other controls such as vessel quotas or taxes) may thus require a highly complex system of regulations.

The disadvantages of license programs could of course be eliminated by using taxes or vessel quotas in addition, in which case the licenses would merely serve the purpose of allocating fishing rights to a particular group of fishermen. Alternatively, privileged vessels could be determined via sale or auction of licenses, and indeed, unless prohibited, licenses will normally be transferred at a price reflecting their scarcity value. License fees and transferable (initially free) licenses are in fact economically equivalent, albeit of limited scope for economic optimization unless the production functions of vessels can be accurately specified and controlled.

FURTHER PROBLEMS

Our basic model is seriously oversimplified in many respects, particularly perhaps from the biological viewpoint. The age, and spatial and genetic structures of fish populations are ignored, as are all matters concerning the role of fish as part of a complex marine ecosystem. Variability and uncertainty (both biological and economic) are likewise ignored. On the economic side, our picture of competition between N individual fishermen overlooks important questions concerning conflict of interest between various groups — fishermen vs. processing companies; commercial vs. recreational fishermen; domestic vs. foreign fleets; international conflicts; and so on.

Regarding international fisheries, it seems clear that regulatory methods involving taxes or fees are out of the question, except for coastal states controlling foreign fishing within economic zones. Taking it for granted that the traditional approach of maximizing (or otherwise optimizing) sustainable yield via total catch quotas (for example) is no longer considered adequate in general, we are left with the method of national quota allocations. These are a matter for negotiation between concerned states (Munro 1979), and various principles (few of which will be established on purely "scientific" grounds) may be involved in allocating total catches. Each participating state is then in a position to regulate its own fishermen on the basis of its catch allocation.

Other conflicts of interest, such as those between commercial and recreational fishermen, which are also subject for negotiation at the political level, will not be further discussed here. Let us remark, however, on the relationship between fishermen and the processing companies to whom they sell their catches. As noted in Clark and Munro (unpublished data), neglecting the processing sector can lead to serious errors in recommendations for fishery management. For example, a perfectly monopsonistic processing firm dealing with a competitive, open-access fishing sector would be motivated to pay prices for fish that would optimize its own returns from the fishery. (In this situation, rent from the fishery would accrue to the processing firm. A limited entry program, or vessel quota system, would then presumably result in the transfer of a portion of this rent back to the fishing sector; modeling this situation might be a worthwhile project.)

The deterministic nature of our models may be considered a serious shortcoming. Appropriate objectives and techniques for the management of fluctuating fish populations are currently the subject of some debate (May et al. 1978). The question of evaluating the relative advantages of the various regulatory techniques in a stochastic setting has yet to be studied. Clearly the various techniques should be assessed in terms of (a) flexibility of management, and (b) riskiness to fishermen. Possible modifications to the basic techniques should also be considered; for example, vessel quotas could be specified in terms of quota "units," with the size of the unit varying from season to season according to stock abundance. But other methods of adjusting total annual quotas can be imagined.

The problem of dealing with bona fide uncertainty (as distinct from quantified variability) in fisheries has received very little theoretical study. In view of the importance of uncertainty in fisheries, it seems clear that flexibility would be a major consideration for any management system. In this regard, the desirability of preventing overexpansion of capacity is especially noteworthy, since the difficulty of achieving necessary changes in management policy is likely to increase with the size of the dependent industry.

Finally, our models are deficient in representing the fish population as a homogeneous mass, isolated from and essentially unrelated to the marine environment. Unfortunately, present understanding of marine ecosystems is too meager in most instances to support operational models involving multiple species and trophic levels, even though important management problems are now arising that involve these matters (May et al. 1979).

One structural characteristic of fish populations that does lend itself to operational modeling is age structure (Beverton and Holt 1957). Theoretical solution of the corresponding economic optimization model presents certain difficulties (Clark et al. 1973; Hannesson 1975), which would also affect the economic theory of regulation. In practice, appropriate harvest rates and size restrictions might be best determined by some ad hoc rule such as "$F_{0.1}$" (Gulland and Boerema 1973), economic efficiency then being achieved by the methods studied in this paper.

Acknowledgments

This work is largely an outcome of my attending two recent conference workshops on the economics of regulating commercial fisheries. It is a pleasure to acknowledge my debt to all of the participants in those workshops. Thanks are due particularly to the following people, who submitted valuable comments on an earlier manuscript: L. G. Anderson, S. J. Holt, D. Huppert, G. R. Munro, P. H. Pearse, G. Pontecorvo. The paper of Moloney and Pearse (1979) was particularly useful in forcing me to notice that allocated quota systems could be proved to be optimal. This research was partially supported by the Natural Sciences and Engineering Research Council of Canada, under Grant A-3990.

ANDERSON, L. G. [ed.]. 1977a. Economic impacts of extended fisheries jurisdiction. Ann Arbor Science, Ann Arbor, MI. 428 p.
          1977b. The economics of fisheries management. The Johns Hopkins University Press, Baltimore. 214 p.
BEVERTON, R. J. H., AND S. J. HOLT. 1957. On the dynamics of exploited fish populations. Minist. Agr. Fish. Food. London. Fish. Invest. Ser. 2(19): 533.
BROWN, G. JR. 1974. An optimal program for managing common property resources with congestion externalities. J. Polit. Econ. 82: 163–174.
CHRISTY, F. T. JR. 1977. Limited access systems under the Fishery Conservation and Management Act of 1976. p. 141–156. *In* L. G. Anderson [ed.] Economic impacts of extended fisheries jurisdiction. Ann Arbor Science, Ann Arbor, MI.
CHRISTY, F. T. JR., AND A. D. SCOTT. 1965. The common wealth in ocean fisheries. The Johns Hopkins University Press, Baltimore. 281 p.
CLARK, C. W. 1976. Mathematical bioeconomics: The optimal management of renewable resources. Wiley-Interscience, New York. 352 p.

1979. Restricted access to common-property fishery resources: a game-theoretic analysis, p. 117–132. *In* P.-T. Liu [ed.] Dynamic optimization and mathematical economics. Plenum, New York.

CLARK, C. W., F. H. CLARKE, AND G. R. MUNRO. 1979. The optimal exploitation of renewable resource stocks: problems of irreversible investment. Econometrica 47: 25–49.

CLARK, C. W., G. EDWARDS, AND M. FRIEDLAENDER. 1973. Beverton–Holt model of a commercial fishery: optimal dynamics. J. Fish. Res. Board Can. 30: 1629–1640.

CLARK, C. W., AND G. R. MUNRO. 1975. The economics of fishing and modern capital theory: a simplified approach. J. Environ. Econ. Manage. 2: 92–106.

COPES, P. 1972. Factor rents, sole ownership, and the optimal level of fisheries exploitation. The Manchester School Soc. Econ. Stud. 40: 145–163.

CRUTCHFIELD, J. A. 1979. Economic and social implications of the main policy alternatives for controlling fishing effort. J. Fish. Res. Board Can. 36: 742–752.

CRUTCHFIELD, J. A., AND A. ZELLNER. 1962. Economic aspects of the Pacific halibut fishery. Fish. Indust. Res. 1(1), U.S. Dep. Interior, Washington, D.C. 173 p.

GORDON, H. S. 1954. The economic theory of a common-property resource: the fishery. J. Polit. Econ. 62: 124–142.

GULLAND, J. A. 1977. The stability of fish stocks, J. Cons. Int. Explor. Mer 37(3): 199–204.

GULLAND, J. A., AND L. K. BOEREMA. 1973. Scientific advice on catch levels. Fish. Bull. 71(2): 325–336.

HANNESSON, R. 1975. Fishery dynamics: a North Atlantic cod fishery. Can. J. Econ. 8: 151–173.

LARKIN, P. A. 1977. An epitaph for the concept of maximum sustained yield. Trans. Am. Fish. Soc. 106: 1–11.

MAY, R. M., J. R. BEDDINGTON, J. W. HORWOOD, AND J. G. SHEPHERD. 1978. Exploiting natural populations in an uncertain world. Math. Biosci. 42: 219–252.

MAY, R. M., J. R. BEDDINGTON, C. W. CLARK, S. J. HOLT, AND R. M. LAWS. 1979. Management of multispecies fisheries. Science 205: 267–277.

MEANY, T. F. 1979. Limited entry in the Western Australian rock lobster and prawn fisheries: an economic evaluation. J. Fish. Res. Board Can. 36: 789–798.

MOLONEY, D. G., AND P. H. PEARSE. 1979. Quantitative rights as an instrument for regulating commercial fisheries. J. Fish. Res. Board Can. 36: 859–866.

MUNRO, G. R. 1979. The optimal management of trans-boundary renewable resources. Can. J. Econ. 12: 354–376.

PEARSE, P. H. [ed.]. 1979. Symposium on policies for economic rationalization of commercial fisheries. J. Fish. Res. Board Can. 36: 711–866.

PELLA, J. J., AND P. K. TOMLINSON. 1969. A generalized stock production model. Bull. Inter-Am. Trop. Tuna Comm. 13(3): 421–496.

QUIRK, J. P., AND V. L. SMITH. 1970. Dynamic economic models of fishing, p. 3–32. *In* A. D. Scott [ed.] Economics of fisheries management — A Symposium, Univ. B.C. Inst. Animal Resource Ecol., Vancouver, B.C.

RICKER, W. E. 1954. Stock and recruitment. J. Fish. Res. Board Can. 11: 559–623.

SCHAEFER, M. B. 1954. Some aspects of the dynamics of populations important to the management of commercial marine fisheries. Bull. Inter-Am. Trop. Tuna Comm. 1: 25–56.

SCOTT, A. D., AND C. SOUTHEY. 1970. The problem of achieving efficient regulation of a fishery, p. 47–59. *In* A. D. Scott [ed.] Economics of fisheries management — A Symposium, Univ. B.C. Inst. of Animal Resource Ecol., Vancouver, B.C.

SMITH, V. L. 1969. On models of commercial fishing. J. Polit. Econ. 77: 181–198.

TURVEY, R. 1964. Optimization and suboptimization in fishery regulation. Am. Econ. Rev. 54: 64–76.

UNITED NATIONS FOOD AND AGRICULTURE ORGANIZATION. 1979. Interim report of the ACMRR Working Party on the scientific basis of determining management measures. FAO Fisheries Circular No. 718, Rome. 112 p.

WILEN, J. E. 1979. Fisherman behavior and the design of efficient fisheries regulation programs. J. Fish. Res. Board Can. 36: 855–858.

[16]

Reviews in Fish Biology and Fisheries **6**, 5–20 (1996)

Individual transferable quotas: theory and practice

R. QUENTIN GRAFTON

Department of Economics, University of Ottawa, PO Box 450, Station A, Ottawa, Ontario, Canada K1N 6N5

Contents

Abstract

The paper examines the theory and practice of individual transferable quotas (ITQs) in fisheries. Using the experience of several countries, a number of ITQ programmes are examined with respect to fisher compliance and their effect upon economic efficiency, employment, the harvesting shares of fishers, cost recovery and rent capture.*

Introduction

Fishery managers regulate fisheries to help ensure sustainability of the resource and to meet socio-economic objectives. Input and effort controls have been the traditional method of regulation along with limits on the total harvest in the form of a total allowable catch (TAC).[†] While such approaches can be effective in preventing biological overfishing, they have often proved ineffective in preventing economic overfishing.[‡]

---

*This paper was prepared for a general readership including fisheries managers and scientists. Readers interested in a more theoretical treatment are referred to Arnason (1990) and Boyce (1992).

[†]Input controls include restrictions on the gear used by fishers while an example of an effort control could be a limited fishing season.

[‡]A good review of the problems of rent dissipation and the economics of fisheries management is provided by Munro and Scott (1985).

6   *Grafton*

Fisheries managed with input and effort controls have all too frequently been characterized by overcapitalization and excess competition among fishers that reduce the net return to fishers and the resource owners.*

The failure of input controls to control fishing effort and help ensure reasonable returns to fishers has stimulated fishery managers to use other forms of management. These management regimes have been based on the allocation and enforcement of property rights in the form of access and use rights in fisheries. These instruments attempt to remedy an externality where the actions of an individual impose costs on others but these costs are not accounted for by the individual. The use of quotas or permits for controlling externalities was proposed by Dales (1968) with respect to pollution, using the insights from Coase (1960). Christy (1973) was among one of the first to suggest their use in fisheries management. Today, such instruments are referred to as individual transferable quotas (ITQs) and allocate a total allowable catch among fishers in the form of individual harvesting rights. ITQs, in contrast to input controls, operate on the principle that incentives rather than controls should be used to manage a fishery.[†]

This paper examines the theory and practice of ITQs in various fisheries. Its contribution is to summarize some of the theoretical advantages of ITQs and to present some experiences of ITQ management with respect to fisher compliance with regulations, economic efficiency, employment, cost recovery and rent capture.[‡] This review should prove useful to fishery managers and scientists contemplating the introduction of transferable harvesting rights into fisheries.

Theory of ITQs

Fishers will harvest provided that expected returns exceed costs. In an open access fishery, fishers do not consider the costs they impose on others from harvesting a fish today (Warming, 1911; Gordon, 1954; Scott, 1955).[§] Thus, in an open access fishery the total harvest will, in general, exceed the level that maximizes the net return from the fishery. ITQs, by ensuring that fishers must pay a price for harvesting an extra fish and providing them with a long-term interest in the resource, can help change fisher behaviour and increase the net return from the fishery.

ITQs may be sold by the owner or regulator of a fishery to fishers or may be allocated on the basis of past harvests and vessel characteristics. In either case, a market for individual quota will arise, provided that fishers differ with respect to the net returns per unit of fish. Those fishers who can earn a net return per unit in excess of the market price for quota will buy quota, while those who have a net return less than the market price will sell their quota. The regulator only determines the initial

---

*See, for example, Dupont (1990) for estimates of the costs of rent dissipation in the salmon fisheries of British Columbia, where input and effort controls have been in place since the early 1970s.

[†]See Squires *et al.* (1995) for a general review of ITQs in fisheries management and Arnason (1993) and Pascoe (1993) for reviews of the experience and potential of ITQs in specific fisheries jurisdictions.

[‡]A brief glossary is appended.

[§]Open access is a situation where a common-pool resource is harvested by many firms or individuals, with no regulation over the amount harvested or entry and exit. Most developed fisheries are more accurately described as limited-user open access, where there may be a conditional but exclusive right to fish but not an exclusive right on how much to fish.

allocation of quota and verifies that fishers are respecting the quota regulations, while it is the fishers through their trades that determine the market price of quota and how much is bought or sold.

Theoretically, if ITQs command a positive price, fishers will take this into account when harvesting extra fish and can thus change the incentives faced by fishers. For example, fishers wishing to increase their harvest under ITQ management must pay for the privilege of leasing or purchasing quota from another quota-holder. Similarly, fishers wishing to harvest only the quota they were initially assigned also face an implicit cost equal to the revenue foregone from not having sold or leased the quota to another. Thus, a share of the externality equal to the quota price is internalized by fishers with ITQs.

ITQs, however, will only maximize the net return from the fishery provided that there are no in-season stock externalities, such that the harvesting costs are invariant to the size and distribution of the biomass; and no congestion externalities, such that harvesting costs are invariant to the amount of fishing effort applied at a given location (Boyce, 1992).* If ITQs do not, in general, lead to a first-best outcome they are, nevertheless, a desirable management tool if they result in a superior outcome to that arising from current practice.

In open access and most limited entry fisheries, each fisher is competing to catch some share of the total harvest. In this race for the fish, it may be profitable for an individual to invest in a faster vessel or in more sophisticated search gear so as to obtain a greater share of the catch. Such investment, although privately beneficial, does not increase the total harvest or total returns from the fishery, but changes the distribution of returns among fishers and increases total harvesting costs. Thus, the race for fish and other inefficiencies will continue if there is only a control on the total harvest and not on the individual outputs of fishers.

In contrast, in an ITQ fishery where the quotas are viewed as providing a durable and exclusive harvesting right, fishers may have a much greater chance of harvesting a given share of the TAC.[†] By ensuring that fishers harvest only their own quota, ITQs provide fishers with every incentive to minimize their costs because their gross revenue is more or less fixed by their quota-holdings. In turn, this can help reduce overcapitalization and racing behaviour between fishers to 'catch the fish before someone else'.

ITQ management may also reduce the need for specific input controls and other types of regulations. It has even been proposed that the price of ITQs be used to set the TAC so as to maximize the net return from the fishery (Arnason, 1990).[‡] An example of an input control which, if removed, may increase the net returns of fishers is a limited fishing season. For example, allowing fishers to harvest year round instead of during a limited fishing season may enable fishers to land a higher-quality product and increase safety at sea. Further, spreading the fishing season over a longer period can prevent sharp falls in prices brought about by a large increase in supply when fishing is restricted to only a few days. Allowing for transferability of the harvesting rights also permits more profitable fishers to harvest a greater share of the TAC. Such transfers

*Smith (1969) reviews the externalities prevalent in fishing.

[†]Scott (1989) reviews the notions of property rights and their characteristics with respect to the fishery.

[‡]Heaps (1993) points out that setting a TAC using only the price of quotas will not, in general, be optimal if vessels are heterogeneous.

should increase the total profits from the fishery and may change the structure of the industry as less profitable fishers exit through the selling or leasing of their quota to others.

One other characteristic of ITQs is that they can provide fishers with an additional interest in the resource. This interest, represented by the value of the quota owned by individuals, may encourage more involvement in management by fishers. Cooperation among fishers in helping to manage the fishery, in collaboration with the owners of the resource, should both improve the management of the fishery and reduce the costs of regulation. Another form of cooperation in the form of mutually beneficial agreements among fishers may also be encouraged with ITQs through assigning quota on both a species and an area basis. When the gains from cooperative behaviour are sufficiently great and fishers can monitor the actions of others, one may observe the pooling of effort among fishers. For example, one fisher may specialize in search gear to find the fish and others in harvesting the fish with the returns shared according to some predetermined arrangement. In turn, this type of cooperative behaviour can reduce total fishing costs and the individual risk faced by fishers.* Scott (1993) has even suggested that ITQs, by 'solving' the distributional question in the fishery, may even serve as a step towards achieving the goal of a joint fisher-ownership structure of the resource.

To ensure that the potential benefits of ITQs are realized, however, it is necessary that fishers view their quota as an exclusive and durable harvesting right. If fishers who are not quota-holders are able to fish with impunity, the quota becomes valueless as a meaningful property right. A share of the externality imposed by harvesting is, therefore, not internalized and the race to fish still remains. The race for fish may also remain in so-called flash fisheries where the TAC is caught in a very short period of time even if there is adequate enforcement.[†]

ITQs, by restricting the harvest of fish that can be legally landed, also provide an incentive to maximize the return per quota unit. In turn, this encourages fishers to land a higher-quality and valued product. It also has the effect of encouraging 'high grading', whereby fishers dump at sea less desirable fish for which they have quota. Both quota-busting and high grading are potentially serious problems for the sustainability of a fishery.[‡] These and other potential problems with ITQs are examined in detail by Copes (1986). The following section examines the experiences of ITQ management and whether in practice ITQs have realized their theoretical potential.

Experiences with ITQs

ITQ programmes have been implemented in a number of fisheries in several countries including Canada, Iceland, Australia, New Zealand and the USA. These programmes differ with respect to their restrictions on transferability, size of the fisheries, number of participants and various other characteristics.

A problem common to all ITQ programmes is how to determine the initial allocation of quota to fishers. In all ITQ programmes to date, fishers have received an allocation gratis based upon on an existing and/or historical participation in the fishery and/or

*An example of cooperative behaviour among fishers is provided by Wilen (1989) in a study of the British Columbia roe herring fishery.
[†]A flash fishery is one where the fishing season is very short for either biological or marketing reasons.
[‡]See Arnason (1994) and Anderson (1994) for a discussion on discarding and high grading in ITQ fisheries.

vessel characteristics. In most jurisdictions, the allocations are denominated as a percentage of the TAC so that fluctuations in harvest are borne exclusively by the fishers.* A notable exception were the initial allocations in New Zealand's ITQ fisheries which were originally denominated on the basis of weight but which were subsequently changed to percentage or share quotas.† Another issue with respect to quota markets is the number and volume of trades relative to the total allocation of quota. Lindner *et al.* (1992) have observed that there have been only a limited number of transactions in some of New Zealand's ITQ fisheries while there has been a substantial amount of trading in other fisheries such as Australia's southern bluefin tuna (*Thunnus maccoyi*, Scombridae) fishery. Trading appears to be more intense where there are significant differences in the gear and vessels used by fishers.

To evaluate the diverse experiences with ITQs, it is necessary to use criteria to judge their successes and failures. In this study, the experiences of some but not all ITQ fisheries are evaluated according to the following criteria:

1. changes in economic efficiency;
2. changes in employment and harvesting shares of fishers;
3. compliance with ITQs;
4. cost recovery, management costs and the capture of resource rents.

CHANGES IN ECONOMIC EFFICIENCY

There is evidence that ITQs have improved economic efficiency in a number of fisheries. In a survey of the rights-based management in Canada, Crowley and Palsson (1992) state that there have been efficiency gains with such regulations. Evidence of improvements in efficiency have also been documented in ITQ fisheries in Australia, New Zealand and Iceland, and have been characterized by reductions in fishing effort and an increase in profitability of fishers. These two issues are examined separately.

Reduction in fishing effort

One of the expected benefits of ITQs is that they can reduce the excess capital employed in the fishery. Depending upon the fishery, however, the removal of excess capital may involve an adjustment period of several years. For example, in fisheries where earnings outside the quota fishery are very low, it will pay a vessel owner to keep fishing with an old vessel so long as the returns cover variable costs. In other fisheries, where alternative employment opportunities exist for displaced labour and capital, the structural change in the fishery brought about by ITQs may be rapid. A listing of the changes in vessel numbers in selected rights-based fisheries in Canada is provided in Table 1. In most cases, the number of vessels employed in the fisheries declined with the introduction of ITQs.

The British Columbia (BC) sablefish, *Auoplopoma fimbria* (Auoplopomatidae), fishery is illustrative of the reductions in the number of vessels that can take place in the short run. In 1989, the year before individual vessel quotas (IVQs) were introduced

---

*For a definitive review of the advantages and disadvantages of quantity and share quotas see Kusuda (1993).
†Less than expected levels of the biomass of some species such as the orange roughy (*Hoplostethus atlanticus*, Trachichthyidae), which would have necessitated costly purchases of quota by the regulator to reduce the TACs, ultimately led to a change from quantity to share quotas in 1989.

Table 1. Changes in numbers of vessels in selected rights-based fisheries

Fishery	Number of vessels		Source
	Before programme	In 1992	
Herring (RTV)	16	11	Crowley and Palsson (1992)
Herring (4WX)	49	40	Crowley and Palsson (1992)
Offshore groundfish	139	115	Crowley and Palsson (1992)
Offshore scallop (4X, 5Ze)	73	61	Crowley and Palsson (1992)
Offshore lobster (4X, 5Ze)	8	8	Crowley and Palsson (1992)
Lake Erie	248	182*	Cowan (1990)
Lake Winnipeg	800–1000	400–600	Crowley and Palsson (1992)
BC sablefish	47	30	Grafton (1992b)

*No. vessels in 1990.

into the fishery, there were 46 vessels harvesting the resource. In 1990, in the first year of IVQs, there were only 30 vessels that actively fished for sablefish (Grafton, 1995). The change in the harvesting pattern arose from transfers of fishing licences with quota among registered vessels in the fleet. This, in turn, resulted in a greater harvest and a higher profit per vessel. A similar adjustment also took place in the Lake Erie ITQ fishery. In 1983, the year before ITQs were introduced, there were some 248 vessels, in 1984 there were 242, and in 1988 there were only 182 vessels actively operating in the fishery (Cowan, 1990). This represented a 27% decline over a 4 year period. It should be noted, however, that both the BC sablefish and Lake Erie ITQ programmes have some of the least restrictive transferability provisions in Canadian fisheries.

In a review of Iceland's fisheries, Arnason (1986) notes that the number of vessels remained virtually unchanged in the first 2 years of the operation of ITQs. There was, however, a decline in the fishing effort as measured by vessel tons per day at sea. On this criterion, aggregate fishing effort fell some 15% in the first year of the ITQ programme in 1984 and then some 6% in the following year. It should be noted, however, that in the Iceland programme vessels were allowed to opt for either quantity quotas or effort quotas. The effort quotas, which were removed in 1990, still provided an opportunity for fishers to compete among themselves and may have limited the reductions in fishing effort that would otherwise have taken place. Nevertheless, Arnason (1993) observed that although herring, *Clupea harengus* (Clupeidae), catches tripled over the period 1977–1990, fishing effort decreased by some 20% with the implementation of ITQs.

A more dramatic effect on the capital employed in a fishery is given with the introduction of an ITQ scheme into Australia's southern bluefin tuna fishery. In 1984, when ITQs were first introduced, there were some 143 quota-holders, while by 1988 there were only 63 quota-holders (Muse and Schelle, 1989). In contrast, in New Zealand there was an increase in the number of vessels in the year following the introduction of ITQs in its inshore fisheries (Clark *et al.*, 1988). Clark *et al.* (1988) suggest, however, that the increase was attributable in large part to the purchase of squid-jigging vessels by New Zealand companies to take advantage of opportunities in the squid fishery. The fact that vessel numbers did not immediately decline in New Zealand with ITQs may also be a reflection of the lengthy adjustment that arises in industries where the value of vessels and chances for employment in other activities are very low.*

Increased profitability

Another benefit attributed to ITQs is that they can increase the profitability of those fishers allocated quota gratis. This increase may arise from reduced racing behaviour, improvement in the quality of the fish and other factors. For those fishers entering the fishery at a later date and who must purchase quota from existing quota-holders, the price paid for quota should reflect the expected discounted profits from harvesting a given share of the resource. For these later participants, any resource rents from the fishery should be capitalized and reflected in the quota price.

Evidence of increased profitability with ITQs for the original quota-holders is available from a number of fisheries. Unfortunately, separating the increased profitability of fishers due to ITQs alone, from other factors such as improved market conditions or increases in the fish stock is often not possible. In the case of Australia's southern bluefin tuna fishery, Geen and Nayar (1988) have simulated the impacts of ITQs and suggest that total net profits in the fishery would, at a steady state, be some A\$6.7 million more with ITQs than with an open access situation. Substantial benefits attributable to ITQs are also found in the Icelandic demersal fisheries. According to Arnason (1986), the value of reduced fishing effort and improved quality of product was US\$15 million in the first year of implementation of ITQs.

In New Zealand, there is qualitative evidence of increased profitability of fishers. In a survey of one group of fishers, Dewees (1989) notes that 23% of respondents claimed that ITQs have led to improved quality of product, 17% have switched to longlining gear, and some 10% of fishers have reduced their fishing effort. In a summary of the study, Dewees (1989) observed that these changes were aimed at maximizing the price received by fishers and minimizing their costs. In a survey of the ITQ programmes in New Zealand, Macgillivray (1990) also credits the programmes with improving the financial performance of the industry. This view is supported by the industry itself (Sharp and Roberts, 1991) in a submission by the New Zealand Fishing Industry Board (NZFIB) to the New Zealand Government Task Force on Fisheries Legislation. The NZFIB states that the ITQ system has been an effective means of moving towards optimal economic benefits while at the same time sustaining the resource. In particular, they single out the most obvious benefit resulting from ITQs: the reduction in the race for fish.

In Canada's rights-based fisheries, there is also evidence of increased profitability due to ITQs. In the BC sablefish fishery it was observed that the value of sablefish licences increased some fourfold with the introduction of individual vessel quotas (IVQs) and the coupling of licences with quota (Grafton, 1992b). Much of this change probably reflects a higher value placed upon quota, which gives a more secure fishing privilege than licences alone. It may also reflect higher profits in the fishery with ITQs. In the sablefish and halibut, *Hippoglossus stenolepis* (Pleuronectidae) fisheries of British Columbia there have also been big changes in the length of the fishing season. For example, before the introduction of IVQs in 1990, the sablefish fleet took only 8

---

*Lindner *et al.* (1992) examine some of the issues with respect to transitions in the New Zealand fisheries brought about by ITQs.

days to harvest the TAC. Under quota management, fishers are harvesting over several weeks and at different times of the year. This allows fishers to coordinate their harvests with market prices and has increased the quality of the fish landed. Both factors have helped fishers to receive a higher return for their product. An added benefit for fishers is improved safety because in the pre-quota regime, irrespective of the weather conditions, it was necessary to fish during the very short designated fishing season.

In a review of the enterprise allocation (EA) programme of Canada's offshore groundfish fishery, Gardner (1988) has noted efficiency improvements in both the harvesting and processing sectors from the introduction of individual quotas. In particular, he notes that individual quotas have allowed for improvements in quality of the product through improved timing of the catch. This has led to improvements in processing, with fewer production bottlenecks, lower inventories, and a higher proportion of higher-valued products being produced.

CHANGES IN EMPLOYMENT AND HARVEST SHARES OF FISHERS

To reap the full benefits of individual quotas, transferability among fishers must be allowed. It is transferability of quota that allows fishers to retire labour and capital. Without the ability to sell quota to other fishers, less profitable fishers may choose to remain in the fishery, preventing more profitable fishers from harvesting a greater share of the TAC.

Transfers of quota also have implications for fishery managers beyond economic efficiency. Concentration of quota may reduce the total employment in the fishery by reducing the number of vessels. It may also create market power for fishers with large quota shares who may be able to manipulate quota and product prices to the detriment of others.* A related concern is that transferability of quota will remove the smaller owner-operator fishers and allow those with larger vessels and processing companies to dominate the fishery. Whether such an adjustment takes place, however, is entirely dependent on the characteristics of the fishery. For instance, a larger and faster vessel may have been at an advantage in a pre-ITQ fishery where fishers were competing for shares of the catch. The same vessel under an ITQ regime may not have a comparative advantage when fishing is spread more evenly throughout the year but is likely to impose higher fixed costs on its owners than older and smaller vessels. For example, in a study of the BC sablefish fishery, the generally smaller and older longline vessels were found to have a higher profit per unit of sablefish landed than the larger trap vessels (Grafton, 1995).

To address the concern over quota concentrations, a number of ITQ programmes have imposed limits on the quota that can be owned by any one individual or company. For example, in New Zealand these limits are 20% for the inshore and 35% for the offshore fisheries (Gibson, 1989). Nevertheless, with these regulations the owned and leased quota-holdings of the ten largest companies in New Zealand increased from 58% of the total in 1986 to 66% in February 1988 (Muse and Schelle, 1988). In other jurisdictions the concern is more with the concentrations of quota on a regional basis. For example, in Iceland, transfer of quota from one region to another requires the authorization of the Minister of Fisheries. In the first year of operation of ITQs in Iceland, there were considerable transfers of quota between regions (Arnason, 1986). A listing of the total

*Anderson (1991) addresses the issue of market power in ITQs.

quota transfers is provided in Table 2 and the percentage change by species by region is provided in Table 3. In the regional transfers, net losses in landings were experienced in the urban areas of the Capital and gains were observed in the smaller fishing villages.

Significant changes in the regional distribution of the harvest of Australia's southern bluefin tuna were also observed with the introduction of ITQs in 1984. For example, Table 4 indicates that by the end of 1987, quota trading had led to the exit of New South Wales fishers from the fishery and reduced by over 50% the quota owned by Western Australia fishers. This has resulted in a concentration of quota in South Australia with the share of the TAC increasing from 66% in 1984 to 91% in 1987 (Geen and Nayar, 1988). Over the same period there was also a change in the share of the catch harvested by gear type, with purse-seine vessels increasing their total share from 16% in 1984 to 42% in 1987.

Another issue in the transfer of quota is its effect on employment in the fishery. The evidence from Canada and other countries suggests that ITQs are likely to reduce employment in the harvesting sector. This is because as vessels are retired from the fishery the labour previously employed on these boats is not re-employed on the remaining vessels. For example, in the year following the introduction of ITQs into the

Table 2. Transfers of quota in cod equivalents (tonnes) in Iceland's ITQ fisheries, 1984–1988*

Year	Rental and leases (t)	Permanent sales (t)	Sales as % of total quota
1984	60 825	14 494	3.3
1985	70 976	19 986	4.7
1986	55 609	20 391	4.4
1987	49 081	21 391	4.3
1988	92 889	28 708	5.8

*Source: *Aegir*, Fisheries Association of Iceland, July 1989, pp. 356–364.

Table 3. Regional redistribution of landings in Iceland's fisheries in 1984*

Species	South-west	West	North	East
Cod[†]	2 20	1 22	1 2	2 7
Haddock[‡]	1 22	1 15	2 60	2 20
Saithe[§]	2 10	1 5	1 5	1 5
Redfish[¶]	1 20	2 1	0	2 40

*Source: Arnason (1986), p. 95. All values are in %.
[†]*Gadus morhua* (Gadidae).
[‡]*Melanogrammus aeglefinus* (Gadidae).
[§]Species not identified in source; assumed to be *Pollachius virens* (Gadidae).

Table 4. Quota-holdings (tonnes) by State in the Australian southern bluefin tuna fishery*

State	1984–85	1985–86	1986–87
Western Australia	2752	1926	1305
South Australia	9272	12 110	13 030
New South Wales	1872	0	0

*Source: Geen and Nayar (1988).

Lake Erie fishery, the labour force fell some 22% (Cowan, 1990). In addition, for those vessels remaining in the fishery, a decrease in the race for fish can mean vessel owners are able to substitute labour for extra time spent at sea. Thus, ITQs may lead to a reduction in the total number of people employed in the harvesting sector but may increase the hours worked of the crew members who remain in the fishery.

An issue related to employment is the remuneration of fishing crew with ITQs. Under quota management, there is less uncertainty faced by vessel owners with respect to their catch and more emphasis on minimizing operating costs. In this scenario and with longer fishing seasons, if the crew does not perform at their best there still remains an opportunity to fulfil the quota at a later date. As a result, the incentive to pay crew a share of the total value of the harvest so as to elicit the crew's best efforts is reduced. There may, therefore, be a change to paying crew on a daily or hourly wage with the introduction of ITQs. For example, in the BC sablefish fishery, some vessel owners have changed the method of paying crew with ITQs and pay a daily wage of $100 instead of the traditional share of 50% of the gross revenue less operating costs. Such a change has led the British Columbia united fishermen and allied workers union to oppose the introduction of ITQs in Canada's Pacific fisheries (Cruickshank, 1991, p. 32). Opposition by crew members and their representatives to ITQs is, however, not untypical in fisheries. A proposal made by Hannesson to address this opposition is to make the initial allocations of quota to fishers and crew collectively (Hannesson, 1988).

Despite likely reductions in employment in the harvesting sector, ITQs may be beneficial to the processing sector. If fish are landed over a greater period of time, the period of employment for processing workers should also increase. Further, spreading the landings over a longer period can enable processors to change the output to obtain a higher-quality and higher-priced product mix. Depending on the labour components in the different products, this may increase or decrease employment in the processing sector. In the case of Canada's Atlantic fishery, there is evidence that individual quotas in the enterprise allocation (EA) programme may have increased employment in the processing industry. According to Gardner (1988), the EA programme increased the quantity of labour-intensive but higher-valued products such as packs of fresh and frozen fillets. This, in turn, increased the labour hours per tonne of fish harvested from 30 h in 1984 to 36 h in 1987.

COMPLIANCE WITH ITQ REGULATIONS

One of the fundamental requirements for a successful ITQ programme is fisher compliance with regulations. Where fishers are able to fish over quota or undertake other prohibited activities with impunity, an ITQ system will not provide the expected benefits and may even be detrimental to the sustainability of the fishery.

A number of schemes have been implemented for monitoring and observing the harvests and landings of fishers to ensure compliance with quota regulations. Often, monitoring fishers' harvest at sea is not feasible in terms of management costs or practical where there are large numbers of small vessels. Instead, many ITQ jurisdictions have set up systems for monitoring fishers' landings. For example, in New Zealand the change to ITQs in 1986 led to a significant change in the monitoring activities of fisheries officers. Instead of an emphasis on dockside monitoring, the fishery managers set up a scheme to monitor the 'paper trail' of quota landings. This monitoring is focused around a catch landing log and a quota management report which must be submitted monthly to the regulator and lists the fish caught by species and area and the quota under which it was landed. In addition, fish processors are required to submit a monthly report that indicates the quantities purchased, the price, the species and the quota identification. This paper record has shifted the focus of monitoring from traditional fishery officers to accountants and persons experienced in fraud investigation. The system has also been helped by the introduction of a goods and services tax (GST) which also ensures that record keeping is consistent. The monitoring programme is aided by the fact that over 85% of New Zealand's harvest is exported. Trade data, therefore, provide additional verification of the quantities landed, and a small domestic market reduces the opportunity for over-the-counter sales that bypass the quota management system.

Another important feature of the New Zealand programme is the legislative power accorded to the Government of New Zealand. Under the 1986 amendment to the *New Zealand Fisheries Act*, powers were defined to enable authorities to confiscate quota and vessels of fishers contravening quota regulations. In imposing such penalties, the onus is on the fisher to prove that he or she is not in contravention of the regulations. The system is not without its problems, and quota-busting does take place, despite the fact that there are severe penalties for contravening the regulations. For example, in 1989 a fishing company lost NZ$4 million of quota and boats following a conviction for deliberate overfishing (Macgillivray, 1990). Another concern is that there has been considerable dumping of fish at sea. This dumping arises from fishers high grading and from discarding of fish for which they do not have quota. To help address the discard problem, New Zealand increased from 10% to 50% the proportion of the market price paid to fishers for those species for which they do not have quota. This change was implemented to encourage the fish to be landed once it is caught but not to provide an incentive for fishers to search actively for species for which they lack quota. The system also allows for overages whereby fishers are allowed to land up to 10% above their quota in a season but the extra landings are deducted from the quota for the following season.

A Canadian example of a relatively successful monitoring of ITQs is provided by the Lake Erie fishery. In the scheme, fishers and processors themselves pay for monitoring of landings with a levy of $\frac{1}{2}$ cent per pound on each group. The system requires fishers to land their product and pack it in ice in designated weights at designated ports. A port monitor then weighs 20% of the total landings to verify that the weights are within 5% of the prescribed weight. The system appears to work effectively and has reduced the enforcement costs of the fishery regulator. Prosecutions have also decreased from 54 in 1980 to none in 1988 (Cowan, 1990). A key component of the programme is that weight and landing conditions have been incorporated into the fishing licence. Procedurally this allows the regulator to suspend a licence, with its corresponding quota, should a condition of the licence be violated, although in practice such suspensions only occur after two warnings. Under the pre-ITQ system, a licence could be suspended only after a court trial.

Less successful monitoring has occurred in some of Canada's Atlantic fisheries. Canada's first ITQ scheme, the 4WX herring fishery in the Bay of Fundy, has been in place in one form or another since 1976 and is noteworthy in its failure to adequately monitor the harvests of fishers at the beginning of the programme. A detailed evaluation of this fishery in its early years is provided by Campbell (1981) and a recent review is given by Stephenson *et al.* (1993). In the original ITQ scheme, quotas were not legally binding regulations but were rather self-imposed and informal arrangements agreed to by vessel owners in collaboration with a fisher organization. Under this scheme, fishers were not required to notify the regulator of the weight of their catch and there was a limited legal framework to prosecute those contravening the regulations.* Not surprisingly, there was a breakdown in observing quota limits such that by one account, 50% of catches were not reported in some years (Mace, 1985). Despite non-compliance by fishers, the monitoring costs for the fishery were substantial and represented about 2.5% of the total landed value. In addition, the ITQ programme did not encompass other gear types, so that in the first 2 years of the programme, increased prices for herring and increased profits for purse seiners increased the fishing effort of drift gill netters in the fishery (Kearney, 1983).

Another example of monitoring of ITQs in Atlantic Canada is provided by the western Newfoundland otter-trawl cod fishery. The programme was originally set up on an experimental basis in 1984. Transferability was initially not allowed and an appropriate system of monitoring the harvests of fishers was not implemented. It is not surprising, therefore, that the fishery was described as being ". . . plagued by discarding and misreporting" (Crowley and Palsson, 1992, p. 7). Recognizing the importance of ensuring adequate monitoring to ensure the benefits of ITQs, fishers are currently funding their own monitoring programme of harvest with a royalty on landings.

The EA programme in Canada has also had some difficulties in its monitoring. A major criticism of EAs was that skippers were high grading to maximize their returns per quota unit and consequently discarding lesser-valued fish at sea. Unfortunately, the extent of the problem is not quantifiable although it is likely that some vessels in the past did discard some of their harvest. To address the problem, the regulator subsequently instituted an observer programme on 100% of the foreign and 50% of the domestic offshore trawler fleet. The costs of such monitoring are substantial but there is no direct charge that is applied to the domestic fleet. Such a solution to discards and high grading is, unfortunately, impossible in many other fisheries where the number of vessels is greater and the landed value per fishing trip is much less.

COST RECOVERY, MANAGEMENT COSTS AND RENT CAPTURE

An attributed benefit of ITQs is that they can reduce the costs of management and that some of the increased profitability of fishers can be collected for the purposes of cost recovery and rent capture. Unfortunately, evaluating ITQ programmes on the basis of their management costs is particularly difficult as many of the costs of regulations are not properly apportioned across fisheries and are difficult to quantify.

In the case of New Zealand a system of checks, such as landing and processing records, was established to monitor the ITQs while many of the previous regulations remained in place. This, coupled with an extensive public relations campaign by the government to persuade fishers to accept ITQs and an arbitration system that ultimately reviewed the quota appeals of some 1500 fishers, undoubtedly increased management costs. In fact, in 1986/87 in the first year of ITQs in the New Zealand inshore fishery, management costs increased in nominal terms by 36% over the previous 12 months (Macgillivray, 1990). In the New Zealand case, however, there has been a systematic

---

*Since 1985 fishers have been obliged to provide the regulator with catch weights.

claim on the resource rent by the owners of the resource to recoup the regulatory costs. Indeed, the industry itself acknowledges its "... willingness to contribute to the costs of fisheries management, where the benefits of that management fall to the industry" (Sharp and Roberts, 1991, p. 86). The resource rentals paid by fishers are determined on the basis of quota values, the expected net returns of fishers, and other factors deemed important by the regulator (Grafton, 1992a). In 1988/89, the New Zealand Government collected some NZ$20 million, a little less than 10% of the total landed value of fish caught by the New Zealand fleet. In total, with foreign fleet access fees and other charges, the revenue generated from the fisheries was some NZ$34 million in 1988/89 or an amount equal to the total operating budget of the regulator (Macgillivray, 1990).

Elsewhere, the cost recovery and the capture of resource rents has a much lower priority. In recent ITQ programmes in Canada provision has, however, been made for some cost recovery with a charge on the landings of fishers. For example, in the BC halibut fishery, fishers currently pay a 9 cents per lb landing charge that fully funds a monitoring and observation programme. In the BC sablefish fishery, a charge of 6 cents per lb is currently assessed to cover the full costs of management and monitoring in the fishery. Similarly in the Lake Erie ITQ programme, the cost of monitoring landings is almost exclusively borne by fishers and processors in a landing charge per pound of fish delivered. In the past, fishers in Atlantic Canada have been charged token access fees. For example, in the EA programme, access fees do not bear any relation to the value of the fishery and in 1988 generated some $2.1 million for the regulator (Gardner, 1988).

In other countries, there are also provisions for some cost recovery. In general, however, there has been little attempt to capture the resource rents for the benefit of the resource owner. The capture of rent also has a number of important implications for the fishery, depending upon the amount and how the rent is collected. In collecting rent, however, a non-trivial problem is to estimate the resource rent and separate it from intra-marginal earnings of highliners.* It should also be noted that any procedure for rent capture should be announced prior to allowing transferability of quota. Without such a provision, the resource owner faces the possibility of capturing rent from fishers who paid the expected resource rent to the original quota-holder. The resource owner also faces a choice of tax instruments including a charge based on the quota price and quota-holdings, a profit charge, a fixed fee charge, an auction of the ITQs, and a landing or royalty charge. The appropriate method of collecting rent may vary across fisheries and may depend upon the level of uncertainty.[†]

Conclusions

The theory behind ITQs suggests that in certain fisheries they can be an effective management tool to prevent rent dissipation and can increase the returns to fishers and the resource owners. ITQs may also be a step towards a joint fisher-ownership structure of fishery resources. To achieve these benefits and to ensure a sustainable fishery, ITQs

---

*Intra-marginal earnings are differential earnings among fishers and the intra-marginal rent per unit of quota represents the difference in the average profit per unit of fish landed between the marginal fisher and all others.
[†]Grafton (1994) examines the issues of rent capture in ITQs under uncertainty while Grafton (1995) reviews the effects of several methods of rent capture in the BC sablefish fishery.

must be viewed as a durable, transferable and exclusive property right. An essential component to the success of ITQs, therefore, is adequate monitoring and enforcement. In jurisdictions where enforcement has been satisfactory, it appears that ITQs have been beneficial to both fishers and the resource owners. There is evidence from Australia, Canada, Iceland and New Zealand that ITQs have improved economic efficiency and increased the returns of fishers.

ITQs are not, however, a panacea for all the problems that arise in fisheries. ITQs will not remedy a decline in stocks or immediately remove excess capital in a fishery. Individual quota management does, however, offer the potential of generating a higher net return from the resource. Ultimately, this offers a long-term advantage to both the fishers and the resource owners.

Acknowledgements

The author is grateful for the valuable comments and suggestions of Halldor Palsson, Jon Conrad, Gordon Munro and Trond Bjørndal.

Glossary

Common-pool resources Natural resources where use is rivalrous – such that one person's harvest affects the harvest of others – and where exclusion is difficult, such that users can only partially exclude others from exploiting the resource.

Communal rights A set of property rights where there are rules governing the exploitation of a common-pool resource by members of a community in the common interest. Often these rules exclude non-members from exploiting the resource.

Discarding The dumping of fish at sea so as to increase the net return from a fishing trip.

Enterprise allocations (EAs) A harvesting right assigned to fishing companies in the offshore fisheries of Atlantic Canada.

Externality A situation which arises when the actions of an individual or firm affects the well-being of others and this is not taken into account in the decision-making of the individual or firm.

Factor of production The inputs into the production of goods and services. These inputs include land, labour, capital and raw materials.

High grading The discarding of smaller and/or lesser-valued fish at sea so as to maximize the net return from a vessel's individual transferable quota.

Highliner A fisher who consistently earns intra-marginal rents.

Individual vessel quota (IVQ) A harvesting right, usually transferable, that is assigned to vessels.

Intra-marginal rent The quasi-rent that accrues to units of a heterogeneous factor of production in excess of that received by the marginal unit. In a fishery, if all participants have the same potential income in an alternative activity, the intra-marginal rent is the earnings of the better fishers in excess of that of the least profitable fisher.

Limited entry fishery A fishery where entry or access is restricted by the licensing of vessels or some other means.

Limited-user open access A situation where a common-pool resource is regulated such that there is a conditional but exclusive right to harvest the resource but not an exclusive right of how much each user can harvest.

Open access A situation where a common-pool resource is harvested by many firms or individuals, with no regulation over the amount harvested or entry and exit.

Property rights The societally accepted rights of individuals, or groups of individuals, to exploit assets for their benefit with at least a partial right to exclude others. A lack of properly defined and enforced property rights is often a cause of externalities.

Quasi-rent The rent that accrues to a factor of production that is temporarily fixed in supply.

Quota-busting The deliberate overfishing above the specified amount entitled by an individual transferable quota.

Quota-market The market for individual transferable quotas which permits buyers and sellers to trade. In most individual transferable quota fisheries, trades take place through traders who match potential buyers with potential sellers for a small commission.

Quota price The market price for individual transferable quotas. The price frequently varies over time and across trades at any point in time.

Rent The amount in excess of the minimum payment to ensure the supply of a factor of production.

Rent capture The appropriation of rent from a factor of production.

Rent dissipation The process by which excess competition by users in a common-pool resource reduces the potential rent.

Total allowable catch (TAC) The total harvest that is permitted in a fishery in a given period of time.

References

Anderson, L.G. (1991) A note on market power in ITQ fisheries. *J. Env. Econ. Manage.* **21**, 229–96.

Anderson, L.G. (1994) Highgrading in ITQ fisheries. *Mar. Resource Econ.* **9**, 209–26.

Arnason, R. (1986) Management of the Icelandic demersal fisheries. In Mollet, N., ed. *Fishery Access Control Programs Worldwide.* Fairbanks: University of Alaska, Alaska Sea Grant Report No. 86-4, pp. 83–101.

Arnason, R. (1990) Minimum information management in fisheries. *Can. J. Econ.* **23**, 630–53.

Arnason, R. (1993) Iceland's ITQ system. *Mar. Resource Econ.* **8**, 201–18.

Arnason, R. (1994) On catch discarding in fisheries. *Mar. Resource Econ.* **9**, 189–208.

Boyce, J.R. (1992) Individual transferable quotas and production externalities in a fishery. *Nat. Resource Modelling* **6**, 385–408.

Campbell, H.F. (1981) The Public Regulation of Commercial Fisheries in Canada: Case Study No. 5, the Bay of Fundy Herring Fishery. Ottawa: Canada Ministry of Supply and Services, *Economic Council of Canada tech. Rep.* No. 20, 43 pp.

Christy, F.T., Jun. (1973) Fisherman quotas: a tentative suggestion for domestic management. *Law of the Sea Institute, Univ. Rhode Island, Occ. Pap.* No. 19, 6 pp.

Clark, I., Major, P.J. and Mollett, N. (1988) Development and implementation of New Zealand's ITQ management system. *Mar. Resource Econ.* **5**, 325–50.

Coase, R. (1960) The problem of social cost. *J. Law Econ.* **3**(2), 1–44.

Copes, P. (1986) A critical review of the individual quota as a device in fisheries management. *Land Econ.* **62**, 278–91.

Cowan, T. (1990) Fisheries Management on Lake Erie. Unpubl. Report, Canadian Dept Fisheries and Oceans, Toronto. 28 pp.

Crowley, R.W. and Palsson, H. (1992) Rights based fisheries management in Canada. *Mar. Resource Econ.* **7**, 1–21.

Cruickshank, D. (1991) A Commission of Inquiry into Licensing and Related Policies of the Department of Fisheries and Oceans, Vancouver. 113 pp. (no doc. no.)

Dales, J.H. (1968) *Pollution, Property and Prices.* Toronto: Univ. Toronto Press. 111 pp.

Dewees, C.M. (1989) Assessment of the implementation of individual transferable quotas in New Zealand's inshore fishery. *N. Am. J. Fish. Manage.* **9**(2), 131–9.

Dupont, D.P. (1990) Rent dissipation in restricted access fisheries. *J. Env. Econ. Manage.* **19**(1), 26–44.

Gardner, M. (1988) Enterprise allocation system in the offshore groundfish sector in Atlantic Canada. *Mar. Resource Econ.* **5**, 389–414.

Geen, G. and Nayar, M. (1988) Individual transferable quotas in the southern bluefin tuna fishery: an economic appraisal. *Mar. Resource Econ.* **5**, 365–88.

Gordon, H.S. (1954) Economic theory of a common property resource: the fishery. *J. Political Econ.* **62**, 124–42.

Grafton, R.Q. (1992a) Rent capture in an individual transferable quota fishery. *Can. J. Fish. Aquatic Sci.* **49**, 497–503.

Grafton, R.Q. (1992b) Rent capture in rights based fisheries. PhD thesis, Dept Economics, Univ. British Columbia. 180 pp.

Grafton, R.Q. (1994) A note on uncertainty and rent capture in an ITQ fishery. *J. Env. Econ. Manage.* **27**, 286–94.

Grafton, R.Q. (1995) Rent capture in a rights based fishery. *J. Env. Econ. Manage.* **28**, 48–67.

Hannesson, R. (1988) Fishermen's organisations and their role in fisheries management: theoretical considerations and experiences from industrialised countries. In Studies on the Role of Fishermen's Organisations in Fisheries Management. *FAO Fish. tech. Pap.* No. 300, pp. 1–27.

Heaps, T. (1993) A note on minimum information management in fisheries. Dept Economics Discussion Paper 93-13, Simon Fraser Univ., Burnaby, BC. 15 pp.

Kearney, J.F. (1983) Common tragedies: a study of resource access in the Bay of Fundy herring fisheries. MES thesis, Inst. Resource Env. Studies, Dalhousie Univ. 441 pp.

Kusuda, H. (1993) ITQ fishery under uncertainty. PhD thesis, Dept Economics, Univ. British Columbia. 134 pp.

Lindner, R.K., Campbell, H.F. and Bevin, G.F. (1992) Rent generation during the transition to a managed fishery: the case of the New Zealand ITQ system. *Mar. Resource Econ.* **7**, 229–48.

Mace, P.M. (1985) Catch rates and total removals in the 4WX herring purse seine fisheries. Canadian Atlantic Fisheries Scientific Advisory Committee, Research Document 85/74. 31 pp.

Macgillivray, P.B. (1990) Assessment of New Zealand's individual transferable quota fisheries management. Ottawa: Canadian Dept Fisheries and Oceans, Economic and Commercial Analysis, Report No. 75. 19 pp.

Munro, G.R. and Scott, A.D. (1985) The economics of fisheries management. In Kneese, A.V. and Sweeny, J.L. eds. *Handbook of Natural Resource and Energy Economics*, Vol. 2. Amsterdam: North Holland, pp. 623–76.

Muse, B. and Schelle, K. (1988) New Zealand's ITQ program. Report CFEC 88-3, Alaska Commercial Fisheries Entry Commission, Juneau. 46 pp.

Muse, B. and Schelle, K. (1989) Individual fisherman's quotas: a preliminary review of some recent programs. Report CFEC 89-1, Alaska Commercial Fisheries Entry Commission, Juneau. 122 pp.

Pascoe, S. (1993) Individual transferable quotas in the Australian South East trawl fishery. *Mar. Resource Econ.* **8**, 395–401.

Scott, A.T. (1955) The fishery: the objectives of sole ownership. *J. Political Econ.* **63**, 116–24.

Scott, A.T. (1989) Conceptual origins of rights based fishing. In Neher, P.A., Arnason, R. and Mollett, N. eds. *Rights Based Fishing*. Dordrecht: Kluwer, pp. 11–38.

Scott, A.T. (1993) Obstacles to fishery self-government. *Mar. Resource Econ.* **8**, 187–200.

Sharp, D.C. and Roberts, P.R. (1991) Task force review of fisheries legislation. A submission prepared for the New Zealand Fishing Industry Board and presented to the New Zealand Government Task Force of Fisheries Legislation.

Smith, V.L. (1969) On models of commercial fishing. *J. Political Economy* **77**, 181–98.

Squires, D., Kirkley, J. and Tisdell, C.A. (1995) Individual transferable quotas as a fisheries management tool. *Rev. Fish. Sci.* **3**, 141–69.

Stephenson, R.L., Lane, D.E., Aldous, D.G. and Nowak, R. (1993) Management of the 4WX herring fishery: an evaluation of recent events. *Can. J. Fish. Aquatic Sci.* **50**, 2742–56.

Warming, J. (1911) Om grundrente af fiskegrunde. *Nationaløkonomisk Tidsskrift* **49**, 499–505.

Wilen, J.E. (1989) Rent generation in limited entry fisheries. In Neher, P.A., Arnason, R. and Mollett, N., eds. *Rights Based Fishing*. Dordrecht: Kluwer, pp. 249–62.

Accepted 10 July 1995

[17]

Minimum information management in fisheries

RAGNAR ARNASON   University of Iceland

Abstract. This paper deals with problems of optimal management of common-property fisheries. It advances the proposition that many fisheries management schemes, which are theoretically capable of generating efficiency, are actually not practicable, owing to their huge informational requirements. This applies, for instance, to management by means of corrective taxes/subsidies. The paper proceeds to show that there exists, under fairly unrestrictive conditions, market-based management systems that require minimal information for their operation but lead nevertheless to efficiency in common-property fisheries. One such system is the Individual Transferable Share Quota system (ITSQ).

La gestion des pêches avec un minimum d'information. Ce mémoire traite des problèmes de gestion optimale des pêches qui sont en propriété commune. On suggère que nombre d'arrangements pour gérer les pêches, qui sont théoriquement capables d'assurer l'efficacité, ne sont pas susceptibles d'être mis en pratique parce que les besoins informationnels sont énormes. C'est le cas des méthodes de gestion utilisant un ensemble de taxes et subventions. On montre qu'il existe des systèmes de gestion fondés sur le marché qui, imposant des conditions vraiment peu restrictives et requérant une information minimale pour fins d'opérations, engendrent une utilisation efficace des pêches en propriété commune. Un tel systéme est ITSQ – un système de quotas individuels transférables.

I. INTRODUCTION

Since the work of Gordon (1954) and Scott (1955) it has been widely recognized that common property fisheries generally operate in a socially suboptimal manner. As suggested by Turvey (1964) this state of affairs can be usefully regarded as stemming from externalities in the harvesting process. Any commercial fishery

A previous version on this paper was presented at the workshop on the Scientific Foundations for Rights Based Fishing in Reykjvaik 1988. I am grateful to the participants in this workshop, especially G.R. Munro and J.E. Wilen, for constructive comments on the paper. I would like to thank the two anonymous referees for their helpful comments.

Canadian Journal of Economics   Revue canadienne d'Economique, XXIII. No. 3
August   août 1990. Printed in Canada   Imprimé au Canada

may exhibit several types of externalities.[1] The fundamental externality of common-property fisheries, however, derives from the resource base itself. The resource stock is a factor in each firm's production function. Thus, by their harvesting activity the firms impose a production diseconomy on each other. The result is a tendency towards excessive fishing effort and overexploitation of the resource.

The fundamental externality problem in common-property fisheries may be treated in various ways. By imposing the appropriate taxes on the production of resource stock externalities, fishing firms can, in principle, be induced to operate optimally. In fact, this was one of the earliest suggestions for management of competitive fisheries (see Smith 1968, 1969). Other management systems that have been proposed include, inter alia, entry limitations, effort restrictions, and individual catch quotas (see, e.g., Lawson 1984; Clark 1985). Given certain conditions, all these management measures can be shown to be capable of restoring economic efficiency in common-property fisheries.

The problem of fisheries management, however, is not merely to devise management systems that are theoretically capable of bringing about efficient use of the resource. A fisheries management system has to satisfy a number of social and economic requirements. Among other things, it must be cost effective. An otherwise efficient system may be unacceptable due to its operating cost.[2] A related problem has to do with the data requirements of the management system and the ability of the resource manager to obtain the information necessary to determine the optimal management.

This paper argues that in most ocean fisheries the data requirements for the calculation of optimal tax rates, catch quotas, etc. greatly exceed the capacity of any resource manager. It follows that management systems based on such approaches are of little practical use. On the other hand, there appear to exist institutional arrangements that allow the resource manager to take advantage of the market mechanism in order to solve the management problem. This paper addresses this issue. It attempts to specify institutional arrangements and management procedures that permit optimal management of fisheries with minimal use of extraneous information. This is referred to in the paper as minimum information management schemes (MIMS). It should be mentioned that although the minimum information management schemes discussed in this paper are formulated in terms of fisheries, many of the principles derived seem applicable to a wide range of economic management problems.[3]

The paper is organized broadly as follows. The second section sets out the basic fisheries model employed in the paper and reviews the fundamental efficiency problems encountered in common property fisheries. Fisheries management by means

1 Thus, Smith (1969) distinguishes between stock, crowding, and mesh externalities. Another important externality in fisheries relates to the search for fish concentrations and the dissemination of information thereon.

2 The issue of management costs has been addressed by Andersen and Sutinen (1985).

3 The exploitation of other common-property resources such as timber and water resources provide obvious examples. Other possibilities concern the optimal money supply, the supply of community building lots, etc.

632   Ragnar Arnason

of taxes and individual transferable quotas are considered in section III. Section
IV presents a variant of the individual transferable quota system that allows the
fisheries manager, under certain conditions, to identify the optimal management of
the resource with very little information. Finally, the main conclusions of the paper
are summarized in section V.

II. THE BASIC FISHERIES MODEL

Consider a fishery in which a number of fishing firms exploit a single stock of fish.
Let the fishing industry consist of N fishing firms, where $N > 0$. At a given point
of time some of these N firms may not be operating in the industry. Thus N refers
to potentially active fishing firms.

The harvesting functions of the fishing firms are

$$Y(e(i;t),\ x(t);\ i),\ i = 1,\ 2,\ ..N,\ \text{for } e(i;t),\ x(t) \geq 0,$$

where $e(i;t)$ refers to the fishing effort of firm i at time t and $x(t)$ represents the
biomass of the fish stock at time t. To simplify the notation, redundant functional
arguments will frequently be suppressed below. The functions $Y(.,.;.)$ are taken
to be twice continuously differentiable, that is, S^2, increasing and jointly concave
in e and x. Moreover,

$$Y(0,.;.) = Y(.,0;.) = Y_e(.,0;.) = Y_x(0,.;.) = 0.$$

The harvesting cost functions are

$$C(e(i);i),\ i = 1,\ 2,\ ..N,\ \text{for } e(i) \geq 0,$$

where the functions $C(.;.)$ are assumed to be S^2, increasing and convex in $e(i)$.
While inoperative fishing firms do not incur costs, fishing firms currently operating
in the industry are assumed to experience costs even if they do not exert any fishing
effort. Thus, $C(0;i) \geq 0$ with the strict inequality applying to operative firms.

Growth of the fish stock is defined by the differential equation

$$x' \equiv \partial x(t)/\partial t = G(x) - \sum_i Y(e(i),x;i),\ \text{all } x \geq 0. \tag{1}$$

The natural growth function, $G(x)$, is assumed to be S^2 and exhibits the following
properties:

$$G(x_1) = G(x_2) = 0,\ \text{where } x_2 > x_1 \geq 0,\ \text{and } G''(x) \equiv \partial^2 G(x)/\partial x^2 < 0.$$

The function $G(x)$ is, in other words, unimodal and concave, and there exists a
biomass level for which growth is positive.

Finally, let p and r refer to the market price of catch and the discount rate, respectively. It is assumed that p is finite and p, $r > 0$. In what follows it will be taken for granted that these prices coincide with social shadow prices.

Given these specifications, t⁺e instantaneous profit function for a representative fishing firm i may be written as

$$\pi(e(i), x, p; i) = p \cdot Y(e(i), x; i) - C(e(i); i), \tag{2}$$

where $\pi(.,.,.;.)$ is S^2 and concave in $e(i)$ and x. The present value of firm i's future profits from the fishery is defined by

$$\mathrm{PV}(\{e(i)\}, \{x\}, p, r; i) = \int_0^\infty \pi(e(i), x, p) \cdot \exp(-r \cdot t) dt, \tag{3}$$

where the braces, $\{\cdot\}$, indicate that the time path of the respective variable is involved. Provided that the economic prices involved in the profit functions, (2) and (3), accurately reflect the respective social values, these functions may be taken as measures of social benefits. Notice, however, that they do not necessarily measure resource rents since some of the profits may be intramarginal ones. (For a discussion of resource rents in fisheries see Copes 1972.)

The efficiency properties of this kind of fisheries model have been extensively investigated in recent years and are now well established (an excellent reference is Clark and Munro 1982). For the purposes of this paper, however, it is helpful to review the essentials of this theory briefly.

The social problem is to find a time path of fishing effort for the fishing firms that maximizes the present value of industry profits subject to the biological and technical constraints of the problem. More formally:

$$\text{Maximize}_{\text{all } \{e(i)\}} \quad \sum_i \mathrm{PV}(\{e(i)\}, \{x\}, p, r; i) \tag{1}$$

Subject to   (a) $x' = G(x) - \sum_i Y(e(i), x)$,

(b) x, $e(i) \geq 0$, all i.

The necessary conditions for a solution to problem (1) include the following (for details see appendix A):

$$(p - \mu) \cdot Y_{e(i)} - C_{e(i)} = 0, \text{ for all } t \text{ and } i \text{ for which } e(i) > 0, \tag{4}$$

where μ represents the current shadow value of an additional unit of biomass along the optimal path. Conditions (4) thus state that to maximize present value of profits each firm's marginal benefits of effort, evaluated at market prices less the shadow value of biomass, should equal its marginal costs of effort.

634   Ragnar Arnason

The movement of μ along the optimal path is given by the differential equation

$$\mu' = \mu \cdot \left(\sum Y_x + r - G_x \right) - p \cdot \sum Y_x. \tag{5}$$

In bionomic equilibrium $x'(t) = e'(i; t) = 0$, all i. Hence, the equilibrium, μ is given by the equation

$$\mu = p \cdot \sum Y_x(e^*, x) / \left(\sum Y_x(e^*, x) + r - G_x(x) \right), \tag{6}$$

where e^* represents the optimal equilibrium effort level of firm i.

So, in equilibrium, the shadow value of biomass, μ, depends directly on the harvesting functions of all active firms, the biomass growth function and the economic prices, p and r. Moreover, since each firm's optimal fishing effort level, e^*, depends on its cost function, so does μ.

Consider now the behaviour of the fishing firms. We take it that each firm seeks to maximize is own profits. In the fisheries economics literature there is some ambiguity concerning the firms' perception of the biomass growth constraint.[4] The most reasonable assumption, however, appears to be that of rationality. This means that the firms take the appropriate notice of all variables and relationships affecting their profit functions including the resource growth constraint and each other's fishing effort. Thus, each firm will attempt to maximize its profits given the fishing effort exerted by other firms. Since, in practice, the fishing effort of other firms cannot be instantaneously observed, the fishing firms must form predictions or expectations concerning this variable. A certain equilibrium, usually referred to as Nash-Cournot equilibrium (see Cournot 1897; Nash 1950), is reached when the firms correctly predict each others fishing effort. Notice, however, that outside biological equilibrium a given Nash-Cournot equilibrium is only momentary, since changes in biomass require adjustments in individual fishing effort.

On these assumptions, the ith firms attempts to solve the following problem:

Maximize $\text{PV}(\{e(i)\}, \{x\}; p, r; i)$ \hfill (II)
$\{e(i)\}$

Subject to   (a) $x' = G(x) - \displaystyle\sum_i Y(e(i), x)$,

(b) $x, e(i) \geqq 0$,

(c) $e(j), j \neq i$ given.

Solving this problem for all the firms yields the following necessary conditions (for details see appendix A):

$$(p - \sigma(i)) \cdot Y_{e(i)} - C_{e(i)} = 0, \text{ for all } t \text{ and } i \text{ for which } e(i) > 0, \tag{7}$$

---

4 For instance Clark (1976, 1985) assumes that competitive firms ignore the biomass growth constraint entirely. For the contrary view see Dasgupta and Heal (1979).

where $\sigma(i)$ is firm i's evaluation of the current shadow value of an additional unit of biomass.

The structure of conditions (7) and the socially optimal ones, (4), above, are identical. The only difference is that private firms modify the market catch price by $\sigma(i)$ instead of the social shadow value, μ. The key question therefore is how the $\sigma(i)$s compare with μ.

The solution to the private profit maximization problem, (II), implies the following movement of the $\sigma(i)$s over time:

$$\sigma'(i) = \sigma(i) \cdot \left(\sum Y_x + r - G_x\right) - p \cdot Y_x, \text{ all } i. \tag{8}$$

Thus, in bionomic equilibrium,[5] $\sigma(i)$, is given by the equations:

$$\sigma(i) = p \cdot Y_x(e(i), x; i) / \left(\sum Y_x + r - G_x\right), \text{ all } i. \tag{9}$$

Therefore, comparing equations (9) and (6) for the same x and $e(i)$s, it is clear that in equilibrium $\mu \geq \sigma(i)$, all i. This means that the social shadow value of biomass is at least as great as the private one. In fact, equality between the social and private shadow values of biomass is attained only when there is a single firm operating in the industry. It follows from conditions (4) and (7) that for a given equilibrium biomass, x, the competitive fishing effort, e, will exceed the optimal one if the number of active fishing firms exceeds one.

This argument is sufficient to establish the fundamental proposition of fishery economics, namely, that competitive utilization of a common fish stock generally yields suboptimal economic results.

The above results, incidentally, also show that the common assertion that competitive fishing firms equate marginal income with marginal costs (see, e.g., Clark 1976, 1985) is not generally valid. Provided the firms are rational, in the sense defined above, their private evaluation of the shadow value of the resource will be positive and marginal income will consequently exceed marginal costs. With rational firms the customary assertion applies only asymptotically, that is, when the number of active firms approaches infinity. However, if there are fixed harvesting costs, that is, $C(0) > 0$, an infinite number of firms is incompatible with profit maximization.[6]

The relationship between the private and social shadow value of biomass becomes particularly simple if the fishing firms are identical. Clearly, in that case $\sigma(i) = \sigma(j) = \sigma$, all i and j, and

$$\mu = N_1 \cdot \sigma, \tag{10}$$

where N_1 denotes the number of active firms in the industry. Thus, in this particular case, we see that private evaluation of the shadow value of biomass decreases monotonously with the number of firms active in the industry.

---

5 Notice that in bionomic equilibrium firms' expectations of each other's fishing effort must be correct. Bionomic equilibrium therefore implies a Nash-Cournot equilibrium.
6 $C(0) > 0$ implies that firms have to reach a finite size to break even.

636   Ragnar Arnason

III. FISHERIES MANAGEMENT

Given the inefficiency of competitive fisheries demonstrated above, it is obviously desirable to devise a regulatory regime that is capable of realizing as much of the attainable economic benefits as possible. Over the years, many management systems have been suggested for this purpose. In this section we briefly consider two of the more respectable of these systems, a tax on catch and individual catch quotas.

1. Taxes on catch
The inefficiency of competitive exploitation of a common fish stocks is due to external diseconomies in production. By reducing the fish stock, each firm's harvesting activity adversely affects the harvesting possibilities of other firms in the fishery. Since the work of Pigou (1912), it has been recognized that many externalities can, at least in principle, be remedied within the market system by imposing corrective taxes or subsidies. In the case of fisheries, the appropriate tax turns out to be analytically elegant but, unfortunately, extremely difficult to apply.

Comparing the social and private conditions for profit maximization, equations (4) and (7), respectively, we see that firm's i imputed net output price is $p - \sigma(i)$ instead of the socially appropriate one, $p - \mu$. It follows that the appropriate corrective output tax for firm i is

$$\tau(i) = \mu - \sigma(i). \tag{11}$$

Equation (11) gives the corrective tax on catch at each point of time. The development of $\tau(i)$ over time is defined by the differential equation

$$\tau'(i) = \mu' - \sigma'(i), \tag{12}$$

where μ' and $\sigma'(i)$ are given in equations (5) and (8) above.

There are two important things to notice about the optimal tax. First, the optimal tax is in general not uniform over firms. Only if the firms are identical will there be a single optimal tax. Otherwise, rational firms will have different evaluations of the shadow value of biomass[7] and this must be reflected in the corrective tax.[8] This result, clearly, has somewhat disturbing socio-political implications.[9]

Second, the informational requirements for determining the optimal tax are immense. To calculate μ and $\sigma(i)$ for all i, the tax authority must solve the social

7 Notice that identical technology does not imply identical firms, since they may be of different sizes. Given the same technology, the bigger firms will generally have a higher evaluation of the shadow value of biomass.
8 Assertions of an identical optimal tax for non-identical firms (see Clark 1985) seem to be based on the tacit assumption of non-rational firms, i.e., firms that do not attach any shadow value to biomass left in the sea.
9 Imposing different output tax rates on firms, not to mention higher rates on the smaller firms as would normally be required, would tend to contradict widely held notions about fairness in taxation.

optimality problem as well as each firm's profit maximization problem. To be able
to do so, the tax authority must have at its command all the data relevant to the
fishing firms. In particular, the tax authority must have full knowledge of the resource growth function, and the harvesting and cost functions of all the firms at
all points of time. Moreover, the tax authority must continuously monitor the state
of the resource and the movement of the relevant economic prices for the optimal
tax must be continuously adjusted to new conditions. Clearly, these tasks would
exceed the capabilities of most tax authorities.

2. An individual transferable quota system

Let us now consider a fishery regulated by means of individual catch quotas. Many
variants of this system are conceivable. Here we restrict out attention to the following.

The catch quotas stipulate the maximum rate of catch permitted to each fishing
firm at a point of time. This is quite restrictive. More generally, a catch quota
limits the catch volume over a period of time which may be of any length. A
quota system constraining the rate of catch implies infinitesimal quota periods and
may be referred to as the *continuous quota system*. Alternatively, a system of catch
quotas with finite quota periods may be referred to as the *discrete quota system*. It
is important to notice that fisheries management on the basis of the discrete quota
system is not generally capable of generating full efficiency.[10]

A central authority, which we may refer to as the quota authority, issues the
catch quotas. The quotas are issued continuously at each point of time. The sum
of the catch quotas constitutes the total quota, Q.

The catch quotas are transferable without any constraints and are perfectly divisible. The quotas thus constitute a homogeneous tradeable commodity. We assume
that there is a market for this commodity and, moreover, that this market is open
to every one interested in trading. Also, to bypass the tedious problems of disequilibrium trades, we assume that all trading takes place at equilibrium prices.
The equilibrium quota price is denoted by s.

The quota authority may allocate quotas to firms free of charge or through the
quota market. Let $q_0(i, t) \geq 0$ represents free allocation of catch quotas from the
quota authority to firm i at time t. The quota authority sells the remainder of the
total quota in the quota market. Let $z(i, t)$ represent firm's i instantaneous quota
purchases at time t. Then the total quota constraint may be written as

$$Q(t) = \sum_i (q_0(i, t) + z(i, t)), \text{ all } t. \tag{13}$$

The individual quota constraint, on the other hand, is

$$Y(e(i), x) \leq q_0(i, t) + z(i, t), \text{ all } i \text{ and } t. \tag{14}$$

10 If the quota period is finite, different exploitation paths will satisfy the quota constraint. Generally, however, not all these paths are optimal.

638   Ragnar Arnason

Let us now consider the behaviour of individual fishing firms within this institutional framework. Their profit maximization problem may be written as

$$\underset{\{e\},\{z\}}{\text{Maximize}} \quad \int_0^{\infty} (p \cdot Y(e, x) - C(e) - s \cdot z) \cdot \exp(-r \cdot r)dt \qquad \text{(III)}$$

Subject to   (a) $q_0 + z \geq Y(e, x)$,

$$(b) \quad x' = G(X) - \sum_i Y(e, x),$$

(c) $e \geq 0$.

Now, it is easy to check (see appendix B) that the solution to problem (III) includes the conditions

$$s > 0 \Rightarrow q_0 + z = Y(e, x), \text{ all } i, \qquad (15)$$

$$(p - s) \cdot Y_{e(i)} - C_{e(i)} = 0, \text{ for all } i \text{ for which } e(i) > 0, \qquad (16)$$

where s, it will be recalled, is the market price of quotas.

The message of (15) is that, provided that the market price of quotas is positive, firms will not leave any quotas unused. It follows that total catch will equal total quotas and $x' = G(x) - Q$. Thus, the quota system under discussion effectively separates individual fishing decisions from the development of the fish stocks. It follows that the basic stock externality imposed by fishing firms on each other in competitive fisheries is eliminated.

Comparing conditions (16) with the socially optimal ones given by equations (4) of the previous section, it is apparent that private harvesting will be optimal if $s = \mu$, that is, if the market price for quotas equals the optimal shadow value of the resource. Now, the market price for quotas will depend, among other things, on the total supply of quotas, that is, Q. To see this, notice that equation (16) defines the following set of instantaneous quota demand functions for active firms:

$$s = p - C_e(e(i))/Y_e(e(i), x), \text{ all active } i. \qquad (17)$$

Equilibrium in the quota market requires

$$Q = \sum_i Y(e, x). \qquad (18)$$

Finally, solving equations (17) and (18) yields the instantaneous quota price

$$s = S(p, Q). \qquad (19)$$

Therefore, by supplying the appropriate total quota, the quota authority can control the quota price and thus ensure optimal utilization of the fish resource.

This argument establishes the important result that it is possible, at least within the framework of this particular quota system, to generate full economic efficiency in the fishery by judicious choice of total quotas. However, just as in the taxation case discussed above, the volume of information needed to accomplish this is daunting. First, since the market price of quotas, s, must be set equal to the optimal shadow value of the resource, μ, the latter must be calculated. This involves solving the social optimality problem (1). Second, in order to select the appropriate total quota, Q, the market price function (19) must be obtained. This involves solving the private profit maximization problems. Both of these tasks require exhaustive and continuously updated knowledge of the biomass growth function, the cost and harvesting functions of all the firms, and the output price, just as in the taxation case discussed above. Compared with the output tax, however, management via catch quotas has one slight advantage. It does not require the calculation of individual firms' shadow value of biomass, that is, $\sigma(i)$. The reason is that, since this quota system eliminates the resource stock externality, $\sigma(i)$ does not influence the behaviour of the fishing firms.

IV. MINIMUM INFORMATION MANAGEMENT IN FISHERIES

The results of the previous section demonstrate the possibility of managing a fish resource optimally with the help of catch quotas. To attain that objective, the quota authority simply has to pick the appropriate time path of the total quota. The snag is that to do this the quota authority must have at its command an immense amount of information about the economics of the fishery. In fact, just as in the taxation case, the quota authority has to know in detail the economic conditions of all the fishing firms.

The unilateral selection of total quotas by the quota authority, does not, on the other hand, exploit the available information efficiently. It may be taken for granted that all information the quota authority can possibly obtain in order to determine the optimal total quota is already available within the fishing industry. After all, the fishing firms have at least as much knowledge about their own cost and harvesting functions as the most determined effort of the quota authority could possibly secure. Moreover, since the state of the fish stocks is a major determinant of their profit function, the fishing firms can be relied on to make efficient use of the available biological data. In fact, given a reasonably competitive environment, only those firms that efficiently collect and interpret all the relevant information will survive. It follows that most of the work necessary for the quota authority to determine optimal total quotas will merely constitute a duplication of work already carried out by private agents in the fishery.

The question thus naturally arises whether there exists a way for the quota authority to harness this market information in order to determine the optimal quota. In this section we shall explore this question.

640   Ragnar Arnason

1. An individual transferable share quota system

Consider a continuous quota system where the quotas are permanent shares in the total allowable rate of catch. In other respects the quota system is as discussed in subsection III.2 above. More precisely, the essentials of this quota system are as follows:

1. The individual catch quotas are shares in the total allowable rate of catch. These quotas are referred to as share quotas.
2. The share quotas impose an upper limit on the firm's permitted rate of catch.
3. The share quotas are permanent in the sense that they allow the holder the stated share in the total quota in perpetuity.
4. The share quotas are transferable and perfectly divisible.
5. There exists a market for share quotas. This market is perfect in the sense that it is open to everyone interested in trading, all the traders are price takers, and the market equilibrates supply and demand instantaneously.
6. The quota authority issues the initial shares and subsequently decides on the total quota at each point of time.

We refer to this system as the individual transferable share quota system or, in short, ITSQ.

The significance of a share quota system compared with a quantity quota system is primarily in terms of the impact of total quota variations on the economics of the firms. Under a share quota system, changes in total quotas are automatically reflected in uncompensated quota increases or decreases for individual firms. Under a quantity quota system, on the other hand, total quota adjustments may be affected by trades in the quota market.[11] Hence, in this system, individual firms are compensated for variations in total quotas, which has important implications for management, as will become clear in subsection IV.3. Otherwise, the practical difference between holding share quotas and quantity quotas, from the point of view of individual firms, is rather trivial. Individual share quotas, for instance, may still be denominated and traded in volume terms.

Within the institutional framework of the ITSQ system, individual quota holdings at time t are given by

$$q(i, t) = \alpha(i, t) \cdot Q(t), \text{ all } i \text{ and } t, \ 1 \geq \alpha(i, t) \geq 0, \tag{20}$$

where $q(i, t)$ stands for the volume of quotas and $\alpha(i \ t)$ the share in total quotas or share quotas held by firm i at time t. As before, $Q(t)$ represents total quotas.

The share quotas held by firm i at time t are given by the equation

$$\alpha(i, t) = \alpha(i, 0) + \int_0^t z(i, \tau) \cdot d\tau, \tag{21}$$

---

11 Uncompensated alterations of quantity quotas by the quota authority are of course also conceivable. If such adjustments are proportional to the quota held, however, the quantity quota system essentially amounts to a share quota system.

where $\alpha(i, 0)$ represents the firm's share quotas at some arbitrary initial point of time and $z(i, t)$ its purchases of share quotas at time t. Notice that a negative $z(i, t)$ is interpreted as a sale of share quotas by firm i at time t.

As in subsection III.2, the instantaneous profit function of firm i at time t is

$$\pi(i) = p \cdot Y(e(i), x) - C(e(i)) - s \cdot z(i), \tag{22}$$

where, it may be recalled, p represents the catch price, $e(i)$ the fishing effort exerted by firm i, and x the fish stock biomass. The market price for a unit of share quota is s.

To simplify the argument let us make the assumption that firms do not hold unused quotas.[12] In other words:

$$Y(e(i), x) = q(i), \text{ for all } i.$$

Therefore, given the properties of the harvesting function, fishing effort must satisfy

$$e(i) = E(q(i), x), \text{ all } i \text{ and } x > 0.$$

And the profit function can be written in a more convenient form as

$$\pi(i) = p \cdot q(i) - C(E(q(i), x)) - s \cdot z. \tag{23}$$

Now, within this particular quota system, the social problem is to pick total quotas and allocate individual quotas to firms so as to maximize economic benefits from the fishery. More precisely:

$$\text{Maximize}_{\text{all}\{\alpha(i)\},\{Q\}} \quad J = \sum_i \int_0^\infty (p \cdot \alpha(i) \cdot Q - C(E(\alpha(i) \cdot Q, x)) \cdot \exp(-r \cdot t) dt \tag{IV}$$

Subject to (a) $x' = G(x) - Q,$

(b) $\sum_i \alpha(i) = 1,$

(c) $\alpha(i) \geqq 0, \text{ all } i,$

(d) $Q \geqq 0.$

The corresponding current-value Hamiltonian function may be written as

$$H = \sum_i (p \cdot \alpha(i) \cdot Q - C(E(\alpha(i) \cdot Q, x)) + \mu \cdot (G(x) - Q), \tag{24}$$

where μ is the costate variable for the resource growth constraint, (a).

The solution to problem (IV), if it exists, must satisfy the following conditions:

$$(p - C_E \cdot E_{q(i)}) = \mu, \text{ for all active firms,}$$

$$(p - C_E \cdot E_{q(i)}(0, x)) < \mu \Rightarrow \alpha(i) = q(i) = 0, \tag{24.1}$$

12 The validity of this common assumption (see, e.g., Clark 1985) is examined in appendix c.

642   Ragnar Arnason

$$\mu' - r \cdot \mu = \sum_i C_E \cdot E_x - \mu \cdot G_x. \tag{24.2}$$

Now, μ is the current shadow value of the resource along the optimal path. In other words, μ measures the marginal contribution of additional biomass to the maximal level of the objective functional, J. The left-hand side (LHS) of (24.1) is the current instantaneous contribution of an additional unit of share quota to the operating profits of firm i. Thus, the message of (24.1) is that firm i should receive additional share quotas while the marginal profits created by these share quotas exceed the shadow value of the corresponding resource units. Those firms for which μ exceeds marginal profits of quotas at all quota levels are not allocated any share quotas.

These results basically reiterate the general optimality conditions for common property fisheries which were derived in section II without any reference to the institutional organization of the fisheries. This is as expected; for the maximum economic benefits attainable from the fishery should not depend on institutional arrangements.

Within the framework of the ITSQ system, firm i's profit maximization problem is

$$\text{Maximize} \quad \int_0^\infty (p \cdot \alpha \cdot Q - C(E(\alpha \cdot Q, x) - s \cdot z) \cdot \exp(-r \cdot t)dt, \tag{v}$$
$$\{z\}$$

Subject to   (a) $\alpha' \equiv \partial\alpha/\partial t = z$

(b) $1 \geq \alpha \geq 0$.

It is worth noticing that the control variable, z, appears linearly in this problem. Consequently, the optimal control will be of a bang-bang character. Since z is unbounded, this means that the actual share quotas of the firms will be instantaneously adjusted to desired levels. Thus, quota holdings will at all times be at their optimal long-run level, given the variables that are exogenous to the firms, that is, Q, x, s, p, and r. When these variables change, however, quota holdings must be adjusted.

The current-value Hamiltonian for this problem may be written as

$$H = p \cdot \alpha \cdot Q - C(E(\alpha \cdot Q, x) - s \cdot z + \sigma \cdot z, \tag{25}$$

where σ is the shadow value of an additional unit of quota to firm i.

The necessary conditions for a solution to (v) include

$$s = \sigma, \text{ for active firms,}$$
$$\tag{25.1}$$
$$s \geq \sigma, \text{ for inactive firms,}$$

$$\sigma' - r \cdot \sigma = -(p - C_E \cdot E_q) \cdot Q. \tag{25.2}$$

According to (25.1), firms should purchase additional units of share quotas in the market as long as their shadow values exceeds their market price and vice versa.

Profit maximization may also require some firms to sell all their share quotas and come to rest with no quota holdings at a point where $s > \sigma$. Equation (25.2) gives the rule of motion for the shadow value of quotas.

Combining (25.1) and (25.2) yields the time path of quota prices.

$$r \cdot s - s' = (p - C_E \cdot E_q) \cdot Q. \tag{26}$$

The LHS of this equation may be interpreted as the cost of holding a unit of share quota. The term $r \cdot s$ represents the opportunity cost of holding a unit of share quota compared with investing its market value. The term s' measures the capital gain/loss of holding a share quota due to its instantaneous price changes. The sum of these two terms is the total cost of holding a unit of share quota. The RHS of equation (26), on the other hand, is the marginal profits of share quota holdings. It represents the economic benefits to the firm of utilizing an additional unit of share quota for fishing.

Equation (26) may be regarded as the fundamental dynamic demand function for share quotas. Provided that the fishery is pursued at all, only the price path defined by (26) is compatible with private profit maximization and, consequently, equilibrium in the quota market. Equation (26) is reminiscent the famous arbitrage rule for asset holdings due to Hotelling (1931). According to the Hotelling rule, the rate of asset price increase must equal the market rate of interest. In this case, however, the assets are potentially productive. Hence the rule is modified by the marginal profits of quota holdings, i.e. by the term $(p - C_E \cdot E_q) \cdot Q$.

Now, let $\alpha^*(i, t)$ denote the share quota holdings by firm i at time t that solve the private profit maximization problem, that is (v). Similarly, let $\alpha^{**}(i, t)$ be the share quota holdings by firm i at time t that solve the social problem, (IV). The following lemma is now available:

LEMMA 1. *For a given initial biomass and time path of total quotas, $x(0)$ and $\{Q(t)\}$, respectively, $\alpha^*(i, t) = \alpha^{**}(i, t)$, all i and t.*

Proof. According to equations (24.1) for the optimal program:

$$C_E \cdot E_{q(i)}(\alpha^{**}(i) \cdot Q, x) = C_E \cdot E_{q(j)}(\alpha^{**}(j) \cdot Q, x), \text{ for all active } i \text{ and } j \text{ and all } t.$$

According to equations (25.1) and (25.2) for the private profit maximization program:

$$C_E \cdot E_{q(i)}(\alpha^*(i) \cdot Q, x) = C_E \cdot E_{q(i)}(\alpha^*(j) \cdot Q, x), \text{ for all active } i \text{ and } j \text{ and all } t.$$

Adding the total quota constraint, $\Sigma_i \alpha(i) = 1$, which must hold in both cases, these conditions yield identical solutions for $\alpha^{**}(i, t)$ and $\alpha^*(i, t)$ as functions of $x(t)$ and $Q(t)$.

Now, biomass at time t is determined by the initial biomass level, $x(0)$ say, and the biomass growth constraint, $x' = G(x) - Q$. In other words, $x(t) = X(x(0), \{Q\})$. QED

644   Ragnar Arnason

Lemma 1 states the important result that, within the framework of the share quota system defined above, the total quota will always be caught in the most efficient manner. It follows that it takes a particularly inept quota authority to escape the generation of some economic rents under this management system.

According to lemma 1, the quota authority can ensure optimal utilization of the fish stock by selecting the appropriate time path of total quotas. Writing the optimal share quotas as

$$\alpha^*(i) = \alpha^{**}(i) = \Gamma(x(0), \{Q\}, i),$$

a formal representation of the problem facing the quota authority is

$$\underset{\{Q\}}{\text{Max}} \sum_i \int_0^\infty [p \cdot \Gamma(x(0), \{Q\}) \cdot Q - C(E(\Gamma(x(0), \{Q\})$$
$$\times Q, X(x(0), \{Q\}))] \cdot \exp(-r \cdot t) dt. \quad \text{(VI)}$$

While this problem has considerably fewer control variables than the general problem, (IV), above, it may not be much easier to solve. Since the solution requires, for instance, full knowledge of the $\Gamma(.,.,.)$ functions, the quota authority has to know each firm's harvesting and cost functions in detail and solve its individual profit maximization problem. However, such a procedure runs afoul of the information problem discussed in subsection III.2.

2. Minimum information management scheme
Above, it has been pointed out that the huge amount of information required to determine the optimal level of total quotas by solving the social optimality problem, renders that approach to the fisheries management problem impracticable in most cases. In this section an alternative approach is proposed that efficiently exploits market information, enabling the quota authority to identify the optimal total quota path with minimal information. The fundamental idea is that within the framework of the permanent share quota system defined above, the prevailing quota market price reflects all relevant information about the current and future conditions in the fishery available to the fishing firms or, more generally, the participants in the quota market. It follows that the quota authority only has to monitor the quota market price to become privy to the same information. We now proceed to clarify this idea.

The dynamic demand function for share quotas, equation (26), above, holds for all active firms at all times. Multiplying this equation for each firm by its private profit maximizing quota level, $q^*(i)$, we find

$$(s' - r \cdot s) \cdot q^*(i) = -(p - C_E \cdot E_q) \cdot q^*(i) \cdot Q, \text{ all } i \text{ and } t.$$

Since $\Sigma_i q^*(i) = Q$, summing over all active firms yields

$$(s' - r \cdot s) = -\sum_i (p - C_E \cdot E_q) \cdot q^*(i), \text{ all } t.$$

And, solving this differential equation for an arbitrary initial time $t = 0$, we obtain (for details see appendix D):

$$s(0) = \sum_i \int_0^\infty [(p - C_E(E_q(q^*(i), \; X(x(0), \; \{Q\})) \cdot q^*(i)] \cdot \exp(-r \cdot t)dt. \quad (27)$$

Now, $s(0)$ is the market price of share quotas at time 0. The term $(p - C_E \cdot E_q)$ on the RHS of (27) is the marginal profits of catch quotas. If all quota units held are as profitable as the marginal one, the expression $(p - C_E \cdot E_q) \cdot q^*(i)$ measures the profits made by firm i. This expression, in other words, represents the resource rents[13] obtained by firm i at a point of time. More precisely, as the future is unknown, $(p - C_E \cdot E_q) \cdot q^*(i)$ must be expected resource rents. In fact, since (27) is based on private profit maximizing behaviour, the variables on the RHS of (27) must be the future expectations of the firms in the fishing industry. Thus, equation (27) states that the current market price of share quotas equals the present value of expected future resource rents in the fishery. In this sense the prevailing quota market price reflects the relevant information about current and future conditions in the fishery.

The share quotas, by definition, sum to unity. It follows that $s(0)$ on the LHS of (27) is also the current market value of all outstanding share quotas. Thus, the following fundamental proposition has been established:

PROPOSITION 1. *Under the individual transferable share quota system,* ITSQ, *defined above the market value of outstanding quotas equals the present value of expected future resource rents generated in the fishery.*

Now, according to lemma 1, $q^*(i) \equiv \alpha^*(i) \cdot Q$ in (27) is in fact identical to the optimal $\alpha(i) \cdot Q$, given the total quota. Hence, the RHS of (27) depends only on the path of total quotas, $\{Q\}$, the initial biomass level, $x(0)$, and the exogenous variables p and r. It follows that adjusting the total quota so as to maximize the value of outstanding quotas is equivalent to maximizing the present value of expected future resource rents in the fishery. This establishes the following corollary to proposition 1:

COROLLARY 1.1

$$\underset{\{Q\}}{\text{Max }} s(0) \Leftrightarrow \underset{\{Q\}}{\text{Max }} \sum_i \int_0^\infty (p - C_E(E_q(q^*(i), \; x)) \cdot q^*(i) \cdot \exp(-r \cdot t)dt.$$

Corollary 1.1 is formulated in terms of expected rents. Expected rents, however, are not necessarily equal to actual rents. On the other hand, future conditions in

---

13 The concept of resource rents was employed by Gordon in his seminal paper on the fisheries problem in 1954. Since then the problem of common-property fisheries has generally been seen as one of dissipation of resource rents and the objective of fisheries management the restoration of these rents. For a fuller discussion of resource rents in fisheries see Copes (1972).

646 Ragnar Arnason

a fishery are normally unknown, and consequently expectations have to be relied upon. The crucial point is that there appears very little reason to give more credence to the expectations of the quota authority than those of the firms. Rather the opposite. After all, the firms are generally more knowledgeable about their own economic conditions than the quota authority is. Moreover, accurate predictions are vital to the profitability of the firms and, for that matter, any player in the quota market. Thus, appealing to the principle of rational expectations,[14] industry expectations may well be the best available predictor of future conditions in the fishery. Formally we adopt the following assumption:

ASSUMPTION 1. *The expectations of the fishing firms are the best available predictor of future conditions in the fishery.*

Given assumption 1, corollary 1.1 suggests an especially efficient way of maximizing resource rents in the fishery. Provided the quota market is reasonably competitive, the quota authority does not need to collect any information about the fishery. It only has to monitor the prices in the quota market and adjust total quotas so that the value of outstanding quotas is maximized. This, according to corollary 1.1, will automatically maximize the resource rents generated in the fishery. Compared with solving the individual optimization problems directly, as the traditional fisheries management systems demand, this task appears relatively tractable.

The fisheries management problem, however, is not the maximization of resource rents. The management problem, more generally, is to maximize aggregate profits in the fishery. The difference between profits and resource rents is the so-called intramarginal rents that are earned by the more efficient firms. Only when the firms are equally efficient will profits coincide with resource rents. On the other hand, at least in competitive economic equilibrium, there are grounds to expect equality between resource rents and profits. If, for instance, capital markets were perfect and all firms had access to the same technology, profit maximization would require them to be equally efficient and no intramarginal rents would be earned. (This argument is developed further in appendix E.)

We proceed by assuming that resource rents and profits are equivalent. In other words:

ASSUMPTION 2. $(p - C_E \cdot E_q(q^*, x)) \cdot q^*(i) = p \cdot q^*(i) - C(E(q^*, x))$, all i and t.

According to this assumption, marginal operating profits, evaluated at the profit-maximizing level of individual quota holdings, equal average operating profits. Alternatively, we may say that marginal operating costs equal average operating costs. Notice also that assumption 2 implies that the firms operate at the point of maximum average profits given their quota holding. Alternatively, they may be said to minimize the costs of filling their quota. These attributes are often associated with the catch quota system.

14 On the principle of rational expectations see, for example, Lucas and Sargent (1981, xi–xvi).

On the basis of assumption 2, equation (27) may be rewritten as

$$s(0) = \sum_i \int_0^\infty [p \cdot q^*(i) - C(E(q^*(i), X(x(0), \{Q\})))] \cdot \exp(-r \cdot t)dt. \qquad (28)$$

Equation (28) states that the market value of all share quotas equals the present value of expected future operating profits in the fishery. This result, it may be noted, is a standard assertion in the theory of productive asset prices (see, e.g., Van Horne 1971, 15–19). Here, however, the proposition has been derived on the basis of individual profit maximizing behaviour and assumption 2.

The expression on the RHS of equation (28) is very similar to the objective function of the quota authority presented in problem (IV), above. By lemma 1, $q^*(i)$ in (28) is identical to the optimal $q(i)$, given the total quota. Thus, the only difference between (28) and the objective function in problem (IV) is that the future variables in (28) are not the expectations of the quota authority. However, by appealing to assumption 1, these expectations may be taken to constitute the best predictor of future conditions in the fishery available to the quota authority.

These arguments have established the fundamental proposition of this paper.

PROPOSITION 2. MINIMUM INFORMATION MANAGEMENT SCHEME (MIMS). *In the individual transferable share quota system,* ITSQ, *defined above, and given assumptions 1 and 2, adjusting current total quotas to maximize the market value of total outstanding quotas at each point of time is equivalent to the maximization of profits attainable from the resource. Formally:*

$$\underset{\{Q\}}{Max}\, s(0) \Leftrightarrow \underset{\{Q\}}{Max} \sum_i \int_0^\infty (p \cdot q^* - C(E(q^*, x)) \cdot exp(-r \cdot \tau)d\tau.$$

The practical implications of this proposition are obvious. To identify the optimal total quota at a point of time, the quota authority does not have to collect data about the fish stocks and the economics of the harvesting firms. It has merely to monitor the share quota price in the quota market and adjust the total quota so as to maximize the total value of the share quotas. Proposition 2, on the other hand, depends heavily on assumptions 1 and 2. It they do not hold, the proposition may not be true. Notice, however, that the essence of 1 is that the firms formulate rational expectations. If that is not the case, it is hard to imagine any market-based management that will attain efficiency in the fishery. Assumption 2, on the other hand, is basically the requirement that firms are equally efficient. Given the existence of quota markets, it is not obvious why less efficient firms would choose to continue fishing operations instead of selling off their share quotas to their more efficient competitors.

3. Alternative transferable quota systems
It will be appreciated that not all individual transferable quota systems or ITQs have the convenient management properties of the share quota system described above. In

648   Ragnar Arnason

particular, quota permanence seems to be a prerequisite for minimum information management schemes of the type described in subsection IV.2. Also, it appears that a system of quantity quotas, even if permanent, requires more information for optimal management than the share quota system.

Turning our attention first to quota permanence, it seems obvious that this will play an important role in any efficiency considerations. Clearly, if the firms hold relatively permanent quotas, they will have a greater interest in maintaining the resource base than if they held only transitory quotas. To examine this issue let us assume the share quota system of the previous session with the modification that the quotas may not be permanent. Thus, let the following equation replace equation (21) of the previous section:

$$\alpha(i, \ t) = \alpha(i, \ 0) + \int_0^t [-\beta \cdot \alpha(i, \ \tau) + z(i, \ \tau)]d\tau, \tag{29}$$

where, as before, $\alpha(i, \ t)$ represents firm i's share quota and $z(i, \ t)$ is share quota purchases at time t. β is the quota non-permanence parameter. $\beta = 0$ denotes permanent quotas as in subsection IV.1 and $\beta > 0$ quota non-permenence. In fact, $\beta \rightarrow \infty$ indicates completely transitory quotas as in subsection III.2. Thus (29) is a fairly general representation of the degree of quota permanence.

Now, repeating the exercise of subsection IV.2, it is easy to derive the following equation corresponding to equation (28):

$$s(0) = \sum_i \int_0^\infty [p \cdot q^* - C(E(q^*, \ X(x(0), \ \{Q\})))] \cdot \exp(-(r + \beta) \cdot t)dt. \tag{30}$$

Thus, it emerges that, if quotas are not absolutely permanent, that is, $\beta > 0$, the market value of share quotas underestimates the present value of expected future operating profits in the fishing industry. A positive quota non-permanence parameter leads to an excessive discounting of future profits. In fact, if $\beta \rightarrow \infty$, then $s(0) \rightarrow 0$. It follows from this that if $\beta > 0$ a management policy of maximizing the total value of share quotas will not yield optimal results. Such a policy would overemphasize present profits at the expense of future ones.[15]

Let us now turn our attention to quantity quotas. Assume the same quota system as defined in subsection IV.1 with the exception that the quotas are now quantity quotas, that is, rights to a certain rate of catch irrespective of the total quota. The instantaneous profit function of a representative firm i remains unchanged as specified in equation (22) above. The state variable, however, is now $q(i)$ instead of $\alpha(i)$, and the individual profit maximization problem is

$$\text{Maximize}_{\{z\}} \quad \int_0^\infty [p \cdot q(i) - C(E(q(i), \ x) - s \cdot z] \cdot \exp(-r \cdot t)dt,$$

---

15 It should be noted that equation (30) and the ensuing arguments also apply to the case of private discount rates differing from social ones by β.

Subject to   (a) $q'(i) = z(i)$,

         (b) $q(i) \geqq 0$.

Proceeding as in the share quota case yields the following quota price equation corresponding to equation (28) of subsection IV.2:

$$s \cdot Q(0) = \sum_i \int_0^\infty [\{p \cdot q^* - C(E(q^*, x))\}Q(0)/Q(t)] \cdot \exp(-r \cdot t)dt. \qquad (31)$$

The current market value of outstanding quotas depends, in this case, on the present value of future expected profits multiplied by the ratio of current to future total quotas. Thus, if total quotas increase over time, the market value of outstanding quotas will underestimate the present value of future expected profits in the fishery and vice versa. This is readily understandable. The quantity quota system implies that such quota increases have to be purchased by the fishing firms. Hence the social benefits of increasing Q are underestimated. Only in equilibrium, when the total quota remains constant, will the maximization of quota values be equivalent to maximizing the present value of profits in the industry.

V. CONCLUSIONS

Fisheries management systems suggested in the literature generally call upon the fisheries manager to calculate the optimal level of certain management controls. These controls may be tax rates, individual catch quotas, access licence prices, etc. In most fisheries the data required to perform these calculations are enormous and exceed, by far, any realistic assessment of the capacity of the fisheries manager. It follows that the traditional fisheries management systems are of limited practical use.

A system of individual transferable quotas appears to alleviate this problem somewhat. Provided the quota market operates reasonably smoothly, any total allowable catch will be harvested efficiently in this system. Consequently, some resource rents will normally be generated under this system.

Moreover, there appears to exist a certain variant of the individual transferable quota system, namely, the individual transferable share quota system, or, in short, ITSQ, that allows the fisheries manager, under certain conditions, to determine the optimal total quota with minimal collection of information. In this system the fisheries manager is essentially required only to monitor the quota market and to adjust the total quotas until the current total quota market value is maximized. This is referred to as minimum information management.

The class of individual transferable quota systems allowing minimum information management is not specified in the paper. However, it is demonstrated that common variants of this system may not have this property. Thus, quota permanence seems to be a prerequisite for minimum information management, and, perhaps more surprisingly, permanent quantity quotas do not appear to allow minimum information management except in equilibrium.

650   Ragnar Arnason

APPENDIX A: BASIC FISHERIES MODEL

1. The social problem
The social optimization problem is

$$\text{Maximize} \sum_i \int_0^\infty (p \cdot Y(e(i),\ x) - C(e(i))) \cdot \exp(-r \cdot t)dt$$
$$\text{all } \{e(i)\}$$

Subject to   (a) $x' = G(x) - \sum Y(e(i),\ x)$,

   (b) $x,\ e \geq 0$, all i.

The Hamiltonian function corresponding to his problem may be written

$$H = \sum [p \cdot Y(e(i),\ x) - C(e(i))] + \mu \cdot (G(x) - \sum Y(e,\ x)).$$

The necessary conditions for a solution to this problem, provided it exists, include:

$$(p - \mu) \cdot Y_{e(i)} - C_{e(i)} \leq 0,\ e(i) \geq 0,\ e(i) \cdot ((p - \mu) \cdot Y_{e(i)} - C_{e(i)}) = 0,$$

$$\mu' = \mu \cdot \left(\sum Y_x + r - G_x\right) - p \cdot \sum Y_x.$$

The constraints (a) and (b), hold.

2. The private problem
The private profit maximization problem for a representative firm i is

$$\text{Maximize} \int_0^\infty (p \cdot Y(e(i),\ x) - C(e(i))) \cdot \exp(-r \cdot t)dt,$$
$$\{e(i)\}$$

Subject to   (a) $x' = G(x) - \sum Y(e(i),\ x)$,
   (b) $x,\ e \geq 0$, all i.

The Hamiltonian function for firm i is

$$H = p \cdot Y(e(i),\ x) - C(e(i)) + \sigma(i) \cdot (G(x) - \sum Y(e,\ x)).$$

The necessary conditions for a solution include

$$(p - \sigma(i)) \cdot Y_{e(i)} - C_{e(i)} \leq 0,\ e(i) \geq 0,\ e(i) \cdot ((p - \sigma(i)) \cdot Y_{e(i)} - C_{e(i)}) = 0,$$

$$\sigma(i)' = \sigma(i) \cdot \left(\sum Y_x + r - G_x\right) - p \cdot Y_x.$$

The constraints (a) and (b), hold.

16 For reference on the optimization techniques employed in this paper see, for example, Takayama (1974).

APPENDIX B: THE CONTINUOUS TRANSITORY QUOTA SYSTEM

The Hamiltonian corresponding to problem (III) is

$$H = p \cdot Y(e, x) - C(e) - s \cdot z + \sigma \cdot (G(x) - \sum_i Y(e, x)) + \mu \cdot (q_0 + z - Y(e, x)).$$

Necessary conditions for solving (III) include

$$s = \mu$$

$$q_0 + z - Y(e, x) \geq 0, \ \mu \geq 0, \ \mu \cdot (q_0 + z - Y(e, z)) = 0.$$

Therefore, if $s > 0$, $q_0 + z - Y(e, x) = 0$.

APPENDIX C: DO FIRMS USE ALL THEIR QUOTAS?

Consider the problem facing firm i:

$$\text{Max}_{\{e, z\}} \quad \int_0^\infty (Y(e, x) - C(e) - s \cdot z) \cdot \exp(-r \cdot t) dt$$

Subject to   (a) $\alpha \cdot Q \geq Y(e, x)$,

　　　　　 (b) $\alpha' = z$.

The present value Hamiltonian function is

$$H = (Y(e, x) - C(e) - s \cdot z) \cdot \exp(-r \cdot t) + \sigma \cdot (\alpha \cdot Q - Y(e, x)) + \mu \cdot z.$$

Necessary conditions for a solution include

$$(Y_e - C_e) \cdot \exp(-rt) = \sigma \cdot Y_e, \text{ for } e > 0$$

$$s \cdot \exp(-r \cdot t) = \mu$$

$$\mu' = -\sigma \cdot Q$$

$$\text{Lim}_{t \to \infty} \mu \cdot \alpha = 0 \text{ (necessary transversality condition)}$$

$$\alpha \cdot Q - Y(e, x) > 0 \Rightarrow \sigma = 0 \text{ (complementary slackness condition)}.$$

Now, assuming $q^* = \alpha^* \cdot Q > Y(e^*, x)$, $\sigma = 0$ by complementary slackness. But then $\mu' = 0$ and μ is a constant. In the time interval of unused quotas, $\sigma = 0$ implies $Y_e - C_e = 0$; that is, marginal operating profits of effort are zero. In a productive fishery, this result occurs for only a relatively low biomass. In fact $Y_e - C_e = 0$ is not compatible with optimal management of the resource. Also, in the time interval of unused quotas, $s \cdot \exp(-rt) = \mu$ implies that $s' = r \cdot s$. In other words, the quota price must increase exponentially at the rate of the interest rate.

652 Ragnar Arnason

APPENDIX D: SOLUTION TO EQUATION (26)

Consider the differential equation

$$s' - r \cdot s = -F(q, x),$$

where $F(q, x) = (p - C_E \cdot E_q) \cdot q$.
The solution to this differential equation is

$$s(t) \cdot \exp(-r \cdot t) = -\int_0^t F(q, x) \cdot \exp(-r \cdot \tau) \cdot d\tau + s(0).$$

Now, since $p(t)$ is finite by assumption, $s(t)$ must also be finite. Therefore $t - \infty \Rightarrow x(t) \cdot \exp(-r \cdot t) \to 0$, and

$$s(0) = \int_0^t F(q, x) \cdot \exp(-r \cdot \tau) \cdot d\tau.$$

APPENDIX E: ARGUMENTS IN SUPPORT OF ASSUMPTION 2

Assumption 2 states that

$$p - C_E(E(q^*, x)) \cdot E_q(q^*, x) = p - C(E(q^*, x))/q^*.$$

Assume that all firms have access to the same technology. According to (26), $r \cdot s - s'$ can be regarded as the cost of holding a unit of quota. The benefit, on the other hand, of holding a unit of quota is $p - C_E \cdot E_q$. Denote the quota holding cost by $S(q^*)$ and the benefit by MP(q^*). Notice that MP(q^*) is the marginal operating profits of quota holdings. Also, let AP(q^*) = $p - C(E(q^*, x))/q^*$, where AP(q^*) is the average operating profits of quota holdings.

Now, assume that $S(q^*)$ = MP(q^*) > AP(q^*), for $q^* > 0$. Then, clearly, the firm will be making an overall loss and will be better off by selling its quota. Therefore, since q can be instantaneously adjusted at no cost, q^* cannot have been optimal to the firm.

Alternatively assume that $S(q^*)$ = MP(q^*) < AP(q^*). This means that the firm will be making overall profits. Therefore, since the firm's technology is available to other firms, these profits are also attainable to them. Consequently, the quota holdings of inactive, namely $q^* = 0$, cannot be optimal.

Therefore, MP(q^*) = AP(q^*).

REFERENCES

Andersen, P. and J.G. Sutinen (1985) 'The economics of fisheries law enforcement.' *Land Economics* 61, 387–97

Clark, C.W. (1976) *Mathematical Bioeconomics: The Optimal Management of Renewable Resources* (Wiley)

— (1985) *Bioeconomic Modelling and Fisheries Management* (Wiley)

Clark, C.W. and G.R. Munro (1982) 'The economics of fishing and modern capital theory: a simplified approach.' L.J. Mirman and D.J. Spulber, eds, *Essays in the Economics of Renewable Resources* (North-Holland)

Copes, P. (1972) 'Factor rents, sole ownership and the optimal level of fisheries exploitation *Manchester School of Social and Economic Studies* 40, 145–63

Cournot A. (1897) *Researches into the Mathematical Principles of the Theory of Wealth* (MacMillan)

Dasgupta, P.S. and G.M. Heal. (1979) *Economic Theory and Exhaustible Resources* (James Nisbet)

Gordon, H.S. (1954) 'Economic theory of a common property resource: the fishery. *Journal of Political Economy* 62, 124–42

Lawson, R.M. (1984) *Economics of Fisheries Development* (Francis Pinter)

Lucas R.E. and T.J. Sargent, ed. (1981) *Rational Expectations and Econometric Practice* (George Allen & Unwin)

Nash, J., Jr (1950) 'Equilibrium points in N-person games.' *Proceedings of the Academy of Sciences* 36, 48–9

Pigou, A.C. (1912) *The Economics of Welfare* (Macmillan)

Scott, A.D. (1955) 'The fishery: the objectives of sole ownership.' *Journal of Political Economy* 63, 116–24

Smith, V.L. (1968) 'Economics of production from natural resources.' *American Economic Review* 58, 409–31

— (1969) 'On models of commercial fishing.' *Journal of Political Economy* 77, 181–98

Takayama A. (1974) *Mathematical Economics* (Dryden Press)

Turvey, R. (1964) 'Optimization and suboptimization in fishery regulation.' *American Economic Review* 54, 64–76

Van Horne, J.C. (1971) *Financial Management and Policy* (Prentice Hall)

[18]

NATURAL RESOURCE MODELING
Volume 6, Number 4, Fall 1992

INDIVIDUAL TRANSFERABLE QUOTAS AND PRODUCTION EXTERNALITIES IN A FISHERY

JOHN R. BOYCE
Assistant Professor of Economics
University of Alaska Fairbanks
Fairbanks, Alaska 99775-1070

ABSTRACT. This paper determines the conditions under which an individual transferable quota (ITQ) system will cause fishermen to engage in cost-decreasing, rather than cost-increasing, competition. If there are production externalities (e.g., congestion or stock externalities) present, the market price of a quota will not be fully reflected in these externalities. Thus, fishermen will not fully internalize the externalities in their effort decisions. Even if there are no production externalities, an individual fisherman imposes costs on others under open access by removing a fish that was available to all fishermen. An ITQ system allows the individual who values that fish most to obtain the right to harvest the fish, so each fisherman must internalize the full social cost. Thus, an ITQ system is capable of solving the common property externality but not the production externalities in a fishery.

KEY WORDS: Transferable quotas, production externalities, fishery.

1. **Introduction.** Open access and common property fisheries each possess the characteristic that ownership of the fish is by a "rule of capture." That is, an individual fisherman can claim ownership only by harvesting a fish. This system of property rights gives each fisherman an incentive to increase effort in order to increase his or her share of the harvest. However, if each fisherman in the fishery responds to this incentive, the result is that each fisherman has higher costs. This is commonly called the "race-for-fish" or "over-capitalization" problem. This causes fishery profits to be dissipated by the cost-increasing competition.

Cost-increasing competition occurs for several reasons. First, if the resource is common property or if access is open to all, there is the problem of the rule of capture. However, this is not the only cause of the cost-increasing competition. In many fisheries, production externalities related to the jointness in production among the fishermen also exist. Ever since Vernon Smith's work [1969] economists have spent considerable energy exploring the effects of "stock" and "congestion" externalities in fisheries. Stock externalities occur because the productivity of a unit of effort depends upon the density of the stock that is being exploited. When people are jointly harvesting from the same resource pool, an individual does not have an incentive to take into account the increase in costs to other fishermen due to his reduction in the stock density. Congestion externalities occur when the return to effort in a particular location depends upon the amount of effort being applied at that location. When choosing to participate in the location, an individual considers the effect of the congestion upon his or her own effort, but does not take into account the effect his or her participation has upon the other fishermen in the fishery.

The question economists and managers have grappled with for a number of years is how to devise a regulatory system that encourages cost-reducing, rather than cost-increasing, competition. The history of this regulation, however, includes few long-term successes (Townsend [1990]). The idea applied to the salmon industry in Alaska and in British Columbia was to limit entry. However, this merely converted an open access fishery into a common property fishery. Those who remained in the fishery continued to engage in cost-increasing competition (Wilen [1979]).[1]

Given the problems of limited entry programs, economists have turned to the idea of creating individual transferable quotas (ITQs) (Christy [1973]). Because ITQs create property rights where formerly no such rights existed, economists often conclude that ITQs will solve the over-capitalization and race-for-fish problems in a fishery (cf., Neher, et al. [1989])[2]. I argue that an ITQ system, as commonly envisioned, is capable of addressing only one of the sources of the cost-increasing competition, namely, the incentive derived from the rule of capture.

Neither a stock nor a congestion externality requires that the resource being harvested be owned by a rule of capture. These production externalities are independent of the ownership of the resource. On the other hand, cost-increasing competition will occur in a common property fishery even without production externalities because of the rule of capture. While there have been many supporters of ITQs, the only major criticism has come from Copes [1986]. Copes argues that an ITQ system will not eliminate the cost-increasing competition in a fishery characterized by production externalities. However, Copes' arguments are in conflict with earlier, more mathematical analysis, most notably by Clark [1980].

This paper presents a formal analysis of some of the arguments made by Copes. I determine how the market price for ITQ permits affects the behavior of the fishermen both with and without production externalities. In either case, relative to open access, ITQs will result in a more efficient allocation of resources. However, the conditions under which ITQs will *eliminate* the incentive for cost-increasing competition are quite restrictive. My principal conclusion is that the market price of a quota will fully reflect the social costs only if no production externalities exist. Thus, a necessary condition for an ITQ program to eliminate cost-increasing competition is that the only external cost of the harvest of a fish by one person is that no one else may now harvest the fish. If there are any other costs imposed, either by the removal of the fish or by the effort required for its removal, these costs will not be reflected in the market price for quotas and will be ignored by fishermen. These arguments are developed in Sections 2 and 3.

In addition, in Section 4 I briefly explore an alternative to ITQs that will cause fishermen to fully internalize the costs of their behavior. The idea is a blending of the ITQs idea and Anthony Scott's [1960] principle of sole ownership: I argue that if the government is going to establish property rights in a fishery, instead of establishing individual transferable rights to harvest the fish, it should create individual transferable shares in the profits of the fishery.

    2. The open access problem. For most fisheries, the status quo is that of open access or common property. Furthermore, fisheries rarely are comprised of homogeneous fishermen. There are typically differences in the physical capital used, as well as differences in the human capital of the fishermen. Most mature fisheries are also managed in some fashion by government regulators. The regulation of the fishery generally includes, at a minimum, season or total harvest constraints. Within a season, the total harvest by the fleet is limited by a fleet-wide quota. Although other regulations are also often imposed, such as limits on gear types and in some cases on the number who may participate in a fishery, I ignore these differences in what follows.

A. *The social planners problem.* To examine the social costs of the open access fishery, I first examine the way the fishery would operate if managed by a benevolent social planner. The in-season optimization problem for the social planner is to choose the effort level (defined by the number of agents and the effort from each agent), and the length of the season to maximize the stream of benefits from harvesting a quantity of fish. Since I am concerned with the in-season optimization problem, the total harvest of fish is treated as exogenous.

Let the rate of harvest by the ith agent be given by the production function

$$y_i(t) = y_i(x_i(t), x_{-i}(t), s(t)).$$

The effort controlled by each fisherman at each moment t is denoted $x_i(t)$, and the production function is increasing and concave both in this argument and in the biomass $s(t)$. In addition, congestion, here measured by the aggregate effort of the other agents participating in the fishery, $x_{-i}(t)$, also enters the production function, although it has a negative first derivative.[3] Heterogeneous in the production capabilities of individual fishermen is indicated by the subscripts on the production functions, y_i.

Individuals may also be heterogeneous in their costs. There are two direct costs associated with fishing. The variable costs are given as an increasing convex function of the level of effort applied by the fisherman,

$$c_i(t) = c_i(x_i(t)).$$

In addition, each fisherman faces avoidable fixed costs of entering, K_i. Both costs may vary across individuals.

The output price P is assumed to be constant over time and, because the output quota is held constant in each model, is assumed to be exogenous. Harvesting reduces the biomass over the course of the season, and there is no in-season replenishment of the stock. Thus, the stock equation of motion is given by

(1) $$s'(t) = - \sum_{j=1}^{N(t)} y_j(x_i(t), x_{-i}(t), s(t)),$$

where $N(t)$ is the number of active fishermen at time t, and y_j is the harvest per fisherman at time t. Without loss of generality, assume that $s(0) = s_0$ and that $s(T) = E$, where E denotes the biological escapement required for future harvests and T is the length of the season. E is determined exogenous to the present problem. Let $Q = s_0 - E$. Q represents the total biomass removed within the season: the fleet-wide quota. The season length is simply the time it takes to harvest the fleet-wide quota; i.e., T is the implicit solution to

$$(2) \qquad Q = \int_0^T \sum_{j=1}^{N(t)} y_j(x_j(t), x_{-j}(t), s(t))\, dt.$$

Since I am concerned with the prosecution of a fishery within a particular season, I simplify the analysis by assuming that the discount rate is zero. When there are no stock effects (i.e., $\partial y_i / \partial s = 0$ for all i), a zero discount rate implies that effort is constant over the course of the season. This will not be the case if the discount rate is positive. While the assumption of a zero discount rate affects the characterization of the prosecution of the fishery, it does not affect the argument regarding the efficiency of ITQs.

The optimization problem for a benevolent social planner is to choose x_i, N and T to maximize

$$V = \int_0^T \left[\sum_{j=1}^{N(t)} P y_j(x_j(t), x_{-j}(t), s(t)) - c_j(x_j(t)) \right] dt - \sum_{j=1}^{N(0)} K_j$$

subject to the equation of motion on the biomass, (1). The variable costs depend upon both the level of effort per fisherman and the number of fishermen at each instant in time. The fixed costs depend only upon the maximum number of fishermen present. Because there is assumed to be no influx of the biological stock during the season, the maximum number of fishermen occurs at the opening of the season.

The Hamiltonian for the variable cost part of the social planner's problem is

$$H_{sp} = \sum_{j=1}^{N(t)} (P - \lambda) y_j(x_j, x_{-j}, s) - c_j(x_j),$$

where the costate variable represents the marginal value of a unit of biomass at each point in time. From the maximum principle, the first order conditions in the variables x_i, s and λ are

$$\frac{\partial H_{sp}}{\partial x_i} = 0 :$$

(3)
$$(P - \lambda)\frac{\partial y_i}{\partial x_i} + (P - \lambda)\sum_{j \neq i}^{N} \frac{\partial y_j}{\partial x_{-j}}\frac{\partial x_{-j}}{\partial x_i} = \frac{\partial c_i}{\partial x_i},$$

$$i = 1, \ldots, N,$$

(4)
$$\frac{\partial H_{sp}}{\partial s} = -\lambda' : \qquad -\lambda' = (P - \lambda)\sum_{j=1}^{N} \frac{\partial y_j}{\partial s},$$

and

(1')
$$\frac{\partial H_{sp}}{\partial \lambda} = s' : \qquad s' = -\sum_{j=1}^{N} y_j(x_j, x_{-j}, s).$$

The economic interpretation of these conditions is straightforward. Equation (1') simply repeats the equation of motion for the biomass. From (3), the effort of the ith agent is chosen so that the net value of the marginal product of effort (the first term) is equated with the marginal factor cost of effort (the third term) plus the sum of the external costs imposed on other participating agents (the second term). Thus, the social planner takes into account the congestion externality in the selection of effort. Equation (4) states that the costate variable declines according to the sum of the net value of marginal production of the biomass.

In the event that the stock effect on production is zero (i.e., $\partial y_i/\partial s = 0$ for all i), then the right-hand side of (4) vanishes, implying that the costate variable is constant over time. Therefore, from (3), each individual's effort is constant. In this case, the length of time that each individual is active in the fishery would be equal. However, if there is a stock effect, (i.e., $\partial y_i/\partial s \neq 0$), then (4) implies that λ will be decreasing over time. This means that effort will not be constant. Thus, an exiting condition is required.

INDIVIDUAL TRANSFERABLE QUOTAS 391

Following Clark [1980], agents are assumed to be ordered so that the lowest cost fishermen enter first, the next lowest second, and so forth. (This ordering may not be unique if fishermen are heterogeneous in both variable and quasi-fixed costs.) The exiting condition for the Nth individual is

$$(5) \qquad (P - \lambda)y_N(x_N, x_{-N}, s) - c_N(x_N) = 0.$$

The interpretation of (5) is simply that the Nth individual exits when variable profits approach zero. The corresponding entry condition is that over the interval $(0, T_N)$ that the Nth individual is active, he must just cover the cost of entry, i.e.,

$$(6) \qquad \int_0^{T_N} [(P - \lambda)y_N(x_N, x_{-N}, s) - c_N(x_N)] \, dt = K_N.$$

B. *The open access equilibrium.* Under open access, the ith individual attempts to maximize his own profits subject to the constraints of the actions of the other agents. Throughout the paper, I assume that the individuals simultaneously solve an *open loop* optimization program. The open loop assumption is made for simplicity of analysis, but it may be defended on the basis of a competitive harvesting sector.[4]

The optimization problem for the ith participating individual is to choose effort x_i to maximize

$$V_i = \int_0^T [Py_i(x_i, x_{-i}, s) - c_i(x_i)] \, dt, \qquad i = 1, \ldots, N,$$

subject to the equation of motion given by (1). The associated Hamiltonian is

$$H_i = Py_i(x_i, x_{-i}, s) - c_i(x_i) - \lambda_i \sum_{j=1}^{N(t)} y_j(x_i, x_{-i}, s), \qquad i = 1, \ldots, N.$$

The first order necessary conditions of interest are the costate equation for λ_i and the derivative of H_i with respect to effort. Assuming an open-loop Cournot solution, the necessary conditions are

$$(7) \qquad \frac{\partial H_i}{\partial x_i} = 0 : \qquad (P - \lambda_i)\frac{\partial y_i}{\partial x_i} = \lambda_i \sum_{\substack{j \neq i}}^{N} \frac{\partial y_j}{\partial x_{-j}} + \frac{\partial c_i}{\partial x_i},$$

$$i = 1, \ldots, N,$$

and

$$
(8) \qquad \frac{\partial H_i}{\partial s} = -\lambda_i' : \qquad -\lambda_i' = (P - \lambda_i)\frac{\partial y_i}{\partial s} - \lambda_i \sum_{j \neq i}^{N} \frac{\partial y_j}{\partial s},
$$

$$
i = 1, \ldots, N.
$$

A comparison of (7) and (8) with (3) and (4) reveals the open access problem. In (7) the second term is multiplied only by the imputed shadow value of the stock λ_i, whereas the same term in (3) is multiplied by $(P - \lambda)$. Thus, even if the individual's marginal valuation of the biological stock λ_i equals the social value λ, the individual still fails to take into account the reduction in revenues to the other participating agents due to the congestion externality the individual imposes on them. Similarly, in (8), the ith agent fails to take into account the gross marginal value of the biomass (P times the summation in the last term) accounted for in the social optimum in (4).

The exiting and entry equations for the marginal entrant under open access are, respectively,

$$
(9) \qquad Py_N(x_N, x_{-N}, s) - c_N(x_N) = 0,
$$

and

$$
(10) \qquad \int_0^{T_N} [Py_N(x_N, x_{-N}, s) - c_N(x_N)]\, dt = K_N.
$$

These conditions imply the Nth individual earns zero profits in gross over the entire planning horizon, and that at the moment of exiting is earning zero variable profits as well. The exiting condition also holds for the infra-marginal individuals. That is, at the moment that an infra-marginal entrant exits, that individual is just earning zero variable profits. However, this does not mean that they earned zero profits over the entire time horizon.

As was the case with the social planner's model in section 2A, when there are no stock effects (i.e., $\partial y_i/\partial s = 0$ for all i), all individuals will exit at the end of the season. This is because neither the costate variables nor effort varies over time when there are no stock effects.

INDIVIDUAL TRANSFERABLE QUOTAS 393

C. *The case of no congestion or stock externalities.* One might believe that the differences in the above two systems of equations is due entirely to the congestion and stock externalities. However, a comparison of the social planner's case with the open access case when neither congestion nor in-season stock externalities exist reveals that this is incorrect. Notice from the costate equations (4) and (8) that if there are no stock effects then the costate variables are constant over the season. Thus, it follows from (3) and (7) that the effort levels of each active agent are constant over the course of the season. If effort is constant over the season, then it also follows that each active individual will participate for the entire season. Given these observations, the problem may be converted into a static optimization problem.[5]

For both the social planner and open access models, the season length constraint (2) may be rewritten as

$$(11) \qquad Q = T \sum_{i=1}^{N} y_i(x_i).$$

The social planner wishes to choose the number of participants N and the effort level by each active participant x_i to maximize

$$V = T \sum_{i=1}^{N} [P y_i(x_i) - c_i(x_i)] - \sum_{i=1}^{N} K_i$$

subject to (11). The problem thus stated is linear in the season length T. For this reason, suppose that the season length has an upper bound, say $T \leq \overline{T}$. Thus the social planner's problem may be restated as[6]

$$L = T \sum_{i=1}^{N} [P y_i(x_i) - c_i(x_i)]$$

$$- \sum_{i=1}^{N} K_i + \alpha \left[Q - T \sum_{i=1}^{N} y_i \right] + \beta [\overline{T} - T].$$

Dropping the effort arguments from the notation, the first order condition for effort by the ith fisherman is

$$(12) \qquad (P - \alpha) y_i' - c_i' = 0, \qquad \text{for } i = 1, \ldots, N.$$

The season length equation is

$$(13) \qquad \sum_{i=1}^{N}(P - \alpha)y_i - c_i = \beta,$$

where $\beta \geq 0$, $\overline{T} \geq T$, and $\beta[\overline{T} - T] = 0$.

The entry condition under the social planner's case is found by solving for the value of N that maximizes V,

$$(14) \qquad Py_N - c_N - \alpha y_N = K_N/T.$$

This condition simply states that the marginal entrant must just cover the fixed costs of entry over the course of the season. It is exactly analogous to (10).

It is easy to show that the social planner will choose the season length to equal $\overline{T}$ in this model. Suppose not; then $\beta = 0$, implying the left hand side of (13) equals zero. However, from (14), the expression on the left hand side of (13) for the Nth fisherman is equal to K_N/T, which is positive. Furthermore, all other individuals in the summation have profits greater than the Nth fisherman. Thus, the summation on the left hand side of (13) cannot equal zero, implying that $\beta > 0$ and the season length equals $\overline{T}$.

Now consider the corresponding conditions for the open access situation. The season length constraint (11) still must hold, but the objective function involves only the individual agent's profits, thus

$$L_i = T[Py_i(x_i) - c_i(x_i)] - K_i + \alpha_i \left[Q - T \sum_{i=1}^{N} y_i \right] + \beta_i[\overline{T} - T].$$

Again, assuming a Cournot solution, the effort equation for the ith fisherman is

$$(15) \qquad (P - \alpha_i)y_i' - c_i' = 0, \qquad \text{for } i = 1, \ldots, N.$$

Differentiating V_i with respect to T gives an equation for the season length,

$$(16) \qquad Py_i - c_i - \alpha_i \sum_{i=1}^{N} y_i = \beta_i, \qquad \text{for } i = 1, \ldots, N.$$

INDIVIDUAL TRANSFERABLE QUOTAS 395

The zero-profits entry condition for open access is (cf. (10))

$$(17) \qquad Py_N - c_N = K_N/T.$$

Unlike the social optimum case, the assumption that the season length is less than $\overline{T}$ does not result in a contradiction. This suggests that the season length is shorter under the open access equilibrium than under the social optimum. This, of course, is as intuition suggests. However, while it is easy to see the differences between the two systems of equations, it does not appear possible to show in general that the season length is in fact shorter under the open access than is socially optimal.

It is possible to show that the two equilibria result in different allocations of resources. Let us compare (12) with (15). Under open access, each entrant, including the marginal entrant, ignores the full social cost of removing addition stock from the fleet-wide quota in their selection of the effort level. Arnason [1989] has shown that under optimal effort levels (solutions to (12)), the social planner's shadow value of the unharvested stock equals the *sum* of the individual shadow values from the open access model, i.e.,

$$(18) \qquad \alpha = \sum_{i=1}^{N} \alpha_i.$$

In the present model, this result is possible only if there is a corresponding relationship between β and the β_i's. Using (12), (15) and (18), and assuming that $x_i^* = x_i^0$ (the solutions to (12) and (15), respectively), it follows that

$$(19) \qquad p - c_i'/y_i' = \sum_{j=1}^{N} p - c_j'/y_j', \qquad \text{for all } i = 1, \ldots, N,$$

which implies

$$(20) \qquad \sum_{j \neq i}^{N} p - c_j'/y_j' = 0.$$

Thus, the open access cannot result in the same level of effort as under the social optimum.

A similar comparison can be made of the entry conditions (14) and (17). In (17), the costate value for the marginal entrant is zero since the marginal entrant earns ero profits for each unit of time. The entry condition for open access implies that the marginal entrant ignores the term αy_N, which is the marginal social value of the stock removed by the Nth entrant at each instant in time.

Since there are no production externalities in the present model, the misallocation of resources is due entirely to the absence of ownership of the resource.

3. Individual transferable quotas. Let us now turn to the central question of whether or not an ITQ system will cause fishermen to internalize the social costs of their behavior. Suppose that all fishermen who participated under the open access are given quota shares. Let each individual in the fishery be given a quota at the beginning of the season, $q_i(0)$, and let $q_{-i}(0)$ refer to the quotas granted to the remaining fishermen.[7] The number of quotas dispersed across the fishery satisfy

$$(21) \qquad\qquad Q = q_i(0) + q_{-i}(0),$$

where Q is the total allowable harvest for the fleet. The number of outstanding or unused quotas held by the ith fisherman changes over time according to

$$(22) \qquad\qquad q_i' = -y_i(x_i, x_{-i}, s) + z_i.$$

The term z_i denotes the ith fisherman's purchase $(z_i > 0)$ or sale $(z_i < 0)$ of quotas at each instant of time. The number of unused quotas held by the remaining $N - 1$ fishermen changes over time according to

$$(23) \qquad\qquad q_{-i}' = -\sum_{j \neq i}^{N} y_j(x_j, x_{-j}, s) - z_i.$$

In (23) the identity that $z_i = -z_{-i}$ is used to substitute z_i in for $-z_{-i}$. This ensures that the adding up condition for sales of quotas is satisfied. In addition, the number of quotas that may be transferred at any moment is subject to the constraints

$$(24) \qquad -q_i(t) \leq z_i(t) \leq q_{-i}(t), \qquad \text{for } i = 1, \dots, N.$$

That is, the ith fisherman can sell at most what he owns, and buy at most what remains on the market.

A. *ITQs in the presence of stock and congestion externalities.* If quotas are transferable, then in a competitive market there will exist a price for these quotas. Let $m(t)$ denote the market price of quotes. To an individual receiving an initial quota of $q_i = q_i(0)$, which may be augmented or reduced by buying or selling quotas, the objective function is to maximize

$$V_i = \int_0^T [Py_i(x_i, x_{-i}, s) - c_i(x_i) - mz_i] \, dt,$$

subject to the equations of motion for the quotas given by (22) and (23), and given the actions of the other fishermen. The associated Hamiltonian is

$$H_i = (P - \nu_i)y_i(x_i, x_{-i}, s) - c_i(x_i) - mz_i$$
$$- \nu_{-i} \left[\sum_{j \neq i}^{N(t)} y_j(x_j, x_{-j}, s) + z_i \right] + \nu_i z_i.$$

For an active fisherman (one who does not sell all of his quota) the necessary conditions for the control variables x_i and z_i for this problem are

(25)
$$\frac{\partial H_i}{\partial x_i} = 0 : \quad (P - \nu_i)\frac{\partial y_i}{\partial x_i} = \frac{\partial c_i}{\partial x_i} + \nu_{-i}\sum_{j \neq i}^{N}\frac{\partial y_j}{\partial x_{-j}},$$
$$i = 1, \ldots, N,$$

and

(26)
$$\frac{\delta H_i}{\delta z_i} = 0 : \quad m = \nu_i - \nu_{-i},$$
$$i = 1, \ldots, N.$$

The costate equations for the quotas held by the ith fisherman and by the others use the fact that the remaining stock equals the escapement plus the outstanding quotas, i.e., $s(t) = q_i(t) + q_{-i}(t) + E$.

$$(27) \qquad \frac{\partial H_i}{\partial q_i} = -\nu_i' : \qquad -\nu_i' = (P - \nu_i)\frac{\partial y_i}{\partial s} - \nu_{-i}\sum_{j \neq i}^{N}\frac{\partial y_j}{\partial s},$$

$$i = 1, \ldots, N$$

and

$$(28) \qquad \frac{\partial H_i}{\partial q_{-i}} = -\nu_{-i'} : \qquad -\nu_{-i'} = (P - \nu_i)\frac{\partial y_i}{\partial s} - \nu_{-i}\sum_{j \neq i}^{N}\frac{\partial y_j}{\partial s},$$

$$i = 1, \ldots, N.$$

First, let us establish that the market price for permits remains constant over the entire season.[8] To see this, differentiate (26) with respect to time to yield

$$(29) \qquad\qquad m' = \nu_i' - \nu_{-i}' = 0.$$

The second equality in (29) comes from (27) and (28). The relationship in (26) also shows that the transfer of permits will instantaneously adjust to the equilibrium value due to the 'bang-bang' nature of the control variable z_i.[9]

Recall the social planner's problem given by (3), (4) and (1'). In the event that

$$(30) \qquad\qquad \nu_{-i} = P - \lambda,$$

and

$$(31) \qquad\qquad \nu_i = \lambda,$$

observe that (25) and (3) are identical. Furthermore, if these conditions hold, then (27) and (28) are equivalent to (4). It is the conditions (30) and (31) that must hold if the ITQ system is to replicate the social planner's problem. Suppose these conditions hold. Substituting (30) and (31) into (26) shows that

$$(33) \qquad\qquad m = P.$$

INDIVIDUAL TRANSFERABLE QUOTAS   399

However, this cannot occur since an individual who buys a unit of the output quota still must incur the costs of harvesting the fish. Since this is a positive cost, the market price for permits must be less than P. This implies the following:

THEOREM ONE. An ITQ system in which transferable quotas allocated at the beginning of the season are traded at a competitive market price is not capable of simultaneously solving the in-season stock externality problem and the congestion externality problem.

Theorem One shows that individual transferable quota programs do not work to cause fishermen to fully internalize the costs of their actions in the presence of stock and congestion externalities. The intuition behind the result is fairly simple. Since the property right associated with a quota of the output carries with it no time specification nor a right to catch fish in an uncongested fishery, there remains a fundamental diseconomy. Even if ITQs are put into place, a race for fish and overcapitalization of the fishery will continue to persist.

It should come as no surprise that an ITQ system is not able to deal with the congestion externality problem, since this result has already been obtained by Clark [1980]. Clark's result in regards to congestion externalities has not been widely discussed, although almost every paper on fisheries ITQs references his paper. Clark, himself, discounted this finding, claiming that congestion externalities probably did not occur in a significant number of fisheries. Of course, whether this is so is an empirical question. Clark obtained his result by assuming that there were no in-season stock externalities. His result is a corollary to Theorem One.

COROLLARY ONE. In the case where only a congestion externality exists, ITQs are not capable of reproducing the social optimum.

PROOF. If only a congestion externality exists, in the system of equations (25)–(28), equations (25) and (26) are unchanged, and equations (27) and (28) each have zero right-hand sides. Again, the relationship that must hold for the market price to fully reflect the social cost of the stock is that equations (30) and (31) hold. Thus, the same problem exists: the market price for *in situ* quotas cannot equal the market price for the harvested output.   □

Clark claims, incorrectly, that when the sole production externality is an in-season stock externality, ITQs are capable of solving the social optimization program. The proof is a corollary to Theorem One.

COROLLARY TWO. In the case where only a stock externality exists, ITQs are not capable of reproducing the social optimum.

PROOF. If no congestion externality exists, then (25) becomes

$$(P - v_i)\frac{\partial y_i}{\partial x_i} = \frac{\partial c_i}{\partial x_i}.$$

All other equations in the system (25)–(28) remain unaltered. Now, suppose that (30) and (31) hold. Again, the quota system would replicate the social optimal system. However, the same problem exists as in Theorem One: the market quota price cannot equal the output price.

Clark's error appears to be in his analysis of equations (25) and (26). While (25) differs from (3) only in that λ in (3) is replaced by v_i in (25), this result is because $\partial y_j/\partial x_{-j} = 0$, not because $v_{-i} = 0$. Thus, $v_i \neq m$. A similar mistake was made by Moloney and Pearse [1979]. However, their claim that ITQs are capable of reducing the social costs of harvesting are correct for the in-season case (which they did not consider), as will be shown in the next section.

B. *ITQs when no congestion or stock externalities exist.* If there are no stock or congestion externalities, ITQs can generate efficient incentives. To see this, consider the system (25)–(28) when no stock or congestion externalities exist:

(25) $$(P - v_i)\frac{\partial y_i}{\partial x_i} = \frac{\partial c_i}{\partial x_i}, \qquad i = 1, \ldots, N,$$

(26) $$m = v_i - v_{-r}, \qquad i = 1, \ldots, N,$$

(27) $$v_{i'} = 0, \qquad i = 1, \ldots, N,$$

(28) $$v_{-i'} = 0, \qquad i = 1, \ldots, N.$$

Consider the shadow values for the two different quotas. The quotas by the fisherman have value to him ($v_i > 0$) because of the profits he may obtain when he harvests his quota and sells his permits on the market. However, by assumption, the remaining stock does not affect the profitability of a fisherman of catching his remaining quota. Therefore, quotas owned by other fishermen do not affect his profits except through his purchases or sales of quotas. This implies that $v_{-i} = 0$. Using (26), we have that $m = v_i$, for all $i = 1, \ldots, N$. It follows that the market price reflects the social value of the stock as long as the market is efficient. Thus we have

THEOREM TWO. When there are no in-season stock externalities and no congestion externalities, then an ITQ system will be capable of generating social efficiency.

It is interesting to examine the entry condition under an ITQ system. The entry condition is

(34) $$\bar{T}[(P - m)y_N - c_N] = K_N.$$

The relationship in (34) can be seen as follows. The permits the fisherman owns could be sold at price m on the market. Thus all harvesting of the fisherman's own permits yields net revenues of $P - m$ per unit harvest. Similarly, for quotas the fisherman purchases, the cost of purchase must be subtracted from the revenues. Thus the net revenues are $P - m$ per unit harvest. The marginal fisherman will earn in net what it costs him to fish. However, this is the return *above* the windfall from being endowed the initial quotas. The proof that the marginal fisherman will utilize the entire season available to him is identical to the argument following equations (12)—(14).[10]

C. *Second-best effects.* The results of Theorems One and Two imply the only case in which an ITQ system is capable of generating the

first-best solution is when neither in-season stock nor congestion externalities exist. However, some combination of these conditions probably
exist in most fisheries. Thus, the question is whether or not there is an
improvement over the open access case. The answer is yes, though it
is not clear what the magnitude of this improvement is.

Suppose that each fisherman's initial quota were exactly the same
share of the fleet-wide quota they would receive under common property. Then it is possible that each fisherman could act exactly as before
the ITQ system was implemented. If so, then they would earn exactly
the same rents as under open access. However, if any fisherman could
improve his situation either by holding off on harvesting his share of the
quota, or by buying a portion of shares from another, or selling part of
his own share, then the ITQ system allows him to do this. Furthermore,
any trade that occurs has to result in either a decrease in congestion
or (possibly and) a decrease in the quantity of fish caught some time
during the season.[11] Thus all other fishermen must also benefit by this
trade. This implies:

THEOREM THREE. ITQs will result in an improvement over the open
access equilibrium.

The question that remains is how much of an improvement over
open access is an ITQ system when production externalities exist?
Unfortunately, there does not appear to be a simple answer to this
question. The system of equations describing the equilibrium in the
presence of ITQs does not admit a simple closed form solution for any
of the variables of interest. Thus, although there will be an incentive
for fewer fishermen to participate than under open access, it is not clear
how many fewer fishermen will be active. The empirical evidence is also
mixed. While some fisheries have witnessed large reductions in effort,
others have had almost no effect (Muse and Schelle [1989], Neher et al.
[1989]).

4. Discussion of the results. The results derived above suggest that economists should be careful in advocating programs that
assign property rights to natural resources. Property rights are very
important—there seems little doubt about this. However, it does not

INDIVIDUAL TRANSFERABLE QUOTAS 403

follows that all property rights systems yield the same social results. In the case of fisheries, a system that assigns property rights to the harvestable output will generate the social optimal conditions if, and only if, the sole source of the externality is due to ownership by rule of capture. If the external costs an individual can impose on others are due to production externalities, then an ITQ system will not be sufficient to obtain the social optimum.

The second-best results suggest that ITQs will result in an improvement over an open access or common property situation. However, Theorems One and Two show that the conditions under which an ITQ system will eliminate cost-increasing competition are quite restrictive. The question of how much improvement will occur under an ITQ system is an empirical question, and one that does not seem to have been adequately studied. This alone should lead economists to be cautious in advocating ITQ programs for fisheries where production externalities are present. As Anthony Scott has noted, institutional changes are costly. Much effort is squandered if the degree of change is relatively small.

Furthermore, the preoccupation in the economics literature with congestion and stock externalities is probably not based purely on theoretical interest. It seems safe to say that these are real phenomena. This suggests property rights should be assigned in a way that takes into account the production diseconomies. The simplest and most direct way to do this is to assign *shares in the profits* from the fishery, i.e., to make the fishermen stockholders in the fishery. If fishermen have shares to the profits, they will be forced to internalize the production diseconomies. If the shareholders have control over effort allocations, they will be motivated to insure that these allocations are efficient. If property rights are assigned to shares of profits, it would require that effort be compensated separately from the compensation of fishery profits shares. Fishermen who are more productive would have to be compensated for their productivity.

This is a restatement of Anthony Scott's [1960] argument about the advantages of sole ownership. The only difference is that the sole ownership is in the hands of a corporation whose (initial) members happen to be the original fishermen. An interesting feature of this approach is how it addresses the problem identified by Karpoff [1987]. Karpoff argued that fisheries regulations are politically acceptable if they do not decrease the number of participants in the fishery. He observed that fishermen prefer regulations that increase employment in the fishery. If the fishermen have transferable shares of the profits to the fishery, then it is possible for them to choose to continue with the same quantity of effort levels as under an open access. However, if they choose employment over profits by encouraging over-capitalization of the fishery, then their choice is, by definition, the social optimum. The reason is that in choosing such an institution (if they do so), they are explicitly choosing to value the employment aspect of the fishery higher than the returns to the fishery. Since the cost of this decision is born entirely by the stockholders who are making the decision, there is no externality. Of course, in the long run, selecting to allow dissipation of the rents should not be expected to occur as long as the shares are transferable.

However, there are both historical and legal problems with ownership of profits being the transferable right. First, as noted by Johnson and Libecap [1982], the Department of Justice has not looked favorably on "monopoly ownership" of fisheries resources. However, most fisheries tend to have very elastic demand for their products. This is true even for fisheries as large as the Alaska salmon fishery. Thus, such criticisms might be overcome in a court of law.

However, a second issue is more difficult to overcome. Historically, especially on the west coast, fisheries were developed by financiers and entrepreneurs who do not live in the local fishing communities. As a result, there has been a mistrust of any scheme that might allow these "outside interests" to obtain control over a fishery.[12] This indicates that there would likely be political opposition to a plan allowing the transfer of ownership to persons outside the fishery. However, this is the same complaint that fishermen have had with regard to ITQs. The difference between shares to profits and shares to the harvest is that fishermen will not *also* oppose the stockholder version of ownership on the grounds that it does not solve the race for fish problem or the over-capitalization problem. Furthermore, even if the ownership shares leave the fishery, the labor market will likely remain.

Finally, granting fishermen property rights to the profits to the fishery, rather than to the quantity of fish they can remove, carries with it the problem of organizational costs. Currently there are almost no costs of running the fishery other than those incurred by regulators.

Under an ITQ system, there would be organization costs only in the transfer of quota permits. Under a corporate fishing fleet, there would be organizational costs that the fishermen themselves would have to incur. On the fact of it, this may appear to be the reason that cooperative arrangements have not been constructed very often in the past.[13] However, as Johnson and Libecap [1982] note, organization costs, even on large fisheries such as the shrimp fishery in the Gulf of Mexico, have not prevented fishermen from developing cooperative agreements. That distinction lies with the Justice Department's use of anti-trust statutes.

Economists have long been enamored with the institution of private property. This, of course, is justly so—no other institution is as capable of causing the individual to internalize social values as is the institution of private property. However, the "creation of property rights," as remarked upon by Anthony Scott [1989, p. 290], "is not something that even monarchs can take lightly." It is important that when economists advocate institutions that create property rights, such as ITQs, that they understand the full ramifications of their proposal. I have shown that while ITQs can solve one type of open access problem in a fishery, it will fall short of eliminating cost-increasing behavior if there exist production externalities. Furthermore, it appears that an alternative property rights system exists which is capable of eliminating cost-increasing competition even in the presence of production externalities.

ENDNOTES

* This paper has benefited from discussions with Diane Bischak, Greg Goering, Bob Logan, and by comments from an anonymous referee. All remaining errors are my own.

1. In Alaska, the state-funded enhancement program, which put more fish into the water, obscured the failure to reduce cost-increasing competition. For a number of years, the increase in quantity increased revenues at a greater rate than the increases in costs due to the cost-increasing competition. However, it appears now that further increases in fish may decrease revenues. Thus, with cost-increasing competition also occurring, the fisheries cannot continue to obtain the same profits as in the past.

2. ITQ programs have in fact been established in several fisheries with varying degrees of success (Muse and Schelle [1989]). In Canada, an ITQ system was recently instituted for the halibut fishery, and in the United States, proposals are rapidly working their way through the bureaucratic system for ITQ programs in the halibut and black cod (sablefish) fisheries in Alaska.

Fisheries Economics I

3. The notation x_{-i} denotes $\sum_{j\neq i}^{N} x_j$, where the summation is over $j = 1, \ldots, N$, for j not equal to i. It follows that $\partial x_{-j}/\partial x_i = 1$, for all $j \neq i$. The use of x_{-i} as a measure of the crowding variable rather than using $x = \sum_{i=1}^{N} x_i$ is a matter of convenience in notation. None of the results are affected by leaving the ith agent's effort out of this argument. However, the interpretation as a crowding variable is somewhat altered. Here, the derivative of y_i with respect to the x_{-i} variable simply denotes the effect upon the ith agent's production by the effort of the other agents participating in the fishery.

4. An alternative equilibrium concept is a closed loop or feedback method. The feedback method has the advantage of being subgame perfect, meaning that agents do not commit to a strategy which they may later regret. The closed loop equilibrium conditions involve an extra term in the costate equation which is, roughly speaking, a conjectural variation term (Negri, [1989]). However, when the number of agents is very large, this term diminishes in value. Thus, although the open loop method does not satisfy subgame perfection in a competitive environment, this is unlikely to cause a difference in the equilibrium paths (Eswaran and Lewis [1985]). Furthermore, Negri [1989] has shown that in the case of renewable ground water models, the feedback solution results in a lower steady state stock than occurs under the open loop equilibrium; thus, the closed loop equilibrium is worse than the open loop feedback. In what follows, we show that the market price for an ITQ cannot eliminate cost-increasing behavior in the presence of production externalities under the assumption of open loop optimization behavior. Since the market price does not cause cost increasing behavior to disappear in this model, it is doubtful that it will work in a model where agents are assumed to be acting more strategically.

5. The reason we convert these problems into static optimization problems is the fixed cost terms. In order to solve for the entry condition in a dynamic problem, we would need to introduce a scrap value function $\sum_{i=1}^{N} K_i$, which contains the control variable N. In order to use the standard results from optimal control theory, we would need to create another state variable measure of the number of vessels which entered the fishery. It is much simpler to turn the problem into a static optimization problem.

6. The multiplier α is analogous to the multiplier λ in the discussion in Section 2A.

7. We are making no assumption about how the quotas are handed out to fishermen. This allocation problem is extremely relevant in terms of allocation of economic rents, but is irrelevant in terms of efficiency (see Johnston and Libecap [1982]).

8. If the discount rate within the season is rate r, then the market price for permits will rise at the rate of interest over the course of the season. A positive discount rate will also imply that even in the case where there is no stock effect (i.e., $\partial y_i/\partial s = 0$ for all i), effort will not be constant over the course of the season.

9. The reader will note that I have not given an explicit statement regarding the optimal number of permits to be traded. The purpose of the model is to show that the market price cannot solve the externality problems. Thus, the actual number of permits traded by an individual is incidental to the problem.

INDIVIDUAL TRANSFERABLE QUOTAS   407

10. In the case of no congestion or stock externalities, the objective function above (15) would be rewritten as

$$V_i = T[Py_i - c_i - mz_i] - K_i + \lambda_i[q_i(0) - T(y_i - z_i)] + \alpha[\overline{T} - T].$$

11. The reason is that there are diminishing returns to effort, so a consolidation of effort has to decrease both the catch and the number of vessels. Both decreases benefit the other fishermen.

12. In Alaska, for example, the early salmon fisheries were dominated by the owners of fish traps. Fish traps were efficient means of catching salmon. The cannery would be located at the first point on a river where it was economical to construct an obstruction across the river. The owners would build a weir that funneled the migrating fish into holding pends where the cannery workers could easily catch them and immediately put them into the processing lines or could store them at low cost during times when the run exceeded the plant capacity. The system allowed for strict control over escapement. However, the fish traps were also an effective means for excluding competitors. Since there was only room for one fish trap per river system, the cannery getting the best spot near the mouth of the river could effectively exclude all other users. Other methods of harvesting were more expensive than using the fish trap. The result was that from the 1920s through the 1950s, the fish trap technology was the dominant means of catching salmon. In practice, this meant that non-Alaska canneries were able to exclude Alaska fishermen from the market. The result was tremendous political pressure to outlaw fish traps. By 1956, when the Alaska State Constitution was drawn up, fish traps were declared illegal. Indeed, one of the main reasons for the Statehood movement was the desire to eliminate the fish traps (Cooley [1963]).

13. There is one contemporary example of which I am aware where fishermen have actually engaged in a profit sharing arrangement. In a herring fishery occurring in the Sitka Sound in Alaska, the fishermen have been organized for several years to share the profits and to restrict the number of vessels actually fishing.

REFERENCES

Ragnar Arnason [1989], *Minimum information management with the help of catch quotas*, in *Rights based fishing* (Neher, Arnason and Mollett, eds.), Kluwer Academic Publishers, Dordrecht, The Netherlands.

Francis Christy [1973], *Fisherman quotas: A tentative suggestion for domestic management*, Occasional Paper Series, Law of the Sea Institute, University of Rhode Island **19**, 1–6.

Colin Clark [1980], *Towards a predictive model for the economic regulation of commercial fisheries*, Canad. J. Fisheries & Aquatic Sci. **37**, 1111–1129.

Richard A. Cooley [1963], *Politics and conservation: The decline of Alaska's salmon*, Harper and Row, New York.

Parzival Copes [1986], *A critical review of the individual quota as a device in fisheries management*, Land Economics **62**, 278–291.

Mukesh Eswaran and Tracy Lewis [1985], *Exhaustible resources and alternative equilibrium concepts*, Canad. J. Econom. **28**, 459–473.

408   J.R. BOYCE

Ronald N. Johnson and Gary D. Libecap [1982], *Contracting problems and regulation: The case of the fishery*, Amer. Econom. Rev. **72**, 1005–1022.

Jonathan Karpoff [1987], *Suboptimal controls in a common resource management: The case of the fishery*, J. Politic. Econom. **95**, 179–194.

David G. Moloney and Peter H. Pearse [1979], *Quantitative rights as an instrument for regulating commercial fisheries*, J. Fisheries Research Bd. Canada **36**, 859–866.

Ben Muse and Curt Schelle [1989], *Individual fisherman's quotas: A preliminary review of some recent programs*, Alaska Commercial Fisheries Entry Commission, 89-1.

Donald H. Negri [1989], *The common property aquifer as a differential game*, Water Resources J. **25**, 9–15.

Philip A. Neher, Ragnar Arnason and Nina Mollett, eds. [1989], *Rights based fishing*, Kluwer Academic Publishers, Dordrecht, The Netherlands.

Anthony Scott [1955], *The Fishery: The objectives of sole ownership*, J. Politic. Econom. **63**, 116–124.

———— [1988], *Development of property in the fishery*, Marine Resource Econom. **5**, 289–311.

Ralph E. Townsend [1990], *Entry restrictions in the fishery: A survey of the evidence*, Land Econom. **66**, 361–378.

James E. Wilen [1979], *Fisherman behavior and the design of efficient fisheries regulation programs*, J. Fisheries Research Bd. Canada **36**, 855–858.

James E. Wilen and Frances Homans [1992], *Market considerations in common property resources*, unpublished paper, University of California, Davis, February.

[19]

A Critical Review of the Individual Quota as a Device in Fisheries Management

Parzival Copes

1. INTRODUCTION

Economists were sensitized to the common property problems of the fishery by H. Scott Gordon's seminal article that appeared in 1954. Ever since they have searched for effective measures of management that would lead to economically rational fisheries exploitation. From the start it was recognized that fishery problems were related to the absence of individual property rights in the fish stocks. Nonexclusiveness of access robbed fishing operators of the incentive to husband the resource, leading almost invariably to excessive levels of exploitation. The fugitive nature of most fish stocks, together with the multiple resource use of their water habitat, made it usually impractical, if not impossible, to solve the problems by dividing fish stocks into discrete units for which effective property rights would be assigned. Consequently, economists' solutions have been limited to devices that bestow at most partial property rights on fish resource users, supplemented by various regulatory mechanisms designed to promote improved resource allocation, perhaps modified by considerations of distributional equity.

Until recently the advice from economists on fisheries management tended to focus on various forms of limited entry licensing. This was designed to restrict inputs of manpower and/or capital in the fishing industry and to circumscribe their use. In addition, "buy-back" programs were advocated to remove existing excess capacity from the fishing industry. Limited entry, indeed, has been used on an increasing scale over the last two decades. The salmon fishery rationalization program in

British Columbia, that was introduced in 1968, was the first scheme of note to incorporate a buy-back program. Initially it was widely acclaimed as a prototype for rational fisheries management.

In the end the British Columbia salmon rationalization program has come to be considered a failure. While there is room for debate on the full analytical detail, there is substantial agreement on the major features of the scheme's failure (cf. Pearse and Wilen 1979; Copes and Cook 1982). The buy-back operation resulted in the removal of vessels accounting for no more than five percent of the catch. But investment in technological improvements of vessels remaining in the fleet (so-called "capital stuffing") caused effective capacity to increase greatly.

Many examples can be given where limited entry licensing appears to have increased the net benefits derived from a fishery and to have helped attain biological conservation objectives. However, the degree of success achieved overall with limited entry schemes has been sufficiently modest for economists to cast about for an alternative approach to fisheries rationalization. If limitation of inputs into the fishery cannot be made foolproof, it was argued, why not go for limitation of outputs?

2. QUOTA ALLOCATION

One form of fisheries output limitation indeed has long been used by both national and international management authorities. They

Land Economics Vol. 62. No. 3, August 1986
0023–7639/86/003–0278 $1.50/0
© 1986 by the Board of Regents
of the University of Wisconsin System

Professor of economics and Director of the Institute of Fisheries Analysis at Simon Fraser University. The author wishes to acknowledge useful suggestions offered by an anonymous referee, as well as helpful comments and valuable research assistance from C.S. Wright.

have frequently sought to meet conservation objectives by imposing a total allowable catch (TAC) in fisheries subject to heavy exploitation pressure. Usually this meant monitoring landings and closing the fishery when the year's TAC was reached.[1] However, the annual TAC helped little to achieve economically rational exploitation. The usual result was a "race for fish" as soon as the annual season opened, with each operator attempting to get as much as possible before the season was closed. Management by TAC has been marked by severe overinvestment in fishing capacity and idleness of manpower and equipment during an often long closed season.

The race for fish could be observed at two levels. In international fisheries it was marked by the competitive build-up of national fleets, often heavily subsidized by their respective governments. At the operational level the race was among individual vessels attempting to outdo one another in fishing power and intensity of operation during the open season. The solution proposed has been to divide the TAC into country quotas at the international level and into individual quotas for fishermen, vessels, or enterprises at the operational level.

With a secure quota, a country would no longer have to strive for its share of the international harvest by increasing its fleet capacity in competition with other countries doing the same. A country could then take its allocated catch in the most economical fashion available to it. The Convention concluded by Canada and the U.S. that established the International Pacific Salmon Fisheries Commission in 1937, provided for a 50/50 quota split of the harvest under the Commission's jurisdiction. This arrangement was a prerequisite for the salmon rationalization program introduced by Canada in 1968.

A country quota in an international fishery is of limited use if a country fails to manage the effort of its own fleet. For it is then still faced with an economically wasteful race among its own flag vessels for shares of the national quota. This, of course, is the same problem any country faces in its domestic fisheries that are subject to a TAC, but that allow unfettered competition for shares of the catch among individual operators. In the final analysis, then, rational exploitation requires

management of the fishing effort put out by individual operators.

3. RATIONALE FOR THE INDIVIDUAL QUOTA

The individual quota may be thought of as a fixed share of the catch allocated in advance to individual operators (i.e., recognized fishermen, fishing units or fishing enterprises). Allocations may be made for a single season (e.g., year), for a longer period, or in perpetuity. While an individual quota may be set as a stated percentage of the total catch, administrative practicality dictates that it will usually involve setting a specific quantity that a fishing operator may take in a particular season. Of course, once a TAC is set for a season, a percentage of the catch translates into a fixed amount in any case.

In the fisheries rationalization debate that economists have conducted over the last few decades, the notion of the individual quota has long been present, though serious consideration of it as a major device for rationalization was initially slow in coming. A considered proposal for the introduction of individual quotas was put forward by Francis T. Christy, Jr., in 1973. Good examples of more recent and more elaborate rationales are those provided by Moloney and Pearse (1979) and by Scott and Neher (1981). In practical management terms, some early instances can be found of the application of individual quotas, e.g., in the Prairie Lakes fisheries of Canada where they have been used since the 1930s. The deliberate application of individual quotas (i.e., quotas at the operational level as distinct from country quotas) to achieve goals elaborated in recent theoretical discussions is still at an early stage of development, though there are now several fisheries—e.g., in Canada, New Zealand, Iceland, Norway and South Africa—where this device is being applied. As yet there appears to be no adequate assessment

---

[1] An early example of such TAC management is that introduced in 1930 in the Pacific halibut fishery by the joint Canada-U.S. International Fisheries Commission (later called the International Pacific Halibut Commission).

of the results on which firm general conclusions may be based.

The purported advantages of management by individual quota allocation lie in the elimination of important external diseconomies, both among those associated with open-access fisheries and those peculiar to fisheries subject to limited entry licensing. The guarantee of an individual quota—it is contended—means that fishing operators do not have to race one another to secure their share of the catch as quickly as possible before the TAC is filled and the fishery is closed. When they are assured of their quota—so it is held—fishermen can take their time, spreading their effort optimally across the entire season and using the most economical configurations of equipment and manpower in the process. Gone will be the need for competitive escalation of speed and fishing power, requiring large capital inputs and driving up costs unnecessarily. As a further advantage operators will find little need to fish in bad weather or under other dangerous circumstances in order to keep up their share of the catch. In addition, harvest gluts can be avoided or reduced and a higher value of sales achieved by meeting optimally the time patterns of demand over the year of both fresh fish consumers and processors.

With no need to race for the fish, operators presumably would be induced to use only the most economical capital and labor input configurations. This would avoid some of the regulatory problems encountered in limited entry licensing. There regulators are caught on the horns of a dilemma. If they allow free play to technological change, it will be used, at least in part, for capital stuffing and thus for a socially inefficient increase in fishing capacity.[2] If, on the other hand, they restrict technological change, they are likely to suppress socially efficient cost-reducing technology along with socially inefficient capacity-increasing technology. The resources used in administering the regulation process, of course, represent a further social loss.

The allocation of individual quotas in fishing has been referred to as "stinting the commons," by analogy with the allocation of quantitative pasturage rights on the medieval commons. Proponents emphasize that stinting introduces a system of property rights or quasi-property rights to the fishery. They imply that this should help solvê the problems of common property resource exploitation which are linked to the absence of property rights in the fish stock.[3] Related to this, advocates of the individual quota usually also emphasize a need to make it an "individual transferable quota" (ITQ). Transferability is an obvious characteristic of ownership. It means that fishing operators may sell either their entire quota, or parts of their quota, to other operators. The sale could involve the quota for a given season only, or for a number of seasons, or it could be in perpetuity.

The evident advantage of transferability is that it further facilitates rationalization. If there is surplus capacity of capital and manpower in a fishery in relation to the TAC, there will generally not be enough fish to keep vessels operating at full capacity throughout the season. Rents could then be generated in the longer term by withdrawal of some fishing units from the fishery. There should be a reasonable expectation that the prospect of rents will lead more efficient operators to buy out the quota entitlements of less efficient operators. Thus quota rights would be consolidated in the hands of the most efficient operators who would be able to fish full time and reduce unit costs of operation. In the process both buyers and sellers of quota rights could share in the net benefits of the rents that would be generated.

Short-term rationalization would be promoted by the flexibility of short-term transfers of quotas, or parts thereof. If in a particular season an operator was unable to use his entire quota (e.g., because of illness or vessel breakdown), he could sell all or part of his quota rights for that season to other operators who

---

[2] Excessive capacity build-up under limited entry licensing is commonly marked by technological advances involving higher capital inputs. Theoretically, of course, capacity could also be increased by "labor stuffing," though this does not appear to occur often.

[3] A thorough exploration of the property rights aspects may be found in Scott (1985). Earlier on Bell and Fullenbaum (1973) interestingly proposed the issuing of "stock certificates" to establish fishermen's property rights.

had exhausted their own quotas before the end of the season, or anticipated doing so.

The introduction of individual quotas may be difficult and controversial. There are many different criteria that could be used in determining the initial allocation amongst individual operators. The quota shares could be auctioned, sold at a fixed price, or given away free. Except for the case of auction, a determination would have to be made as to how large a quota each operator would receive. Operators could be given equal shares, or shares based on historical catch performance, or on fishing vessel and gear capacity, or numbers of crew, or various combinations of these and/or other criteria related to considerations of equity, rationality or practicability. A discussion of the options is beyond the purpose of this paper. Suffice it to say that the introduction of any new scheme of fisheries management is likely to be controversial, but that this has not necessarily robbed new schemes of their inherent merit or prevented them from being implemented.

4. WHAT CAN GO WRONG

One of the arguments that is often used in promoting individual quota management is that limited entry licensing is inherently deficient as a management device because of the skill fishermen show in circumventing the rules or defying the intent of entry limitation. The capital stuffing process, whereby additional capacity enters the industry despite, or because of limited entry, is mentioned in evidence. Certainly, the externalities inherent in the common use of a pool resource drive fishermen to act in accordance with their individual interests, where often this is contrary to their collective interest. That they show great ingenuity in doing so is beyond doubt. As a result one may well proclaim that fisheries are exceptionally vulnerable to Murphy's Law: "If anything can go wrong with a new fisheries management scheme . . . it will."

Ironically, when it comes to promoting individual quota management, its proponents often fail to apply the sharp insights gained in exposing the deficiencies of limited entry licensing. There is no reason to assume that fishermen, where confronted with the rules of individual quota management, will lose either their ingenuity at circumvention or their incentive to promote individual interest at the expense of collective interest. Recognizing such to be the case, this paper will explore a variety of problems that should be anticipated with the introduction of individual quota management. Without claiming to be exhaustive, problems will be identified in fourteen areas under the following headings: (1) Quota Busting, (2) Data Fouling, (3) Residual Catch Management, (4) Unstable Stocks, (5) Short-Lived Species, (6) Flash Fisheries, (7) Real Time Management (8) High-Grading, (9) Multi-Species Fisheries, (10) Seasonal Variations, (11) Spatial Distribution of Effort, (12) TAC Setting, (13) Transitional Gains Trap, and (14) Industry Acceptance.

Quota Busting

For many fisheries, enforcement is likely to be one of the most difficult problems with an individual quota system. Obviously, there is a material incentive for fishermen to engage in "quota busting," i.e., catching a larger amount of fish than the individual quota allows. The extent of compliance with quota limits will be influenced by such factors as individual conscience, community culture and social sanctions, effectiveness of official monitoring and enforcement efforts, severity of penalties on conviction for infractions, and extent of gain from cheating on quotas. Analysis of the trade-off between potential gains and losses from legal infractions may be found in the literature on the economics of crime (see e.g., Polensky and Shavell 1979).

In different fishing communities distinctly different attitudes towards enforcement of fisheries conservation and management regulations prevail. Thus in the South Australia rock lobster fishery there is strong pressure from fishermen for rigorous enforcement of limits on the number of traps allowed to be fished, including severe penalties for infractions (Copes 1978). In contrast, in the lobster fisheries of the Canadian Maritime Provinces infringement of the regulations on size limits and permitted number of traps is known to be endemic. Attempts by fisheries officers to enforce the regulations have provoked violent reactions (e.g., Anon. 1983a). In the face of community pressure the courts there have dealt leniently with violators. Obviously, enforcement of fisheries regulations is impeded if there is a noncooperative attitude towards such enforcement by the community at large. Where, on the other hand, the community favors regulations, enforcement is enhanced and may be reinforced by social sanctions applied against violators, such as ostracism or reporting to the authorities of observed infractions.

Many of the factors underlying community attitudes towards fisheries regulations are most competently analyzed by anthropologists and sociologists. But one important factor should be mentioned here. The attitude of fishermen is evidently influenced by the credibility of enforcement, including particularly the likelihood of detection of infractions. The early experiments with individual boat quotas in the Bay of Fundy herring fishery were abandoned under pressure of skippers who knew that colleagues were cheating on their quotas without being caught (Anon. 1983b). The Director-General of Fisheries for the region reported (Crouter 1983) that "attempts to enforce vessel quotas proved to be largely unsuccessful." He added that the experience was "that each and every fisherman will attempt to 'cheat' on his quota and processors promote that attempt through collusion in falsifying records."

The chance of detection of quota busting is enhanced where, in relation to the size of the catch, the number of vessels and the number of points of landing is small. Thus a small number of inspectors can easily monitor the catches from the fleet of a few hundred large groundfish trawlers operating on Canada's Atlantic coast. They must land their fish at one or another of a limited number of processing plants. It is difficult to hide a trawler-load of fish and penalties can easily be made prohibitive for large fishing companies with substantial investments at risk. The cost of a few inspectors can be met easily from the public revenues generated by a high volume fishing operation.

In contrast, consider the British Columbia salmon fishery with over 5,000, mostly small boats. They can potentially land their catches at hundreds of places along an indented coastline that measures thousands of miles. While most salmon fishermen now sell to a few large companies, there are substantial numbers of smaller fish handlers eager to take their catch and they can sell also directly to the public at numerous wharves. Monitoring and enforcing individual quota limits under those circumstances would appear well-nigh impossible. While the individual boat quota has been proposed for the British Columbia salmon fishery, it is no wonder that it has not been accepted and implemented. It is easy to conclude that the individual quota will be very difficult to enforce in a fishery characterized by many small vessels, numerous actual and potential marketing channels, and geographically widely dispersed activity.

Data Fouling

Fisheries managers require reasonably accurate reports on catch and effort from vessel operators as a basis for their estimation of stock strengths and optimal exploitation rates. But if the individual quota system results in fishermen taking catches in excess of their quotas, they are almost certain to underreport their catches in order to evade detection. They may also falsify their reports on effort in order to make these appear compatible with their incorrect catch reports. It has already been observed by fisheries scientists that the introduction of quotas in some places has led to severe deterioration in the quality of data that fisheries managers have to work with (Gulland 1985). In the common fisheries zone of the EEC country quotas are allocated. It is claimed that an internal EEC Commission report has found that Dutch fishermen have systematically cheated on Common Market catch limits with the connivance of some Dutch officials (Lichfield 1984). As a result of this and other suspected transgressions, EEC fisheries scientists have started to add in an often large adjustment factor for "unreported catches" in their calculations (Brander and Gulland 1984). Needless to say, such reliance on guesswork will result in unreliable stock estimates and fishing effort controls.

If an individual quota system results in eva-

sive reporting on catches and effort by fishermen, it is equally likely to lead to distorted
reporting on cost and earnings data in order to
complete the cover-up. Thus socioeconomic
studies on the condition of the fishing industry
may also lose in reliability with the data fouling that accompanies the introduction of individual quotas. This will impair efforts to
monitor the effects of the individual quota system and to determine relevant social policy.

Residual Catch Management

The greatly varying nature of different fish
stocks and of fishing operations on those
stocks, calls for distinctly different techniques
and regulations in fisheries management. In
most fisheries, managers each season, implicitly or explicitly, must determine a desired division of the stock into catch and escapement.
According to the nature of the fish stock and
the fishery, managers may find it distinctly
more effective in one case to set a TAC for the
catch, so that escapement becomes the residual, and in another case to set escapement as
the target, thus making catch the residual.

As an example, salmon management in the
Northeast Pacific clearly requires the latter
technique, as getting the "right" number of
fish of each stock to escape up-river in the
annual spawning run is the crucial factor in
achieving optimal reproduction. Consequently, when it has been determined that the
right number of fish have escaped, the fleet
must be encouraged to quickly mop up the
remainder, both to provide for a better catch
and to prevent a deleterious overloading of the
spawning grounds. At that point it may be essential to utilize the full extent of fishing capacity available in order to mop up the residual
of the spawning run in the available time,
which may be a matter of a few days or hours.
To parcel out individual boat quotas at such a
point in time is patently absurd. With an unknown size of residual catch the quotas could
not be calculated in the first place. And the
need of the moment obviously is to use all
available fishing effort and not constrain any
operator. Individual quota management is inherently unsuited to fisheries where the catch
is residual to a managed escapement target.

Unstable Stocks

Relating to the preceding point, it may be
more generally observed that individual quota
management does not work well when the
TAC cannot be determined with certainty at
the beginning of a fishing season. With relatively slow-growing stocks of relatively long-
lived species an appropriately set annual TAC
is likely to amount to a rather modest part of
total biomass. There is then some leeway in
setting the size of the TAC without any serious
danger to conservation. Given relative stock
stability there is ample time to adjust estimates
of permissible TACs from one season to the
next. It is then possible to set a firm TAC at
the beginning of an annual fishing season and
stick to it. Most groundfish stocks in the North
Atlantic are probably susceptible to firm TAC
determinations in this fashion.

Species that are prone to producing highly
variable year-class strengths are characterized
by serious stock instability, particularly when
they are short-lived. Several pelagic species
(e.g., herring) fall into this category. Determining annual TACs for stocks of these species is a hazardous undertaking. It is frequently essential to set only a tentative TAC
at the beginning of the fishing season, to monitor stocks and catches constantly during the
season, and to adjust fishing plans and allowable catches at short notice, accordingly. But
adjusting the TAC, and thus the individual
quotas based on it, in mid-season is incompatible with the rationale for the individual
boat quota. Fishermen must be confident that
they have the entire season in which to decide
where and when to fish without fear of losing
any part of their quota. Uncertainty as to
whether the initially allocated quota will still
be allowed later in the season, will cause fishermen to "race for the fish" at the beginning
of the season with as much equipment as they
can muster. The presumed advantages of the
individual quota are lost. Cancellation of an
initially allocated quota just once is enough to
induce many fishermen never to trust fisheries
managers again. It may be concluded that relatively unstable stocks, even where they are
managed by catch targeting, will require frequent fishing plan adjustments which make
them ill suited to an individual quota system.

Short-Lived Species

In some fisheries there is no observable relationship between size of catch and subsequent recruitment, i.e., the accession to the stock of catchable fish. This appears to be the case particularly with species of high fecundity where even a small number of surviving spawners is sufficient to fully restock the available ecological niche each season. The adequacy of a spawning stock for this purpose in some cases may be aided by gear selectivity that leaves a sufficiently large stock beyond the reach of the fishery. Or it may be helped by a spawning stock in a sanctuary not touched by the fishery.

Many stocks of crustaceans are believed to have recruitment that for practical purposes is independent of parent stock size. Among these are tropical prawn and shrimp stocks that are short-lived and that are each year available to the fishery as mature individuals for a short period of a few months or weeks only. Obviously, during this period it is important to fish the stock hard in order to mop it up before it succumbs to natural mortality. Under such circumstances it would be irrational to impose individual quotas. In the first place it would be impossible to determine the TAC to be divided, when all of an unknown biomass should be taken. It would also be counter-productive to force any vessel to quit fishing because its quota is taken, when all of the available fishing capacity should be utilized to secure quickly a catch that will otherwise be lost to natural mortality.

Flash Fisheries

Some specialized fish products can be obtained only if the fish yielding the product is caught in a particular condition that occurs over a very short period only. The British Columbia herring fishery provides an example. The priority product of that fishery is roe for the lucrative Japanese market. To meet the exacting standards of that market the roe must be taken as the fish are moving inshore to spawn. In each spawning location there is a window of opportunity that may be as short as a few hours only. Fisheries officers sampling the fish at each spawning site determine

the right moment to open and close the fishery at each site and signal the waiting fleet accordingly. Under these circumstances a "race for the fish" is a necessity. There is no time to fuss with quotas. The fishery cannot wait for the unlucky or unskilled fisherman who is not able to take his full share of the catch before the fishery is closed again.

Real Time Management

Improved knowledge of fish stock dynamics, together with up-to-date monitoring techniques, sophisticated data processing and advanced fleet communication networks, are allowing the introduction of "real time" management in increasing numbers of fisheries. Real time management may be effected by continuous stock monitoring in conjunction with quick-response time and area closures or gear restrictions. This would be in pursuit of objectives such as the protection of critical spawning activity, the avoidance of immature stock components, and targeting on high yield stock components. Real time management responses may be particularly important in fisheries where optimal exploitation is sensitive to the precise timing and pattern of stock migrations and to the variable mixing of targeted and protected stock components.

In the conditions referred to, an individual quota regime would not likely be suitable. It could not avoid a race for fish, for fishermen would not want to be caught with an unfilled quota at the time of a sudden closure. In those cases where closures did prevent some fishermen from filling their quotas, there would be protests and pressure for a reopening in the name of equity. Also, when managers identify isolated stock components that are suitable fishing targets, optimal harvesting strategy may call for the rapid application of all available fishing power to such stock components before they are lost through migration or natural mortality. The constraints of individual quotas at such a time would only lead to waste of available fish.

High-Grading

A fishing operator whose catch is confined to a given individual quota will wish to obtain the greatest net value from that quota. Usually this means that he will want to fill the quota with the best quality of fish only. If fish of a particular size or condition (e.g., with or without spawn) fetch a significantly better price, he may well be induced to "high-grade" his catch by discarding fish of lesser quality.[4] As mortality of discarded fish tends to be high, this practice may be expected to lead to a waste of fish that diminishes the aggregate net revenue obtainable from the fishery. As discards normally are not reported it will also lead to data fouling, depriving fisheries resource managers of accurate data on fishing mortality.

To some extent high-grading is an alternative strategy to quota busting for fishermen. Once a fisherman is retaining fish in excess of quota, he might as well retain and sell both his higher quality and lower quality catch that is in excess of quota and therefore stop high-grading. It is possible, however, that a fishing operator will wish to engage in both practices. When a catch excess can be eliminated by discarding lower quality fish, the risk of retaining it may not be worthwhile. When an excess catch of high quality fish is taken and retained, lower quality fish may be dumped to make more hold room for the better fish.

An "enterprise quota" has been introduced in the Canadian Atlantic groundfish trawler fishery, with individual quotas for fish processing companies that operate the trawlers. Landings are easily monitored at the limited number of plants where the fish is processed. Relatively stable stocks allow firm TACs to be set at the beginning of each season. The fishery, consequently, has been considered well suited to the individual quota system. Recently, however, fisheries managers have discovered that there is a significant discard problem, though they are unable to determine how serious it is.[5] The initial optimism regarding the suitability of the individual quota in this fishery may now be called in question.

Multi-Species Fisheries

Multi-species fisheries are notoriously difficult to manage. An effort level that is optimal for one species in the mix is likely to be too high or too low for other species. A directed

management of effort on one species inevitably involves by-catches of other species. A general discussion of multi-species problems is beyond the scope of this paper, but it is appropriate to comment here on some extra problems that individual quotas may create for management of multi-species fisheries.

Fisheries managers may attempt to set separate sets of individual quotas for different species in a mixed-stock fishery. The chances that a fishing operator's catch would conform precisely to the proportions of the various species quotas allotted to him are almost nil. Inevitably he will fill some quotas before others and will find himself with excess catches of some species when he continues to fish in order to fill all his species quotas. He may either retain the excess catches, which would be illegal quota busting, or discard them, either of which would interfere with rational management and lead to socially undesirable results. Some improvement might be effected where quotas are allowed to be exchanged or traded among operators, but even then it is unlikely that a precise match could be obtained, as chances that the aggregate catch mix would precisely reflect the TACs for the various component species would be remote. Managers will likely be induced to be tolerant to an extent with regard to excess by-catches of species that are not primary targets. But the more tolerant they are in order to prevent discard waste or quota busting, the more fishing operators will contrive to "accidentally" take larger excess by-catches, particularly of the more valuable species in the mix.[6] In a fishery managed by seasonal closure a stop can be put to this when the aggregate catch for all species is about right. But to retain management cred-

---

[4] The term "high-grading" here is appropriately descriptive of the fishery phenomenon to which it refers. It has a somewhat different meaning in the mining industry.

[5] Information obtained from (non-quotable) sources in the Canadian Department of Fisheries and Oceans.

[6] The problem could possibly be attenuated by allowing by-catches, while taxing them at a level where it was just worth bringing in a by-catch, but not worth targeting on it. However, such precision would be almost impossible to achieve, particularly as the tax level that would induce the desired behavior would vary among fishermen.

ibility in an individual quota fishery, the season must be left open for all operators who have not filled all of their quotas.

It is possible, alternatively, to manage a mixed-stock fishery through a single, all-species, individual quota for each operator. But this is likely to result in an extensive effort at high-grading, with operators racing for the higher-value species, while discarding lower-value species along the way. In a fishery managed by seasonal closure, racing for the fish would also take place, but there would be less incentive to high-grade as there would be no individual quota limit to induce discarding.

Seasonal Variations

As discussed above, a major advantage claimed for the individual quota is that it obviates the need for fishermen to "race" for fish at the beginning of the season. But as Christy (1973) has acknowledged, this advantage might not materialize, or not fully so, if a stock is naturally subject to significant intra-seasonally declining yields. It is generally more profitable to the individual operator to fish when stocks are concentrated and the catch per unit of effort is high, than when they are dispersed or thinned out later in the season. All participants in the fishery then may attempt to fill their quotas from denser stocks at the beginning of the season, engaging in capital stuffing to prepare themselves better for the early-season race to the fishing grounds.

Of course, some operators may still keep part of their quota for the late season in order to benefit from price advantages at a time when landings are down. Nevertheless, the tendency for operators competitively to concentrate effort in the season with highest yields is bound to be excessive (i.e., socially nonoptimal) and thus to find expression in external diseconomies. While the individual quota may attenuate the tendency to race for fish, it is unlikely to eliminate the practice entirely.

Spatial Distribution of Effort

Many fisheries are characterized by different grounds, with different revenue yields per

unit of effort when initially exploited. This may result from different stock densities on the various grounds, from different qualities of sub-stocks, or from their more or less advantageous location relative to ports and markets. In an open-access fishery the tendency is for the most profitable grounds to be exploited first, resulting in declining revenue yields per unit of effort on these grounds. Additional grounds are brought under exploitation when their revenue yields per unit of effort match the declining yields of the grounds first exploited (Gordon 1954). The intramarginal grounds are thus inevitably overexploited resulting in dissipation of the rents they could yield.

This pattern of spatial maldistribution of effort is not broken by an individual quota regime. For the boats with unfilled individual quotas still have open access to any grounds within the fishery. They will still tend to overexploit the higher-yield grounds, fully dissipating any rents available there in excess of those on the grounds last to be brought under exploitation. There will also be a tendency towards capital stuffing as operators prepare to race each other to the best grounds. This racing, of course, will also contribute to a seasonally nonoptimal concentration of effort. Even so, with an appropriately restricted TAC there may be a margin of rent left on all exploited grounds. But the aggregate rent will be below that attainable because of the socially nonoptimal spatial and time distribution of effort. Indeed, if an individual quota regime is effective in reducing aggregate effort, it will tend to sharpen the concentration of effort on the higher yield grounds. Exhaustion of quotas in fishing those grounds would cause other grounds to go unexploited even when they are capable of yielding at least a low level of rent. In this case again, the individual quota might reduce the loss of rent that occurs in open access fisheries, but not eliminate it.

TAC Setting

An individual quota system hampers the targeting of a precise annual total catch (Clark 1985). No fisherman is allowed to take more than his quota. However, for various reasons some fishermen may not be able or willing to take all of the quota allocated to them. Of course, if their quotas are transferable they may sell or otherwise dispose of them. However, they will not always have the time and opportunity to transfer unused portions of their quotas to other fishermen. If in an individual quota fishery the authorities succeed in suppressing quota busting, they will likely experience an opposite problem. With no fishermen exceeding his quota, while some fail to fill theirs, the total catch will fall below the TAC. If the TAC is set to mark the optimal catch, any shortfall will result in a nonoptimal catch. In anticipation of a shortfall, the authorities could set a TAC above the optimum—but obviously with an uncertain outcome, as the size of the shortfall cannot be determined in advance.

Transitional Gains Trap

As discussed above, the individual quota cannot achieve its full purpose of rationalization unless it is a transferable one (ITQ). However, transferability may lead to another problem. A common social and political purpose of fisheries rationalization for government is to solve a chronic problem of income deficiency that is exhibited by the fishing industry in many localities. After all, economists going back to Gordon (1954) have remarked on the propensity of common property resource exploitation to lead to deficient income levels.

Tullock (1975) demonstrated that where a government applies a measure of long-term assistance to an industrial sector in which it wishes to improve income levels, the gains to the class of people thus favored tend to be transitional. At least this is so where the right to the benefits is transferable. In that case the initial generation of beneficiaries is able to capitalize the stream of future benefits and extract them from those succeeding them, who must purchase these rights at their full value. As a result succeeding generations enjoy no net benefits, as their gross benefits will be offset by the purchase price they were required to pay. If initial circumstances have not changed, the succeeding generations will fall to the lower levels of net income that government action was designed to overcome in the first place.

This scenario may be expected to be acted out in the fishing industry if rationalization produces net benefits, the rights to which are transferable (i.e., saleable). Indeed there is a relevant example in the British Columbia salmon rationalization program, where limited entry licenses were made transferable (Copes 1978; Copes and Cook 1982). The initial reduction in capacity through buy-back (before capital stuffing took off), together with a fortuitous rise in fish prices, produced some rents and expectations of further rents. Their value was capitalized in the price charged for a license on transfer. The consequent precarious financial position of many new fishermen who bought licenses, and the seizure of their boats for nonpayment of loans, is well known.

In the case of some fisheries rationalization schemes (e.g., limited entry licensing), transferability of rights is not an especially crucial ingredient. But as explained above, it is rather important in the case of individual quota management. If, however, it is also important to bring about a long-run improvement in income levels for succeeding generations in a particular fishery, the ITQ approach is likely to prove unsuitable.

Industry Acceptance

To bring about successful reform in fisheries management, it is usually important to secure the approval and cooperation of the fishing industry. Indeed many governments will not attempt any major changes in their system of fisheries management unless such cooperation is assured in advance. Inevitably in any new scheme, participants will be affected. Some participants may expect to benefit more than others, and some may indeed consider it likely that they will lose. It is therefore difficult enough to develop a consensus for change.

The ITQ has some special psychological drawbacks that are likely to diminish its acceptability. The common property condition tends to make fishermen a particularly individualistic and competitive breed. They tend to be relatively risk-prone. Believing in themselves, they are often convinced of their inate ability to outfish their rivals and earn the status of "highliner." Most seem affected by the "prospector's syndrome," believing that tomorrow's luck will bring the big catch. Having a fixed quota diminishes the opportunity for fishermen to show their mettle and to better themselves by superior performance or to benefit from a lucky big catch they feel they deserve. A large catch becomes a matter of the financial resources to purchase a large quota, rather than a matter of fishing skill or serendipity. There is likely to be a latent fear also that quotas will become concentrated in the hands of a small number of operators with substantial financial resources and that fishing companies, either directly or by proxy, will end up with substantial control over fishing rights. At this stage there is not much evidence that the ITQ will be widely favored by fishermen.

5. CONCLUSION

The use of the individual quota (and particularly the ITQ) as a major regulatory device has received much attention and support in recent years from academic fisheries economists,[7] and increasingly also from officials in management agencies. The individual quota, indeed, seems to have replaced limited entry licensing as the new "conventional wisdom." Canada has often taken the lead in fisheries management experimentation and is already applying the individual quota in a number of fisheries—though generally not (or not yet) on a fully transferable basis. The Economic Council of Canada has pronounced itself in favor of the ITQ. With some qualifications, two major fisheries commissions that brought out reports on the Canadian Atlantic and Pacific fisheries, respectively, have also endorsed it (Kirby 1982; Pearse 1982). In New Zealand the individual quota is being used in a few fisheries and being proposed for more. The device is also being advocated strongly for some fisheries in Australia (e.g., Campbell 1984) and the United States (e.g., Stokes 1983). The number of experiments with the individual quota, so far, has been limited and mostly of short duration, so that a general assessment of its effectiveness is not yet available.

From a theoretical perspective, supported by the necessary simplifying assumptions, the ITQ may be presented as an ideal management device, leading to the generation of maximum net economic returns. What should be recognized is that other management schemes theoretically may produce equally ideal results, if the practical problems that are likely to arise with them are assumed away. Not so many years ago limited entry licensing was presented in its simple theoretical purity, with the promise that it would solve the problems of the fishery and generate maximum returns in the form of rent. The real-life experience with limited entry regimes has been sobering. But many of the problems encountered have been so straightforward, that it is difficult to conclude that with a modest application of foresight they could not have been anticipated and contained, circumvented or ameliorated (Copes and Cook 1982).

It does take considerable time for new management schemes to be planned, accepted and put into place. Therefore it is now still possible to make a careful assessment of the practical feasibility of the individual quota before a wider application is attempted. As this paper has tried to show, caution is warranted as much can go wrong with the individual quota. This is not to say that it is a necessarily inferior device for fisheries management and that its application ought not to be considered. Scrutiny of the record of fisheries management reveals no alternative scheme that is free of significant problems. It should be clear, however, that the individual quota is also prone to quite serious defects and that there should be no rush to embrace it as the new panacea—for which there seems to be a present tendency.

The advocates of the individual quota probably have made too much of the property rights aspects of the scheme. The rights to the fish stock bestowed by the individual quota— even in the form of the ITQ—are still far from fully specified property rights. For that matter, the now maligned limited entry licensing schemes also conferred partial property rights, which didn't save them from serious problems. What really counts in rationalizing

---

[7] For a notable exception see McConnell and Norton (1980, 193–94).

the fisheries is not what property rights have been installed, but what externalities remain or are newly created by the particular form of partial property rights introduced.

There is a near-infinite variety of biological, economic, social, and political circumstances affecting different fisheries around the world. Each set of circumstances will create its own unique set of actual and potential problems, requiring a particularized management approach for its solution. At the outset it must be realized that what is to be considered a problem depends in part on the objectives of a society's fisheries policy. Almost always this will include a goal to improve net economic returns from the fishery, however, this is qualified by considerations of distributional equity, lifestyle preferences, employment needs and community viability. Therefore the capacity of a management device to generate rents remains a general requisite and a touchstone of success.

It is difficult to be categorical about the merits of the individual quota in relation to those of other fisheries management devices, because so much depends on the vastly different circumstances that pertain to different fisheries. A review of these could fill several volumes. In this paper, a few generalizations will be offered regarding the practical applicability of the individual quota. Given the considerable advantages offered by it under ideal circumstances, it is perhaps best to consider the individual quota a generally attractive management device, except where circumstances leave it vulnerable to serious problems. Drawing on the preceding analysis of this paper, an attempt will be made in the following paragraphs to identify the circumstances under which use of the individual quota should be avoided or approached with particular caution.

There are some fisheries in which individual quota management could not be adapted to biological circumstances or would be in serious conflict with biological management imperatives. Thus the individual quota would be wasteful in a fishery on short-lived species that should be fished up quickly with all available capacity. It would also be irrational in an escapement targeted fishery, where a residual surplus needs to be mopped up with deliberate

speed. However, in such a fishery it might be feasible to assign individual quotas for the earlier part of the season, while allowing free fishing on the available surplus at the end of the season. In flash fisheries, also, it will usually be impractical to operate with individual quotas. In general, fisheries with quickly changing, unstable or unpredictable stock levels are ill-suited to an individual quota regime and are better managed by time and area closure in conjunction with limited entry. Fisheries on widely distributed, long-lived species, with stable and slowly changing stock levels, on the other hand, may be better suited to individual quota management. Even where stock levels are not stable, it may prove possible to manage a fishery through individual quotas, if the season is long enough to allow the quotas to be parcelled out piecemeal as the season progresses and knowledge of stock levels is refined.

Because of the strong incentive to engage in quota-busting, enforcement can be a serious problem in individual quota management. The problem is likely to be the more serious, the larger is the number of fishing units involved, the more extensive is the geographical area over which they are dispersed, and the greater is the number of possible marketing channels for the catch. The problem may not be a serious one in an industrial fishery with few landing ports and processing facilities. The problem may be insuperable in a widely distributed small-boat fishery for a luxury species that may be sold over-the-side to a wide range of potential customers. In a small-boat fishery confined to a local community the feasibility of individual quota management may depend on the community's culture and attitude towards self-policing.

In some fisheries the discard problem may be sufficiently serious to rule out individual quota management as being too wasteful. However, in certain other cases it may be feasible to reduce discarding to a tolerable level by fine tuning regulations. Separate quotas might be given for different species or for different fish sizes that have different values per unit weight. If, at the same time, the trading of quotas among fishermen were kept easy and were allowed to take place after the landing of surplus catches, much of the incentive to discard might be removed. It might also be possible in some situations to devise value quotas, which could be filled by catches of any designated species or fish size according to landed value.

Many fisheries in the developed world, and most in the Third World, produce average incomes that are considered undesirably or unacceptably low from a public policy perspective. In these cases a major objective of improved management is likely to be the permanent raising of average incomes in the fishery. But a higher income level is likely to be undermined in the long run under a management regime that provides for transferable fishing rights, because of the transitional gains trap. Many of the benefits of individual quotas depend on their transferability, which makes their suitability in the case of "social" fisheries doubtful.

The various problems that could arise with individual quotas, and their often serious nature, suggest that great caution should be exercised when considering the introduction of individual quota management in any fishery. But the same warning should be heeded when contemplating any alternative management scheme. One might simply say that every fisheries resource manager should be required to reflect carefully on Murphy's Law before attempting any new move. Experience so far suggests that we should be nondogmatic in our choice of management technique and that we should select from the array of available fisheries management devices, the combination that is most beneficial and least deficient in any particular set of circumstances. Above all, we must reconcile ourselves to the fact that the best *possible* solutions will still be flawed.

References

Anon. 1983a. "The Language of Fire." *Canadian Fishing Report* 5(6): 4.

Anon. 1983b. "The Trouble with Boat Quotas." *Canadian Fishing Report* 5(10): 4.

Bell, F. W., and R. F. Fullenbaum. 1973. "The American Lobster Fishery: Economic Analysis of Alternative Management Strategies." *Marine Fisheries Review* 35(8): 1–6.

Brander, Keith, and John A. Gulland. 1984. Personal communications in November 1984 from Keith Brander, scientist at the Fisheries Laboratory in Lowestoft of the Directorate of Fisheries Research in the U.K. Ministry of Agriculture, Fisheries and Food, and John A. Gulland, fisheries consultant.

Campbell, David. 1984. *Individual Transferable Catch Quotas: Their Role, Use and Application.* Fishery Report No. 11. Darwin: Department of Primary Production, Northern Territory.

Christy, Francis, T., Jr. 1973. "Fisherman Quotas: A Tentative Suggestion for Domestic Management." Occasional Paper No. 19 of the Law of the Sea Institute, University of Rhode Island.

Clark, Colin W. 1985. "The Effect of Fishermen's Quotas on Expected Catch Rates." *Marine Resource Economics* 1(4): 419–27.

Copes, Parzival. 1978. *Resource Management for the Rock Lobster Fisheries of South Australia.* Adelaide: Government of South Australia.

Copes, Parzival, and B. A. Cook. 1982. "Rationalization of Canada's Pacific Halibut Fishery." *Ocean Management* 8:151–75.

Crouter, R. A. 1983. "Quotas by Fishing Gear for the Herring Fishery of the Bay of Fundy." Unpublished paper, Canada, Department of Fisheries and Oceans, Atlantic Fisheries Service, Scotia-Fundy Region.

Gordon, H. Scott. 1954. "The Economic Theory of a Common-Property Resource: The Fishery." *Journal of Political Economy* 62(2): 124–42.

Gulland, J. A. 1985. Fisheries Management Problems: An International Perspective." In T. Frady (ed.), *Proceedings of Conference on Fisheries Management: Issues and Options.* 33–56. Fairbanks: University of Alaska.

Kirby, Michael J. L. (Chairman). 1982. *Navigating Troubled Waters: A New Policy for the Atlantic Fisheries.* Ottawa: The Task Force on Atlantic Fisheries.

Lichfield, John. 1984. "EEC Accuses the Hague in Fish Scandal." *Daily Telegraph,* October 8, 1984. London.

McConnell, Kenneth E., and Virgil J. Norton. 1980. "An Evaluation of Limited Entry and Alternative Fishery Management Schemes." In *Limited Entry as a Fishery Management Tool,* R. Bruce Rettig and Jay J. C. Ginter, eds. Seattle: University of Washington Press.

Moloney, David G., and Peter H. Pearse. 1979. "Quantitative Rights as an Instrument for Regulating Commercial Fisheries." *Journal of the Fisheries Research Board of Canada* 36:859–66.

Pearse, Peter H. (Commissioner). 1982. *Turning the Tide: A New Policy for Canada's Pacific*

Fisheries. Vancouver: The Commission on Pacific Fisheries Policy.

Pearse, Peter H., and James E. Wilen. 1979. "Impact of Canada's Pacific Salmon Fleet Control Program." *Journal of the Fisheries Research Board of Canada* 36:764–89.

Polensky, A. M., and S. Shavell. 1979. "The Optimal Trade-Off Between the Probability and Magnitude of Fines." *American Economic Review* 69(5): 880–91.

Scott, Anthony. 1985 (in press). "Catch Quotas and Shares in the Fishstock as Property Rights."

In *Essays in Honor of James Crutchfield*. Seattle: University of Washington Press.

Scott, Anthony, and Philip A. Neher. 1981. *The Public Regulation of Commercial Fisheries in Canada*. Ottawa: Economic Council of Canada.

Stokes, Robert L. 1983. *Limited Entry in the Pacific Halibut Fishery: The Individual Quota Option*. Council Document #20. Anchorage: North Pacific Fishery Management Council.

Tullock, Gordon. 1975. "The Transitional Gains Trap." *Bell Journal of Economics* 6:671–78.

[20]

Entry Restrictions in the Fishery: A Survey of the Evidence

Ralph E. Townsend

For twenty-five years, policies to restrict access to fisheries have been dominant topics in fishery economics. During this period, economists have accumulated an extensive theoretical literature on restricted access. The adoption of new restricted access programs, and the maturing of existing programs, has generated an increasing body of practical experience with restricted access. The present work conducts an empirical survey of this experience and attempts to find themes that support or contradict the economic theories of restricted access.

Restricted access, which is also known as limited entry or license limitation, has gone through a cycle of attitudes among economists. When proponents (such as Sinclair [1961]) first suggested that license limitation could replace the missing property rights, the economics profession enthusiastically adopted restricted access as a keystone of its recommendations on fishery policy. As economists found themselves dealing with various inadequacies of restricted access programs, this initial consensus within the profession was gradually eroded. Some economists attributed these inadequacies to faulty design or administration; others suggested more fundamental problems with the institutions. As economists experienced a variety of successes and failures, the assessments of limited entry came to reflect these various experiences. Although the work by a number of authors could illustrate the evolution of this thought, the recent assessment of restricted access by Copes (1986) is representative of the more critical perspective.

The present paper seeks common themes from among this range of experi-

ence. Rather than simply concluding that some limited entry programs work and some do not, the analysis tries to identify features of limited entry that seem to contribute to success and to failure.

A review of the experiences with limited access may be especially timely. Individualized transferable quotas (ITQs) may have replaced limited entry as the "consensus" within the discipline. As Copes (1986) warned, our experiences in limited entry should not be discarded when approaching ITQs. The two institutions share features that make the experiences of limited entry directly relevant to ITQs. More fundamentally, economic thinking about ITQs is likely to experience changes not unlike the evolution of ideas about limited entry.

I. METHODOLOGY

The analysis proceeds first by reviewing the relevant experiences with about thirty limited entry programs from around the world. Building upon this review, the analysis then extracts seven themes from these experiences.

A working definition of restricted access is necessary to conduct such a review. In the economic literature, there is clearly a wide variation in how the terms "limited entry" or "restricted access" are used. For present purposes, a program is considered "restricted access" if some institution establishes administrative pre-conditions that determine who may or may not fish. This definition would exclude indirect limitations, such as taxes or royalties on catch

Department of Economics, University of Maine.

This work is the result of research sponsored by NOAA, National Sea Grant Program, Department of Commerce, under Grant NA86AA-D-SG047 through the Maine–New Hampshire Sea Grant Program. The author also wishes to acknowledge the very useful comments of two anonymous referees.

0023-7639/00/-0001 $1.50/0

(which some authors define as restricted access). The vast majority of restricted access programs are some form of license limitation. Most commonly, these licenses are issued to a group of historic participants. These licenses may apply to the fisher or to the vessel; they may be transferable or nontransferable. A wide array of ancillary features apply to these programs, such as total gear restrictions, areal limitations, or seasonal restrictions. A few of the more recent programs include some form of individualized quota.

This review, of necessity, covers only programs for which some evaluative evidence on economic performance is available. Evaluative evidence is defined broadly here and even includes judgmental evidence by non-economists. Even with this broad criteria, many restricted access programs are not reviewed here because no evidence of effectiveness is available. Individualized transferable quotas are especially affected. Because individualized transferable quotas are a relatively new phase of license limitation, *ex post* evaluations have yet to be published for many of these programs.

Evidence of the economic benefits of management may take several forms. Direct tax collections or license fee revenues offer clear evidence. Governments usually do not collect rents by taxation, so rents appear as income in excess of the opportunity cost of labor and capital. The willingness of fishermen to accept wages below opportunity costs (Anderson 1980) complicates the calculation of such rents, however. Higher physical yields per unit of fishing effort (relative to similar unmanaged fisheries or relative to pre-management yields) are indirect evidence of income in excess of opportunity cost. Market prices for licenses, which represent capitalized rents, are most frequently cited as evidence. A number of market imperfections may limit the quality of this information. In particular, the need to purchase a license may reduce the value of equipment or human capital. Initially at least, license values may reflect a reduction in quasi-rents

earned by immobile factors (Johnson and Libecap 1982).

Stock size and landings levels may provide indirect evidence of the impact of limited entry. For example, if a stock managed by limited entry falls below the maximum sustainable yield (MSY) stock size, then *a fortiori* the stock must be below maximum economic yield. Therefore, stock reductions are one indicator of economic failure of limited entry. Conversely, a limited entry program that maintains stocks and landings at or near MSY levels has some positive economic benefits. Consumer surplus is increased (relative to over-exploited fisheries). For communities with excess labor, the increase in employment associated with higher stocks and landings may also be a benefit.

The establishment of limited entry frequently has profound distributional effects. The difficulty of evaluating equity implications is self-evident. For a few programs, some limited data on redistribution of income and the social and political implications of that redistribution are available.

Largely absent from this evaluation is any discussion of the costs of management. With rare exceptions, the costs of implementing and enforcing limited entry are almost never tabulated. Obviously, these costs are difficult to measure. These costs may be difficult to separate from the expenses a government must bear to discharge its general responsibilities. For example, the costs of courts or of scientific research may be incurred whether or not limited entry is implemented. Still, it is disappointing to find not even the most rudimentary estimates of costs. If, as Pearse (1980) argued, the costs of public enforcement determine how closely management should approximate sole property, the absence of cost estimates is a serious deficiency.

In presenting this review of restricted entry, I found it convenient to arrange the programs roughly by the apparent degree of economic success. While this ranking is clearly inexact and subject to interpretation, it does help identify the factors that contribute to success or failure of restricted access programs. We now turn to these programs and work from the most successful to the least successful.

II. REVIEW OF PROGRAMS

Alaska applied limited entry in textbook form in 1973. The state created a commission with authority to administer limited entry in over-exploited fisheries. Limited entry was first applied to salmon fisheries and later to some herring, crab, and sablefish fisheries. The salmon plan, which has attracted the greatest attention, froze effort in each area at historic fishing levels. The effect was generally to reduce the number of licenses. A formula that weighted various "dependence" factors then allocated the fixed number of licenses. The plan created a large number of different licenses, because a license was specific to a particular area (e.g., Bristol Bay) and to a particular gear type (e.g., gillnets). An individual could hold licenses for different area/gear type combinations but could not hold more than one license for a given area and gear type. Otherwise, licenses were freely transferable. Licenses could not, however, be leased or used as collateral for private loans.

Initial evidence from Alaska (Owers 1975) indicated that incomes remained low immediately after implementation of limited entry. For the period 1977 to 1979, however, license values increased threefold. According to data collected by the Alaskan Commercial Fisheries Entry Commission, license values in 1979 varied from $5,000 to $175,000 (Langdon 1980). License values have tended to fluctuate dramatically in response to year-to-year variations in catches. Young (1983), in a generally critical article, argued that the high license values have been the product of historical good luck. Salmon catches go through long cycles, and limited entry began in one of the most depressed stock conditions in 1974, when only 22 million salmon were caught. Harvests increased steadily, reaching 111 million in 1981. Young (1983) implicitly argued that economic conditions will deteriorate as the cycle reverses itself.

Although restrictions on gear that predated limited entry were continued, Schelle and Muse (1986) reported rather dramatic increases in capital and labor per license. For example, they reported a doubling of capital and a 33 percent increase in labor between 1973 and 1982 both in the Cook Inlet gillnet fishery and in the Prince William Sound seine fishery. Also, some fishermen qualified for licenses for multiple gears and/or areas. When they sold one or more licenses, one diversified vessel became two or more specialized vessels.

For herring, Alaska's limited entry program has not entirely ended the race to catch fish. Tillion (1985) reported that the 1982 herring fishery in Prince William Sound lasted only four hours and still exceeded the area quota.

The Alaskan program has also been criticized for its distributional impacts. Native Alaskans often sold their rights very early and have been largely excluded from the limited entry fisheries subsequently. Koslow (1979) argued that this exclusion has increased the social problems of native villages. A report to the Alaskan Legislature by Langdon (1980) eased fears that limited entry licenses would be transferred to nonresidents, but raised the fear that limited entry licenses were being transferred from rural residents to urban residents. Schelle and Muse (1986) discussed a variety of studies that address the sale of Native Alaskan licenses and the rural to urban transfers. These continue to be political issues for Alaskans.

Western Australia encouraged the initial development of its prawn fishery by granting exclusive concessions for exploratory fishing. Twenty-five concessions were granted in Shark Bay in 1963; 17 were granted in Exmouth Bay in 1965; and 13 were granted in Nickol Bay in 1971. Each concession could operate a single vessel of whatever design it chose. Processors generally held multiple concessions. The number of concessions was subsequently increased administratively to 35 in Shark Bay and 22 in Exmouth Bay. By strictly limiting concessions. Western Australia guaranteed rents to the concessionaires. Meany (1978) reported returns to capital in excess of 25 percent and license values of $A 150,000 to $A 200.000 ($US 165,000 to $US 220,000).[1] Kailis (1982) reported even higher license values of $A 375,000 ($US 367,000) for Exmouth Gulf.[2]

Experience in the Southern Australian prawn fishery has been similar. License control was implemented in March 1968, two months after discovery of the fishery. Twenty-six permits were initially authorized for Spencer Gulf and 5 for Gulf St. Vincent. Subsequent administrative actions increased those numbers for 1979 to 39 in Spencer Gulf and to 14 in Gulf St. Vincent. New licenses are awarded by random drawing to qualified fishermen. The government of South Australia has attempted to reduce capital-stuffing by restrictions on overall boat length, on double-rigged trawls, on net headline, and on engine power and by occasional closures. Licenses are nominally not salable, but boat sales routinely reflect the license value. Byrne (1982) reported such license values at $A 100,000 to $A 200,000 ($US 98,000 to $US 196,000); Bain (1985) reported license values in excess of $A 200,000 ($US 133,000); and Kailis (1982) reported $A 350,000 ($US 343,000). Byrne (1978) reported annual rents per vessel of $A 17,000 ($US 20,000) in Spencer Gulf and $A 49,000–$A 59,000 ($US 58,000–$US 70,000) in Gulf St. Vincent. Byrne (1978) suggested that Spencer Gulf was exploited above maximum economic yield, while exploitation in Gulf St. Vincent was below maximum economic yield. Byrne (1982) also reported an increase in fishing power per boat of 50 percent in Spencer Gulf and of 100 percent in Gulf St. Vincent. License fees in the Spencer Gulf fishery are now based upon a 3.5 percent royalty, so the government did obtain some $A 500,000 ($US 333,000) of the rents in 1984–85 (Lilburn 1986). Bain (1985) reported that license fees for Gulf St. Vincent have also been increased, to $A 7,000 ($US 4,700).

By the time entry was limited in the Western Australian lobster fishery in 1963,

a dramatic increase in fishing effort had already occurred. For instance, Meany (1978) reported a 66 percent increase in pots from 1958–59 to 1961–62. The 830 licenses issued in 1963 were each permitted three pots per foot of boat length, subject to a 200 maximum. Meany (1978, 1979), Morgan (1980a, 1980b), and Rogers (1982) all have reported significant problems in managing this fishery. Effort has continued to increase through an increase in unregulated boat features, such as engine size and electronics. Administrative reductions in the length of the fishing season have been necessary to protect the fishery. Moreover, regulations on pot/boat length combinations have created incentives to build unseaworthy boats. Despite these difficulties, license values have reached $A 1,000 ($US 1,100) to $A 2,000 ($US 2,200) per pot licensed, for a total license value that can range to $A 200,000 ($US 220,000) (Meany 1978; Kailis 1982).

Individual transferable quota (ITQ) management was introduced in 1984 in the Australian southern bluefin tuna fishery. Prior to 1984, the individual states had made sporadic efforts to control entry. In 1984, the Australian Fisheries Council, which includes all state fisheries ministers plus the Commonwealth minister responsible for fisheries, implemented ITQs that reduced total catches by one-third. The impact was dramatic. Although quotas were issued to 143 fishermen based upon prior landings, quotas were rapidly transferred so that only 85 vessels ultimately owned quotas and only 57 actually landed fish (Robinson 1986). Approximately one-quarter of the total catch was sold for sashimi, which commanded prices double the old cannery prices. Average fish size increased from 74.4 cm to in excess of 80 cm. Robinson

---

[1] For comparison, all currencies are converted to U.S. dollars using annual average exchange rates reported in OECD (1984).

[2] Note that Kailis has been a critic of limited entry and has considered high license values to be evidence of the failure of limited entry. Kailis's estimates of license values are consistently higher than other estimates and should be interpreted with some caution.

(1986) reported that ITQs sold for $A 2,200 per ton ($US 1,900) at the end of 1984, which implies a total asset value of $A 30 million ($US 25 million) for the ITQs. Hagan and Henry (1986) confirmed prices near $A 2,500 ($US 2,160) per ton at the end of 1984, although prices fluctuated widely d..ring the season. It should be noted that the export orientation of the industry and the concentration of processing facilities makes enforcement more feasible.

ITQs were introduced to the Icelandic demersal fisheries in 1984. Although the quotas are freely transferable within a season, they are issued annually only to vessels that continue to fish. Because license renewal requires continued fishing, the 578 vessels licensed in 1984 essentially all continue to fish. Arnason (1986) estimated economic rents for 1984 at between $US 15 million and $US 30 million. These are compared to average annual landing values of about $US 300 million and an estimated rent of $US 155 million under optimal effort reduction. Effort fell modestly after ITQ introduction: 15 percent in 1985 and 6 percent in 1986. On net, the program seems modestly successful in more efficient use of existing capital, but it does not reduce the excess capital.

California limited entry to its herring roe fisheries in San Francisco Bay and Tomales Bay in 1974 to the 17 vessels already engaged in the fishery. Limited entry was suspended in 1977, and reintroduced in 1978, when 155 gillnet and 5 purse seine or lampara nets were licensed. Quotas were established for each gear type. Gillnetters faced restrictions on total net length, alternate-week fishing assignments, and a maximum vessel catch of 91 metric tons per season (in 1980). Roundhaul vessels faced a trip limit of 36 metric tons (in 1980). Huppert (1982) estimated that participants in the fishery in 1979 earned rents of $5.6 million, and that rents were near the theoretical maximum of $6.3 million. Among the larger, more efficient purse seine vessels, a race to exhaust the quotas has developed. In the 1980 season, purse seine vessels took 50 percent of their seasonal quota in 30 hours. As Huppert and Odemar (1986) explained, California has pursued a policy of avoiding excessive rents in its limited entry fisheries. Therefore, the number of licenses has steadily increased over time: 17 in 1974 and 1975; 57 in 1976; 267 in 1977; and 447 in 1982.

In the oyster fishery of the eastern United States, there have existed both privately leased oyster bars and open access public bars. In general, the public bars were naturally productive areas, while private bars were developed by private cultivation of naturally barren bars. Agnello and Donnelley (1975, 1976) examined the relation of private leaseholds to production using data from 16 eastern coastal states (Massachusetts to Texas) for 1950 to 1972. They demonstrated that the yield per acre for oyster beds was correlated to the percent of private oyster leaseholds across the sample of states. Because the natural, public bars are thought to be inherently more productive than the artificially created private bars, this result indicated substantial benefits to private leases.

Maine law since 1913 has permitted local communities to restrict access to clam flats. About half of the towns have exercised this option. Until very recently, most towns simply excluded non-residents. Those towns that did restrict access had 15 percent higher yield per unit of effort (Townsend 1985). Furthermore, among the towns that do restrict access, yields were positively correlated with more restrictive management. The evidence also suggested that this mild form of restriction only captured a small fraction (about one-quarter) of maximum rents (Townsend 1986).

In Maine's lobster fishery, cultural barriers have served to create de facto restricted access. Although arguably not restricted access, these restrictions provide some useful comparisons.[3] Acheson (1972, 1975)

---

[3] There is, in fact, a broad anthropological literature on culturally based management of the commons. Two good references are McGoodwin (1984) and McCay and Acheson (1987). Most of this anthropological literature relates to pre-capitalist or developing societies, and is therefore of limited relevance to the present review.

has documented cultural barriers that determine who can and who cannot fish for lobsters along the Maine coast. Virtual property rights exist for the right to fish on the grounds adjacent to certain isl. ids. In other areas, a local "gang" of fishermen collectively defend certain areas by a variety of extra-legal devices. Where the power of the gang is significant, it determines the conditions for entry to the gang. Wilson (1977) found that fishermen in the most restricted areas had 39 percent higher annual incomes, had 69 percent higher catch per trap haul, and landed slightly larger lobsters as compared to essentially open access areas.

The South African government began limited entry to its west coast shoal fishery for pelagic species (primarily pilchard and maasbankers) in the early 1950s (Gertenbach 1962, 1973). The government limited licenses for processing facilities (including canning, fish meal, and fish oil plants) in 1949, and hold capacity for the fleet was frozen and quotas were established in 1953. Hold capacity certificates and processing tonnages were transferable. The system gave licensed processors monopsony power, and the industry restructured itself to integrate fishing and processing activities. Although the resultant industrial concentration has been criticized for its monopoly elements, the system generated resource rents for a significant period (Gertenbach 1962, 1973).

Recent evidence (Butterworth 1981; Cram 1981; Troadec, Clark, and Gulland 1980) indicated a severe collapse of this fishery that was hastened by excessive exploitation. Both Cram (1981) and Troadec et al. (1980) reported that the political decision to permit two large offshore vessels to enter the fishery in 1966 destroyed industry support for the restraints on licensing. Even without this new entry, the system was creating significant over-capacity. Cram (1981) reported a tripling of fleet hold capacity between 1950 and 1970. Troadec et al. (1980) also suggested that the harvesting efficiency increased even faster because of technological innovations. When the government finally tried to restrain harvests

by reducing quotas, enforcement was undermined by misidentification of species and by falsified scale readings (Cram 1981).

British Columbia introduced limited entry in its salmon fishery in 1969. The program reduced the number of licenses from 7,500 to 5,500 (roughly) in its first four years. Part of this license reduction was accomplished by government purchase of 362 licenses (and boats) during 1971–74 and another 26 vessels in 1981. A complicated process of attrition continues for the "Class B" licenses, which were granted as nontransferable licenses. The program has encouraged substantial investment in construction of new vessels by license holders. Fraser (1979) concluded that fishing power grew as rapidly after regulation as before, while Pearse and Wilen (1979) estimated that the rate of new investment moderated slightly. The management authority has had to shorten the legal fishing season repeatedly to compensate for this increase in fishing effort. Pearse (1982, 13–14) contended that over-fishing has remained a problem in the fishery and suggested that catches were one-third below the level possible under maximum sustainable yield. Appraisals by Newton (1978) and Rettig (1984) are similar. Some rents are evident in the value of licenses. Fraser (1978) estimated the value of all licenses at $C 12.4 million ($US 12.4 million) in 1973, $C 79.0 million ($US 79.3 million) in 1974, and $C 43.3 million ($US 42.6 million) in 1975. Increasing real prices and stock improvement from government-financed habitat restoration were largely responsible for these rents. These license values must be compared to government costs. Regulation and habitat restoration cost the Canadian government $C 20 million ($US 20 million) per year in the 1970s, while license fees generated only $C 1 million ($US 1 million).

Some of the gear used in the British Columbian salmon fishery is also used in another limited entry fishery: herring roe. The entire catch is sold to Japan. Canada restricted entry for herring roe in 1974 to 252 seiners and 1,579 gillnetters. These were still substantial increases over the 161 active seiners and 223 active gillnetters in

1973, but well below the expected 5,900 new entrants in an unregulated fishery (Fraser 1982). Licenses were non-transferable, personal licenses and contained an "owner-operator" requirement. License fees were substantial: $C 2,000 ($US 2,010) for seiners and $C 200 ($US 201) per gillnetter. To deal with the very hectic pattern of many boats converging on each opening, the government restricted each license to one of three areas in 1981. In 1982 and thereafter, vessels were allowed to lease or buy licenses from idle vessels in other areas.

As Fraser (1982) reported, this fishery has been subject to wildly erratic external changes since its inception in 1971, and therefore any evaluation is difficult. Fraser concluded that limited entry did partially stem the inflow of capital and fishermen that would have resulted from the price increases. Fraser also reported above-average fishermen wages for the late 1970s. Pearse (1982, 38–43) reported that the fleet has enormous over-capacity and that open seasons are often measured in hours or even minutes. Both Fraser (1982) and Pearse (1982) were very critical of the personal license, which permitted unrestricted increases in capital per license and which made enforcement difficult under the hectic conditions of the herring roe fishery. MacGillivray (1986) suggested that area licenses reduced costs of fishing and of processing. A 30 percent reduction in fleet size occurred after transfers of licenses were permitted in 1982. MacGillivray implied that the lease value of licenses is substantial, but did not provide quantitative data.

Japan has had a comprehensive and complex system to limit entry (Comitini 1967; Asada 1973, 1985). The legislative basis for this system was passed in 1949, although its origins go back to the early 1900s. Responsibility for managing Japan's fisheries is shared among the national government, prefectural governments, and cooperatives. Because the management system is actually a broad collection of different but interrelated programs, evaluation is not simple. There seemed to be agreement among several commentators (Com-

itini 1967; Kasahara 1971; Asada 1985) that the system maximizes fisheries investment, gross sales, and gross income, but that economic rents are generally absent. The Japanese have historically tried to divert excess effort into high-seas fisheries by favorable licensing and subsidy programs (Comitini 1967; Kasahara 1971). Comitini (1967) cited specific evidence that rates of return are higher in the smaller, near-shore fisheries than in larger, distant-water fleets. Herrington (1971) more critically suggested that many stocks are over-fished. The tonnage restrictions on licenses have also encouraged the construction of vessels to meet licensing rather than technological needs. Keen (1973) specifically described the construction of unconventional and often unsafe boats in the tuna fleet.

Evaluations of the role of cooperatives have varied from Shima's (1985) rather positive view to Herrington's (1971) quite negative evaluation. The differences may be attributable to which cooperatives are familiar to specific authors. There is agreement that cooperatives emphasize "equitable" (and by inference equal) distribution of income. Fishing rights are often rotated among members, so that everyone enjoys approximately equal access over time. Both Herrington (1971) and Asada, Hirasawa, and Nagasaki (1983) indicated that cooperatives often restrict the use of efficient harvesting technologies in the pursuit of equity.

In the Victorian abalone fishery of Australia, limited entry and voluntary retirement have decreased the number of licenses from 200 in 1968 to 91 in 1979 (Stanistreet 1982). Fishermen, rather than boats, are licensed, and licenses are non-transferable. The Victorian Commercial Fisheries Management Committee has pursued a policy of maintaining an essentially artisan fishery and uses an average cost formula to insure adequate incomes when setting the number of licenses. Although abalone diving is an arduous job best suited to young men, divers tend to remain in the fishery well into middle age. The state is reluctant to exercise its nominal authority to require inactive fishermen to "show cause" for continued use of the license, because replacement with younger men who dive longer hours might threaten the resource. In general, Stanistreet (1982) described a license system that encourages inefficiency to pursue conservation goals.

Cicin-Sain, Moore, and Wyner (1978) provided a more general review of the abalone programs of Australia, which include state-run programs in South Australia, Victoria, New South Wales, Tasmania, and Western Australia. Their review suggested that attrition has reduced the number of divers in these fisheries. In fact, because licenses are non-transferable in South Australia, Victoria, and New South Wales, the replacement of retiring divers is a thorny political issue. In Tasmania, licenses are transferable and bring a price of $A 8,000 ($US 8,800) (Cicin-Sain et al. 1978). Also, the states capture modest rents through license fees of $A 200–$A 250 ($US 220–$US 275). Rather conflicting evidence was suggested by Kailis (1982), who cited license values of $A 30,000 ($US 29,500) for Western Australia, which also permits market transfers. Bain (1985) suggested that abalone management is feasible in Australia primarily because Australia controls a large percent of the world market. Restrictions on production raise world prices, and higher prices tend to offset any short-run reduction in income due to smaller catches.

Regulation has created similar results in the Victorian scallop fishery of Australia. License limitation was introduced in 1968 for Port Phillip Bay and in 1971 for Lakes Entrance. Licenses were offered to all fishermen then engaged in the fishery, although a few declined to purchase the license. Licenses are nominally not transferable, but the government routinely issues a license to the new owner of a boat. Dredge size is specified on the license, and daily bag limits, closed seasons, and proscribed fishing hours all restrict fishing options. The resulting fleet is very inefficient, and at least part of the fishery is over-capitalized. Again, a kind of average cost calculation is used to determine the number of licenses. Sturgess, Dow, and Belin (1982) estimated

relatively low de facto license values: $A 2,470 ($US 2,420) for the Port Phillip Bay license; $A 3,350 ($US 3,280) for Lakes Entrance; and $A 4,570 ($US 4,480) for a combined license. Sturgess et al. (1982) suggested that annual rents on the order of $A 44,000 ($US 43,100) per license are theoretically possible for Port Phillip Bay.

The evidence is less compelling for economic rents in the southern Australian rock lobster program than in the western Australian lobster fishery examined earlier. Limited entry was enacted in 1966 in Tasmania, in 1967 in Victoria, and in 1968 in South Australia. Approximately 1,000 licenses were issued by the three states. Each license received a "pot allocation" based upon boat length, crew size, and zone fished. Licenses are transferable, but pot allocations are not divisible. In the southern zone of the South Australia lobster fishery, a 15 percent across-the-board pot reduction was ordered in 1984. Staniford (1988) estimated that this action reduced effort only by 9 percent because of adjustments in the use of other inputs.

Copes (1978) estimated rents at $A 2,647 ($US 3,152) per fisherman in the southern Australian lobster fishery for the 1971–72 season, in comparison to estimates of rents per fisherman of $A 23,980 ($US 28,560) at maximum economic yield (MEY). These figures can be compared to average annual earnings of $A 7,000–$A 8,000 ($US 8,300–$US 9,500) in the fishery. MEY would require a 70 percent reduction in fishing effort by Copes's estimates. Also, Copes estimated negative rents for the relatively poor fishing season 1975–76. As in many limited entry fisheries, individual fishermen have increased fishing effort by increasing the use of unregulated inputs. More recent analysis by Sudmalis (1982) reported declines in profitability, in real income, and in license values in the fishery. Sudmalis also argued that current regulations on pots per vessel significantly reduce the efficiency of the industry.

Experiences have been similar in northern Australia's prawn fishery (Lilburn 1986). By the time entry was limited in 1977, 292 boats qualified for licenses. Prior to 1977, no more than 160 ever fished in any year. In fact, effort was so excessive that only 195 boats operated in 1977 and 150 in 1978 (MacLeod 1982). The season for banana prawns, the primary species, was progressively shortened over the period 1974–77 (MacLeod 1982). Lilburn (1986) reported that the Australian Bureau of Agricultural Economics estimated that 55 percent of the boats in 1980–81 and 1981–82 were unable to earn a positive return on investment. A liberal boat replacement policy coupled with boat-building subsidies caused capacity to continue to grow. A more restrictive boat replacement policy was implemented in 1984. A license buyback scheme funded by the government purchased 17 licenses in 1985. These new policies, coupled with favorable stock conditions, seem to have improved economic conditions in the fishery (Lilburn 1986).

Limited entry management of Norway's purse seine fleet began with a moratorium in 1970. This fleet originally pursued Atlanto-Scandian herring, but switched onto Barents Sea capelin and other pelagic stocks when the herring collapsed. In 1973, hold capacity restrictions were added. Hannesson (1986) argued that fleet capacity continued to increase even with hold restrictions. Licenses command a price of 800 to 1,000 Kroner ($US 140–$US 180) per hectoliter of hold. Hannesson (1986) attributed this value to the cost advantages of larger boats, rather than to resource rents. The number of licensed purse seiners declined from 321 in 1973 to 253 in 1979, but total hold capacity increased about 5 percent. Substantial increases in other inputs were also suggested by the increasing costs of new vessels. Beginning in 1979, the government purchased 18 percent of fleet capacity through vessel buy-back programs that paid up to 6 million Kroner ($US 1.2 million) per vessel. Hannesson (1986) calculated that the rate of return on the social investment in the buy-back program was near 14 percent. Brochmann's (1984a, 1984b, 1985) and Hannesson's (1985) discussion of Norwegian fishery policy indicated that income subsidies continue to attract new capital and labor to all fisheries,

even as the government buys up existing vessels.

Wisconsin introduced limited entry to its Lake Superior fishery in 1968. Limited entry does not cover Native Americans, who in 1972 won court cases establishing special rights. The number of licenses issued fell from 68 in 1968 to 38 in 1970–71 via voluntary exit. Reductions to a maximum of 20 were accomplished largely by excluding part-timers. The key species, lake trout, was subjected to individualized quotas. Licenses eventually contained restrictions on total gear used. Despite the apparent reduction in effort, Bishop, Johnson, and Samples (1978) reported that incomes remained low—perhaps too low to attract replacements for retirees.

Michigan's limited entry program, introduced in 1968 for fisheries on Lakes Michigan, Huron, Superior, Erie, and St. Clair, was based upon a clear recreational bias in state policy. Licenses were issued to individuals and have been transferable, but the Michigan Department of Natural Resources apparently has discouraged such transfers (Talhelm 1978). Commercial licenses were reduced from 300 in 1968 to 140 in 1978, primarily by the exclusion of part-timers. The primary benefits of the limited entry program have been increased opportunities for recreational fishermen (Talhelm 1978). Borgeson (1972), in an early evaluation, concluded that the total benefits from commercial fishing were less than Michigan's costs of administering the fisheries.

Ontario's freshwater commercial fisheries, which include both Great Lakes and northern inland lakes, have been subject to limited entry since the 1960s. Under this program, the purchase of more efficient gear by the permit holders created overcapitalization and over-exploitation. In 1984, a system of transferable vessel quotas was introduced. Cowan (1986) reported that after implementation of the system, total investment fell 10 percent, employment fell 20 percent, and total income increased by about 20 percent. Both Cowan (1986) and Haxwell (1986) reported good compliance with the quotas. Haxwell (1986) acknowledged some minor problems, including vertical integration of fishermen into processing and into direct sales, as well as incomplete and tardy reports. Quota compliance is clearly the Achilles' heel of ITQs, so longer experience may be necessary before this auspicious start can be thoroughly evaluated.

In the Isle of Man herring fishery, entry was limited to 100 British vessels, 24 Irish vessels, and the local Manx fleet in 1977. The plan was designed to maintain the status quo while more comprehensive management was studied. Very favorable stock conditions in the first limited entry season generated large incomes for license holders, and McKellar (1977, 1982) reported that the government acquiesced to the resulting political pressure to lift the entry moratorium.

Riley (1982) provided some evidence on limited entry in New Zealand. After repeal of a restricted license program that had lasted from 1936 to 1963, limited entry was reintroduced on a fishery-by-fishery basis beginning in 1969. Riley (1982) cited the experience of the Foveaux Strait oyster fishery, where limited entry was reestablished in 1969, as a successful limited entry program. It would seem that the quota and not limited entry preserved the resource. The non-transferable permits created a fleet of old vessels with old captains, and this fleet has had abnormally high maintenance expenses. Waugh (1985) seemed to report that New Zealand's inshore fisheries have gone through the classic cycle of over-fishing followed by ineffective moratoria on entry. Clark and Duncan (1986) criticized the over-capitalization and inflexibility under the rock lobster limited entry plan of 1978 and the wetfish limited entry plan of 1980. The inshore wetfish fishery, in particular, they characterized as showing signs of "over fishing, over capitalization, potential biological damage to some commercial fish species, and a significantly declining economic performance."

In the Bay of Fundy herring fishery in Atlantic Canada, government assistance in 1976 permitted purse seiners to establish a "club," the Atlantic Herring Fishermen's Marketing Cooperative (AHFMC) (Kear-

ney 1984; Peacock and MacFarlane 1986). This club adopted vessel quotas, negotiated prices with both domestic and foreign buyers, and established rules for the fishery. The club successfully increased the income of fishermen, primarily by negotiating higher prices and by increasing sales of herring for food (which command higher prices). This collective control weakened when a second club, South West Seiners Co. Ltd., formed in 1979, and collective action completely disintegrated by 1981. The loss of a European market, which had provided higher prices, was a major contributor to this disintegration. Government decisions to stop over-the-side sales to Poland in 1979 and 1980 also severely weakened AHFMC's position. Government efforts to reimpose vessel quotas after 1980 proved largely unenforceable. Crouter (1985) summarized that experience: "Our experience with vessel quotas has been that each and every fishermen will attempt to 'cheat' on his quota and processors will promote that attempt through collusion in falsifying records." Peacock and MacFarlane (1986) also reported that vessel quotas for 1983–86 were largely ignored. They estimated that 40 percent of the catch was unreported in 1984, despite enforcement that cost $C 500,000 ($US 400,000) in a fishery valued at $C 18 to $C 20 million ($US 14 to $US 16 million).

The management of fisheries in the state of Washington has been overshadowed by Judge Boldt's 1974 decision that awarded half the salmon catch in most rivers to Native Americans. Bell (1978) indicated that the Boldt decision was the driving force for the moratorium on entry, which was enacted almost immediately. Despite the implementation of limited entry and a buyback program, the non-Native commercial fishing industry · emained very large relative to its 50 percent share of the catch. Jelvik (1986) reported that idle licenses accounted for about 8 percent of all licenses in 1983, about 80 percent in 1984, and 25 percent in 1985.

The Boldt decision has negatively affected other Washington fisheries as salmon fishermen turned to other species.

By the time limited entry was enacted in the herring-roe fishery, the entire quota could be taken in as little as two days (Trumble 1977). Trumble did suggest that economic returns in the bait herring fishery, which is financially less important, may have been positively affected by limited entry.

Huppert and Odemar (1986) reported similar effects in California's salmon limited entry program. By the time limited entry was established in 1980, effort was already excessive. From 1960 to 1978, the number of licenses grew from 1,365 to 4,919 and the constant dollar value of landings per boat fell from $8,290 to $3,460. At the time of the initial moratorium in 1980, 5,119 vessels landed salmon. Under an interim limited entry system for 1983–85, the number of vessels landing salmon dropped from 3,223 to 2,308. This exit was voluntary and uncompensated, and was due to poor stock and economic conditions.

Rettig (1984) would generalize this negative evaluation of Washington's and California's salmon programs to include the program in Oregon. Rettig summarized his evaluation of all three states' salmon programs:

Nevertheless, a general impression emerges that the high profile of fishermen participating in program design, ease of entrance qualifications, minimal reduction policies and low license fees have tended to lead to high political feasibility, acceptable administrative feasibility, low program costs, contentious but relatively satisfactory equity treatment, and minimal impact on either resource conservation or economic efficiency goals. (1984, 242)

When limited entry was established in 1967 for the lobster industry of Atlantic Canada, individual fishermen were licensed. In 1969, a two-tiered vessel license system was introduced. Transferable Class A licenses were issued to "full-time" fishermen and non-transferable Class B licenses were issued to "part-time" fishermen. Retirement of Class B licenses to date has been minimal. The number of Class A licenses alone was very large relative to the resource. Moreover, fishermen increased

fishing power of licensed boats by adding pots, pot-haulers, electronics, and larger engines. DeWolf (1974) evaluated the program as generally ineffective. He found that total traps increased 5.6 percent between 1968 and 1972, that catch per trap haul fell, and that incomes remained generally low. He concluded his analysis by predicting

license limitation and trap limits . . . will redistribute income more equally, will not affect total effort substantially, may lead to an increase in total number of traps, may have an adverse effect on economic efficiency, and will increase the value of boats as the right to fish lobsters becomes capitalized. (DeWolf 1974, 53)

Note that DeWolf's prediction of capitalized access value is somewhat inconsistent with his generally pessimistic view.

The groundfish fisheries of Atlantic Canada are the largest and most important in the region. Fishery policy for this fishery has been a potpourri of license limitations (in 1973), quota allocations and reallocations, quality incentives and restrictions, and subsidies to both harvesters and processors. Shifts in policy emphasis have been the rule rather than the exception (Macdonald 1984). Regulation of this fishery has been hampered by the general policy of using fisheries as the employer of last resort in Atlantic Canada (Copes 1982). In 1981–82, "enterprise quotas" were negotiated for the four large vertically integrated companies. Subsequently, the government provided a significant financial infusion into these four firms, and in the process, restructured the four firms into two.

The Kirby Commission (1982) acknowledged that the complicated system of allocations, subsidies, and regulations in the fisheries of Atlantic Canada has produced a generally inefficient and stagnant industry. Fraser (1986) characterized the situation in 1981 as one of "heavy competition among the individual vessels," "inconsistent quality," "gluts," and a generally "bleak situation." Macdonald (1984) has argued that stocks have been preserved by the regulations. Both Macdonald (1984) and Fraser (1986) have argued that the enterprise quotas for the two major companies have permitted more rational harvesting, and they expressed optimism for the future of these enterprise quotas.

The Canadian share of the Pacific halibut catch was reduced in 1979 from 75 percent of catch to 60 percent by the unilateral withdrawal of the United States from the International Pacific Halibut Commission. Management reduced effort by about 20 percent, but much of the effort simply moved into related fisheries, such as salmon, herring, and sablefish (Copes and Cook 1982). Pearse (1982, 121–26) was very critical of the halibut licensing program. He concluded that "in spite of a limited entry system intended to prevent it, the licensed fishing capacity has expanded alarmingly." He reported that 422 licenses were granted in 1981, compared to only 33 in 1979 and well above his estimate of 100 "mainly halibut" vessels prior to 1979. For the related sablefish (blackcod) fishery, Pearse (1982) also reported capacity that is three to four times the level needed to take current harvests.

A limited entry program for Atlantic surf clams was introduced by the U.S. Mid–Atlantic Fishery Management Council in 1977. In addition to limiting new entry, the program set annual and quarterly quotas, and it limited days and hours fished by license holders. The limits on days fished were an attempt both to reduce effort and also to provide stable landings for processing facilities. Fishermen have especially disliked the designation of fishing days, because fishermen often have faced the choice of fishing in poor weather or giving up allocated days. Strand, Kirkley, and McConnell (1981) concluded that the moratorium was largely irrelevant because independent economic factors caused the fishery to contract: Ocean quahogs were substituted for surf clams; the 1979–80 recession reduced demand; and processors exercised monopsony power to drive prices down. These lower prices caused total fishing effort to fall. Also, 60 new licenses were granted under the limited entry plan. Nichols (1985), in a more positive evaluation, argued that overall quotas have promoted

stock recovery. He did agree that weekly quotas served to promote orderly processing and that fishermen have seen little economic benefit from the plan. Nichols (1986) noted that vessels continue to retire voluntarily from the fishery.

A moratorium on new licenses was established for lobsters in Massachusetts in 1975. Licenses were non-transferable and subject to cancellation for inactivity. Smith (1978) has shown that effort continued to increase after the moratorium, because the amount of gear per license increased from 162 pots per license in 1975 to 182 in 1977. Catch per license has shown no clear trend. In the nearby Maine lobster fishery, attempts to limit entry in 1974 backfired when fishermen rushed to qualify for licenses while the courts invalidated the initial law on constitutional grounds.

In the mackerel fishery of southwest England, company-owned freezer trawlers were limited to those that fished in 1980. They are subject to a "sectoral quota" that is essentially administered by a federation of companies (Derham 1985). All other vessels were subjected to weekly vessel quotas. Derham (1985) reported that these vessel quotas were increasingly ignored as the fishermen realized that legal penalties were impossible to impose. Widespread misreporting by fishermen was actively aided and even encouraged by buyers. Derham suggested that an effective monitoring program would require an observer at every point of sale and could easily require resources equal to 25 percent to 80 percent of the landed value of the resource.

III. TENTATIVE CONCLUSIONS

Any conclusions drawn from the existing body of empirical work on limited entry must be tentative. While the number of plans reviewed here is moderately large, the evidence from no plan is unassailable. Economists have devoted much less effort to the *ex post* evaluation of these programs than to a priori prescriptions. A clear conclusion must be that further empirical research is absolutely necessary. Despite this caveat, several broad hypotheses receive general support from the empirical evidence assembled to date.

First, the restrictiveness of a program is correlated to its economic success. The most restrictive programs have either reduced effort significantly (as in Alaska's salmon fishery) or closed entry before effort reached rent-dissipating levels (as in the Australian prawn fisheries). The success of property rights systems, as exemplified by the American oyster fishery, also illustrates the effectiveness of restrictive rights.

Less restrictive plans have been marginally successful. Moratoria on entry that included a phased reduction in fishing effort have been marginally successful. Relevant examples include the salmon fishery of British Columbia and the Australian lobster programs. The least restrictive programs, which generally established simple moratoria with no effort reduction, have not generated economic rents. The clear examples here include the U.S. Pacific salmon programs and the lobster and groundfish programs of Atlantic Canada.

Unfortunately, the most restrictive programs are also potentially the most expensive. Effective management may require prohibitive enforcement expenditures by the government or high compliance costs for fishermen. The experiences of the Canadian government in the Bay of Fundy, of the English government in its mackerel fishery, and of the southwest African pelagic shoals fisheries all demonstrate the difficulties created by misreporting under vessel quotas. For schooling pelagic species, the difficulty of monitoring catches when a large part (or all) of the annual catches can be taken in hours or minutes presents serious problems. This has been especially evident in a number of herring-roe fisheries. These problems are especially noteworthy in the context of the current interest in individual vessel quotas (cf. Copes 1986).

A recurrent issue in limited entry has been control of growth in inputs per license. When a management program reduces capital-stuffing by tight gear restrictions, management costs are shifted in part

to fishermen in the form of higher operating costs. In Japan's tuna fleet, in Western Australia's lobster fleet, in the New Zealand Foveaux Strait oyster fishery, and in Australia's Victoria scallops, restrictions on gear design or on renovation have created fleets of boats that were not only inefficient but perhaps even unseaworthy.

Very restrictive programs are most often implemented by administrative "blitzkrieg," with little concern for equity aspects. The ITQ program for Australian bluefin tuna is exemplary. On the reverse side, protracted deliberations over legislation or administrative rules invite preemptive actions by fishermen. By the time the Maine Supreme Court had invalidated limited entry legislation in the lobster fishery, total licenses had increased 50 percent. Richards and Gorham (1986) reported a 72 percent increase in licenses while the North Pacific Fishery Management Council debated limited entry for halibut.

Second, there is an inverse relationship between the complexity of a fishery and the success of management, *ceteris paribus*. The complexity of a fishery is determined by the ability of a fishery to respond to outside forces, such as management. Fisheries that harvest multiple species are more complex than single species fisheries, because fishermen can adjust to management by changing the target species. Fisheries that use multiple harvesting technologies, such as otter trawling, gillnetting, and longlining, are more complicated than single technology fisheries, because management may induce technological responses. Fisheries that extend over wide geographical areas and over international boundaries are more complex than geographically localized fisheries. In general, the greater the geographic mobility of the species and the greater that technological mobility of fishermen, the more complex a fishery is.

Because the regulation of complex fisheries invites changes by fishermen that undermine the regulations, complex fisheries require more complicated regulations. *Ceteris paribus,* limited entry has been more successful in simple fisheries. Sedentary shellfish resources, such as Maine clams and the American east coast oyster fisheries, have proven easier to manage than more complex fisheries. The lobster fisheries of Australia are also relatively simple fisheries. Salmon fisheries are more complicated, because several species are harvested and because several harvest technologies are used. The British Columbian salmon programs have had great difficulty in restraining capital growth because of the technological options available to the license holder. At the far extreme, the limited entry program for groundfish in Atlantic Canada has been little more than a catch allocation scheme.

Third, the social and political environment affects the success of limited entry plans. When these plans are socially and politically unpopular, their success tends to be limited. Strictly extra-legal social institutions, such as the lobster fiefs of Maine, require no overt enforcement for success. When management co-opts social attitudes, as in the residency requirements in Maine clam ordinances, the costs of enforcing management are also low. As Ciriacy-Wantrup and Bishop (1975) discussed, many apparently open-access-fishery fisheries are in fact subject to rules of common access.

Effective management is very difficult when the political environment insists upon government guarantees of fishermen's incomes. Subsidies, such as those enacted in Norway and in Atlantic Canada, make the necessary exit from the fishery virtually impossible to obtain. When management must contend with basically antagonistic social attitudes, which is typical in U.S. fisheries, management is even more difficult. The current interest in "co-management" reflects a growing recognition of the importance of supportive social institutions to management (Pinkerton 1989).

Fourth, limited entry has generated economic benefits more often by reducing short-run externalities than by eliminating long-run stock externalities. This deduction is based upon a seeming contradiction: License values and/or income data indicate economic rents in a number of fisheries where evidence of stock depletion is also

clear. This reduction of stocks below MSY levels indicates failure to solve stock externalities. Therefore, these rents must be generated by easing short-run externalities. Note that as long as limited entry keeps the entry rate below the open access entry rate, short-run externalities will be moderated. A number of programs simply slow the rate of effort expansion, rather than freezing effort levels.

At one level, this is a unexpected conclusion, because the classic model of fisheries economics (Schaefer 1957, Gordon 1954) models only long-run, stock externalities. Short-run, or "crowding," externalities have received much less attention from economists. One can argue, however, that it should have been clear (and perhaps was clear to some) that limited entry does not solve the underlying incentives of individual fishermen to catch as many fish as possible as soon as possible. Under virtually all limited entry programs, no fisherman can invest in future catches by delaying current catches. The destructive effects of this inherent competition are constrained by the limits on effort, but the fundamental incentives for individual fishermen are unchanged. Even under ITQs, the fisherman cannot reduce today's catch in return for higher catches tomorrow.

Failure to solve stock externalities is evident even in the most restrictive plans. The South African program, which was criticized as monopsonistic, did not avoid stock collapse. A number of programs, including the Australian rock lobster fisheries and the U.S. and Canadian Pacific salmon fisheries, have faced continual capital growth that threatens the resource. No fishery has recovered from stock collapse via limited entry alone. Rather, very restrictive quota systems have been the major tool for stock restoration. The gap between a priori economic theory and *ex post* experience is large here. The practical significance is also great: The empirical evidence suggests that biologists correctly insist that limited entry is, at best, one component of an effective management program.

In fact, limited entry has often generated rents not because effort was decreased but rather because prices increased. That is, limited entry created rents not by reducing the existing externalities but by delaying more severe short-run externalities when prices increased. This is essentially the case for the salmon fisheries of British Columbia and for the rock lobster fisheries of Australia. Australia's abalone programs also may have been possible only because of Australia's monopsony power. For the Canadian Bay of Fundy herring clubs, higher prices were the glue that held the clubs together. When prices weakened, the clubs disintegrated.

Fifth, there is little evidence that weak limited entry plans evolve into strong, successful plans. Experience belies the political arguments for a gradual approach to an efficient system. Economists have frequently supported such policies, in part because phased implementation reduces the present value of losses imposed upon disfavored fishermen. Any program becomes entrenched, because management creates vested interests to oppose further change. The moratoria in Massachusetts lobsters, Washington salmon, Atlantic surf clams, Canadian lobsters, and Canadian herring have all stagnated as ineffective programs. The government buy-back program in the British Columbia salmon fisheries has been halted because of the cost of license acquisition, which the industry is unwilling to fund. The effective programs, such as Alaskan salmon, Australian prawns, and South African pelagic shoal fisheries, were instituted initially in essentially their final form.

When limited entry proves to be a clear failure, management may progress to individualized quotas. This seems to have been the case in Ontario's freshwater ITQs and in New Zealand. The political economy of this transition from unmanaged to ineffective limited entry to ITQs may be problematic. Do marginal limited entry programs delay the implementation of effective management by stabilizing the economic climate and by creating vested interests? Is a failed limited entry program a political step necessary to create a favorable environment for the eventual switch to an effective alternative? If limited entry is a necessary

stepping stone, should the Machiavellian economist encourage creation of badly flawed limited entry programs to speed the transition to effective management? These are difficult questions that address the appropriate role of the economist in such overtly political decisions. Understanding the political economy of this evolution might well be a high priority for economists.

Sixth, the rents created by limited entry are often politically problematic. Because fisheries tend to be rural, artisan industries, specific taxes (or very high license fees) on license holders are rarely imposed. Leaving rents with fishermen, however, creates strong pressures to increase the number of licenses so others may share in the profits. The experience in a large number of fisheries around the world (Japan, Britain's Isle of Man herring, British Columbia herring, California herring-roe, and Australia's Victoria scallops) all indicate that political realities tend to favor those who want more licenses issued. For the British Columbian salmon fishery, the dissatisfaction with high license values resulted in union demands that the subsequent licensing program for herring-roe be non-transferable. This seriously complicated enforcement in the fishery. Quite possibly, government expropriation of part (or most) of the rents by higher license fees, by royalties, or by license transfer taxes may actually improve the long-run political prospects of a system. Higher license fees and royalties are increasingly common in limited entry programs.

Finally, the rights under restricted access have proven to be quite different from the terrestrial property rights. Early proponents of limited entry had suggested that the establishment of marketable rights would reduce the government's role in fisheries to something analogous to its limited role in the legal system for other property rights. Limited entry licenses are inherently imperfect rights, because the restricted license holders continue to exploit the resource in common. The government must intervene at least periodically to manage these inherent imperfections. A number of these weaknesses have been identified already in these conclusions. The government must still establish overall quotas. The problem of capital-stuffing seems endemic. Even the race to harvest is not ended in all fisheries, as in the Alaskan and Californian herring fisheries. If the government tries to reduce effort by a policy of attrition, it must eventually deal with the problem of allocating replacement licenses. This has become an issue in the Australian abalone fisheries and in Wisconsin's Lake Superior chub fishery. In general, the parallels drawn between limited entry rights and terrestrial property rights are weak.

These empirical conclusions support some elements of economic theory and contradict others. As we would expect, more aggressive management produces greater economic efficiencies. Also, the net benefits of management are greater when the costs of management are lower, either because of the simplicity of the managed fishery or because of a favorable social and political environment. Unexpectedly, most benefits of limited entry flow from reductions in short-run externalities, which economists have tended to ignore, rather than from the long-run externalities usually modelled by economists. The political and economic case for gradualist policies is contradicted by the evidence, which suggests initial plans become permanently entrenched, or even weakened by pressures to "share the wealth." Finally, limited entry has not proven to be a one-shot panacea for fisheries management. Limited entry is, at best, one element in a broader program of fisheries management.

References

Acheson, J. M. 1972. "The Territories of the Lobstermen." *Natural History* 81 (Apr.): 60–69.

———. 1975. "The Lobster Fiefs: Economic and Ecological Effects of Territoriality in the Maine Lobster Industry." *Human Ecology* 3(3):183–207.

Agnello, R. J., and L. P. Donnelley. 1975. "Property Rights and Efficiency in the Oyster Industry." *Journal of Law and Economics* 18 (Oct.):521–33.

———. 1976. "Externalities and Property Rights in the Fisheries." *Land Economics* 52 (Nov.):518–33.

Anderson, L. G. 1980. "Necessary Components of Economic Surplus in Fisheries Economics." *Canadian Journal of Fisheries and Aquatic Sciences* 37 (5):858–70.

———, ed. 1981. *Economic Analysis for Fisheries Management Plans.* Ann Arbor, MI: Ann Arbor Science.

Arnason, R. 1986. "Management of Icelandic Demersal Fisheries." In Mollett (1986). pp. 83–101.

Asada, Y. 1973. "License Limitation Regulations: The Japanese System." *Journal of the Fisheries Research Board of Canada* 30 (Dec.):2085–95.

———. 1985. "Licence Limitation Regulations: The Japanese System." In FAO (1985a), pp. 307–17.

Asada, Y., Y. Hirasawa, and F. Nagasaki. 1983. *Fishery Management in Japan.* Rome: FAO Fisheries Technical Paper 238.

Bain, R. 1985. "Vessel Licenses, Gear Control and Fishermen Licensing in Australia: Australian Experience of These and Related Management Measures." In FAO (1985a), pp. 329–42.

Bell, D. M. 1978. "Gear Reduction/Buyback Programs in British Columbia and Washington State." In Rettig and Ginter (1978), pp. 353–57.

Bishop, R. C., G. V. Johnson, and K. Samples. 1978. "Wisconsin's Limited Entry Experience." In Rettig and Ginter (1978), pp. 317–22.

Borgeson, D. P., ed. 1972. *Status of Michigan's Fisheries Management—1971.* Lansing, MI: Department of Natural Resources, Fisheries Division.

Brochmann, B. S. 1984a. "Regulation of Fishing Effort through Vessel Licenses." In FAO (1984), pp. 149–51.

———. 1984b. "Financial Measures to Regulate Effort." In FAO (1984), pp. 167–72.

———. 1985. "Fishery Policy in Norway—Experiences from the Period 1920–82: Case Study." In FAO (1985b), pp. 108–22.

Butterworth, D. S. 1981. "The Value of Catch-Statistics-Based Management Techniques for Heavily Fished Pelagic Stocks with Special Reference to the Recent Decline of the Southwest African Pilchard Stock." In *Applied Operations Research in Fishing,* ed. K. Haley, 441–64. New York: Plenum Press.

Byrne, J. L. 1978. "The South Australian Prawn Fishery: A Case Study of Limited Entry Regulation." Unpublished masters thesis, Simon Fraser University.

———. 1982. "The South Australian Prawn Fishery: A Case Study in License Limitations." In Sturgess and Meany (1982), pp. 205–24.

Cicin-Sain, B., J. E. Moore, and A. J. Wyner. 1978. "Limiting Entry to Commercial Fisheries: Some Worldwide Comparisons." *Ocean Management* 4:21–49.

Ciriacy-Wantrup, S. V., and R. C. Bishop. 1975. " 'Common Property' as a Concept in Natural Resources Policy." *Natural Resources Journal* 15 (Oct.):713–27.

Clark, I. N., and A. J. Duncan. 1986. "New Zealand's Fisheries Management Policies—Past, Present and Future: The Implementation of an ITQ-Based Management System." In Mollett (1986), pp. 107–40.

Comitini, S. 1967. "Economic and Legal Aspects of Japanese Fisheries Regulation and Control." *Washington Law Review* 43:179–96.

Copes, P. 1978. *Resource Management for the Rock Lobster Fisheries of South Australia.* Report to Steering Committee for the Review of Fisheries of the South Australian Government.

———. 1982. "Implementing Canada's Marine Fisheries Policy: Objectives, Hazards, and Constraints." *Marine Policy* 6 (July):219–35.

———. 1986. "A Critical Review of the Individualized Quota as a Device in Fisheries Management." *Land Economics* 62 (3):278–91.

Copes, P., and B. A. Cook. 1982. "Rationalization of Canada's Pacific Halibut Fishery." *Ocean Management* 8:151–75.

Cowan, T. 1986. "Recent Adjustments in Ontario's Fisheries." In Mollett (1986), pp. 245–49.

Cram, D. 1981. "Hidden Elements in the Development and Implementation of Marine Resource Conservation Policy: The Case of South West Africa/Namibian Fisheries." In *Resource Management and Environmental Uncertainty: Lessons from Coastal Upwelling Fisheries,* eds. M. H. Glantz and J. D. Thompson, 137–56. New York: Wiley.

Crouter, R. A. 1985. "Quotas by Fishing Gear for the Herring Fishery of the Bay of Fundy." In FAO (1985a), pp. 409–14.

Derham, P. J. 1985. "The Problems of Quota Management in the European Community Concept." In FAO (1985a), pp. 241–50.

DeWolf, A. G. 1974. *The Lobster Fishery of the*

Maritime Provinces: Economic Effects of Regulations. Ottawa: Fisheries Research Board of Canada Bulletin 187.

FAO. 1984. *Papers Presented at the Expert Consultation on the Regulation of Fishing Effort (Fishing Mortality).* Rome: FAO Fisheries Report 289 Supplement 2.

———. 1985a. *Papers Presented at the Expert Consultation on the Regulation of Fishing Effort (Fishing Mortality).* Rome: FAO Fisheries Report 289 Supplement 3.

———. 1985b. *Case Studies and Working Papers Presented at the Expert Consultation on Strategies for Fisheries Development.* Rome: FAO Fisheries Report 295 Supplement.

Fraser, C. A. 1986. "Enterprise Allocations in the Offshore Groundfish Fishery in Atlantic Canada: 1982–1986." In Mollett (1986), pp. 207–13.

Fraser, G. A. 1978. "License Limitation in the British Columbia Salmon Fishery." In Rettig and Ginter (1978), pp. 358–81.

———. 1979. "Limited Entry: Experience of the British Columbia Salmon Fishery." *Journal of the Fisheries Research Board of Canada* 36 (July):754–63.

———. 1982. "License Limitation in the British Columbia Roe Herring Fishery: An Evaluation." In Sturgess and Meany (1982), pp. 117–37.

Gertenbach, L. P. D. 1962. "Regulation of the South African West Coast Shoal Fisheries." In *Economic Effects of Fishery Regulation,* ed. R. Hamlisch, 423–58. Rome: FAO Fisheries Report 5.

———. 1973. "License Limitation Regulations: The South African System." *Journal of the Fisheries Research Board of Canada* 30 (Dec.):2077–84.

Gordon, H. S. 1954. "The Economic Theory of a Common-Property Resource: The Fishery." *Journal of Political Economy* 62 (Apr.):124–42.

Hagan, P., and G. Henry. 1986. "Potential Effects of Differing Management Programs on the Southern Bluefin Tuna Fishery." *Marine Resource Economics* 3 (4):353–89.

Hannesson, R. 1985. "Inefficiency through Government Regulations: The Case of Norway's Fishery Policy." *Marine Resource Economics* 2 (2):115–41.

———. 1986. "The Regulation of Fleet Capacity in Norwegian Purse Seining." In Mollett (1986), pp. 65–82.

Haxwell, C. 1986. "Management Measures to Control Commercial Fish Harvest: The Ontario Experience." In Mollett (1986), pp. 231–44.

Herrington, W. C. 1971. "Operation of the Japanese Fishery Management System." University of Rhode Island, Law of the Sea Institute Occasional Paper 11.

Huppert, D. D. 1982. "California's Management Programmes for the Herring Roe and Abalone Fisheries." In Sturgess and Meany (1982), pp. 89–116.

Huppert, D. D., and M. W. Odemar. 1986. "A Review of California's Limited Entry Program." In Mollett (1986), pp. 301–12.

Jelvik, M. 1986. "Washington State's Experience with Limited Entry." In Mollett (1986), pp. 313–16.

Johnson, R. N., and G. D. Libecap. 1982. "Contracting Problems and Regulations: The Case of the Fishery." *American Economic Review* 72 (5):1005–22.

Kailis, T. G. 1982. "Limited Entry—An Industrial View." In Sturgess and Meany (1982), pp. 77–86.

Kasahara, H. 1971. "Japanese Distant-Water Fisheries: A Review." *Fishery Bulletin* 70 (2):227–82.

———. 1973. "Management of the Fisheries in the North Pacific." *Journal of the Fisheries Research Board of Canada* 30 (Dec.): 2348–60.

Kearney, J. 1984. "The Transformation of the Bay of Fundy Herring Fisheries, 1976–1978: An Experiment in Fishermen-Government Co-management." In Lamson and Hanson (1984), pp. 165–203.

Keen, E. A. 1973. "Limited Entry: The Case of the Japanese Tuna Fishery." In Sokoloski (1973), pp. 146–58.

Kirby, M. J. L., Chair. 1982. *Navigating Troubled Waters: A New Policy for the Atlantic Fisheries.* Report of the Task Force on Atlantic Fisheries. Canadian Department of Fisheries and Oceans.

Koslow, J. A. 1979. "Limited Entry Policy and the Bristol Bay. Alaska Salmon Fishermen." Institute of Marine Resources IMR Reference No. 79-7, Scripps Institution of Oceanography.

Lamson, C., and A. J. Hanson, eds. 1984. *Atlantic Fisheries and Coastal Communities: Fisheries Decision-making Case Studies.* Dalhousie Ocean Studies Programme. Halifax, Nova Scotia.

Langdon, S. 1980. "Transfer Patterns in Alaskan Limited Entry Fisheries." Final Report for Limited Entry Study Group of the Alaska State Legislature.

Lilburn, B. 1986. "Management of Australian Fisheries: Broad Developments and Alternative Strategies." In Mollett (1986), pp. 141–87.

Macdonald, R. D. S. 1984. "Canadian Fisheries Policy and the Development of Atlantic Coast Groundfisheries Management." In Lamson and Hanson (1984), pp. 15–75.

MacGillivray, P. 1986. "Evaluation of Area Licensing in the British Columbia Roe Herring Fishery: 1981–1985." In Mollett (1986), pp. 251–74.

MacLeod, N. D. 1982. "Limited Entry Management for the Northern Prawn Fishery: A Review of its Development." In Sturgess and Meany (1982), pp. 225–74.

McCay, B. M., and J. M. Acheson, eds. 1987. *The Question of the Commons.* Tucson: University of Arizona.

McGoodwin, J. R. 1984. "Some Examples of Self-Regulatory Mechanisms in Unmanaged Fisheries." In FAO (1984), pp. 41–61.

McKellar, N. B. 1977. "Restrictive Licensing as a Fisheries Management Tool." Sea Fish Industry Authority, Fisheries Economics Research Unit Occasional Paper No. 6.

———. 1982. "The Political Economy of Fisheries Management in the North East Atlantic." In Sturgess and Meany (1982), pp. 349–64.

Meany, T. F. 1978. "Restricted Entry in Australian Fisheries." In Rettig and Ginter (1978). pp. 391–415.

———. 1979. "Limited Entry in the Western Australian Rock Lobster and Prawn Fisheries: An Economic Evaluation." *Journal of the Fisheries Research Board of Canada* 36 (July):789–98.

Mollett, N. 1986. *Fishery Access Control Programs Worldwide: Proceedings of the Workshop on Management Options for the North Pacific Longline Fisheries.* Alaska Sea Grant Report 86-4.

Morgan, G. R. 1980a. "Population Dynamics and Management of the Western Rock Lobster Fishery." *Marine Policy* 4 (Jan.):52–60.

———. 1980b. "Increases in Fishing Effort in a Limited Entry Fishery—The Western Rock Lobster Fishery 1963–1976." *Journal du conseil international pour l'exploration de la mer* 39 (1):82–87.

Newton, C. H. B. 1978. "Experience with Limited Entry in British Columbia Fisheries." In Rettig and Ginter (1978), pp. 382–90.

Nichols, B. 1985. "Management of the Atlantic Surf Clam Fishery under the Magnuson Act, 1977 to 1982." In FAO (1985a), pp. 431–47.

———. 1986. "The Past, Present, and Future of

Magnuson Act Surf Clam Management." In Mollett (1986), pp. 275–99.

OECD. 1984. *Main Economic Indicators, 1964–1983.* Paris.

Owers, J. E. 1975. "An Empirical Study of Limited Entry in Alaska's Salmon Fisheries." *Marine Fisheries Review* 37 (July):22–25.

Peacock, F. G., and D. A. MacFarlane. 1986. "A Review of Quasi-Property Rights in the Herring Purse Fishery of the Scotia-Fundy Region of Canada." In Mollett (1986), pp. 215–30.

Pearse, P. H. 1980. "Property Rights and the Regulation of Commercial Fisheries." *Journal of Business Administration* 11:185–209.

———. 1982. *Turning the Tide: A New Policy for Canada's Pacific Fisheries.* Final Report of the Commission on Pacific Fisheries Policy.

Pearse, P. H., and J. E. Wilen. 1979. "Impact of Canada's Pacific Salmon Fleet Control Program." *Journal of the Fisheries Research Board of Canada* 36 (July):764–69.

Pinkerton, E., ed. 1989. *Cooperative Management of Local Fisheries.* Vancouver: University of British Columbia Press.

Rettig, R. B. 1984. "License Limitation in the United States and Canada: An Assessment." *North American Journal of Fisheries Management* 4 (Summer):231–48.

Rettig, R. B., and J. C. Ginter, ed. 1978. *Limited Entry as a Fishery Management Tool.* Seattle: University of Washington.

Richards, H., and A. Gorham. 1986. "The Demise of the U.S. Halibut Fishery Moratorium: A Review of the Controversy." In Mollett (1986), pp. 33–54.

Riley, P. 1982. "Economic Aspects of New Zealand's Policies on Limited Entry Fisheries." In Sturgess and Meany (1982), pp. 365–84.

Robinson, W. L. 1986. "Individual Transferable Quotas in the Australian Southern Bluefin Tuna Fishery." In Mollett (1986), pp. 189–205.

Rogers, P. P. 1982. "Boat Replacement Policy in the West Coast Rock Lobster Fishery—A Historical Review and a Future Option." In Sturgess and Meany (1982), pp. 189–204.

Schaefer, M. B. 1957. "Some Considerations of Population Dynamics and Economics in Relation to the Management of the Commercial Marine Fisheries." *Journal of the Fisheries Research Board of Canada* 14 (Oct): 669–81.

Schelle, K., and B. Muse. 1986. "Efficiency and Distributional Aspects of Alaska's Limited Entry Program." In Mollett (1986), pp. 317–52.

Shima, K. 1985. "The Role of Cooperatives in the Exploitation and Management of Coastal Resources in Japan." In FAO (1985b), pp. 243–47.

Sinclair, S. 1961. *License Limitation—A Method of Economic Fisheries Management.* Ottawa: Canada Department of Fisheries.

Smith, L. J. 1978. "Case Studies on Economic Effects of Limiting Entry to the Fisheries." In Rettig and Ginter (1978), pp. 416–28.

Sokoloski, A. A., ed. 1973. *Ocean Fishery Management: Discussions and Research.* Department of Commerce NOAA Technical Report NMFS Circ-371. Seattle, Washington.

Staniford, A. 1988. "The Effects of the Pot Reduction in the South Australian Southern Zone Lobster Fishery." *Marine Resources Economics* 4:271–88.

Stanistreet, K. 1982. "Limited Entry in the Abalone Fishery of Victoria." In Sturgess and Meany (1982), pp. 139–52.

Strand, I. E., J. E. Kirkley, and K. E. McConnell. 1981. "Economic Analysis and the Management of Atlantic Surf Clams." In Anderson (1981). pp. 113–38.

Sturgess, N. H., and T. F. Meany, eds. 1982. *Policy and Practice in Fisheries Management.* Canberra: Australian Government.

Sturgess, N. H., N. Dow, and P. Belin. 1982. "Management of the Victoria Scallop Fisheries: Retrospect and Prospect." In Sturgess and Meany (1982), pp. 277–315.

Sudmalis, R. 1982. "Economic Aspects of Limited Entry in the Southern Rock Lobster Fishery." In Sturgess and Meany (1982), pp. 153–65.

Talhelm, D. R. 1978. "Limited Entry in Michigan Fisheries." In Rettig and Ginter (1978), pp. 300–316.

Tillion, C. V. 1985. "Fisheries Management in Alaska." In FAO (1985a), pp. 291–97.

Townsend, R. E. 1985. "An Economic Evaluation of Restricted Entry in Maine's Soft-Shell Clam Industry." *North American Journal of Fisheries Management* 5 (Winter): 57–64.

———. 1986. "Evidence from Controlled Harvests for Potential Economic Benefits from Management of Soft-Shell Clams (*Mya arenaria*)." *North American Journal of Fisheries Management* 6 (Fall):592–95.

Troadec, J.-P., W. G. Clark, and J. A. Gulland. 1980. "A Review of Some Pelagic Fish Stocks in Other Areas." *Rapports et procès-verbaux des réunions du conseil international pour l'exploration de la mer* 177 (Nov.): 252–77.

Trumble, R. J. 1977. "Effects of Limited Entry Legislation on Management of Washington State Commercial Herring Fisheries." Washington Department of Fisheries Progress Report.

Waugh, G. D. 1985. "Regulation of Fishing Effort (Fishing Mortality) in New Zealand." In FAO (1985a), pp. 343–54.

Wilson, J. A. 1977. "A Test of the Tragedy of the Commons." In *Managing the Commons*, ed. G. Hardin and J. Baden, 96–111. San Francisco: Freeman.

Young, O. R. 1983. "Fishing by Permit: Restricted Common Property in Practice." *Ocean Development and International Law Journal* 13 (2):121–70.

[21]

Contracting Problems and Regulation: The Case of the Fishery

By RONALD N. JOHNSON AND GARY D. LIBECAP*

The inefficiencies of common property fisheries are of continuing concern to economists.[1] The early work by Scott Gordon (1954) and Anthony Scott (1955) outlined the problem and later studies by James Crutchfield and Giulio Pontecorvo (1969) and Frederick Bell (1972) provided empirical estimates of the losses that result. Those studies were followed by the dynamic models of Colin Clark (1976), and James Quirk and Vernon Smith (1970) of optimal harvest rates and the use of corrective taxes or quotas to achieve them. But in spite of a large and growing literature and the persuasiveness of the outlined efficiency criteria, most fisheries retain common property aspects with overcapitalization and excessive labor input. Why those conditions persist and the failure of the regulatory response to them are the issues addressed in this paper.

We examine a number of fisheries, but focus on the Texas shrimp industry, which is one of the nation's most valuable fisheries for a single species and which shares the common property characteristics observed elsewhere. It is considered overcapitalized and catch per unit of effort is falling.[2] Ex-

amination of the fishery reveals the many margins along which rent dissipation occurs and the nature of the regulations necessary for controlling fishing effort to avoid those losses. The regulatory environment in Texas is complicated by conflict within the fishery between inshore and offshore fishermen. The latter assert that bay shrimping reduces the number of shrimp that successfully migrate to the Gulf. The inshore fishery is highlighted by another issue—the recent resettlement of some 30–45,000 Vietnamese refugees, including many fishermen, along the Texas Coast (Marine Advisory Service, Paul Starr). Their entry has been met by hostility and violence from existing shrimpers who recognize that they are in an environment characterized by the absence of property rights. Bay shrimpers have lobbied the Texas Legislature for broad limits on new entrants. Yet, ironically, the sale of additional boats by individual shrimpers to the Vietnamese has facilitated entry of the refugees into the fishery.

Regulations in the Texas shrimp and other fisheries are incomplete, leaving many options for rent dissipation uncontrolled, because of high contracting costs among fishermen and political factors that mold government actions. Contracting costs are high among heterogeneous fishermen, who vary principally with regard to fishing skill.[3]

*Montana State University and Texas A&M University, respectively. We benefitted from comments by Terry Anderson, Raymond C. Battalio, Gardner Brown, Oscar Burt, Micha Gisser, Wade Griffin, John R. Moroney, Anthony D. Scott, Peter Temin, and participants in workshops at Texas A&M University and the University of Washington. Research assistance was provided by Scott Barnhart and Phil Mizzi. Funding was provided by Sea Grant, Texas A&M University.

[1] Throughout this paper we use the terms common property and open access interchangeably. Here the terms describe a situation where no property rights, group, or individual exist(s) for the resource.

[2] Catch per unit of effort in the inshore Texas Gulf and Federal Gulf waters is presented for 1963–77 by W. L. Griffin, C. G. Tydlacka, and W. E. Grant. They show catch per unit of effort generally falling in inshore and offshore waters. While catch per unit of effort fluctuates from year to year, it falls from approximately 360 kg shrimp per unit of effort to 220 kg from 1963 to 1977. The common property nature of the fishery is reflected in the following statistics. Since 1975 the number of

vessels has grown by 23 percent (Nelson Swartz). For fish houses the record of entry is similar. Between 1970 and 1977, the number of firms grew from 259 to 287, though during that period, many left the industry as others entered. The number of fish houses was calculated from license data supplied by the Texas Parks and Wildlife Department. The record of entry by fish houses suggests that monopsony as discussed by Colin Clark and Gordon Munro is absent in the fishery.

[3] There may be minor differences in labor-leisure choices and capital, but our statistical evidence and discussions with fishermen indicate that catch variations are largely due to skill. Government regulations in the Texas bays restrict capital, and our empirical evidence shows boats to be relatively homogeneous.

The differential yields that result from heterogeneity affect the willingness of fishermen to organize with others for specific regulations. In developing this point, we deny the traditional assumption of zero economic rents in an open access fishery. Regulations that recognize existing rankings of fishermen, while increasing total yields, will be supported. By contrast, regulations that pose disproportionate constraints on certain classes of fishermen will be opposed by those adversely affected. This suggests that fishermen are unlikely to readily agree to individual quotas such as those described by Clark (1980) and David Moloney and Peter Pearse (1979).

Political constraints affect the solutions offered by governments. Both federal and state governments emphasize the right of all citizens to access fisheries and other wildlife. They refuse to assign private territorial rights to areas large enough to cover migratory species. Moreover, informal voluntary efforts to control entry are opposed by the Justice Department and the Federal Trade Commission as violations of the Sherman Act. Further, antitrust actions have been taken against fishermen unions along the Gulf and Pacific Coasts when they attempted to regulate prices and to limit entry. The regulations generally adopted by governments are visible, yield-enhancing policies that avoid more controversial restrictions. Accordingly, the arrangements that achieve consensus from both heterogeneous fishermen and politicians allow many of the relevant margins of fishing effort to continue unregulated.

In the following section we describe the private agreements that have been reached in a number of fisheries in the absence of government support. Those agreements reflect the contracting problems that also must be overcome by either a governmental regulatory agency or by a firm, if sole ownership were a viable political alternative. The issue of heterogeneity and its implications are outlined in Section II. In the third section we consider the specific case of the Texas shrimp fishery in detail. Finally, we offer some concluding thoughts on why satisfying certain marginal conditions in fishery management is likely to remain an elusive goal.

I. Contracting in the Absence of State Support

Models of fishery exploitation generally begin by noting that property rights to the resource stock are absent. Entry under open access conditions is described as continuing until the average cost of catching a standard unit of fish equals the market price. At the limit, then, the rental value of the fishery is dissipated. But if the absence of property rights to the stock is the source of the problem, the assignment of such rights would appear to solve it. Yet, sole fishing rights have been historically rejected by the federal and state governments as leading to monopoly control of the fishery.[4] Where they have existed, private territorial rights to fugitive fisheries have been dismantled in the United States and elsewhere in response to egalitarian pressures.[5] Thomas Lund describes the elimination of fishery rights arrangements in U.S. inshore waters in the early nineteenth century. Similarly in the twentieth century, George Rogers shows that aboriginal use rights to Alaskan salmon were outlawed by the 1924 White Act which provided that "no exclusive or several right of fishery shall be granted therein..." (1979, p. 784). White Act provisions were later incorporated in the Alaskan constitution to prevent feared, nonresident control of the salmon fishery.

Government ownership of fisheries for their common use by all citizens and the associated rejection of alleged monopoly controls have been repeatedly emphasized by federal and state courts.[6] For example, the Texas Supreme Court in 1950 rejected state legislation to limit entry in Texas coastal waters: "...If allowed to stand, the statute and action already taken under it are rea-

---

[4]An exception is the occasional granting of private leases to oyster beds. See Richard Agnello and Lawrence Donnelley. Also in inshore Japanese fisheries customary rights are recognized and enforced by the state. See Salvatore Comitini.

[5]Breton describes the rejection of fishing rights in the Caribbean off Venezuela. Established in 1821, they were broken up after a 1928 revolution in response to redistribution pressures.

[6]*McCready v. Virginia* 94 U.S. 391 (1887), *Toomer v. Witsell* 334 U.S. 385 (1948), *Stephenson v. Wood* 34 S.W. 2nd 246 (1931), *Dodgen v. Depuglio* 209 S.W. 2nd 588 (1948).

sonably calculated to perpetuate in effect a monopoly of commercial fishery for the favored class" (*Dobard v. State*, 233, S.W. 2nd 440). The courts have not allowed state governments to discriminate against out-of-state residents in devising regulatory schemes. Limited entry arrangements for inshore and state territorial waters must include all U.S. citizens. That requirement no doubt reduces the incentive of state legislatures to effectively regulate fisheries, since any resulting gains must be shared with outsiders and cannot be restricted to voting residents.

Even though formal private rights are absent, fishermen have frequently resorted to informal contracting and the use of unions and trade associations to mitigate open access conditions. The record reveals, however, that both have provided limited gains because informal arrangements lack enforcement, and because of government opposition to union attempts to restrict fishing effort. Examination of informal contracting and fisherman unions reveals the types of regulation to which fishermen can agree and the nature of the government response.

A. *Informal Contracting*

In Texas, bay shrimpers and Vietnamese refugees have attempted to informally restrict the entry of additional boats into Galveston and San Antonio bays where refugee resettlement has been most intense. On Galveston Bay an agreement included: "The Vietnamese agree to discourage other Vietnamese against moving into Seabrook or buying any more boats. The Vietnamese agree to sell their shrimp for the same price as the native shrimpers or within 10 to 15 cents of that price. The Vietnamese would also not drag one net with two boats..." (News release, January 7, 1981).[7] The agreement has not been binding and conflict has resulted as additional boats have entered the bay. Moreover, the agreements are considered by the

FTC to be in violation of antitrust laws.[8] There is related, but limited, contracting among shrimpers for sharing information regarding the location of shrimp. Such information is closely held and exchanged only within small cliques. Limited sharing of information may increase costs to newcomers who are typically denied access to it, and hence retard entry. Knowledge of the location of shrimp is valuable. Not only do locators get first opportunity for the shrimp, but they get to them before they are scattered by repeated trawling. (See also Raoul Andersen, 1972, and David White.)

Anthropological studies (see Yvon Breton; John Cordell; Shepard Forman), while pointing to the existence of territorial rights and hostility toward outsiders, reveal the breakdown of informal agreements, particularly as competition for the resource increases. Significantly, the studies show few intragroup controls on effort. For example, James Acheson discusses voluntary contracts among Maine lobstermen for territorial rights which are enforced by surreptitious violence. Yet, he shows that many areas are not adequately defended and approach common access conditions. James Wilson's study (1977, p. 109) of the same locality notes that voluntary contracting is absent in over 90 percent of the fishery. Bell's examination of the northern U.S. lobster fishery confirms that the agreements are incomplete, since he estimates that economic efficiency (as he defines it) could be achieved with half the observed level of effort. Similar territorial schemes in the early inshore Newfoundland cod fishery have been investigated by Andersen and Geoffrey Stiles, Andersen (1979), and Kent Martin. Private fishtrap sites were established on unused spots following informal arrangements. They, however, broke down in the face of disputes and the advent of off-shore fishing.

---

[7]Seabrook, Texas Police Department. A similar document titled a "Statement of Concensus" dated 10 May 1980 was written at Palacios, Texas, and signed by represented shrimpers. Copy provided by the Marine Advisory Service, Texas A&M University.

[8]In December 1980, John Townsend, chairman of the Texas Governor's Task Force for Indochinese Resettlement, was warned by the FTC that voluntary agreements to limit the number of boats in the Texas bays was in violation of the Sherman Act (personal communication with the authors, January, 1981).

B. *Fishermen Unions*

As a more structured arrangement for restricting outsiders and for policing compliance of members, fishermen unions and trade associations are an alternative to more nebulous informal agreements. They emerged along the U.S. coasts to limit entry and to negotiate price agreements with wholesalers and canneries. Unions were particularly active from the 1930's through the 1950's, and they implemented policies to increase member incomes. But they were subsequently dismantled by the federal government as violations of the Sherman Act.[9]

The *Gulf Coast Shrimpers and Oystermens Association v. U.S.*, 236 F. 2nd 658 (1956) case is of particular interest because of its relation to shrimping and because of the detailed regulations imposed by the union. The union was organized in the 1930's to regulate shrimping and to set prices along the Mississippi coast. The 5th Circuit Court of Appeals affirmed earlier convictions of the association and its officers for violation of the Sherman Act. The court found that the union had not merely attempted to fix prices, but also had excluded from the market those not complying with association rules. Practically all commercial shrimp and oyster fisherman operating from the five major ports in Mississippi were members.[10] They were permitted to sell only at or above the association's floor price and to packers who agreed to its rules. In its opinion the court denied the association protection from antitrust prosecution as provided for labor unions by the Norris-La Guardia Act (29 U.S.C.A. 113). Further, the group's actions were found to have exceeded the exemptions provided by the Fisheries Collective and Marketing Act

(15 U.S.C. Sec. 521). Crucial in the denial of union status in this and other cases was the finding that fishermen either owned their own boats or worked for shares and were hence independent entrepreneurs.[11] Conflicts over negotiated prices were not considered a legitimate labor dispute; nor, ironically, was conservation a justification for group action: "A cooperative association of boat owners is not freed from the restrictive provisions of the Sherman Anti-trust Act, section 1-7 of this title, because it professes, in the interest of the conservation of important food fish, to regulate the price and the manner of taking fish unauthorized by legislation and uncontrolled by proper authority" (15 U.S.C.A. Sec. 522).

By fixing prices to control the fishing of certain size classes of shrimp and restricting entry, the union could provide some increase in member income even though shrimp were sold in a national market.[12] A study of the transcript in the *Gulf Coast* case and interviews with individuals knowledgeable of the union indicate that price fixing had the objective of increasing the value of total catch by directing effort toward larger, more valuable shrimp.[13] Minimum price lists based on shrimp size (tails per pound) were distributed among packers and members. By setting a minimum price for smaller shrimp that generally exceeded prices elsewhere, the association reduced the quantity demanded by packers.[14] Indeed, testimony in the case shows that whenever the market price for small shrimp fell below the association floor price, as was frequent, packers closed down

[9]Major cases included *Columbia River Packers v. Hinton* 315 U.S., 520 (1942), *Manaka v. Monterey Sardine Industries* 41 F. Supp. 531 (1941), *Hawaiian Tuna Packers v. International Longshoremen's and Warehousemen's Union*, 72 F. Supp. 562 (1947), *McHugh v. U.S.* 230 F. 2nd 252 (1956), and *Local 36 of International Fishermen and Allied Workers of American et al. v. U.S.* 177 F. 2nd 320 (1949).

[10]Captain Joe Ross, who was a union member, estimated membership at 1,800 men with 600 boats (phone conversation with authors, March 24, 1981).

[11]On the issue of share payments to crew members Crutchfield notes: "The roots of the legal problem of fishermen's unions lie in the nearly universal practice of compensating fishermen on a share or 'lay' basis" (1955, p. 542).

[12]In 1951, Mississippi accounted for 3.8 percent of total Gulf catch (U.S. Department of Commerce, 1951).

[13]Transcript of Record, *Gulf Coast Shrimpers and Oystermen's Association v. United States*, pp. 51–53, Government Exhibit #6; phone interviews with Captain Joe Ross, Biloxi, and J. Y. Christmas, Gulf Coast Research Laboratory, Ocean Spring, Mississippi, March 24, 1981.

[14]Although the prices were occasionally changed up or down, they were apparently fairly rigid (Transcript of Record,..., pp. 65–67, 90).

and shrimpers stopped fishing.[15] The higher price for smaller shrimp, then, acted as a conservation measure by reducing catches of smaller, immature shrimp, thereby increasing the yield of higher-valued, larger shrimp later in the season. The price per pound for larger shrimp was at least double that for smaller shrimp, and union minimum prices for larger shrimp were generally at or below the market price in adjoining states.[16] With the price floor reducing fishing effort for small shrimp and with Mississippi's small share of the national market, the association's actions established a quasi quota for small shrimp on the Mississippi coast.

The price-setting efforts of the union coincided with the establishment in 1934 of a legal minimum size count for Mississippi shrimp of 40 per pound, larger than the 4 inch or 68 per pound requirement in neighboring Louisiana.[17] The union was apparently an advocate of the legislation, and it attempted to enforce it. If shrimp smaller than the minimum were brought to a packing house, union peelers refused to peel them.[18] To further enforce the rules, the union required that all captains fishing for small shrimp carry a purchase contract from a buyer at the association price. Fines and suspension from fishing were levied for failure to comply.[19]

The union's efforts were directly aimed at obtaining larger shrimp. Its efforts were seemingly successful. Louisiana shrimpers were attracted to Mississippi waters, but union members opposed entry from outsiders.[20] The association pressured packers

not to buy below the union price, and to deny ice and fuel to nonunion shrimpers. Testimony from a packer clearly illustrates when and how the union enforced its rules against nonmembers:

Q. And the crews of ships were from where?
A. The majority of them were from Louisiana.
Q. Did you have any difficulty with the defendant association...?
A. We did in July, 1951....We were buying the shrimp on the same basis that we would buy shrimp in Louisiana —different sizes and different prices ...On the large shrimp we were paying, I believe $70.00 a barrel, and I think the association price was around $45.00 or $55.00, I don't recall; but on the smaller shrimp, they were $35.00 a barrel; we were paying $25.00 or $30.00 for that particular size shrimp because the market wouldn't justify it. During that period I was contacted by members of the Association because we were not paying union price.

[Transcript of Record,..., p. 95]

The confrontation in July 1951 between association members and out-of-state crews led to eviction of Louisiana shrimpers from Mississippi waters as the testimony reveals:

Q. During this period of time did the boats that were at dock in Pascagoula leave?...
A. In other words, all those boats...we didn't own those boats.... Those boats went back to the points in Louisiana and all of them never have been back to this day.

[Transcript of Record,..., p. 108]

The association's emphasis on protecting small shrimp is currently repeated by every Gulf Coast state through minimum size limits, though significantly none include limited entry as part of the regulatory

[15]Testimony of Joe Castigliola, a packer (Transcript of Record,..., pp. 78–99).

[16]Oliver Clark, a packer, testified that his prices for large shrimp were above the association floor price (Transcript of Record,..., p. 174).

[17]Louisiana size regulations, Louisiana Department of Wildlife and Fisheries, New Orleans. For Mississippi size regulations, see Food Commission, State of Mississippi. Ordinance #3, 1934.

[18]Phone interview, Captain Joe Ross, union member, March 24, 1981.

[19]Gulf Coast Shrimpers and Oystermen's Association By-Laws, p. 11 (Transcript of Record,...).

[20]Captain Joe Ross, Biloxi, phone conversation March 24, 1981. He emphasized the importance of protecting small shrimp to increase the catch of larger

shrimp. He also argued the union was effective, and that union members were hostile to any outsiders and denied them fuel and ice.

scheme.[21] In the absence of such controls on entry, some rent dissipation will occur. Mississippi's limit is now 68 whole shrimp per pound, corresponding to neighboring states (Gulf of Mexico Fishery Management Council, Section 3, pp. 31, 32).

The *Gulf Coast Shrimpers* case illustrates the nature of union agreements to restrict fishing effort. They were designed to increase the value of the total catch for members, but, notice, none of the court cases examined showed that unions implemented individual effort constraints on their members. In the following section we document the heterogeneity of fishermen with respect to skill, and argue that under such conditions, limits on individual effort are extremely costly to agree to and enforce. Those costs not only limit the type of voluntary agreements that can be reached within fishing groups, but they reduce the ability of fishermen to act as a cohesive political force in seeking government regulation.

II. Heterogeneity and Contracting

Following Gordon, standard analyses of the fishery problem generally assume homogeneous fishermen. The supply curve of aggregate fishing effort is then infinitely elastic, and under open access conditions the resource rent is totally dissipated. Effort is added until the value of the average product equals the opportunity cost of labor. The usual remedy is either a call for taxes on catch, or for individual quotas instituted and enforced by the government. Such facile solutions and the assumptions on which they are based abstract too far from actual fishery conditions. In particular, the assumption of

homogeneity leads to neglect of the high costs of contracting, either through political channels for government regulation or through private arrangements among fishermen to limit catch. Indeed, if fishermen had equal abilities and yields, the net gains from effort controls would be evenly spread, and given the large estimates of rent dissipation in many fisheries, rules governing effort or catch would be quickly adopted. With the ease of contracting implied by homogeneity, the rules selected would not only include restrictions on entry, but also regulations covering other margins where significant dissipation occurs. For example, total effort could be restricted through uniform quotas for eligible fishermen. But if fishermen are heterogeneous, uniform quotas will be costly to assign and enforce because of opposition from more productive fishermen. Without side payments (which are difficult to administer), uniform quotas could leave more productive fishermen worse off than under common property conditions. As A. Adasiak (1979) has shown, egalitarian pressures are likely under government quota schemes, and even if rent maximization calls for equal quotas, resistance is probable.[22]

Recognition of differential abilities among fishermen has been noted in much of the descriptive literature on fisheries, though the implications for contracting have not been drawn. For example, Scott notes: "Fisheries experts repeatedly speak of durable groupings of skippers, vessels, and crews according to the size of their catch or earnings, year in and year out" (1979, p. 733). Our own research reveals that Texas shrimpers categorize fishermen on the basis of their fishing ability.[23] Repeated success by some fishermen (higher than average catches) is primarily attributed to knowledge of how to set nets

---

[21] There is evidence that the association was effective in restricting entry and increasing the value of yield. Average shrimp price data from 1948, when data are first available, through 1959 in Mississippi and Louisiana show significant differences between the two states (U.S. Department of Commerce). If Mississippi catch had a greater proportion of larger, more valuable shrimp, the average price would be higher. A one-tailed *t*-test of the difference in the means of the ratios of Mississippi to Louisiana prices for 1948–53 (the period the union was active) and 1954–59 (the post-union period) shows that they are significantly different from zero at the 95 percent confidence level.

[22] Adasiak shows that the allocation of licenses in Alaska was based on notions of "social considerations," "minimum social dislocation," and "excess profits occurring to some," (pp. 775). We are not arguing such considerations should be ignored, but rather are saying that they increase contracting costs.

[23] Based on interviews with fishermen and county extension marine agents, conducted July 1980 at Galveston, Dickinson, Seadrift, Port Aransas, Port Lavaca, and Bay City, Texas.

and regulate their spread, correct trawling speed, and the location of shrimp. While skill is apparently the most important determinant of catch, more productive fishermen also tend to have somewhat better equipment. Capital in the Texas bays, though, is relatively homogeneous because state regulations restrict the size and number of nets which can be pulled by each bay vessel.

Heterogeneity of fishermen can be seen from an analysis of daily catch data from one fish house for the fall 1978 bay shrimp season. A fish house is a packing house that buys shrimp from fishermen and sells it to dealers and processors. The data covered the period September 15–December 14, 1978, and were for commercial food shrimp. Twenty-nine full-time shrimpers were included in the sample, where full time applied to all shrimpers who fished at least three days a week for at least five weeks.[24] Daily catches were listed by fishermen, and they were regressed against an intercept term, identification dummy variables for each of the fishermen, and dummy variables for each day of the season. The dummy variables for each day were used to account for day-specific effects such as those of weather and tidal changes:

$$(1) \quad Catch_{ij} = a + \sum_{i=1}^{n-1} b_{1i} Day_i$$

$$+ \sum_{j=1}^{k-1} b_{2j} Fishermen_j + e_{ij},$$

$$n = 86; \quad k = 29.$$

The joint F-value for the day variables is $F(85, 1107) = 3.01$, and the joint F-value for the fishermen identification variables is $F(28, 1107) = 12.81$; both are significant at the .01 level.[25] Thus, differential performance of individual fishermen is a crucial feature of the sample.

To further illustrate the differences that exist among fishermen, as well as to show the existence of consistently successful fishermen, the sample was divided into good, average, and poor shrimpers on the basis of average catch. Those categories, though, mask wide variation within the classes. Mean daily catches for each shrimper and a sample mean were calculated; we classified as good those fishermen with catches more than one standard deviation above the sample mean; as poor those having mean catches below one standard deviation. The remaining fishermen are labeled average.[26] Weekly catch means for the three categories were then calculated for each of the thirteen weeks of the season. The catch of good fishermen ranged from a weekly mean of 1,098 pounds to 485 pounds. For average, the range was 652 to 286, and for poor ranges were 515 to 150. These results show persistent differences in catch, underscoring the regression results, with better shrimpers routinely catching more than their less-skilled counterparts.

The observed differences in fishing ability are largely attributed to acquired knowledge and innate skills. Since those skills are unlikely to be readily transferable assets, economic rents exist in the fishery, even under open access conditions (see Richard Bishop).[27] The aggregate supply curve for fishing effort is thus positively sloped and inframarginal fishermen receive rents. The likelihood of upward-sloping effort supply curves was previously noted by Steven Cheung, Parzival Copes, and Colin Clark (1980), but they did not develop the implica-

---

[24] The full-time criteria was arbitrarily selected to avoid weekend shrimpers and those fishing during vacations from other occupations with one month assumed to be the vacation limit. There were 29 full-time shrimpers, and 86 days out of the possible 91 were fished due to poor weather for 5 days.

[25] $R^2 = .36$ with $F(113, 1107) = 5.47$.

[26] The means used were least squares means calculated for unbalanced designs in analysis of variance tests. The classification, based on one standard deviation from the sample mean, resulted in 2 good fishermen, 6 poor fishermen, and 21 average fishermen.

[27] Fishing skills are not the same as managerial talents, and it does not follow that better fishermen are good managers. Hence, there may be no advantages to skilled fishermen in forming fleets. Since fishing skills are not easily transferable and fishermen are paid on a share basis, the higher yields of good fishermen may not be captured by fleet owners. Ship-shore operations are also costly to coordinate for fleets as pointed out by Andersen (1972, pp. 124–26).

tions of heterogeneity for contracting. Such heterogeneity limits both the nature of voluntary agreements and the effectiveness of fishermen as political lobbyists.

To illustrate the impact of heterogeneity on contracting we introduce heterogeneous fishermen into the H. Scott Gordon model. In that model the average and marginal products of fishing effort decline as effort increases. The stock of fish enters into the catch function, and it declines as aggregate effort increases. Hence, there is a direct relationship between catch and aggregate effort in the fishery. A regulatory scheme establishing catch quotas accordingly implies commensurate levels of effort, and effort quotas commensurate catches. Following Gordon, we abstract from more dynamic considerations that require the use of time and appropriate discount rates since for our purposes, they are unnecessary.

If there are N fishermen, the catch for each can be expressed by

$$(2) \qquad h_i = f(e_i, X); \; i = 1, 2, 3, \ldots N.$$

Here e_i denotes the units of standard fishing effort exerted by fisherman i; X is the stock of fish. The stock of fish is affected by total catch $(\Sigma_{i=1}^N h_i)$, and we assume h_i alone has no significant impact on X. As total effort (E) is increased the stock of fish is reduced, and average catch per aggregate unit of effort falls. Aggregate fishing effort (E) is usually defined as the sum of the number of fishing vessels times their individual catching power (the proportion of the stock each vessel is capable of catching per unit of time).[28] Here, we define (E) to be the sum of indi-

vidual efforts e_i, where each e_i is a function of capital, labor, and the specific abilities of fishermen that affect catching power. Individual effort, then, as shown in equation (2), is an input in the catch function. With individual effort measured in standard units the partial derivative of equation (2) with respect to e_i is a constant and the same for all fishermen. Hence, the value of the marginal and average products of effort as viewed by each individual fishermen are equal. The cost per standard unit of effort, however, varies across fishermen because of differences in fishing abilities.[29] The cost of supplying individual effort is independent of total effort in the fishery. The net rents received by fishermen i after considering all opportunity costs, $C_i(e_i)$, is given by

$$(3) \qquad rents_i = Pf(e_i, X) - C_i(e_i),$$

where P, the price of landed fish, is constant and exogeneously determined. We further assume $C_i'(e_i) > 0$, and $C_i''(e_i) > 0$ in the relevant range. Only for the marginal fishermen will rents be equal to zero such that the value of the marginal product of individual effort equals average cost

$$(4) \qquad P \cdot \partial f / \partial e_i = C_i(e_i)/e_i.$$

Figure 1 illustrates the complications differential skills provide for fishery regulation. Panel (a) of the figure shows the effort supply curves for two distinct categories of fishermen: Good fishermen $\hat{S}$ and less productive fishermen $\bar{S}$. Fishermen in each category are identical, and the supply curves $\hat{S}$ and $\bar{S}$ are the sums of the marginal costs for fishermen in the two categories. We assume that the supply functions are equal until effort level H is reached, and diverge thereafter. That assumption clarifies the discussion, but the main results do not depend on it. Panel (b) of the figure shows aggregate fishing effort with MR_E and AR_E the monetized marginal and average product curves. Both curves de-

---

[28] The concept of fishing effort is vague, as noted by James Wilen. Fishermen employ various combinations of capital (vessel length, tonnage, well configuration) and labor inputs. Our concept of effort embodies those choices and is consistent with the usual definition (see Scott, 1979, p. 727). We present our arguments using the cost of producing effort, rather than the cost functions for catching a unit of fish. The latter, as indicated by equation (2), contains the cost of producing effort but would necessarily shift with changes in the stock levels. On the other hand, the cost curves for producing effort are independent of stock levels, thus simplifying the presentation and allowing us to show equilibrium conditions.

[29] Note that cost functions can also vary across fishermen if their opportunity costs are different. We return to this point when discussing the role of part-timers in the fishery.

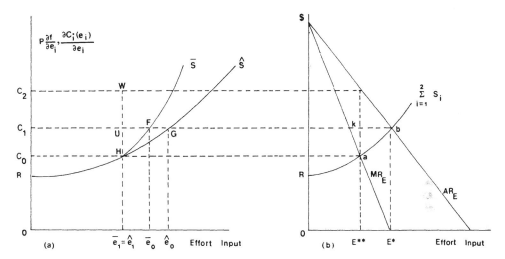

FIGURE 1. QUOTAS AND DIFFERENTIAL RENTS

cline as the stock X falls in response to greater aggregate fishing effort.[30] The supply of total effort is the sum of the effort curves for both classes of fishermen. Under open access conditions, total effort equals E^* (arbitrarily chosen to equal maximum sustained yield). The level of total effort E^* and a corresponding equilibrium stock of fish result in a value of the average catch per unit of effort equal to C_1. In panel (a), fishermen in the two categories provide $\bar{e}_0$ and $\hat{e}_0$ units of effort, equating the relevant supply curves with C_1. To maximize rents in the fishery, however, aggregate effort must be reduced to E^{**}. Following Adasiak's point that egalitarian pressures are common in quota schemes, we have constructed an example where maximization calls for equal quotas $\bar{e}_1 = \hat{e}_1$. Those quotas lead to aggregate effort E^{**} and provide a net gain in rents equal to abE^*, where

$$(5) \quad abE^* = (C_2 WUC_1 - UGH)$$

$$+ (C_2 WUC_1 - UFH).$$

The first right-hand term in (5) is the net

[30] The MR_E and AR_E functions as shown in Figure 1 are derivable from biological models with steady-state yields; see Clark (1976, pp. 14–15).

gain or loss to better fishermen, while the second applies to the less productive fishermen. With the cost curves assuming arbitrary slopes beyond H, there is no reason why the first term in (5) cannot be negative, if the second term is sufficiently large to leave abE^* positive.[31] In that case, without side payments, better fishermen would oppose quotas.

Opposition to quotas could be mitigated if they were assigned on the basis of historical effort or catch and made transferable, thereby lessening the constraints on more productive

[31] With the cost curves assuming arbitrary slopes beyond H, the only constraints are

(a) $UG\hat{e}_0\hat{e}_1 + UF\bar{e}_0\bar{e}_1 < 2C_1 U\hat{e}_1 0,$

(b) $\hat{e}_1 HG\hat{e}_0 < UG\hat{e}_0\hat{e}_1,$

(c) $\bar{e}_1 HF\bar{e}_0 < UF\bar{e}_0\bar{e}_1.$

The first condition follows from the upward-sloping supply function in panel (b) and the linear average catch function, implying that $C_1 k = kb$ and $(E^* - E^{**}) < kb$. The last two conditions follow from the upward-sloping individual supply functions. For instance, if $\hat{e}_1 HG\hat{e}_0 = 4.9$, $UG\hat{e}_0\hat{e}_1 = 6$, $\bar{e}_1 HF\bar{e}_0 = .8$, $UF\bar{e}_0\bar{e}_1 = 1$, $2C_1 U\hat{e}_1 0 = 8$, and $2C_2 w\hat{e}_1 0 = 10$, the above conditions will be met. abE^* equals .7, but the net gain to better fishermen is $-.1$. Without compensation, better fishermen in this case will oppose quotas.

fishermen.[32] Further, the individual effort curves in Figure 1 show only one of many possible outcomes. All parties in some cases may gain even if the relative gains vary. Nevertheless, the assignment of effort or (more common) catch quotas are an assignment of wealth. Competition for that wealth is unlikely to produce a unified, collective effort by fishermen for a quota system. Even assuming that a regulatory bureaucracy were concerned with efficient outcomes, the problem remains. With heterogeneous fishermen and limited knowledge of individual effort supply functions, quota systems are costly for regulatory agencies to devise, and, as a result, disputes over the distribution of gains is likely. In addition, the fish stock is rarely known with precision and is subject to changing biological conditions. In that setting, individual quotas (generally thought of as a fixed percent of the total allowable catch) would have to vary across seasons. The assignment of individual seasonal quotas, however, will not control all rent dissipation. With the stock of fish included in the catch function each fisherman has incentive to harvest early, when stocks are high. The resulting rush raises the aggregate cost of landing the allowable catch. Optimality, then, calls for variable quotas during the season.[33] These complications cannot be taken lightly. Not only are the costs high of managing a quota system in a stochastic environment, but fishery managers will not be able to specify the potential gains of the system to each fisherman. The reluctance of those who have adapted well to open access conditions and are earning rents to support quotas is predictable.

Empirical evidence from the Bay of Fundy herring fishery reveals the difficulty of designing quota arrangements which achieve an efficient allocation of effort and minimize political opposition. Limited entry quotas based on historical catch patterns, and fishermen subsidies were instituted in 1977 after persistent falling catch. Nevertheless, conflict emerged and some fishermen attempted to leave the government cooperative around which the regulations were based. Further, preliminary data indicate that catch per unit of effort continued to fall under the regulatory arrangement (Harry Cambell, 1980, Table 3.3a).

The Bay of Fundy herring fishery was biologically overfished at the time quotas were introduced. That condition reduces opposition to quotas. To see why, consider Figure 1. If aggregate effort is initially beyond the maximum sustainable yield point ($MR_E = 0$ in panel (b)), then total catch could increase with the imposition of the quota system and provide a net gain to more productive fishermen.[34] Quotas or other property rights arrangements will also be less costly if they are assigned when the fishery is undeveloped. In that case, there will be no preexisting claims that must be reconciled in the new system. T. F. Meaney provides an example of the Shark Bay fishery in Australia where the government has permitted a single firm to operate the majority of the shrimping vessels in a previously undeveloped area for over twenty years. Catch and earnings have risen dramatically with no evidence of excessive rent dissipation.

Another commonly advocated management tool is a corrective tax on either effort

---

[32] Scott (1979) and David Moloney and Peter Pearse have presented arguments in favor of transferable catch quotas. Although they consider various ways that initial quota assignments can be made, their discussion focuses on the efficiency aspects of quotas once they are in place.

[33] Paul Bradley offers one of the earliest discussions of controlling effort during the season and across seasons.

[34] Essentially, we are arguing that gains from contracting increase as the fishery becomes progressively overfished. In a dynamic setting, that would imply falling average catch over time as a falling stock eventually reduces recruitment. A system of quotas could increase the stock and subsequently raise average catch. To see the argument within the context of our static model, however, allow for a new, steeper AR_E function to pass through point b in panel (b) of Figure 1. The corresponding new MR_E function pivots at point k, intersecting the old MR_E function from above. Accordingly, the new sustainable yield point will be to the left of E^*. The level of aggregate effort under open access remains the same (E^*). While the new rent-maximizing level of effort would necessarily decrease, moving $\bar{e}_1$ and $\hat{e}_1$ to the left in panel (a), average catch will increase. The gains of a quota system to more productive lower-cost fishermen must eventually overcome any losses if the AR_E function is made sufficiently steep.

or catch, but taxes have even less chance of being supported than do quotas. With heterogeneous fishermen and upward-sloping effort supply schedules, a corrective tax would lower rents received by fishermen (for example, a unit tax equal to WH in Figure 1). Hence, corrective taxes negatively affect the group most likely to seek fishery regulation, an ironic result since fishery programs are generally aimed at raising the income of fishermen rather than economic efficiency (Scott, 1979, p. 729). A system of lump sum payments to fishermen from tax revenues is possible; but it would encounter the same problem of the distribution of gains among heterogeneous fishermen. It is not surprising that fishery experts are unable to point to a single example where regulatory taxes or royalties on either effort or catch are in use.[35]

Other forms of regulation are more common, such as season closures, a total allowable catch for the entire fleet, and controls on entry. These more limited arrangements are often criticized because each regulates only one of the many options for rent dissipation. (for example, see Wilen). The 1950 Texas Supreme Court ruling, *Dobard v. State*, is instructive because it shows that the court early recognized that rent dissipation would occur at other margins if only entry restrictions were enacted: "...It cannot be said with the least certainty that reduction or increase in the number of boats, especially without any provision as to the size or other characteristics of the boats, would reduce or increase the total number of shrimp taken..." (233 S.W. 2nd 440). Meaney also shows that limited entry in the western Australia rock lobster fishery has been accompanied by dramatic increases in horsepower and other attributes of lobster boats as existing fishermen compete to raise catches. Alex Fraser, and Pearse and Wilen reveal similar reactions to limited entry in the British Columbia salmon fishery. Crutchfield (1979, p. 746), however, argues that although dissipation occurs at unrestricted margins, it is not complete as exhibited by positive values for fish-

ing licenses. Further, with heterogeneous fishermen and a fixed level of fishing knowledge and skills limited entry can be expected to provide some gains to existing fishermen.

Those potential gains and the fact that few internal constraints are typically involved, suggest that limited entry will be supported by fishermen as a general objective. Once again, however, heterogeneity will lead to conflict over the details of any limited entry program. For example, part-time fishermen are often candidates for exclusion by means of entry restrictions; however, without adequate cost data one cannot conclude which group should be removed from the fishery.[36] It is possible that rents may be higher for less-skilled fishermen, who catch fewer fish, because of generally lower opportunity costs. State-sponsored buy-back programs such as in British Columbia support the objectives of limited entry by retiring vessels and licenses through government purchases (see Fraser). The program, though, is costly and controversial. Finally, if limited entry involves the assignment of transferable fishing licenses, disputes may arise over the issue of transferability.

To see why, we illustrate the problem by using three distinct categories of fishermen, where fishermen are identical in each category. Further, group contracting does not take place. Panel (a) in Figure 2 shows the aggregate effort supply curves for each of the three categories. Panel (b) shows the aggregate effort supply S_0 and value of catch per unit of effort AR_E. Under initial open access conditions and fish prices equal to P_0, only individuals in categories 1 and 2 with marginal costs $C_1'(e_1)$ and $C_2'(e_2)$ are actively fishing. The imposition of a limited entry program that grants transferable licenses to only active fishermen does not in itself change the equilibrium conditions. If for exogenous reasons, however, the price of fish rises to P_1, the value of the average catch per standard unit of effort shifts to AR_E^1. Under those conditions, a third category of fishermen (with the same number of fishermen as

---

[35] "So far as I know, no regulatory tax or royalty on catch is anywhere in effect today" (Scott, 1979, p. 735).

[36] Scott (1979, p. 731) also questions the arbitrary elimination of part-timers arguing that the literature on fishing has little to say on the issue.

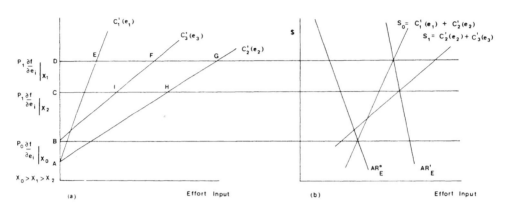

FIGURE 2. TRANSFERABILITY OF RIGHTS AND AGGREGATE RENTS

licenses in category 1) with costs represented by $C_3'(e_3)$ will enter the fishery through purchase of licenses from those in category 1, if area CIB exceeds ADE; that is, transactions will occur if individual rents for fishermen in category 3 exceed those for fishermen in category 1. With the transfer of licenses to more productive fishermen, however, aggregate effort expands along supply curve S_1 in panel (b). We assume that fishermen in category 3 recognize the effect of increased aggregate effort in calculating gains from purchasing the licenses. The greater effort resulting from the transfer of licenses reduces the rents earned by fishermen in category 2 who were not party to the transaction. Aggregate rents increase only if $CIB > ADE + DGHC$, but with individual contracting we only know that $CIB \geq ADE$. Hence, fishermen in category 1 will support license transfers, while in the absence of compensation, those in category 2 will oppose it.

The example employs considerable foresight by fishermen on the effect of entry by different groups, more foresight than they may possess. But uncertainty over the details of limited entry such as transferability, contributes to dissention and raises the political costs of the program. The example also points to the importance of well-defined property rights for efficient resource use. Limited licensing assigns only a right to fish, not a property right to the fish resource; hence dissipation of rents still occurs. The example

shows that transferability of licenses in the absence of well-defined rights to the resource need not provide any improvement in aggregate wealth. Indeed, it can reduce it.[37]

Figures 1 and 2 illustrate the effects of heterogeneity on the success of fishery management. To illustrate our argument, the figures simplify the problem by focusing on well-defined categories. Empirical data, however, reveal wide-ranging differences among fishermen, which raise the costs of internal agreements among fishermen to regulate individual effort. Reasonably cohesive groups may form if there are sharp intergroup distinctions, such as those between commercial fishermen employing different types of capital (see Robert Higgs) or as in Texas between bay and Gulf shrimpers. To the extent well-defined groups exist, intergroup conflict and bargaining result as each attempts to impose differential constraints on other fishermen. For the successful group, the total available catch is at least temporarily increased and internal controls avoided.

The discussion demonstrates the hazards for existing fishermen of quota assignments, taxes, and limited entry with transferable licenses. If enforcement costs are comparable, the uncertainty of government activities, largely outside the control of fishermen,

---

[37] This conclusion contrasts with the beneficial effects of transferability described by Crutchfield (1979, p. 746).

would lead them to prefer regulation by fishermen unions or trade associations. Since that has been denied, government controls remain the only possibility to formally limit effort. Yet heterogeneity among fishermen suggests they will not lobby for an efficient allocative scheme until pressure on the fishery is sufficiently intense to raise the net gains from lobby group formation. Until that occurs, fishermen are more likely to support arrangements that do not affect status quo rankings and that increase their total catch—season closures, hatcheries, gear restrictions to protect juvenile fish, and controls on fishing by members of other groups.

The desire of the fishermen to enhance total catch coincides with the interests of politicians and bureaucrats. Since property rights have been historically denied to fisheries, access is determined by the political process. Catch-enhancing policies are popular among politicians because they forestall the implementation of controversial allocative schemes. Accordingly, the heterogeneity of fishermen and the self-interest motives of politicians and bureaucrats lead to similar predictions for the types of regulations sought and provided for in the Texas bay shrimp fishery. Those regulations are examined in the following section.

III. Empirical Evidence of Regulation in the Texas Shrimp Fishery

A. *Limited Entry*

Shrimp are an annual crop with no clear relationship between population size and annual recruitment (Lee Anderson, 1977, p. 103). Even though the fishery cannot be biologically overfished, catch per unit of effort is falling. Despite the potential gains offered by limited entry for existing fishermen, we predict it will not be advocated by shrimpers until economic conditions in the fishery become critical. Even then, the emphasis will be on constraining outsiders rather than established fishermen. The reluctance of shrimpers to support limited entry stems from the uncertain and possibly negative effects of government administration of the regulation—who will be restricted, how will

entry limits change, and how will rules on transferability affect existing fishermen? The prediction is supported by the evidence. There are currently no formal, limited-entry programs in the fishery. Efforts to restrict access have focused on outsiders—against Louisiana shrimpers in 1947 and 1949 (both statutes declared unconstitutional), and against Vietnamese in 1981.[38]

The lack of support for limited entry is also reflected by the absence of informal access controls. The bay shrimp fishery is divided into three separate bay systems: Galveston, Matagorda, and Aransas. Within a bay, shrimpers could deny dock space, ice, and fuel to outsiders to reserve local waters. Yet, there is no evidence of efforts within communities to define territorial waters or to deny outside shrimpers support facilities. The incentive for such actions is reduced by the need for frequent migration to other areas in response to local catch fluctuations due in part to pollution and fresh water discharge. Access to other areas requires reciprocal privileges to local waters to outside shrimpers. Analysis of annual catch data for the three bay systems reveals the need for temporary migration and access to distant waters. The catches do not move together, so that a poor season in one area is not replicated elsewhere.[39]

[38] The first law levied a $2,500 license fee on out-of-state vessels and a $200 fee on individual fishermen. While upheld in *Dodgen v. Depuglio* 209 S.W. 2nd 588 (1948) by the Texas Supreme Court, the U.S. Supreme Court rejected discriminatory licensing fees in *Toomer v. Witsell* 334 U.S. 385 (1948). The Texas Legislature responded with the limited licensing scheme, ruled unconstitutional by the Texas Supreme Court, in *Dobard v. State* 233 S.W. 2nd 435 (1950).

[39] Based on interviews conducted July 1980. Shrimpers argued that catch fluctuations were not shared across bays, and they accordingly wanted temporary access to areas of better catch. Local controls on entry were viewed as bringing retaliatory restrictions on access to other bays. We collected annual catch data for the three bays from the *Annual Reports* of the Texas Parks and Wildlife Department for 1939–49. Annual catches for each bay were regressed against a constant and time, and the residuals examined to determine the extent to which catches fluctuated together. Correlations of the residuals show no statistically significant relationship:

	Galveston	Matagorda
Matagorda	− .57	
Aransas	.14	.30

B. *Season Closures, Gear Restrictions and Minimum Shrimp Size*

We predict support from fishermen for these measures because they increase the value of the total catch by protecting juvenile shrimp and do not allocate individual effort. Further, the conservation of small shrimp is consistent with the biological management goals of the administrative agency, the Texas Parks and Wildlife Department. Their joint efforts are reflected in the following passage from the agency's *Annual Report*:

"After conferring with the canners, wholesale dealers and fishermen, the Coastal Division's Marine Biologist prepared a discussion on the shrimp situation and recommended certain seasonal restrictions.... The suggestions were adopted and agreed upon by a mass meeting of the shrimp fishermen and dealers held at Port Lavaca and were later presented to the Legislature and enacted into law.
[1940, p. 39].

The support among shrimpers for closed seasons is also shown by voluntary, local extensions of closed periods at Palacios, Texas in 1973, and at Seadrift and Rockport, Texas, in the late 1970's to allow immature shrimp to grow.[40]

Gear restrictions are of two types: Minimum net mesh size and limits on the number and size of nets that can be used in the bays. The former reinforce the effect of season closures by allowing small shrimp to escape the pull of the nets. The latter, however, are of a different nature; they are used in the bays to reduce the competitive advantage of large Gulf vessels over inshore boats. There are no restrictions on the number or size of trawl nets used in the Gulf, but in the bays only one net, 25 feet in width in the spring and 95 feet in the fall is allowed per vessel

(Texas Parks and Wildlife Department). That restriction effectively keeps the larger Gulf vessels out of the bays.[41]

Support for minimum shrimp size controls is divided along industry lines and depends upon the type of shrimp involved. There are two types at issue: white shrimp, which remain in the inshore bays and are harvested in the fall, and brown shrimp, which migrate as juveniles from the bays to the Gulf. Bay shrimpers have incentive to protect immature white shrimp, since they have access to them after they have grown. There are corresponding minimum size limits for the fall white season. Bay shrimpers do not have that incentive for brown shrimp, and there are no minimum size limits for the spring brown shrimp season in the bays. Group conflict between the bay and Gulf fisheries over the brown season has influenced regulatory efforts in Texas since 1959, when the state's principal shrimp management law was enacted.[42] In that year the Gulf lobby successfully closed the spring season from March 1–July 15. In 1963 bay shrimpers secured an amendment to allow a limited season from May 15–July 15.[43] To additionally restrict fishing in the bays, the Gulf lobby was able to impose individual catch limits on bay shrimpers during the spring brown season. The limits were first 250 pounds per vessel per day and later 300 pounds. No similar restrictions exist for bay white shrimp or brown shrimp on the Gulf. These catch limits are not quotas in the usual efficiency sense, but are the outcome of competition by bay and Gulf shrimpers for brown shrimp. Repeated evasion of the limits was tolerated

---

[40] Based on interviews conducted July 1980 with shrimpers at Seadrift, Rockport, and Palacios, Texas. A hazard to local extension of a closed season is that shrimpers from other areas may not comply. At Seadrift, when local shrimpers withdrew effort, outsiders appeared; further evidence of the precarious nature of voluntary agreements under open access conditions.

[41] Most Gulf vessels are 55 feet or longer, weigh over 60 tons, have crews of at least 3, and fish for 2 or more weeks per trip (Robert Maril). Bay boats are smaller; during 1975–80, 88 percent were 50 feet or less with most of the newly entering boats 25 feet and under 5 tons (Swartz). Boats are commonly operated with a crew of 2, and fish only in day trips. The common use of gear restrictions as a means of blocking entry by outsiders is described by Higgs, and by Crutchfield and Pontecorvo.

[42] *Statutes of the 56th Texas Legislature* (1959, pp. 407–18). The political pressures behind the 1959 Shrimp Conservation Act are described in the *Corpus Christi, Texas, Caller*, February 10, 1959.

[43] *Statutes of the 58th Texas Legislature* (1963, pp. 895–907).

from their enactment in 1959 until depressed conditions in the Gulf in the late 1970's led to political pressure from the Gulf industry for improved enforcement. In 1980, the Parks and Wildlife Department added fifty additional wardens to the normal ninety to enforce quotas and other restrictions during the bay spring season. The total number of shrimp violations in the bays correspondingly rose from 331 in 1977 to 439 in 1980 (Texas Parks and Wildlife Department, Coastal Law Enforcement Division).

C. *Taxes, Quotas, and Other Internal Effort Controls to Protect the Stock of Shrimp*

Because these regulations pose the greatest possibilities for rent redistribution and reduction, we predict little support for them among fishermen. There are no taxes levied to restrict effort in the fishery, and license fees are minimal, $40 for the bay and $50 for the Gulf. Moreover, there has been no discussion of effort or catch quotas to reduce fishing pressure to meet conservation goals. The fishery remains relatively unregulated.

The empirical record of regulation in the Texas shrimp fishery is consistent with the arguments of the paper: Regulations that expand total catch and are not aimed at redistributing rents have been supported by both fishermen and the regulatory agency; group effort has increased as fishing pressure has grown; and those efforts have focused on outsiders—the Vietnamese in the case of the bay fishery, and bay shrimpers in the case of the Gulf fishery. No internal effort controls have been implemented, except those imposed exogenously by other groups.

IV. Concluding Remarks

Our predictions of the types of agreements heterogeneous parties can voluntarily agree to are likely to apply to similar contracting situations elsewhere in the economy. Our focus here is on persistent common property conditions in fisheries and the nature of the regulatory response to them. Currently, government regulation is the only means of increasing fishery rents, since sole ownership and other private efforts to control entry and

effort have been rejected as illegal. Under government regulation numerous options remain for rent dissipation for two reasons: first, heterogeneous fishermen do not form cohesive lobby groups for government controls on individual effort. We have emphasized the hazards facing heterogeneous fishermen from effort or catch quotas, corrective taxes, and transferable licenses and why group agreement in such regulation is costly, at least until the fishery is intensively depleted. Fishermen can be expected to rally for general regulations to raise total yields such as season closures or entry controls on outsiders. Those programs raise rents for existing fishermen above open access conditions, even though dissipation continues along other margins. In the absence of political support from fishermen, politicians, and bureaucrats facing periodic reelection and budget review will not pursue efficiency goals in regulation if the programs are controversial, as is likely. Second, information and measurement costs for regulating migratory species are high. Biological knowledge of the fish stock is uncertain, as is knowledge of the nature of its interdependency with other species, and the impact of environmental changes. Additionally, with heterogeneous fishermen costly measurement of individual effort and catch is necessary for devising quota, tax, and limited entry schemes. Further, enforcement of regulations across large territories is costly, particularly if a general consensus among fishermen regarding the regulations has not been reached. While government regulation may be incomplete with considerable rent dissipation continuing, it does not immediately follow that more regulation is called for. As Ronald Coase pointed out, "But the reason why some activities are not the subject of contracts is exactly the same as the reason why some contracts are commonly unsatisfactory—it would cost too much to put the matter right" (1960, p. 39).

Many of these same costs apply also to voluntary contracting and private arrangements for regulating the fishery, and would be encountered in union or trade association arrangements. Agreement on internal catch or effort restrictions is costly for heterogeneous fishermen—hence the broad regulations adopted by the Gulf Coast Shrimp and Oystermen's Association: restrictions on outsiders; price setting to protect small shrimp; no individual specific controls on members. Additional supportive evidence is the absence of voluntary, informal controls on individual effort among Texas shrimpers. One cannot conclude that private contracting will lead to consensus, and that all margins of dissipation will be regulated. This mirrors the cartel literature which is replete with examples of the abrogation of contracts within cartels. There is, though, one crucial advantage offered by sole ownership and trade associations over government regulation: they will internalize the costs of regulation. Accordingly, they should not arbitrarily be denied consideration in the selection of management policies if the elusive goal of maximizing the rental value of the fishery is to be achieved.

REFERENCES

Acheson, James M., "The Lobster Fiefs: Economic and Ecological Effects of Territoriality in the Maine Lobster Industry," *Human Ecology*, July 1975, *3*, 183–207.

Adasiak, A., "Alaska's Experience with Limited Entry," *Journal of the Fishery Research Board of Canada*, July 1979, *36*, 770–82.

Agnello, Richard J. and Donnelley, Lawrence P., "Property Rights and Efficiency in the Oyster Industry," *Journal of Law and Economics*, October 1975, *18*, 621–34.

Andersen, Raoul, "Hunt and Deceive: Information Management in Newfoundland Deep-Sea Trawler Fishing," in his and Cata Wadel, eds, *North Atlantic Fishermen: Anthropological Essays on Modern Fishing*, Toronto: University of Toronto Press, 1972; 120–40.

_____, "Public and Private Access Management in Newfoundland Fishing," in his *North Atlantic Maritime Cultures*, New York: Mouton Publishers, 1979; 299–336.

_____ **and Stiles, R. Geoffrey,** "Resource Management and Spatial Competition in Newfoundland Fishing: An Exploratory Essay," in Peter H. Frickle, ed, *Seafarer and Community*, London: Croom Helm, 1973; 44–66.

Anderson, Lee G., *The Economics of Fishery Management*, Baltimore: Johns Hopkins University Press, 1977.

Bell, Frederick W., "Technological Externalities and Common-Property Resources: An Empirical Study of the U.S. Northern Lobster Fishery," *Journal of Political Economy*, January/February 1972, *80*, 148–58.

Bishop, Richard C., "Limitation of Entry in the United States Fishing Industry: An Economic Appraisal of a Proposed Policy," *Land Economics*, November 1973, *49*, 381–90.

Bradley, Paul G., "Some Seasonal Models of the Fishing Industry," in A. Scott, ed, *Economics of Fisheries Management: A Symposium*, H. R. MacMillan Lectures in Fisheries, Vancouver: University of British Columbia, 1970; 33–44.

Breton, Yvon D., "The Influence of Modernization on the Modes of Production in Coastal Fishing: An Example from Venezuela," in M. Estelle Smith, ed, *Those Who Live From the Sea*, New York: West Publishing Company, 1977; 125–38.

Cambell, Harry, "The Impact of Government Regulation on the Bay of Fundy Herring Fishery," Programme in Natural Resource Economics, Vancouver: University of British Columbia, 1980.

Cheung, Steven N. S., "The Structure of a Contract and the Theory of a Nonexclusive Resource," *Journal of Law and Economics*, April 1970, *13*, 49–70.

Clark, Colin W., *Mathematical Bioeconomics*, New York: John Wiley & Sons, 1976.

_____, "Towards a Predictive Model for the Economics Regulation of Commercial Fisheries," *Canadian Journal Fishery and Aquatic Science*, July 1980, *37*, 1111–29.

_____ **and Munro, Gordon R.,** "Fisheries and the Processing Sector: Some Implications for Management Policy," *Bell Journal of Economics*, Autumn 1980, *11*, 603–16.

Coase, Ronald H., "The Problem of Social Cost," *Journal of Law and Economics*, October 1960, *3*, 1–44.

Comitini, Salvatore, "Marine Resources Exploitation and Management in the Economic Development of Japan," *Economic Development and Cultural Change*, July 1966, *14*, 414–27.

Copes, Parzival, "Factor Rents, Sole Ownership and the Optimum Level of Fisheries Exploitation," *Manchester School of Economics and Social Studies*, June 1972, *40*, 145–62.

Cordell, John, "Carrying Capacity Analysis of Fixed-Territorial Fishing," *Ethnology*, January 1978, *17*, 1–24.

Crutchfield, James A., "Collective Bargaining in the Pacific Coast Fisheries: The Economic Issues," *Industrial and Labor Relations Review*, July 1955, *8*, 541–56.

_____, "Economic and Social Implications of the Main Policy Alternatives for Controlling Fishing Effort," *Journal of the Fishery Research Board of Canada*, July 1979, *36*, 742–52.

_____ and Pontecorvo, Giulio, *The Pacific Salmon Fisheries*, Baltimore: John Hopkins Press, 1969.

Fraser, Alex G., "Limited Entry: Experience of the British Columbia Salmon Fishery," *Journal of the Fishery Research Board of Canada*, July 1979, *36*, 754–63.

Forman, Shepard, "Cognition and the Catch: The Location of Fishing Spots in a Brazilian Coastal Village," *Ethnology*, October 1967, *6*, 417–26.

Gordon, H. Scott, "The Economic Theory of a Common Property Resource: The Fishery," *Journal of Political Economy*, April 1954, *62*, 124–42.

Griffin, W. L., Tydlacka, C. G. and, Grant, W. E. "A Comparison of Real and Nominal Fishing Effort in Policy Evaluation," Department of Agricultural Economics, Texas A&M University, no date.

Higgs, Robert, "Legally Induced Technical Regress in the Washington Salmon Fishery," in *Research in Economic History*, 1982, *7*, 55–86.

Lund, Thomas A., *American Wildlife Law*, Berkeley: University of California Press, 1980.

Maril, Robert L., "Shrimping in South Texas: Social and Economic Marginality Fishing for a Luxury Commodity," paper presented at Association for Humanist Sociology, Johnston, Pennsylvania, 1979.

Martin, Kent O., "Play by the Rules or Don't Play at All: Space Division and Resource Allocation in a Rural Newfoundland Fishing Community," in Raoul Andersen, ed., *North Atlantic Maritime Cultures*, New York: Mouton Publishers, 1979; 277–98.

Meaney, T. F., "Limited Entry in the Western Australian Rock Lobster and Prawn Fisheries: An Economic Evaluation," *Journal of the Fishery Research Board of Canada*, July 1979, *36*, 789–98.

Moloney, David G. and Pearse, Peter H., "Quantitative Rights as an Instrument for Regulating Commercial Fisheries," *Journal of the Fishery Research Board of Canada*, July 1979, *36*, 859–66.

Pearse, Peter H. and Wilen, James E., "Impact of Canada's Pacific Salmon Fleet Control Program," *Journal of the Fishery Research Board of Canada*, July 1979, *36*, 764–69.

Quirk, James and Smith, Vernon, "Dynamic Economic Models of Fishing," in A. Scott, ed, *Economics of Fisheries Management: A Symposium*, H. R. MacMilliam Lectures in Fisheries, Vancouver: University of British Columbia, 1970, 3–32.

Rogers, George W., "Alaska's Limited Entry Program: Another View," *Journal of the Fishery Research Board of Canada*, July 1979, *36*, 783–88.

Scott, Anthony, "The Fishery: The Objectives of Sole Ownership," *Journal of Political Economy*, April 1955, *63*, 116–24.

_____, "Development of Economic Theory on Fisheries Regulation," *Journal of the Fishery Research Board of Canada*, July 1979, *36*, 725–41.

Starr, Paul D., "Vietnamese Fisherfolk on the Gulf Coast: A Case Study of Local Reactions to Refugee Resettlement," *International Migration Review*, Spring-Summer 1981, *15*, 226–38.

Swartz, A. Nelson, "The Character of the Texas Shrimp Fleet," Staff Paper, Department of Agricultural Economics, Texas A&M University, 1980.

White, David R. M., "Environment, Technology, and Time-Use Patterns in the Gulf Coast Shrimp Fishery," in M. Estelle Smith, ed, *Those Who Live From the Sea: A Study in Maritime Anthropology*, New York: West Publishing, 1977; 195–214.

Wilen, James E., "Fishermen Behavior and the Design of Efficient Fisheries Regulation Programs," *Journal of the Fishery Research Board of Canada,* July 1979, *36,* 855–58.

Wilson, James A., "A Test of the Tragedy of the Commons," in Garrett Hardin and John Baden, eds, *Managing the Commons,* San Francisco: W. H. Freeman and Co., 1977; 96–110.

Gulf of Mexico Fishery Management Council, *Draft Environmental Impact Statement and Fishery Management Plan for the Shrimp Fishing of the Gulf of Mexico, United States Waters,* Tampa: U.S. Department of Commerce, 1979.

Marine Advisory Service, "Strangers in a Strange Land—A Delicate Balance," Texas A&M University, 1980.

Texas Game, Fish, and Oyster Commission, *Annual Reports,* Austin: 1920–62; **Texas Parks and Wildlife Department,** *Annual Reports,* Austin: 1963–79.

Transcript of Record, *Gulf Coast Shrimpers and Oystermen's Association v. United States,* No. 15680, 5th Circuit, U.S. Court of Appeals, 1956.

U.S. Department of Commerce, *Fishery Statistics of the United States.* Washington: USGPO, 1977, 1948–59.

Name Index